CHARLES RICKETTS

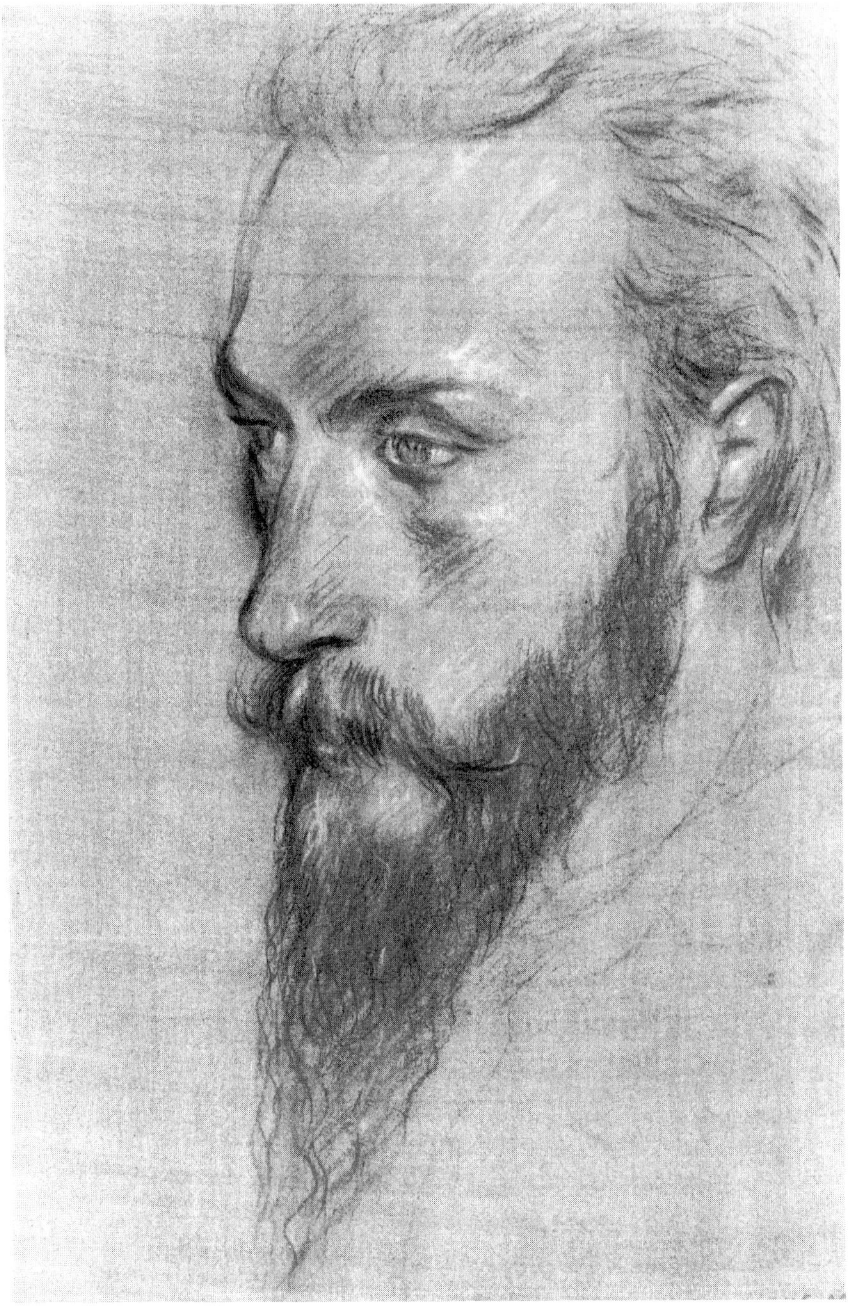

Charles Ricketts, (1899) drawing by Charles Shannon (National Portrait Gallery)

Charles Ricketts

A Biography

J. G. P. DELANEY

CLARENDON PRESS, OXFORD

1990

Oxford University Press, Walton Street, Oxford OX2 6DP
Oxford New York Toronto
Delhi Bombay Calcutta Madras Karachi
Petaling Jaya Singapore Hong Kong Tokyo
Nairobi Dar es Salaam Cape Town
Melbourne Auckland
and associated companies in
Berlin Ibadan

Oxford is a trade mark of Oxford University Press

Published in the United States
by Oxford University Press, New York

British Library Cataloguing in Publication Data
Delaney, Paul
Charles Ricketts: a biography.
1. English visual arts. Ricketts, Charles.
Biographies
I. Title
709'.2'4
ISBN *0-19-817212-5*

Library of Congress Cataloging-in-Publication Data
Delaney, J. G. P.
Charles Ricketts: a biography/J. G. P. Delaney.
p. cm.
Bibliography: p.
Includes index
1. Ricketts, Charles S., 1866-1931.
2. Artists—England—Biography. I. Title.
N6797.R5D45 *1989*
709'.2'4—dc19
[B]
ISBN *0-19-817212-5*

Typeset, printed and bound in Great Britain by
Butler & Tanner Ltd, Frome and London

For
Sir Rupert Hart-Davis
&
in memory of
George Warner Allen

Acknowledgements

IT is gratifying to review and record all the help I have received in my research. Miss Henriette and the late Mr Daniel Sturge Moore and all the members of the Sturge Moore family have offered help and hospitality most generously, and given me permission to quote copyright material by Ricketts, Shannon, 'Michael Field', and Thomas Sturge Moore. My earliest debt (and one of my greatest) is to Sir Rupert Hart-Davis, who suggested that I write this book. An international group of Ricketts enthusiasts has helped me in innumerable ways: Mr Peter and Mrs Angela Astwood, Mr Michael Barclay, Mr Stephen Calloway, Mr Joseph Darracott, Mr David Gould, Professor Charles Gullens, Mr Ton Leenhouts, M. Henri Locard, Mr Simon Reynolds, Mr Albert Sperison, Mr Charles Stewart, Mr Paul van Capelleveen, and Professor Carl Woodring. Members of the Ricketts family have given me every kind of assistance: Dr and Mrs Vincent von Sinner, Mrs Marguerite and Mr Benjamin de Seigneux (all of whom were most generous hosts), Mr and Mrs Robert Friedenrich, Mr Ralph Ricketts, and Lt.-Col. J. M. Ricketts. Ricketts's closest surviving friend, Mr Cecil Lewis, answered many questions in discussion and in letters: I am grateful for his co-operation.

I am also indebted to the following: Sir Harold Acton, Baron Pierre Abbadie de Madières, Mr Alan Anderson, Mr Stephenson B. Andrews, Ms Susan Ashbrook, Mr Malcolm Baker, Mr Vincent G. Barlow, Dr and Mrs François Barras, Mr Alistair Beaton, Mr Alan Bowness, Mr Quentin Bridge, Lady Brown, the late Mr Richard Carline, Mrs Florian Carr, Mr Turner Cassity, Mr Max Chapman, Mr John Christian, Mrs Betty Clark, the late Lord Clark of Saltwood, Mr Stephen Clifford-Wilson, the late Mr Francis Cooper, Ms Lili Couvée-Jampoller, Miss Gabrielle Cross, Mr Timothy d'Arch Smith, Mr Roger de Grey, PRA, Mr Jan Dekkerd, Viscountess Eccles, Professor Dennis Farr, Mr Richard Finny, the late Mrs Barbara Fitzgerald, the late Professor Ian Fletcher, Mr Colin Franklin, Miss Mary Laura Gibbs, Dr Myles Gleeson White, Dr Warwick Gould, Mr Basil and Mrs Nicolete Gray, Mr Peter Greenham, Miss Cecily Hacon and the Hacon family, Ms Julia Hanson, the late Mr Keith Henderson, Mrs Marina Henderson, Dr Desmond Henry, the late Mr Christopher Hewett, Mr Malcolm Hillier,

Mr Ian Kenyur-Hodgkin and Mr Ian Hoy, Mr Martin Holmes, Mr Michael Holroyd, Mr H. Montgomery Hyde, Miss Karin Jaspers, Mr Stephen Jones, Professor Norman Kelvin, Lord Kennet, Mr Frederick R. Koch, Dr G. Krishnamurti, Dr Mary Lago, Mrs Marie-Jacqueline Lancaster, the late Mr and Mrs Roger Lancelyn Green and Mr Richard Lancelyn Green, Mrs Helen Langdon, Miss Sylvia Legge, Dr John Lipscomb, Ms Elizabeth Loizeau, Mr Justin Lowinsky, Ms Helen MacGregor, the late Mrs Dorothea Maclagan, Ms Elizabeth Masterman, the late Mr Raymond Mortimer, the late Mrs Violet Nyberg, Mr Peter Nahum, Mr William H. O'Donnell, Mr Stanley Olson, the late Miss Pamela Ovens, Mr Terence Pepper, Mr Geoffrey Perkins, the late Mr Arnold Pienne, Mr J. R. Piggot, Mr Bernard Pile, Mrs Christine Raikes, Professor John Rewald, the late Professor Giles Robertson, Professor Martin Robertson, Mr Peter Rose and Mr Albert Gallichen, Mr John Paul Ross, Sir John Rothenstein, Mr Terence Rowe, Professor Ann Saddlemyer, Mrs Diana Scarisbrick, Professor Ronald Schuchard, Ms Pamela E. Selwyn, Fr Brocard Sewell, O. Carm., Mr Donald Slee, Mr Michael and Mrs Patricia Snodin, Mr John Russell Taylor, Mrs Lance Thirkell, Mr Brian D. L. Thomas, OBE, the late Dame Sybil Thorndike, Mr Julian Trevelyan, Mr Reginald Tye, Mr Peter Vernon, Mr R. W. Vint, Mr Alex Werner.

The staff of the following institutions answered enquiries and provided information: Mrs Clark, Art Gallery and Museums Dept., Aberdeen; Mr I. M. de Grott, Rijksmuseum, Amsterdam; Mr David Farmer and Ms Cathy Henderson, Harry Ransom Humanities Research Center, Austin, Texas; Dr Mary Holbrook and Ms Joan Roston, Holburne of Menstrie Museum, Bath; Mr H. J. Turner, Bedford Museum, Bedford; Miss Halina Grubert, Cecil Higgins Art Gallery, Bedford; Mr Martyn Anglesea, Ulster Museum, Belfast; Ms Claudine Lemaire, Bibliothèque Royale de Belgique; Dr Sigrid Achenbach, Staatliche Museen Preussicher Kulturbesitz, Berlin; Dr Werner Schade, Staatliche Museen zu Berlin; Mr George Breeze and Miss Andrea Rose, City Art Gallery, Birmingham; Mr Paul Sykes, Director of Recreation, Blackburn; James K. Burkett and P. Dunderdale, Grundy Art Gallery, Blackpool; Mr Graham Teasdell, Russell-Cotes Art Gallery, Bournemouth: Mrs Caroline Krzesinska and Frances S. Clarke, Cartwright Hall, Bradford; Mr Conner, Art Gallery, Brighton; Mr Francis Greenacre, City Art Gallery, Bristol; Mr Eric Chamberlain, Miss Jane Munro, and Mr Duncan Robinson, Fitzwilliam Museum, Cambridge; Dr Michael Halls, Kings College Library, Cambridge University; Ms Ann R. Milstein, Fogg Art Museum, Cambridge, Mass.; Mr W. H. Bond and Mr Rodney G. Dennis, The Houghton Library, Cambridge, Mass.; Ms Lynn McClelland, South African National Gallery, Cape Town; Mr A. D. Fraser-Jenkins, National Museum of Wales, Cardiff; Ms Laura Hamilton and Mr Robert Hugg, Carlisle Museum and Art Gallery; Mr David Addison and Miss Gaynor Andrews, Cheltenham Art Gallery; Mr R. A. Clarke, Herbert Art Gallery and Museum, Coventry; Mr David Fraser, Museums and Art Gallery, Derby; Mr T. J. Manby, Museum and Art Gallery, Doncaster; Ms Gertaute Lippold, Staatliche Kunstsammlungen, Dresden; Mr John K. Hutchinson, National Gallery of Ireland, Dublin; Ms Felicity O'Mahony, Trinity College Library, University of Dublin; Mr Alan Bell, Librarian of Rhodes House, Oxford; Mr Douglas Hall and Ms Valerie A. Edmonds, Scottish National Gallery of Modern Art, Edinburgh; Miss C. Jane Baker, Royal Albert Memorial Museum, Exeter; Ms Anne Donald, Glasgow Museums and Art Gallery; Mr Jack Baldwin and Mr P. K. Escreet, University of Glasgow Library; the late Mr Wilfrid Blunt, Watts Gallery, Guildford; Mr K. W. Brown, Museum Service, Harrogate; Mr John Boyden, Museum of Art, Hove; Ms Patricia M. Butler, Museums, Ipswich; Mr J. Pemble, Borough Amenities Officer, Kettering; Mrs Dunn, Ferens Art Gallery, Kingston-upon-Hull; Miss Dilys Davies, Museum and Art Gallery, Kirkcaldy; Mr Stanley T. Dibnah, Library and Museum Service, Kirklees; Mr David Masson, Brotherton Library, University of Leeds; Mr Alexander W. Robertson, Temple Newson House, Leeds; A. R. Hinshelwood, Museums and Art Galleries, Leicester; Mr R. H. Wood, Usher Gallery, Lincoln; Mr Colin Bailey and Ms Marjorie E. Crook, Walker Art Gallery, Liverpool; Mr John Abbott, Mr Philip Athill, and Mr Eric Holder, of Abbott & Holder, London; Ms Nicole Weil, Service Culturel, Ambassade de France, London; Miss Janet Backhouse, The British Library, London; Ms Frances Carey and Mr Nicholas Turner, British Museum, London; Mr. T. G. Bunce, The City and Guilds Institute, London; the late Mr James Ingram, P. & D. Colnaghi & Co. Ltd., London; Mr Peyton Skipwith, Fine Art Society, London; Mr M. A. Ford, J. S. Maas & Co., London; Miss Judith Prendergast, National Portrait Gallery, London, Miss Sarah Boys-Stone and Miss Helen Valentine, Royal Academy of Arts, London; Mrs J. Agnew, Royal Commission on Historical Monuments, London; Miss Lucy Sutton, Society of Authors, London; Ms Jennifer Booth, Mr Christopher J. M. Johnston, and Ms Mandy Green, Tate Gallery, London; Miss Young, University of London Library; Mr John Sunderland, Witt Library and Mr Benedict Read, Courtauld Institute of Art, London; Miss Gillow and Mr Peter Cormack, William Morris Art Gallery, London; Mr John Bidwell, Mrs Edna C. Davis and Mr Thomas F. Wright, Williams Andrews Clark Memorial Library,

University of California; Ms Glenise Matherson, The John Rylands University of Manchester Library; Mr A. J. N. Arber, Ms Jane Edwards, Mr Derek Lawson, Ms Treena Lucas, and Mr Julian Spalding, City of Manchester Art Galleries; Ms Angela Bolger and Mr Robert Hopper, Whitworth Art Gallery, Manchester; Miss C. M. Parry, County Art Galleries, Merseyside; Ms Judith A. Schiff and Mr Vincent Giroud, Yale University Library, New Haven, Conn.; Ms Gill Hedley, Laing Art Gallery, Newcastle-upon-Tyne; William McNaught, Archives of American Art, New York; Ms Jean R. McNiece, Mr Kenneth Oberembt, and Ms Elizabeth E. Roth, New York Public Library; Mr W. N. Terry, Central Museum, Northampton; Ms Norma Watt, Castle Museum, Norwich; Mr David Phillips, Castle Museum and Art Gallery, Nottingham; Mr Myron Lasin, jun., Ms Monique McCaig, Miss Elaine Phillips, and Mr Brian Stewart, National Gallery of Canada, Ottawa; Ms Margaret Crum, Ms Margaret Green, and Mr D. G. Vaisey, Bodley's Librarian, Oxford; Mr K. J. Garlick, Mr Christopher Lloyd, Mr Dan Lowe, Dr Nicholas Penny, and Ms Noelle Brown, Ashmolean Museum, Oxford; Mme Caroline Godfroy and Mlle France Daguet, Durand-Ruel & Co., Paris; Mlle Anne Roquebert, Musée d'Orsay, Paris; Mr James Blair, Perth Art Gallery and Museum; Mrs Lindsey Hoole, Museums and Art Gallery, Portsmouth; Ms Jean E. Preston, Princeton University Library; Ms Karen T. Campbell, Museum of Art, Rhode Island School of Design, Providence, R.I.; Eric J. Stanford, Museum and Art Gallery, Reading; Mr Alan Jutzi and Ms Sara S. Hodson, The Huntington Library, San Marino, California; Ms Anne L. Goodchild, City Art Galleries, Sheffield; Mr H. B. Ratcliffe, Atkinson Art Gallery, Southport; Mr A. R. Mountford, City Museums and Art Gallery, Stoke-on-Trent; Mr R. F. Took, Sunderland Museum; Mr John S. Bunt, Glynn Vivian Art Gallery, Swansea; Miss Margaret Machell and Ms Christine Ardern, Art Gallery of Ontario, Toronto, Canada; Mrs Margaret A. Slater and Mr J. R. Watkin, Art Gallery and Museum, Warwick; Ms Cynthia Ott, Archives of American Art, Washington, D.C.; Mr J. Duxbury, Memorial Library, University of Wisconsin.

I am also indebted to the following for permission to quote excerpts from copyright material: the estate of the late M. Jacques-Émile Blanche, from *Portraits of a Lifetime*; Mr Richard Lancelyn Green, from Gordon Bottomley's letters; the estate of the late Lord Clark of Saltwood, from *Another Part of the Wood*; Mr Martin Holmes, from *Self and Partners (Mostly Self)*; the estate of Mr Laurence Housman, from *The Unexpected Years*; Lord Kennet, from Kathleen Bruce's *Self Portrait of an Artist*; Cecil Lewis, from the introduction to *Self-Portrait*;

the estate of the late Miss Lillah McCarthy, from *Myself and my Friends*; the Pissarro family, for letters from Camille and Lucien Pissarro; Fr Timothy Radcliffe, OP, from letters of Fr. John Gray; the estate of the late W. Graham Robertson, from his letters; Sir John Rothenstein, from Sir William Rothenstein's *Men and Memories* and his own *The Artists of the 1890s*; the Society of Authors, from George Bernard Shaw (© 1989 The Trustees of the British Museum, the Governors and Guardians of the National Gallery of Ireland and Royal Academy of Dramatic Art).

For more general and long-term help, I would like to thank the staff of the London Library, the British Library, the Bodleian Library, the Art Library at the Victoria and Albert Museum, the Cambridge University Library, the New York Public Library, and the Bibliothèque Nationale, Paris.

Dr Barbara Heinzen read the manuscript and made useful suggestions, as did the anonymous reader and my editor Dr Jane Wheare at OUP. Mr Paul Chipchase's sharp eye detected some factual errors and many embarrassing slips and solecisms; for his patient and meticulous reading of the typescript I am deeply grateful. With his long experience of publishing, Mr Peter Day has been an assiduous guide and counsellor. Mrs Stephanie Haynes and Miss Pat Deacon transformed my scribbles into lucid typescript. The late Miss Annabel Sprigge, Miss Kayoko Akagi and Sig. Stefano Marostica have helped with photocopying. My single greatest debt is to the late Mr George Warner Allen, an admirer of Ricketts since he saw the Ricketts Memorial exhibition in 1933. Any section of this book that might show a deep knowledge of the history and techniques of painting can safely be attributed to him; without his help in this and in many other ways and his unfailing encouragement I doubt if I could have finished this biography. I much regret that he did not live to see its publication.

Finally, my agent Caroline Dawnay deserves my heartfelt gratitude for her support and persistence. So do Sir Rupert Hart-Davis, Paul Chipchase, and Stefano Marostica who read the proofs. My editors at OUP, Frances Whistler and Anne Ashby, have been models of patience and sympathy.

J.G.P.D.

Marske-in-Swaledale, 10 December 1973
Brightwell-cum-Sotwell, 11 June 1988

Contents

List of Illustrations xv
Note on Sources and Quotations xix
List of Abbreviations xxi
Prologue 1
1. 'My young youth' 5
2. Ricketts and Shannon: 'as others see us' 19
3. The aesthetes of the Vale, Chelsea 38
4. The 'Book Beautiful' 94
5. The 'Seven Devils of the Collector' 114
6. An 'undiscovered Master' of the nineteenth century 142
7. The 'House Beautiful' 181
8. The Theatre Designer 224
9. Egypt and the *Ballets Russes* 251
10. 'I died during the war' 277
11. 'Shannon's latest folly' 309
12. 'My Canadian berth' 337
13. 'I'm tired of Art' 372
Epilogue 399
Bibliography 402
Index 415

List of Illustrations

1. Charles Robert Ricketts (1838–83) 4
2. Edward Woodville Ricketts (1808–95) 9
3. Charles Shannon (?1880s) 20
4. Charles Ricketts (early 1880s) at 164 Kennington Park Road 23
5. The Vale, Chelsea (*c.*1888) 39
6. *Had Zimri Peace Who Slew his Master?* After a Pen illustration by Charles Ricketts for 'Jezebel', in the *Universal Review* (1888) 42
7. *Oedipus and the Sphinx* (1891). Pen drawing by Charles Ricketts 47
8. *Mr Ricketts and Mr Shannon in the Enjoyment of Popular Success* (1907), by Max Beerbohm 50
9. Cover design by Charles Ricketts for *A House of Pomegranates* (1891), by Oscar Wilde 61
10. *Love in the Snow.* Wood-engraving by Charles Ricketts for *Daphnis and Chloe* (1893) 74
11. Cover design by Charles Ricketts for *In the Key of Blue* (1893), by John Addington Symonds 77
12. *The Two Old Kings.* After a Pen illustration for *Poems Dramatic and Lyrical* (1893), by Lord de Tabley 79
13. Cover design by Charles Ricketts for *The Sphinx* (1894), by Oscar Wilde 83
14. Portrait miniature of Edith Cooper (1901) by Charles Ricketts (Fitzwilliam Museum) 89
15. Katherine Bradley (1846–1914; photograph) 90
16. Pen illustration for *The King's Quair* (1895; collection of Frederick Koch) 99
17. Border and frontispiece designed and cut in the wood by Charles Ricketts for *Nimphidia*, by John Drayton (Vale Press, 1896) 105
18. *The Man in a Black Shirt* (self-portrait). Oil painting (1897) by Charles Shannon (National Portrait Gallery) 111

19. *The Man in the Inverness Cape* (i.e. Charles Ricketts). Oil-painting (1898) by Charles Shannon (National Portrait Gallery) 112
20. Lansdowne House, showing *Lapiths and Centaurs* by Piero di Cosimo 120
21. Renaissance border designed and cut in the wood by Charles Ricketts for *Julia Domna*, by 'Michael Field' (Vale Press, 1903) 129
22. The Sabatai ring (1904) designed by Charles Ricketts for 'Michael Field' (Fitzwilliam Museum) 139
23. *The Lost Piece of Silver*. Proof woodcut designed and cut by Charles Ricketts for *The Parables* (Vale Press, 1903; Colnaghi's) 172
24. The Vale, dining-room (*c.*1890) 182
25. Lansdowne House, dining-room 183
26. Charles Ricketts (photo: 13 September 1903) by Beresford 189
27. *The Descent from the Cross*. Oil-painting (*c.*1905) by Charles Ricketts (William Morris Gallery) 206
28. *Don Juan in Hell*. Oil-painting (*c.*1908) by Charles Ricketts (Courtauld Institute, University of London) 208
29. *Don Juan in Hell*. Oil-painting (1931) by Charles Ricketts (Grundy Art Gallery, Blackpool) 209
30. *Heliodorus Driven from the Temple*. Oil-painting (1906) by Charles Ricketts (photo: Fine Art Society, London) 212
31. *The Head of John the Baptist*. Oil-painting (1928) by Charles Ricketts (Witt Library) 215
32. *Chimaeras* (or *Chacun sa chimère*). Oil-painting (1917) by Charles Ricketts (photo: Sotheby's) 217
33. Stage setting for *King Lear* (1909) by Charles Ricketts (Victoria and Albert Museum) 236
34. *The Plague*. Oil-painting (1911) by Charles Ricketts (Musée d'Orsay, Paris) 245
35. Charles Ricketts in Greece (photo: 1911) 256
36. *Silence*. Bronze statuette (1905) by Charles Ricketts (William Andrews Clark Library, University of California, LA) 273
37. Poster in lithography by Charles Ricketts for *The Dynasts* (1914), by Thomas Hardy 293
38. Costume design for Lillah McCarthy as Judith (1919), in

Arnold Bennett's play. Inscribed 'To her serene highness the Princess of Monaco' (private collection) 296

39. *The Man with the Greek Vase* (i.e. Charles Ricketts). Oil-painting (1917) by Charles Shannon (Art Gallery and Museum, Leamington Spa) 307

40. Costume design for *Parsifal* (1921) by Charles Ricketts (British Museum) 313

41. Life drawing said to be of Cecil Lewis (1920s) by Charles Ricketts (collection of Albert Sperison) 317

42. *Montezuma Sacrificing to the Sun.* Oil-painting (1915) by Charles Ricketts (Manchester City Art Gallery) 320

43. *The Wise and Foolish Virgins.* Oil-painting (1914) by Charles Ricketts (private collection) 323

44. *Orpheus and Eurydice.* Statuette in bronze (1905–7) by Charles Ricketts (Tate Gallery) 324

45. *Orpheus and Eurydice.* Oil-painting (1923) by Charles Ricketts (Peter Nahum Gallery) 325

46. *Don Juan Challenging the Commander.* Oil-painting (1928) by Charles Ricketts (Royal Academy of Arts, London) 329

47. *Joan and her Voices.* Design by Charles Ricketts for Shaw's *St Joan* (1924; Carlisle Museum and Art Gallery) 343

48. *The Vision of Queen Katharine.* Production photograph with Sybil Thorndike as Katharine of Aragon in *Henry VIII* (1926) 348

49. Costume designs for female chorus figures for *The Mikado* (1926; Ulster Museum, Belfast) 350

50. Plaster bust of Cecil Lewis (1926) by Charles Ricketts (Witt Library) 354

51. Henning Nyberg (1903–64) (photo: collection of the late Mrs Violet Nyberg) 382

52. Charles Ricketts painting *Faust and Mephistopheles* (*c.*1930) 394

53. Charles Ricketts, bust (1902), by F. R. Wells, at Arolo, Lago Maggiore, Italy 398

Note on Sources and Quotations

THE main manuscript sources to which the footnotes refer are listed in the Bibliography; others, less important, are given more fully in the notes. Letters and diary entries which are quoted in part or whole in *Self-Portrait: Taken from the Letters and Journals of Charles Ricketts, RA*, collected and compiled by T. Sturge Moore, ed. Cecil Lewis (1939), are indicated by *SP*, but the versions given in the text are taken from the original manuscripts where possible. Dates of letters given in square brackets are my own suggestions. For books and articles the place of publication is London unless otherwise indicated.

Ricketts's spelling was idiosyncratic: although he had his published writings corrected, he was unrepentant in his private letters. 'Why scold me for [not] using 2 g's in exagerate,' he wrote to Robbie Ross in June 1911, 'these small imperfections secure my popularity. Too perfect, I should be a subject for *controversy*.' Since his misspellings are extremely frequent, I have dispensed with the convention of acknowledging these with '[sic]'. I have, moreover, tidied up Ricketts's baffling punctuation, in the interest of fluency, without indicating my 'improvements'. Where an ampersand is used in the original manuscripts, it has been retained in the quotations. Ricketts's misspellings of proper names in the text have been corrected in the index.

List of Abbreviations

Affectionate Cousins	Sylvia Legge, *Affectionate Cousins* (Oxford, 1980).
All for Art	Joseph Darracott, *All for Art: The Ricketts and Shannon Collection* (Cambridge, 1979).
Bibliography of Vale Press	*A Bibliography of Books issued by Hacon and Ricketts* (1904).
Birnbaum	Martin Birnbaum, *The Last Romantic* (1960).
Bliss Catalogue	*Catalogue of the Bliss Collection of Alphonse Legros*, introduction by Charles Ricketts (1923).
Bookselling	Temple Scott, 'Mr Charles Ricketts and the Vale Press', *Bookselling*, 2 (December 1896), 501–12.
Calloway	Stephen Calloway, *Charles Ricketts: Subtle and Fantastic Decorator*, foreword by Kenneth Clark (1979).
CHS	Charles Shannon.
Clark	Kenneth Clark, *Another Part of the Wood* (1974).
CR	Charles Ricketts.
CR, *Wilde*	Charles Ricketts, *Oscar Wilde: Recollections* (1932).
Defence	Charles Ricketts, *A Defence of the Revival of Printing* (1899).
Dial	*Dial*, 1–5 (1889–97), ed. Charles Ricketts and Charles Shannon.
Field	Charles Ricketts, *Michael Field: A Memoir*, ed. [J.G.] Paul Delaney (Edinburgh, 1976).
Friend of Friends	*Friend of Friends: Letters to Robert Ross*, ed. Margery Ross (1952).
Friends of a Lifetime	*Friends of a Lifetime: Letters to Sydney Carlyle Cockerell*, ed. Viola Meynell (1940).
Furst, *Apollo*	Herbert Furst, 'Charles Ricketts, ARA, and his Stage Work', *Apollo*, 1 (1925), 329–54.
GBS	George Bernard Shaw.
'Heirs'	[J. G.] Paul Delaney, 'Heirs of the Great Generation: Yeats's Friendship with Charles Ricketts and Charles Shannon', in Warwick Gould (ed.), *Yeats Annual*, 4 (1986), 53–73.

Holmes	C. J. Holmes, *Self and Partners (Mostly Self)* (1936).
Lady Kennet	Lady Kennet (Kathleen Bruce), *Self-Portrait of an Artist* (1949).
Letters CR to MF	*Letters from Charles Ricketts to 'Michael Field', 1903–14*, ed. [J. G.] Paul Delaney (Edinburgh, 1981).
Letters of W. B. Yeats	Allan Wade (ed.), *The Letters of W. B. Yeats* (1962).
'Marginal Recollections'	Charles Ricketts, 'Some Marginal Recollections of the "Nineties"' (review of W. Rothenstein's *Men and Memories*, i), Observer (15 Mar. 1931).
McCarthy	Lillah McCarthy, *Myself and Friends* (1933).
Men and Memories	William Rothenstein, *Men and Memories*, 2 vols. (1931–2).
MF	'Michael Field'.
'Musical Recollections'	Charles Ricketts, 'Desultory Musical Recollections', *Music Bulletin*, 5 (1923), 10–12, 43–6, 88–90, 97, 160.
nd	no date
Pages on Art	Charles Ricketts, *Pages on Art* (1913).
pc	postcard
pm	post mark
Prado	C. S. Ricketts, *The Prado and its Masterpieces* (1903).
Rewald	John Rewald (ed.), Camille Pissarro, *Letters to his son Lucien*, 4th edn. (1980).
RNRH	R. N. R. Holst.
Rothenstein, *Artists*	John Rothenstein, *The Artists of the 1890s* (1928).
Rutter	Frank Rutter, *Art in my Time* (1933).
Schuchard	Ronald W. Schuchard, 'W. B. Yeats and London Theatre Societies, 1901–1904', *RES*, NS, 29 (1978), 415–46.
Since Fifty	William Rothenstein, *Since Fifty* (1939).
65 Ill.	Charles Ricketts, RA, *Sixty-five Illustrations*, introduction T. Sturge Moore (1933).
Some Letters	*Some Letters from Charles Ricketts and Charles Shannon to 'Michael Field', 1894–1902*, ed. [J. G.] Paul Delaney (Edinburgh, 1979).
SP	*Self-Portrait: Taken from the Letters and Journals of Charles Ricketts, RA*, collected and compiled by T. Sturge Moore, ed. Cecil Lewis (1939).

TEL	T. E. Lowinsky.
Theatre Arts Monthly	Gordon Bottomley, 'Charles Ricketts, RA', *Theatre Arts Monthly*, 16 (May, 1932), 377–95.
Theatre Notebook	Ifan K. Fletcher, 'Charles Ricketts and the Theatre', *Theatre Notebook*, 22 (1967), 6–23.
Titian	Charles Ricketts, *Titian* (1910).
TSM	T. Sturge Moore.
TSM, original introd.	T. Sturge Moore, original introd. to his collection of Ricketts's letters (BL Add. MS 61713).
Unexpected Years	Laurence Housman, *The Unexpected Years* (1927).
Vale Prospectus	Prospectus for the Vale Press, 1895–8.
WBY	W. B. Yeats
Wilde, *Letters*	*The Letters of Oscar Wilde*, ed. Rupert Hart-Davis (1962).
Yeats and Moore Correspondence	*W. B. Yeats and T. Sturge Moore: Their Correspondence, 1901–1937*, ed. Ursula Bridge (1953).

Prologue

'HE was a painter, sculptor, book illustrator and stage designer of real distinction, and much of his work, especially his sculpture and his costume designs, will be "rediscovered" with astonishment.' So wrote Kenneth Clark about his friend Charles Ricketts, RA (1866–1931). He went on to compare him with another key artistic figure of the day, Bernard Berenson, who 'had the greater mind' (he was Clark's own master), while Ricketts 'knew how things were made'.[1] In another place, Lord Clark calls Ricketts and Roger Fry 'the leading critics of art of their time', contrasts Ricketts's exquisite taste and lively conversation with the dull paintings on Fry's wall and his 'sonorous monologue', and concludes that Ricketts's friends would first remember him for 'his laugh, destroying all pomposity and encouraging us to say outrageous things!'[2] There are numerous such tributes to Ricketts. Wilde described him as 'cultivated and interesting';[3] Yeats as 'one of the greatest connoisseurs of any age';[4] G. B. Shaw as 'the noble and generous Ricketts . . . a natural aristocrat as well as a genuine and devoted artist'.[5] For these three, Ricketts designed both books and plays (though Wilde's *Salome* was a posthumous production). Having collaborated with such notable writers, and with Lillah McCarthy, Sybil Thorndike, Harley Granville-Barker, and Sir Thomas Beecham in his theatre work, Ricketts's name crops up in many biographies and memoirs of the 1890s and early twentieth century, but his achievements have never been considered in their own right. No biography of Ricketts has ever been attempted before now, though there were two books on him in the 1930s and three more have been published in the last decade. A biography has long been overdue, not only for the light it can cast on Ricketts's distinguished friends, but so that he himself can be better understood.

Today, Ricketts is probably best remembered for his book designs,

[1] Clark, pp. 178–9.

[2] Calloway, p. 6.

[3] Wilde, *Letters*, p. 249.

[4] Quoted in Ann Saddlemyer, 'The Heroic Discipline of the Looking Glass', in R. Skelton and A. Saddlemyer, *The World of W. B. Yeats* (Seattle, 1967), 70.

[5] *65 Ill.*, p. [v] = *SP*, p. 18.

particularly those done in the 1890s. John Gray's *Silverpoints* and Wilde's *The Sphinx* are now seen, along with Beardsley's *Salome*, as the quintessence of that fascinating period, and command prices that would have amazed and delighted Ricketts. His designs have had enormous influence in Europe and America. The Vale Press he founded was among the most important early private presses. His sculpture and theatre designs, especially the superb finished costume-drawings he liked to present to friends, are becoming better known but still deserve attention. His most famous productions, Shaw's *St Joan* with Sybil Thorndike, the D'Oyly Carte *Mikado*, *The Coming of Christ* at Canterbury Cathedral, were among the most brilliant theatrical successes of his day. Yet the story of these and many other productions to which he contributed has never really been told. As a painter Ricketts is still largely neglected. With the revival of interest in Victorian painting, and particularly in Symbolism, his aims and values are likely to be better appreciated. Particular attention has been given here to his painting, which has received summary treatment to date. Ricketts was also a superb writer. His books on the Prado and Titian are among the most lively and evocative works on painting ever written in English, and deserve to be read by anyone interested in art. Finally, Ricketts was a connoisseur and collector of a very high order. On an income that rarely exceeded £1,000 a year, he and his lifelong friend Charles Shannon amassed a superb collection which, particularly in drawings, rivalled those of contemporary millionaires and was mostly left to the Fitzwilliam Museum, Cambridge. Such a wide range of activities might suggest that Ricketts was a dilettante, but he was not, as Kenneth Clark indicates. His knowledge of how things were made included not only an impressive scholarship (always worn lightly) but also an intrinsic understanding of most materials, similar to that of William Morris. His versatility has, however, counted against him with art historians, making him difficult to pigeon-hole and prompting suspicion and distrust.

All these elements must be woven into a life that was full of contradiction. Ricketts was secretive and rather reclusive, yet led a life full of feverish activity. He was passionately devoted to art in almost any form, yet as fiercely antagonistic to Post-Impressionism as he was to philistinism (of which this 'French disease' was in his view a form). His conversation was full of wit and gaiety, yet inwardly he was tormented by fears and anxieties. In a published account, his old friend Thomas Sturge Moore described Ricketts's friendship with Shannon as 'the most marvellous human relationship that has come within my observation ... but neither easily indulged the other; their union was more bracing than comfortable'.[6] Privately, however, he acknowledged that their friendship was 'full of strains and nearly broke down under it'.[7] Indeed, it has been largely misunderstood, and often summed up misleadingly in the arch clause 'They did everything together!' The story of their relationship, which almost foundered at least twice, can be built up from hints in his diaries and letters, even though Ricketts vigorously censored the former, deleting passages, tearing out sections and pages, and even destroying volumes covering whole years. The entries (spanning a period of eleven years) that survive are only a fraction of what there once was.

Finally, Ricketts himself has been an enigma. People now, as in his lifetime, are strongly either for him or against him. His wit, his ebullience, his generosity, his fine critical mind won him allies; his powerful personality, his love of influencing people, his violent anger and antagonisms, his 'machiavellian' scheming to further his own artistic values earned him as many enemies. He fascinated friend and enemy alike. Paradoxically, underneath his friendship with Shannon and his great vitality lay a loneliness, a tragic childhood, a deep pessimism about life which has never been analysed but which is the key to his character. His experience of life led him beyond the exclusive aestheticism of his youth to more humane values.

[6] *65 Ill.*, p. [v] = *SP*, p. 17.

[7] TSM to Owen Lewis (n.d.), transcript in Sturge Moore Papers, Univ. of London, Box 20/21.

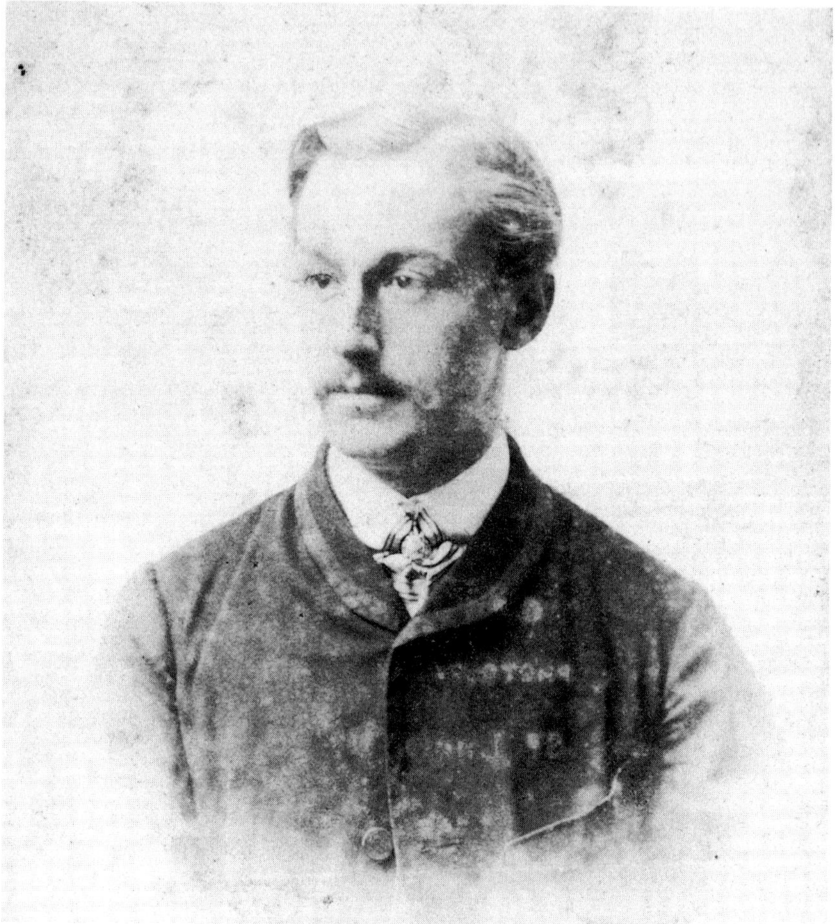

FIG. 1. Charles Robert Ricketts (1838–83)

1 *'My young youth'*

Charles de Sousy Ricketts was born at 57 rue du Rhône, Geneva, on 2 October 1866, the second child of an English father and a French mother. Their first child, a daughter, had died in infancy. His father, Charles Robert Ricketts (Fig. 1), was studying painting in Geneva with the help of an allowance from his own father. After a short career in the Royal Marine Light Infantry, Charles Robert had risen to be a First Lieutenant on *HMS London* by 1859. However, he had been invalided out of the service by 1863, aged only twenty-five, and retired on half pay, suffering from the effect of wounds.[1] With his naval background, he specialized in painting marine subjects, and eventually exhibited his seascapes at the Royal Academy and the Royal Society of British Artists in London.

In November 1862 he was in Naples, where he is said to have met his future wife, Hélène Cornelie Jouhan, a widow three years his senior, born Hélène Cornelie de Soucy.[2] Her background has proved elusive: Ricketts said that he was 'born of a French mother, who was bred Italian & had Spanish blood by her father', and her English marriage certificate gives her father as Louis, Marquis de Soucy.[3] On other early documents her name is spelt Sousy, Soussy, Souchey, Soucey; Ricketts used the form 'Sousy' on his birth certificate for his middle name. Unless her father's title was bogus, the only person he could be was Armand Louis de Fitte, Comte de Soucy, who, as the last male heir, presumably had some claim to the title of Marquis borne by his grandfather. As Hélène was not listed among his daughters at his death in 1862, she must have been illegitimate, a hypothesis which is not only supported by some circumstantial evidence but which would explain a great deal about Ricketts's character.[4] She seems

[1] Navy List, 1859.

[2] Mrs Ursula Bridge attempted unsuccessfully to trace the marriage record in Naples. See BL Add. MS 61713.

[3] MF Journal 1898, fo. 97^{v}. See also TSM, 'Charles Ricketts' [obituary], *Observer* (11 Oct. 1931).

[4] Printed Obituary Notice in file of Général de division Armand Louis Joseph de Fitte, Comte de Soucy, Service historique de l'Armée de Terre, Château de Vincennes, Paris.

to have had no title, no money, no inherited possessions. Ricketts, moreover, had no contact at all with her family and only once referred to it, when he mentioned that his grandfather had known Rossini, but gave no name or achievements.[5] This may have been deliberate, out of a desire to shield his mother from any hint of scandal or shame. Ricketts had a secretive side to his nature and in Victorian times the fact of having an illegitimate mother would not only have strengthened this tendency but also increased his insecurity and defensiveness.

The Comte de Soucy, of a family which had belonged to the *noblesse de robe* since the sixteenth century and is now extinct, had been a general in the French army, Inspector-General of French colonial troops, and Governor of Guadeloupe and, later, Martinique.[6] Curiously, the Ricketts family also had connections with the West Indies where they had large holdings, one of their ancestors having accompanied Admiral Penn and General Venables in the Conquest of Jamaica in 1655. However, much of their wealth had been lost with the abolition of slavery, so that nineteenth-century descendants, having returned to the mother country and settled in Hampshire, sought careers in the civil service, the navy, or the Indian army.[7]

On 20 January 1868, when Ricketts was just over a year old, his parents were married in a civil ceremony in London, with two of the groom's brothers as witnesses,. This was probably a remarriage because they had failed to register with British consular officials their wedding in a Neopolitan church. Hélène de Soucy was a Roman Catholic, though lapsed, and it seems likely that there had been a church wedding in Italy.

The family settled in Norwood, south of London, where the elder Ricketts's brother Arthur lived. Nearby was the Crystal Palace, the magnificent erection of iron and glass built in Hyde Park for the Great Exhibition of 1851 and then re-erected three years later south of the Thames. With its many courts demonstrating the culture and architecture of different civilizations, its museum and art gallery, it was, Ricketts remembered, 'a paradise for children, and one of the most comprehensive art museums in the world (this I knew later)'. It

[5] 'Musical Recollections', No. 1: pp. 10–12, No. 2: pp. 43–6, No. 3: pp. 88–90, 97; No. 5: p. 160. This article is the main source of information on Ricketts's childhood and the basis of this account, unless otherwise indicated. Part of it is in *SP*, pp. 1–3.

[6] Roman d'Amat, *Dictionnaire de biographie française*, 13 (Paris, 1971), 1415.

[7] Ralph Ricketts to author, 14 Feb. 1982. See *Burke's Landed Gentry,* i (1952), 605–6.

was also, for the 1870s, 'the home of music in England . . ., with daily concerts, a small local opera, crashing brass bands, a mammoth organ, great Saturday classical concerts, and huge Handel festivals'. Its attraction to children, and especially to a sensitive and imaginative child like Ricketts, was broad: 'The place was not only full of appeals to the imagination from the toy stalls to the great intimidating groups of statuary, it was a world full of sound.' Ricketts always claimed that his early aesthetic sense—and Shannon's—was stimulated by the 'iced cakes and monumental wedding cakes' that he used to see there.[8] Together with his continental background, it helped to provide him very early on with a base for that astonishing breadth of culture, ranging through the art, music, and literature of many nations, that marked him as an adult.

The particular bent of his character revealed itself early: his earliest memories centred around music and visual images. From his father he inherited his strong visual sense; from his mother his love of music. Almost his first memory was of the rainbow reflections of a Victorian crystal chandelier on the walls as he lay on the floor. His first 'scandal' occurred when he rushed into a lady's skirt, shouting 'blue, blue, blue!' Out of this visual fascination grew his great love of flowers. The first ones he remembered grew by the Thames, poppies and scabious, and he was so fond of the look of pansies, red daisies, and geranium leaves that they used to be sent by friends and relatives to 'little Charlie'.[9] Another very early memory was of the cadences of his mother's beautiful voice as he sprawled on the floor and she, frowning slightly and entirely self-absorbed, practised vocal scales and trills at the piano. Rossini had declared that it was a pity that her voice could not be heard upon the stage; an opinion, Ricketts admitted, that might have been influenced by truffles and Chambertin. Still, she had a very good voice, and sang arias from 'dear old Italian Operas of the worst type', so that the young Ricketts grew more familiar with Italian operatic music than with the other European classics. Though he had seen Verdi conduct his *Requiem* with a chorus of 1,200 at the Albert Hall in 1875, his most significant early musical memory was of being taken, aged six, to *Don Giovanni*; he was impressed by the

darkened stage, a white figure, dishevelled and clinging to a man in violet velvet, an incomprehensible duel, a ballroom scene with a threat of pistol

[8] CR to Gordon Bottomley, p.m. 17 Dec. 1922 and 24 Dec. 1927.

[9] MF Journal, 1902, fo. 47^{v}.

shots, a familiar serenade, then the awful ghostly figure of a man on a horse, and many other dim fateful things moving to the beck of the conductor

His vivid memory of this first opera shows not only how responsive he was but also the emergence of a fascination with Don Juan that was to haunt him all his life.

One of his father's sisters, Eleanor, who was musical and played the harp as well as the piano, introduced him in his seventh or eighth year to a composer who became one of the passions of his life. His aunt was practising behind a closed door: the music was unfamiliar and the boy became 'dimly conscious of something new and strange, unlike the cadenced questions and replies of most music I then knew'. He appeared, inquisitive and anxious, to ask 'What was that, Auntie? Play it again.' She complied, and for years afterwards he would ask her to play the 'immortal and unsurpassed *Fantaisie Impromptu*' by Chopin. Along with Schumann and Wagner, Chopin was to remain one of Ricketts's musical gods.

His Aunt Eleanor lived at his paternal grandfather's house at Ryde on the Isle of Wight. Ricketts had reservations about her: 'she was over-feminine & without imagination'—indeed, he hated most of his relatives and later recorded their visits in his diary as 'cinders, ashes, and dust'—but he was quite fond of his Aunt Flora because she 'was more like a man and very generous', and very fond of his grandfather Edward Woodville Ricketts, who had been Chief Clerk to His Majesty's Treasury (Fig. 2). Ricketts remembered him as 'kindness & sweetness incarnate', with 'beautiful voice & diction'.[10] From him Ricketts inherited his love of beautiful books, for the old man was a connoisseur of fine editions; his bequests to friends reflect his delight in a fine volume, such as 'the beautiful Pine's Horace given to my father by his brother'.[11] He had also collected pictures in a small way on a tour of Italy with Lord Northesk, and Ricketts had a vivid recollection of the Italian paintings which had belonged to his parents and grandfather. On visits to the London auction rooms in later years he sometimes recognized these. In 1909 he found four 'primitive' pictures which had

[10] See *SP*, p. 57, and CR to Florence de Sinner, Townshend House [after May 1923] and Lansdowne House [before May 1923].

[11] Supplementary List, The Will of Edward Woodville Ricketts, 6 June 1892. (In the coll. of Dr Vincent von Sinner.)

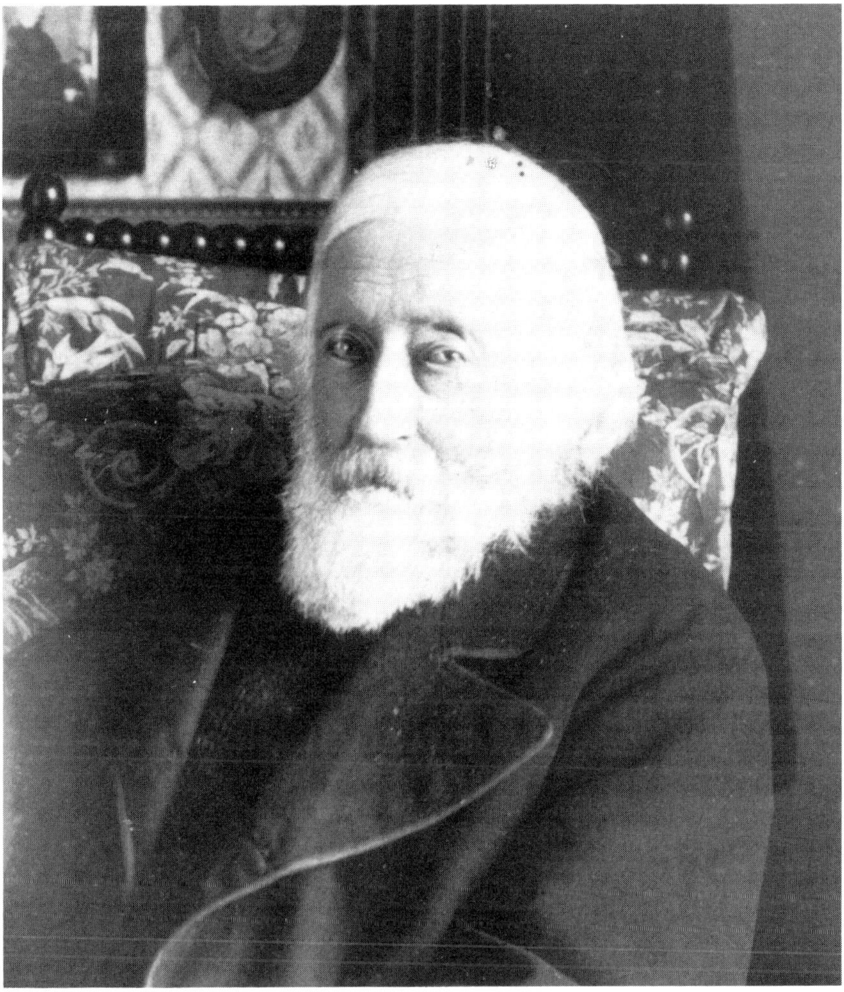

FIG. 2. Edward Woodville Ricketts (1808–95)

once hung at home (by an imitator of Crivelli rather than by Vivarini as he had supposed), which he had not seen since 1888, when he was twelve, but he was able to state that originally they had not been varnished.[12] A few years later some of his grandfather's collection turned up at Sotheby's.

> I went there with some emotion [he recorded] as I knew the old 'Bassano' once in the Ryde dining room was there ... My emotion was increased beyond endurance when I recognised two panels of saints by Masaccio skied, and a predella piece by Pesselino lying on the floor. My heart started beating so violently that I had to leave the room and I felt faint on the stair case.[13]

Ricketts, who could not remember a time when he was not interested in pictures, had an astonishing visual memory which, he claimed, he had 'not found equalled by any expert or director of museum, but, Alas! no memory for names & dates'.[14]

By the mid 1870s the family had moved to Kensington, to 79 Harcourt Terrace, in a row of 'damned houses' which his father jovially described as 'The Catacombs'. Here they had musical Sundays, when friendly Italians sang loudly and gossiped louder still about the great divas and the great operas. Astonished neighbours came out on to their balconies to listen. His mother had joined the Choral Society of the Albert Hall, whose organist would visit and show young Charlie the stops 'with delightful names' that produced sounds so different from the 'horrid collective hymn-singing of St. Luke's', the local Anglican church which they attended, and for which Ricketts held, even after many years, a 'shuddering dislike'. He believed that 'the vocal horrors of our Anglican church service and woolly anthems' had blighted the musical taste of many an Englishman.

The move to Kensington coincided with attempts at a more intensive education from a more capable governess. On walks with her he would stop to see a cutter-yacht made of sugar in the window of a shop opposite Holland House. Ricketts later associated it with

> a period of unaccountable melancholy and dawn of the inner life which began with me at the age of 8, when I brooded and actually ventured on suicide with emerald green watercolour paint. Earlier in life my character had been

[12] CR to Mr Laing, 15 Jan. 1909.
[13] CR Diary, 28 June 1915 = *SP*, p. 242.
[14] CR to Mr Laing, 15 Jan. 1909.

sunny and outward, given over to visible things from the first, flowers, colours, music, art, the gardens and courts of the Crystal Palace.[15]

Whether the cause of this was the new governess or the more intensive schooling is not known, but worries over his mother's ill-health must also have contributed to it. The jovial outward and the tortured inward aspects were to remain permanent features of his character.

Lively, sensitive, highly intelligent, but at the same time wilful, arbitrary, and highly strung, he must have been 'a bewitching child, the joy of the governess', a friend concluded. When he misbehaved he was warned that he would be sent out to become a bootblack and once, persisting in his naughtiness, was actually taken to the front door, where his pride quailed before the horrible fate that awaited him. He was so impressionable and hypersensitive that he sobbed inconsolably at the sight of the funeral of a total stranger, when his mother explained that the person was 'going away': so susceptible was he at an early age that he could not bear the thought of loss. Though he was fascinated by the piano, he had little aptitude for such mechanical inventions and his attempts to learn to play it often ended in angry bursts of tears, until his lessons were wisely given up, but he would steal back to the instrument when alone, to be enthralled by single notes, random chords or 'delicious shuddering crashes'. He often caught his fingers when slamming the lid when someone approached, 'a guilty resonance still in the air', but this was ignored, till once his father, discovering him, laughingly dubbed him 'the piano tuner', and 'from that day this habit ceased'. There was an element of steely pride in his character.

Ricketts usually speaks of his childhood as if he were the only child, and indeed had something of the egotism and isolation of one, but a sister, Blanche Isabella, had been born shortly after their arrival in England. No doubt he resented her arrival from the first, and even more when she proved to have little in common with him. After their childish squabbles the young Charlie, since he was not permitted to beat her, used to beat her dolls to relieve his frustration.[16] Throughout his life, he would always need to give vent to his anger, and there was, moreover, a streak of ruthlessness in his nature.

A friend recorded that Ricketts 'had nothing of his Anglo-Saxon

[15] CR to WBY, p.c. (n.d.), [after 1914; possibly 1926]. See WBY, *Autobiographies: Reveries over Childhood and Youth* (1914); collected edn, 1926), 37.

[16] MF Journal, 1899, fo. 122.

father but the name'.[17] Even though they shared an interest in the visual arts, Ricketts commented pointedly 'we liked totally different things in the National gallery'.[18] Their different taste in painting was a sign of their differing characters. By not going in for athletics and field sports he disappointed his father, who sometimes used crushingly to address his only son as 'Mademoiselle'.[19]

He was much closer to his mother. He was her confidant, and shared with her not only a passion for music but also that animation and quickness of mind that seem peculiarly French. His dependence on her was increased by the family's frequent moves, from Switzerland to England, from Norwood to Kensington, and later from England to France. No continuity of friendship with his peers was possible. As he felt little rapport with his father or sister, his mother must have been the one constant, close attachment in his early life.

Concern over her health, however, cast a shadow of fear and anxiety over the last years of his childhood. She suffered from a tumour.[20] When she became an invalid Italy was recommended, Ricketts observed, 'by doctors short of advice'. On the journey to Boulogne both mother and son fell sick, and she was saved from a sudden crisis by a local practitioner under whose care she stayed for two years. Boulogne, where they lived in a dark house in the rue de l'Écu, Ricketts recalled as 'a dusty, dirty stinky old place', where he visited the '*unromantic* churches' with a Catholic governess and loved the ramparts and the gateway with the Virgin in a boat in which crocuses used to bloom in the spring. Here he went to school for the first time at the Collège Communal, and, as well as Latin, he later recalled (mischievously mocking Victorian morality), 'I learnt filthy French words and other practices—hence I view it with affection'. At school, however, he was an alien, and made to fight because he was a 'sale gosse Anglais'.[21] Again and again in his youth he found himself an outsider.

Their next move, when Ricketts was twelve, was to Asnières, a boating centre on the left bank of the Seine near Paris, 'that ridiculous

[17] MF Journal, 1898, fo. 97^v.

[18] CR to Hamilton Minchin, 29 June 1914. Quoted in Edward D. Nudelman, *Catalogue*, 9 (Seattle, Wash., 1985), 34.

[19] Holmes, p. 163; MF Journal, 1904, fo. 38.

[20] MF Journal, 1902, fo. 48.

[21] CR to TEL [*c.*1919]; CR to MF [12 Aug. 1910]; CR to WBY, p.c. (n.d.) [after 1914; possibly 1926].

old place in which I spent two of the happiest years of my boyhood'.[22] This meant of course Paris, the Louvre, the salons, two 'banal' visits to the opera, grand masses at the churches, the crowding of 'countless precocious perceptions, and the dawning of personality'. In the Paris of 1879 and 1880 he began to formulate the taste and artistic principles that were to guide his life. He fell under the spell of Puvis de Chavannes, whom he regarded as Europe's greatest living artist, and of Gustave Moreau, who more than any other influenced his painting; but his tastes were quite wide. After a visit to the Metropolitan Museum in New York in 1927, he wrote, 'I was delighted to see old favourites of my boyhood, Bastien Lepage's *Joan of Arc*, Gustave Moreau's *Oedipus*, Manet's *Girl with the Parrot*, Regnault's *Salome*, looking very vulgar but perfectly preserved.'[23] What these all have in common is their human subjects, their relevance to human experience; for art to Ricketts was a comment on life, more often than not on human tragedy. What is absent from this list is any interest in the Impressionist landscapes that were then making their appearance, came to be so highly prized in his lifetime, and are so highly priced today.

The years in France laid the foundation of his wide knowledge of French literature. Flaubert and Zola were at the time the leading novelists; Anatole France was commencing the career that would lead to his being regarded as the greatest French novelist of the age, an opinion that Ricketts shared. In his boyhood he developed an interest in Flaubert, 'whose pessimism', he noted later, 'is quite unlike anything of the kind known [in England], where the pessimist—if he exists—is a sparse creature with a long upper lip and nothing in his trousers'. This more earthy French tragic sense—'an element in all things, almost a central fact'—answered something in his own nature, and its influence contributed to the tragic character of his paintings.[24] A generation before, Baudelaire (Ricketts's favourite poet, along with Catullus and Keats) had initiated that fascination with decadence that was so to enervate the comfortable bourgeoisie in an increasingly prosperous France. Verlaine, after his early paganism, was about to turn to religious subjects, with which Ricketts had little sympathy. Although he had at first been interested in Verlaine's younger followers, he gave them up

[22] CR Diary, 21 Mar. 1915 = *SP*, p. 232.

[23] CR to Mary Davis, Sunday, 20 [Nov. 1927].

[24] CR to Laurence Haward, 8 Mar. 1915.

with Henri Regnier; only Verlaine, he decided, 'came to stay'.[25] In 1915 he wrote, 'I probably know Baudelaire, Flaubert, Verlaine, Fromentin better than most Frenchmen and, I am sure, better than any living Englishman.'[26]

The winter of 1879–80 was the severest of the century in France, and the Seine froze solid.[27] That December the family set out once again for Italy. On their way to Rome they stopped off at Genoa, but three days before the intended departure for Rome, on 30 December 1880, Hélène Ricketts suddenly died; her age, forty-six. She was buried on a hillside in the city. Ricketts was fourteen: 'I felt old as the earth', he recalled.

He later confessed that if he showed little affection for his friends it was because it was all with his mother.[28] Even allowing for some dramatization, this is essentially true. Her loss at such an impressionable age could only have exacerbated that vein of pessimism in his character that grew out of his reading and his fears and anxiety over her health. Throughout his life, he always took the worst for granted. 'With me', he said, 'pessimism is a survival of the religious instinct, and a sort of subconscious hint to Fate that I know the worst, and that it would be unsportsmanlike to act on my premonition.'[29] The worst in fact had already happened to him with his mother's death; happiness and security were ever after to seem precarious. This fear and pessimism characterized the inner man, but it had its effects on the outer man too. It explains not only his fits of nerves and temper but also 'a certain amount of despair', which, he observed, 'counts always in my personal energy and gaiety'.[30]

The family returned to England, where he spoke English with a French accent and found himself again an outsider. As he was considered too delicate for school, attempts to give him a conventional education now ceased. For the next two years he 'basked' in museums and read day and night. He was allowed to read what he liked, so that it must have seemed hard when he was forbidden to see the first performance of Wagner's *Ring* in London in 1882. Because of the incest

[25] MF Journal, 1901, fo. 7^{v}.
[26] CR to Mary Davis, [Apr. 1915].
[27] 'Paris', *Encyclopaedia Britannica*, 11th edn, 20 (1911), p. 821.
[28] MF Journal, 1899, fo. 24^{v}.
[29] CR to Cecil Lewis [*c.*1926]; *SP*, p. 373.
[30] CR to Laurence Binyon [*c.*1915].

in *Die Walküre* there had been some 'idiotic call for censorship in the Press', which had startled his father. All he saw of this famous production was the two property goats attached to Fricka's chariot outside the stage door. Countless Sundays were spent in two friendly houses, where the works of Beethoven, Chopin, and Schumann became even more familiar to him. One of these, near Plumstead church, was the home of an eccentric spinster with untidy hair, who had 'a distrust for her sex, and a liking for exaggerated character in men, or for that matter, for men without character'. Her playing was as untidy as her hair and rather limited by rheumatic wrists and the 'tedious stress of an anxious life', but she fired Ricketts's imagination with her memories of performances by Rubinstein and Liszt, and her admiration for Chopin. The piano, however, was better played at a house at Sydenham, where another older woman friend opened to him the range of the art of Chopin and Schumann. This second family may have been called Gray, friends of the Rickettses who had three children, Norman, Saxon, and Brenda. The latter turned up again in 1904, now married and called Mrs Wiggins.[31]

Ricketts felt that these two elderly women alone influenced him in his early days. His youth was spent largely among grown-ups, whose influence tended to strengthen his own rather conservative outlook. Indeed, his tendency to form friendships at this age with people much older than himself is indicative of a backward- rather than forward-looking mentality. But these two women, who were highly intelligent and interested in finer things, helped form his character and interests. They were free thinkers; the first was a declared atheist. By the time he was fifteen Ricketts too had declared himself a convinced atheist and materialist whose sole religion was beauty.[32] (As if to prove his convictions, he once walked downstairs 'to meet and challenge a ghost which came nightly to a part of my father's chambers' in Victoria Street.[33]) He never abandoned this view of life. The practice of religion simply passed his understanding. Nearly fifty years later, after reading *The Golden Bough*, he wrote: 'I feel more than ever convinced that Religions have done more to harm the world than other forms of stupidity & that "Humanity" has been a slowly evolved ideal running

[31] CR Diary, 27 Oct. 1904 = SP, p. 111.

[32] MF Journal, 1904, fo. 176.

[33] CR to WBY, p.c. (n.d.) [after 1914; possibly 1926]; CR to Florence de Sinner [Jan. 1931].

counter to the horrid placation ceremonies & taboos invented in fear of Nature who seems awful & purposeless enough.'[34]

Yet Ricketts's character had its contradictions. Whilst courageous enough to set out to challenge a ghost, he could be extraordinarily timid in dealing with much more ordinary situations. On his first visit to Oxford, where he went to see some relations, he was so upset by the bleak day and the gaudy, commercial side of the city that he nearly burst into tears in the court of St John's College, went back to the Clarendon Hotel, gave up the room which had been booked for him, and returned to London.[35] His nerves often got on top of him.

He lived for a time at his grandfather's at Ryde, and renewed contact with his Ricketts relations. Among these was his cousin Florence, the daughter of his father's brother Arthur. She and her mother were among the few relatives that Ricketts liked, and over the years the two cousins kept in touch, though it was always Florence who took the initiative in writing letters and paying visits. Two early letters to her, the only ones found from this period of his life, show something of his boyish character. They are full of exuberance, of bad puns, awful jokes and adolescent teasing, and of amusing marginal sketches. 'Norman Grey called,' he informed her when she was away in Scotland, 'and he and Papa had a game of Billiards, so I had a real live greyvance, as it prevented my going to the Mascotte, and, greycious, if he comes again he and I shall disagrey.' In the same letter he included a ballad which probably drew on an experience of hers, for she had attended a Quaker School. It was written, he claimed, in 'very early English to the tune of *The Old Cow Died Of*':

> A Quaker School to the Academy went,
> And there an afternoon in talking spent,
> All went happily un till By and By,
> The picture of Phryne, the Leighton they spy.
> Miss Sharp nearly fainted
> To se[e] such a thing painted.
> Some one speaks of the picture's combination of colour,
> This rises of the young ladies the rightious choler,
> 'Combination indeed!' is the unanimous bawl
> 'Why Phryne has got no combination at all'.[36]

34 CR to G. Bottomley, p.m. 16 Nov. 1926.

35 MF Journal, 1897, fo. 109.

36 Combinations were a form of Victorian underwear worn by both sexes for warmth.

He concluded in a tone that sounds more like the adult Ricketts, 'On second thought, I think a pare of gloves might have been suggested, not to speak of a kilt.'[37] The second letter is full of marginal decorations of nuns, because his cousin was boarding at a convent in France.[38]

Though he often did drawings simply to amuse his family, he was deeply serious about art. Even at this early age he was determined, he told his cousin, to follow what he believed to be real art and not just what would achieve popularity. He would prefer his work to be appreciated later, even after his death, rather than to betray his artistic principles. At fifteen he drew up a thoughtful and sensible list of rules for a proposed art exhibition. With more youthful egotism he announced to his cousin that he would one day make a name for himself and that she would write his biography.[39]

His visits to Ryde ceased when his grandfather, after the death of his first wife Isabella Gibson in 1883, married his cousin Frederica Ricketts and moved to her house in Wales. This marriage evidently caused a scandal, and contributed to the break-up of the family. In 1885 Aunt Nellie married Alfred de Mutach, of a patrician Bernese family, and went off to live in Switzerland. She was followed there in 1896 by cousin Florence who had married de Mutach's nephew Edward de Sinner, a baron of the Holy Roman Empire. Ricketts's father went, apparently for his health, to live in Rothesay on the Isle of Bute in Scotland, where he was able to carry on his sketching of marine subjects. In 1870 he had his half pay commuted for the sum of £1,121 17*s.* 6*d.* He died on 29 June 1883, at the age of forty-five. His personal estate, amounting to only just over £133, was held in trust for Blanche, who is mysteriously described as the 'only next of kin'. Ricketts never understood what had happened to several thousand pounds from the estate of his grandmother which his father had received.[40]

Blanche, only fourteen, was placed under the guardianship of her uncle Arthur. In 1895 she married a German doctor, Wilhelm Sachs,

The painting was Leighton's *Phryne at Eleusis* (now untraced). See Leonie and Richard Ormond, *Lord Leighton* (1975), pl. 182.

[37] CR to Florence Ricketts (afterwards de Sinner) [summer 1882].

[38] CR to Florence Ricketts, 6 Jan. 1882.

[39] Florence de Sinner to F. E. Jackson, 1 Nov. 1931. The 'Notice to Exhibitors', is in the coll. of Dr Vincent von Sinner.

[40] CR to Florence de Sinner [after May 1923]. Ricketts's father was buried in plot 244 in the new burial ground, Rothesay.

and moved to Germany. She kept in touch with her brother, but he kept his distance. When she invited him to visit he commented, 'It would be as absurd as for a dove to be among sea-gulls.'[41] He was never shy of showing a fine opinion of himself. In his diaries he makes only two references to her, the bald notes of her sickness and death (in 1903 at the age of thirty-five). No mention is ever made of her two sons, one of whom died fighting on the German side in 1915.[42] This was the chilling side of Ricketts's character.

By sixteen, then, he was an orphan. He had spent his childhood moving from place to place, constantly under the shadow of the illnesses of both his parents. Even more than most of us, he was a mixture of contradictory qualities; richly gifted but deeply flawed. In some fields he was so precocious and sure of himself as to be brilliant. Yet he could be touchy, difficult, a prey to terror and worry. For good and ill, he had a lively dramatic imagination. He was generous but malicious, sensitive but ruthless, bold but in some ways timid, sure of his opinions but basically insecure, exceptionally well read but not well educated. Having largely escaped conventional schooling, he had never been moulded in any conventional pattern, but allowed to develop his own tastes and interests in tune with his own nature. Finally, he was also lucky. With the death of his father, he would have been alone in the world, had he not fortunately already embarked on the friendship that was to occupy him for the rest of his life.

41 MF Journal, 1901, fo. 38v.

42 Information from the von Sinner family; CR Diary, 15 and 20 May 1903.

2 *Ricketts and Shannon: 'as others see us'*

TWO Cs interlocked was the early collector's mark used by the two Charleses, Charles Ricketts and Charles Shannon. This coupling of their initials signified a commitment, but one whose nature is not so simple and obvious as has generally been supposed.

They met on Ricketts's sixteenth birthday, 2 October 1882. He had just begun an apprenticeship to a wood-engraver at the City and Guilds Technical Art School, in Kennington Park Road, South London. Shannon (Fig. 3), who was nineteen, had been at the school for a year when Ricketts arrived. His background was totally different from Ricketts's. Born at Quarrington, Lincolnshire, on 26 April 1863, he was the second son of the Revd Frederick William Shannon, rector of Quarrington with Old Sleaford, and his first wife, Catherine Emma Manthorpe. The rector, who had 8 children by 2 wives, had been in this parish since 1861, thus providing for his son a family circle and a foundation of stability and continuity that the young Ricketts never knew. Shannon's rural upbringing in such conditions produced a very steady base to his character. At the age of twelve he had been sent to St John's, Leatherhead, a school for the sons of clergymen, where he was a keen cricketer and footballer. When his father recognized his artistic gifts and ambitions he enrolled him at Kennington. Shannon was in fact the kind of son that Ricketts's own father would have liked, a factor which may have drawn Ricketts to him.

Their friendship had the attraction of opposites. With his blond hair, rosy complexion, tall, elegant figure, and quiet, friendly manner, Shannon was personable and popular. As an artist, he soon had a great reputation among his fellow students. Ricketts, on the other hand, was small in stature, not much over five feet tall, with pale skin drawn over a finely modelled face. 'He had so swift a mind,' Laurence Binyon recalled after Ricketts's death, 'pouncing on essentials, that in his company others of bright wit seemed slow. He flushed and quivered, now with admiration, now with anger; his laughter was frequent and

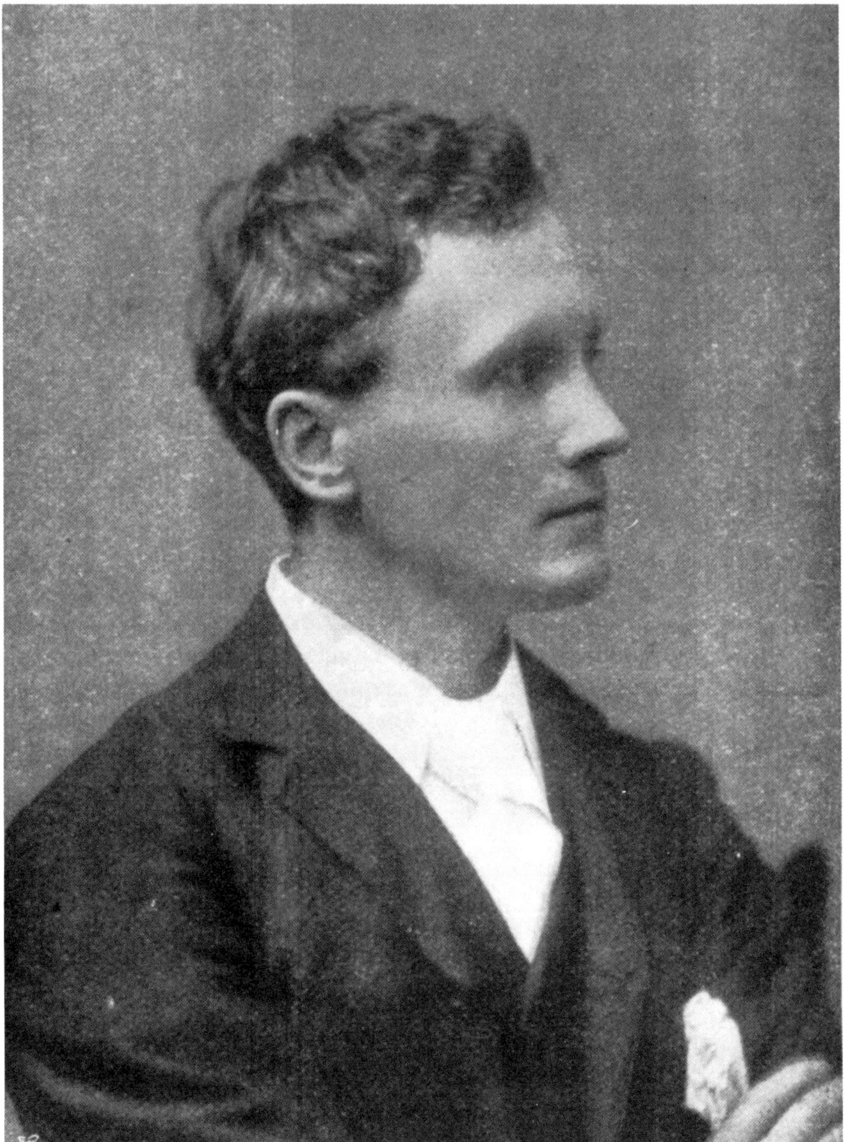

FIG. 3. Charles Shannon (? 1880s)

tempestuous.'[1] His strange self-sufficiency went hand in hand with a sense of isolation. Shannon understood this: in his portraits of his friend there is a haunting loneliness about the eyes. Shannon, moreover, soon came to replace Ricketts's mother in providing him with the one strong relationship that he needed.

It is not difficult to see what drew him to the younger man. The richness and the strange mixture of Ricketts's gifts were to attract people to him for the rest of his life. Shannon fell under his spell, and spoke to his friend Thomas Sturge Moore, who had not yet met Ricketts, of a prodigy who had an extraordinary ability to 'evolve compositions, to pick out good work, and to remember all he had seen'. Sturge Moore described his first meeting with the twenty-year-old Ricketts:

> At last I saw him, a short, ramshackle youth in a cloud of extremely fine, tow-coloured hair which stood round his head like a dandelion puff. At 11 a.m. he had just turned out of bed and had no collar, no socks, and but just enough on to enable him to cross the two hundred yards from their rooms to the school. When he entered he was imitating an orchestra with gestures as well as sounds, a characteristic mode of progression from room to room, all his life. Somebody began describing a picture of young girls with hay-rakes: he pretended to be shocked, 'Young girls with rakes!' and danced about hiding his face in his hands and laughed even more than we did.

With his French background, he differed from the others not only in character but also in knowledge, interests, and experience. He could talk of artists and art movements on the Continent of which they knew nothing. The contents of every major gallery in Europe seemed to be familiar to him, though in fact the Louvre was the only one he knew well. In 1881 he had visited the Brussels Museum, and he may have seen the Genoese collections and one or two galleries when as a boy he had visited Germany, but most of his knowledge of these came from books.

However, if he depended on books for such information, he did not derive his opinions from them. His responses to art were fresh and original. Whilst his first reaction was largely instinctive, for his mind was primarily intuitive rather than intellectual, he was also excellent at analysis, at divining the artist's intentions, at exposing merits and faults. What he felt about a work of art did not derive from established reputation or current fashion. To rely on these was in Ricketts's view

[1] Review of *Self-Portrait*, *Observer* (10 Dec. 1939).

to 'see through your ears'. Because he had largely been left to educate himself, he never started from the generally accepted view, from what one was supposed to think. What mattered to him was his own immediate impression. However, as one's impressions can sometimes change, he was apt to change his views rather suddenly, and praise one day what he had condemned the day before. This did not worry him, though it often bewildered those who did not know him well.

Together with this precocious assurance and erudition, there were odd gaps in his knowledge, due to his unusual background. Only when, as an art student, he was confronted with the dangling bones did he realize that there was a skeleton in each of us. 'But how did you suppose you were made?' he was asked. 'Oh, out of a kind of india-rubber', he answered, ready to improve the joke against himself.[2] Ricketts, moreover, never learnt to spell or punctuate properly. Whether for his amazing learning or his amazing ignorance, he was the sort of student that was talked about.

The room in which Sturge Moore met Ricketts was the same one that Shannon had met him in some five years before, a first-floor front room in a terraced house, 164 Kennington Park Road, where he remained until 1887. The year before, Shannon had joined him there, having previously shared lodgings in Lavender Sweep, Clapham, with another student, Leonard Raven-Hill, later to become one of the *Punch* illustrators. The following year Ricketts and Shannon moved, together with another friend, Reginald Savage, to 12a Edith Terrace, Edith Grove, Brompton. In this and in subsequent moves their address got better and better, and they shared a house for the rest of their lives.

It is almost universally assumed nowadays that they were lovers, but there is no conclusive evidence of this, no acknowledgement in diary, letter, or reported conversation that this was the case. Ricketts himself was secretive, and theirs was the discretion of an age which did not discuss such things. A close friend, Gordon Bottomley, observed that if Ricketts and Shannon had never met, Ricketts's life would have been substantially the same, but Shannon's would have been entirely different.[3] Shannon would have married and had a family. His works show how sensitive he was to female beauty, and the recurrent themes of mother and child suggest a yearning for these ordinary joys that his life with Ricketts denied him. Shannon several times became involved with his female sitters, and wanted to marry at

[2] *65 Ill.*, p. [1]. [3] G. Bottomley to William Rothenstein, 19 May 1944.

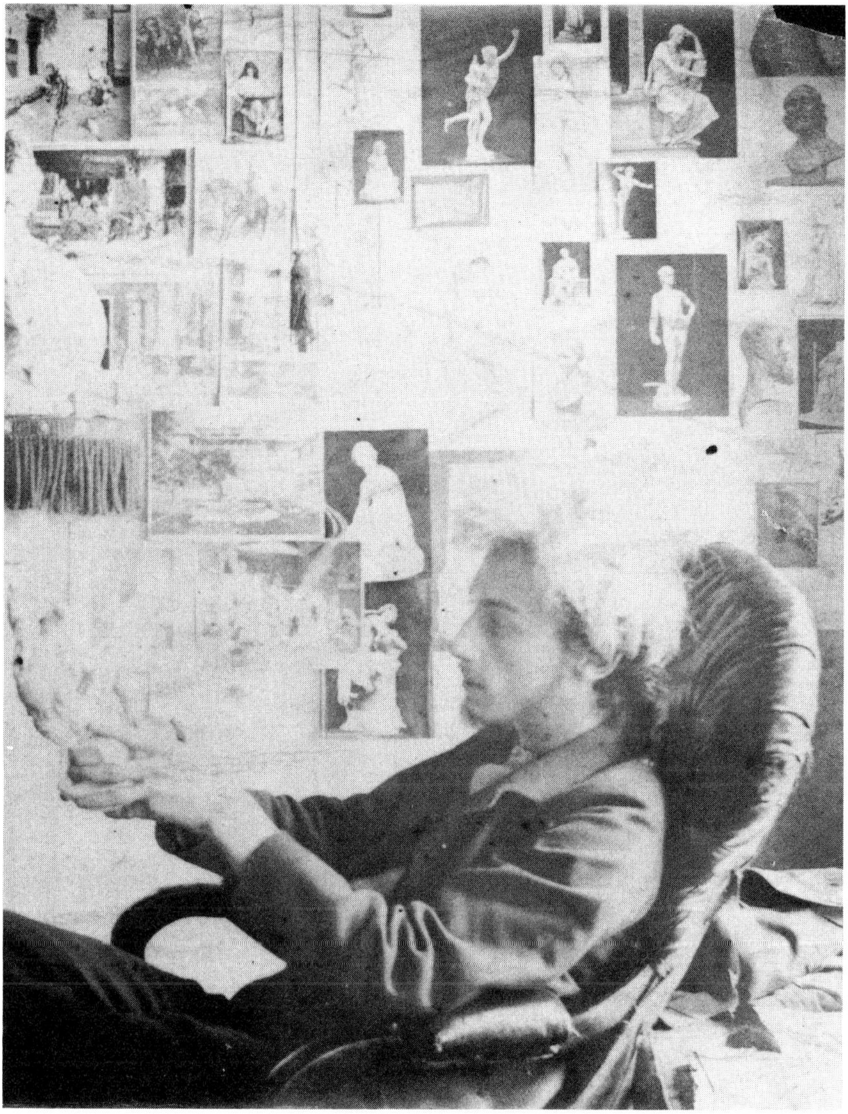

FIG. 4. Charles Ricketts (early 1880s) at 164 Kennington Park Road (Fitzwilliam Museum)

least one of them. Though Ricketts was certainly homosexual, Shannon was mainly—if not entirely—attracted to women. Ricketts often made jokes about Shannon's sexual preferences: putting £15 on a painting by Etty, famous for his female nudes, was, he remarked, 'a concession to C. H. Shannon's sex instincts; for me the money represents only the moustaches of the captive hairdresser & the lap dog in the corner of the picture'.[4] Ricketts rigorously expurgated his diary of the more personal revelations. Among the few that have survived is the comment that 'the night had also given me that which I have never possessed', but this in isolation remains opaque.[5] We do not know either who or what it refers to. If to a sexual matter, as seems very likely, then it is significant that it occurs nearly twenty years after Ricketts met Shannon. Whoever it refers to, it suggests that Ricketts was slow to find full sexual expression. Ricketts's letters to Shannon, though deeply affectionate, are far more prim than those to Robbie Ross. It is also significant that Lucien Pissarro only came to the conclusion that they were 'a couple of old aunties' after many years of close association.[6] Their oldest friend, Sturge Moore, always got angry at this question, but it was never clear whether his anger was because it was true or because it was not true.[7] Though a visitor recorded their twin bedsteads in a shared cubicle, a note on Lansdowne House by 'Bengy' Lowinsky says they had separate bedrooms.[8]

After a quarrel Ricketts noted that he was 'grateful that we do not see each other as others see us': this seems to mean that they regarded themselves more as individuals sharing common interests than as the unit others considered them to be; they were two celibates sharing an ideal, as William Rothenstein's phrase for them—'the sisters of the Vale'—suggests.[9] The 'we' that Ricketts always used referred mainly to their shared artistic values and opinions.[10] Certainly they never saw themselves as a homosexual couple, never belonged to any homosexual circle or organization. Nor were they prepared to make any revelations.

4 CR to R. B. (Robbie) Ross, p.c. [*c.*1905].

5 CR Diary, 6 Feb. 1900.

6 W. S. Meadmore, *Lucien Pissarro: Un Coeur simple* (1962), 76.

7 Miss Henriette Sturge Moore in conversation with the author.

8 Jacques-Émile Blanche, *Portraits of a Lifetime*, trans. and ed. Walter Clement (1937), 130; a note by TEL, attached to his letters from CR.

9 *Max and Will: Max Beerbohm and William Rothenstein: Their Friendship and Letters, 1893–1945*, ed. Mary Lago and Karl Beckson (1975), 28.

10 William Rothenstein, *Men and Memories*, i (1931), 173.

Shortly before his death, John Addington Symonds visited Ricketts and Shannon at the Vale in the company of another literary homosexual, Charles Kains-Jackson. In the course of a spirited conversation, Symonds began asking increasingly impertinent questions. Finally, Ricketts angrily told them to leave. As he stood on the landing, Symonds looked up from the bottom of the stairs and exclaimed 'But you are, aren't you? You do, don't you?' Many years later Ricketts repeated this story with amusement, but still did not answer Symonds's questions.[11]

Being homosexual in the Victorian era, especially after the Wilde trial, made it necessary for Ricketts to camouflage his true nature and his real feelings for many of the handsome young protégés he helped throughout his life. Most of these relationships were almost certainly platonic, even to the extent that the object of Ricketts's devotion was probably not aware of the nature or intensity of his passion. Such a necessity can only have enhanced his inclination to be secretive. ('I have moments of secretiveness', he confessed, 'which look blacker than they are.'[12]) Later, he appears to have eased this strict reserve, a development which was probably first encouraged by his friendship with Robbie Ross, the first friend (even including Shannon) with whom he could be completely open, in letters and probably even more so in conversation. It is doubtful whether he ever broached this subject with Wilde. That he could tell the story about Symonds against himself is one indication of how he changed, as is the fact that he could much later openly declare his love to one young man.

What seems very obvious to us was not so to their contemporaries. Their generation was much less 'knowing' than ours. Many men remained bachelors, or shared a house for company and convenience, without any of the raised eyebrows or speculation that have become common since the popularization of psychological theories. Among Ricketts's and Shannon's contemporaries, there were many artists and writers who never married, like Alfred Stevens, Leighton, Swinburne, Tonks, Wilson Steer, Ronald Gray, George Moore, Sargent, and Henry James; in earlier days, Reynolds, Lawrence, Turner, and Etty among others; the great majority of these were not homosexual. The fact that

[11] Told by CR to Prof. Sir J. D. Beazley, who told Prof. Martin Robertson, who told the author. See also the anonymous obituary of Symonds, by Kains-Jackson, *Pall Mall Budget* (27 Apr. 1893), 658.

[12] CR to Sydney Cockerell [Dec. 1910].

Steer always went on painting trips with his friends Fred Brown and W. C. Coles, called 'The Inseparable Three', did not arouse the least suspicion. Nor did the ménage of Swinburne and Watts-Dunton at The Pines, Putney, or even the all-male 'monastery' in Lewes headed by the American Edward Warren, which Ricketts visited twice to examine their collection of classical antiquities.

That Ricketts and Shannon loved and cared for each other is indubitable; that their relationship was sexual at one stage or other seems likely, but we shall probably never know what if anything went on behind closed doors. This is how they would have wanted it.

Whatever the exact nature of their relationship, its real foundation was their love of art. This first drew and then kept them together. Ricketts and Shannon were of the last generation of the British Aesthetic Movement, and in some ways its purest and most consistent adherents. Their whole lives were devoted to art, to studying it, to creating it, to propagating it, to collecting it. 'Such a little delights me', Ricketts confessed in 1911, 'outside the world of art, where I am omnivorous and insatiable.'[13] Aestheticism filled their lives more completely than it did Whistler's or Wilde's, who were its more public advocates. To find a parallel for them we must turn to France, to the Goncourt brothers, who shared a similar life devoted to their art and their collecting. Art to Ricketts and Shannon was not an embellishment of life but an austere and demanding commitment. 'Art', Ricketts noted, 'is the evidence of man's opposition to mere necessity and chance and the nearest approximation to a sense of immortality.'[14] Their devotion to it was derived from a coherent code which regarded beauty as the highest value. That this creed was spurned by the British public was to increase Ricketts's sense of isolation and put him at bitter odds with society.

It was probably Ricketts who had the greater part in determining their values and life-style. Shannon, however, was not a mere cypher. He was 'a very quiet & reserved man', Sybil Waller—a sitter—recorded, '& his extreme gentleness of manner was very fascinating. He was a delightful talker, when with one or two companions, but in a large gathering, his low voice & modest manner kept him rather in the background.' This was especially the case 'when his friend

[13] CR Egyptian Diary (1911), p.[6].

[14] This is a sentiment Ricketts often expressed. See *SP*, pp. 426–7 and CR, 'The National Gallery of Canada', *Observer* (19 Nov. 1930).

Ricketts was present, whose extreme vivacity & lively conversation quite overshadowed his more serious friend'.[15] For this reason, many friends of the two artists made sure they occasionally saw Shannon on his own. Though less brilliant and assertive than Ricketts, he provided Ricketts with a foundation of which he was deeply in need. His relationship with Ricketts was, one could say, the platform on which Ricketts could perform. In fact, Ricketts undoubtedly needed him far more than he needed Ricketts. Moreover, Ricketts respected his taste and judgement, and would often defer to him. When they disagreed, Shannon's view often prevailed: Ricketts, for instance, attributed a drawing of a baby they had acquired in 1904 for twenty-five shillings to Lorenzo di Credi, but after studying photographs of Leonardo's drawings he re-attributed it to this master in 1914: 'so Shannon is right and I was wrong', he noted; it is now given to Verrocchio.[16] Shannon was always given the better studio in their houses. Shannon performed many of the more practical tasks that Ricketts found hard or irksome. In the early days, he often wrote out Ricketts's letters for him. He made travel preparations and soothed Ricketts's nerves during the journey, especially when crossing the Channel. In Ricketts's book work, Shannon did the ruling ('I'm a fool with a ruler', Ricketts admitted) and later prepared the grounds and did the glazes for Ricketts's paintings. Throughout their lives Shannon wrapped any parcels containing books, prints, or drawings that they sent to friends: 'this will reach you . . .', Ricketts often promised, 'when Shannon, cardboard, paper, and string can be brought into conjunction.'[17]

These years at art school gave both Ricketts and Shannon a sound basis for their achievements, particularly of the next decade. The choice of wood-engraving as a career, and of the City and Guilds Technical Art School, may at first seem strange for boys of their social background. The school had been founded to 'make Art subservient to industrial purposes' and to provide training for those 'engaged in the art industries of the locality' (like Doulton's pottery), and the wood-engraving class especially attracted middle-class girls (such as

[15] Sybil Waller, *Memoir*, p. 2 (TS in collection of Mr Malcolm Hillier, who owns a portrait of her by Shannon).

[16] CR Diary, 20 Feb. 1914. See *Friends of a Lifetime: Letters to Sydney C. Cockerell*, ed. Viola Meynell (1940), 342, and *All for Art: The Ricketts and Shannon Collection*, ed. Joseph Darracott (1979), pp. 80–1, pl. 39.

[17] CR to G. Bottomley, p.m., 13 July 1907.

Clemence Housman, the poet's sister) who wished to earn their own living.[18] Yet there were good reasons for both these choices. Wood-engraving was still the main source of book and magazine illustrations, gave promise of well-paid work for illustrated magazines, and provided a basic training from which one could branch out into other art forms. What drew Ricketts to the genre was his love of the wood-engravings of the 1860s, designed by Rossetti, Millais, Boyd Houghton, and others, especially Rossetti's illustrations for the Moxon *Tennyson* and Millais's illustrations for *The Parables of our Lord* (1863)—which Ricketts always considered Millais's masterpiece.[19] At Kennington their master was Charles Roberts, the foremost wood-engraver of the day, in whose studio many of the students later went on to work. He was a reproductive engraver, who cut designs made by other men, as earlier wood-engravers had cut designs by Rossetti, Millais, and the others. As for the school, in addition to being close to Ricketts's relatives and friends in South London, it was one of the leading art schools of the decade, under the dynamic supervision of John Sparkes, who was also Principal of the National Art Training School at South Kensington. Sparkes, testifying before a Royal Commission on Technical Education in 1882, pointed out that the difference between Lambeth Art School and the others was that 'we have always had life classes, always had designing classes, and always had modelling classes, three things which are not common'.[20]

In 1882 there were fourteen students, only three of whom were male, in the wood-engraving class. Just up the road from Ricketts's lodgings, the school was housed in two buildings, 122 and 124 Kennington Park Road. Following the custom of the time, male and female students worked in different studios, each group of four students working at a circular table which had an elevated block in the centre for the lamps and the glass water-lenses which concentrated the light on the wood-block. Wood-engraving was a slow, painstaking art, and it is not difficult to imagine that the quick and restless Ricketts did not submit easily to its limitations. He was 'an Ishmaelite', a rebel, not

[18] 'Some notes about the Beginning of City and Guilds Art School' and other papers in City and Guilds Archives; and Laurence Housman, *The Unexpected Years* (1937), 106–7.

[19] CR to RNRH [1916] and Easter 1919.

[20] 'Evidence of Mr Sparkes', [*Second*] *Report of the Royal Commission on Technical Instruction, 1882–4*, Reports Commissions in House of Commons Sessional Papers 31(ii) (1884), 101. I am indebted to Mr Alex Werner for this.

only against the masters and the teaching methods, but against other students as well, but none the less became the leading artistic influence among his fellows.[21]

As well as wood-engraving, they also attended the life and composition classes, where they developed such powers as draughtsmen that they, along with all the other men, abandoned wood-engraving for the more lucrative and congenial work of drawing original designs on the wood for others to cut. What also stimulated this migration to the life class was the development in the early 1880s of new photographic processes to replace wood-engraving as a medium of reproduction. Ricketts's and Shannon's earliest work to be published indeed consisted of pen-and-ink drawings reproduced by line block. Only later did they begin to make use of their training as engravers and as designers. Finally, they must also have attended the advanced design class, where instruction was given in 'higher decoration and design, especially in the Italian and classical styles' by Hugh Stannus, who had worked as assistant to Alfred Stevens, arguably the greatest exponent of classical design in British art.[22] Here they must have imbibed not only their great admiration for Stevens, whose work they later collected avidly, but also something of the particular Italianate quality of their early wood-engravings.

Ricketts and Shannon also attended classes in the school in Miller's Lane in South Lambeth, where John Sparkes was superintendent of studies. The work of the Senior Sketch Club was criticized by Royal Academicians, among them Alma-Tadema, Frank Dicksee, H. Stacy Marks, and W. F. Yeames, but these famous names of the time left far less impression than the biting but salutary comments of Sparkes. One RA however, criticizing the work of Ricketts and Shannon, 'admitted that he only wished that he could have produced their exhibits'. They won prizes. One of Ricketts's successes was on the subject of Athaliah, the Queen of Israel, who ascended the throne only after the massacre of all the legitimate heirs, her grandchildren, except one, who later deposed her.[23] Even as early as this, Ricketts (like Delacroix, whom he admired) was attracted to the subject of massacres.

[21] MF Journal, 1902, fo. 48; and M. H. Spielmann, 'Our Graphic Humorists: Mr L. Raven-Hill', *Magazine of Art* (1896), 494.

[22] South London Technical Art School, *Prospectus* for 1882–3 (1882).

[23] TSM, original introd., 25 (BL Add. MS 61713). Only part of this was published in *SP*, pp. 14–18.

Such successes were, however, not treated with reverence, and Ricketts's most celebrated water-colour sketch was cut up soon after it had been singled out for a prize. What pieces contented him were retained, and the others were distributed among eager friends. Soon, even the bits kept back were disposed of. Ricketts, commented Sturge Moore, 'advanced all the time, destroying his past'. His high sense of standards precluded any sentimental attachment to his own student work, and he and Shannon, having carried off 'every prize for sketching & free work' one year, disdained to try the next.[24]

At the same time, Ricketts and Shannon were laying the foundation of that formidable scholarship that was to underlie their work and their collecting. Ricketts particularly was forever classifying, passing judgement on the works of art that came to his notice, and he gathered together photographs and reproductions of works he admired. Having covered the hideous wallpaper of his room with brown paper, he pinned up copies of works by Baudry, Puvis de Chavannes, Gustave Moreau, Rossetti, Watts, and many others (Fig. 4). Bulky albums and portfolios were filled with the work of illustrators, not only those of the sixties, but also of artists as various as Menzel, Willette, Howard, Pile, and Abbey. Eclectic though his tastes were, he retained his high standards and was continually weeding out the inferior work.

The same can be said of his treatment of his fellow students. Leonard Raven-Hill, who had shared digs with Shannon, became a close associate for a short time, when Ricketts, who was himself interested briefly in satirical drawings, influenced him considerably. A closer friend and collaborator was Reginald Savage, who accompanied Ricketts and Shannon in their moves over the next decade. However, all his life Ricketts's relationships with almost everyone were prone to strain, and this was especially true during his youth. As well as being tortured by inner fears and anxiety, he was perhaps too aware of his superiority, too apt to express his quick and vivid responses decisively, too given to producing collisions and hurt feelings among those who did not know or understand him well. If his company was exciting and stimulating, he could also be very tactless and abrasive. W. B. Yeats later told him, 'You are always in open revolt against yourself,' and it is true that (despite his wit, liveliness, and generosity) his swift shifts and reactions, his fervent, intolerant loves and hates were disturbing; he

[24] *65 Ill.*, p. [i]; MF Journal, 1902, fo. 48v.

rarely left peace and tranquillity behind him.[25] Whilst his ability to laugh at himself when he realized he was being absurd made some forget such differences as easily as he did himself, there were others who could not cope. Shannon recalled that at art school he himself had 'heaps of friends until Ricketts quarrelled with them'.[26]

To some extent, therefore, they kept themselves aloof, as they were always to do. This was due as much to their high notion of the artistic vocation and a growing estrangement from current art movements as to Ricketts's quarrelsomeness. They already had their own artistic values, based on the study of the past, an interest in figure subjects, and a commitment to the imagination as a primary quality in the artist. Such a scholarly view of art was often expressed with an unscholarly passion or an overscholarly narrowness. Once, Ricketts was being questioned about his early life and said, 'Hadn't you always heard that in those days we were both pedantic and disagreeable?' When told this was not so he continued genially, softly striking the table with flattened palm, 'Yes, I *insist* that we were pedantic and disagreeable.'[27]

Weekly, they visited the art library at the Victoria and Albert Museum in South Kensington to see the art journals and the expensive art publications. On the way Ricketts would hold forth, making outrageous remarks: 'Flaubert is God the Father, Zola, God the Son, and Sarah Bernhardt, God the Holy Ghost.'[28] He enjoyed the absurdity, the blasphemy, in such observations, but there was also a solid basis in appreciation: Flaubert was remote and serene, Zola more down-to-earth, and Bernhardt transformed by passion and inspiration.

Such throw-away comments were carefully recorded by his new friend Thomas Sturge Moore, who sometimes accompanied Ricketts and Shannon to the museum library. A pupil of Shannon's at the Croydon Art School, where the latter now taught several days a week, he had been persuaded to transfer to Kennington. Though he met Shannon in 1885, it was two years before he was introduced to Ricketts, who had, from Shannon's stories, assumed an almost legendary quality. Moore, an intelligent but vague young man, and brother of the Cambridge philosopher G. E. Moore, applied all the gravity of his Nonconformist upbringing to the new aesthetic credo he imbibed from

25 *SP*, p. 127; MF Journal, 1908, fo. 7^v.

26 Holmes, p. 163.

27 Rothenstein, *Artists*, p. 181.

28 TSM, original introd., p. 29.

Ricketts and Shannon. They were often amused by his vagueness and impracticality: Ricketts held that Moore had '*timbre*, but a muddled brain', and described him affectionately as 'an Ass, but a Golden Ass'.[29] He felt there were 'two Tommies'; one he delighted in—'the simple creature of naïve sensation as fearless as it is naïve'—and one he hated who, as Shannon said, went 'about in flannel drawers'.[30] However, they genuinely admired his wood-engravings and writings on art, and Ricketts considered some of his poetry 'the best work of his generation and perhaps of the present time'.[31] A friend commented that they hoped to 'train up Tommy to be a sort of Ruskin to them. He has elements of the preacher & therefore of success in England.'[32] Indeed, he was their closest associate and, after their deaths, their faithful apologist.

Shannon's part-time teaching, which brought him forty pounds a year, was his principal source of income at this time, while Ricketts received a quarterly cheque of twenty-five pounds from his grandfather. To bring in more money, he did hack illustrating. One of his jobs was to draw ladies' stockings in a shop in Bond Street for a catalogue. When the pattern of one specimen did not show clearly, the manageress asked, 'Would you like to see it on?' The timid young man was seized with embarrassment until he realized that the stocking was to be demonstrated on the leg of a dressmaker's dummy.[33] He also did satirical pen drawings in the style of Edwin Abbey, some of which were published in the *Alarum*, a short-lived 'panorama of the times' issued between October 1886 and March 1887. For such tasks he and Shannon turned to their collection of the work of illustrators, not only as a crib but also for a standard to aim at and measure themselves against.

Not all this period was spent in work or study. Aunt Flora Ricketts, who lived nearby, must have received visits, as did Ricketts's grandfather. Ricketts still frequented the houses in Plumstead and Sydenham, and it was probably at the latter that he was commissioned to paint a ceiling, in about 1887, which

[29] MF Journal, 1901, fo. 121: 1905, fo. 105v.

[30] Ibid., 1905, fo. 109v.

[31] CR to G. Bottomley, 19 July 1919. This letter is not in Ricketts's hand.

[32] MF Journal, 29 Mar. 1902, fo. 41v. See Sylvia Legge, *Affectionate Cousins* (Oxford, 1980), 187.

[33] Holmes, p. 164.

represented the four seasons; each occupied a quarter of the rectangle to be joined together by mouldings of wood or plaster. Each unstretched canvas was in turn painted on the floor. The sky and amoretti varied seasonably and there were suggestions of architecture on which some of the children stood while others flew or floated on smartly feathered wings. Ricketts treated this work only half-seriously; it was done to please and, if paid, only sufficiently to cover costs.[34]

Ricketts still maintained his connections with France. Like many people of mixed background, he was never to feel at home in either country. He lived in England, but he hankered after things in France, especially the greater seriousness with which, he felt, art and artists were regarded. In 1885 Ricketts and Shannon visited France together. Ricketts was eighteen, but still unbelievably timid, and, in spite of his flawless French, would walk Shannon (who spoke no French) off his feet before he could pluck up enough courage to enter a restaurant and order a meal. Though in 1883 they had gone to Antwerp, no doubt to study the magnificent collection of Flemish painting, Paris was their real mecca throughout the 1880s and 1890s, as it was for many young artists. Yet Ricketts and Shannon went for entirely different reasons from others. Most were attracted by what was new and revolutionary in painting, the theories and techniques of Impressionism, the ateliers like the celebrated Académie Julian, whereas Ricketts and Shannon were drawn by the past and living masters of the classical tradition of figure painting. In the Louvre they could study one of the greatest collections of Renaissance paintings in the world. Moreover, in France lived the greatest modern exponents of classical art, Puvis de Chavannes and Gustave Moreau. While these artists employed classical or traditional techniques of design and painting, their subjects were largely Romantic. In fact, the Symbolism of Puvis and Moreau represented the last phase of romanticism, which found almost its final expression in the work of Ricketts and Shannon. While in Paris, the two friends also studied the work of the Impressionists and of other modern artists. Ricketts never liked to reject or accept artists, movements, or periods wholesale, and his antipathy to Impressionism and its aftermath did not harden into outright opposition until years later. Lodging in the rue Racine in the Quartier Latin, he was able to buy the latest books of his favourite French authors in the galleries of the Odéon and, in

[34] TSM, original introd., p. 22.

the evening, go to the Comédie Française, where he saw all the great actors of the late 1880s.

In 1885 they saw Puvis de Chavannes standing in the Louvre in front of a piece of antique statuary and then (one of the last people to leave the gallery at closing time) before Poussin's *Le Déluge*, works showing 'the trend of his partialities', his classicism.[35] On this occasion they did not speak to him, but two years later they wrote to ask the master if they could visit him. He suggested between nine and ten the next morning. So excited were they and so afraid of oversleeping that they remained up the whole night before seeking out his studio, near the Place Pigalle. Ricketts considered him one of the most remarkable men he had ever met, 'most curiously apart, speaking sometimes to you, sometimes to himself, without affectation'.[36]

> I remember him as the man of his work, simple, grave and genial, touched and charmed by our raw and uncultivated admiration for his painting.... From time to time his speech became admonitory, and he launched forth into disapproval of current tendencies—the photographic drawing of many, 'la perfection bête qui n'a rien à faire avec le vrai dessin, le dessin expressif!' and against 'les pochades d'atelier et de vacance'.[37]

The stress on design and the dismissal of what was merely photographically realistic, hasty or unconsidered appealed greatly to Ricketts and Shannon with their love of imaginative painting. Because of the growing ascendancy of naturalism there, he dissuaded the two young English artists from studying in Paris, as many of their contemporaries were doing. 'Paris, c'est une foire', he commented. Afterwards, he remarked of them to a friend, 'Quelle charmante jeunesse'.[38] However, when they returned to Paris the following year the master was unable to see them.

On one such trip to Paris Ricketts had gone on ahead and rented the studio of a friend, who had decorated it with an old fishing net. It was a particularly hot summer. 'In the course of a night or so', Ricketts recalled, 'I became covered with rash and bumps and spots of blood on my night-shirt—it was in the pre-pyjama days. I met Shannon at the station with a long face covered with hectic spots and told him I feared I was stricken with fearful disease of the kind of which Herod

[35] *Pages on Art*, p. 69.
[36] MF Journal, 1900, fo. 55. See also Rothenstein, *Artists*, p. 181.
[37] *Pages on Art*, pp. 69–70.
[38] Rothenstein, *Artists*, pp. 181–2.

and other sinful persons had been smitten.' Shannon was less alarmed and, after investigating the guilty bedding, discovered the cause of the plague: fleas.[39] We may speculate why Ricketts was so ready to attribute his condition to a punishment for sin: he may have been improving the story. What is important here is Shannon's calm, sensible, practical approach, while Ricketts had given way helplessly to imaginary fears and anxiety. Soothing and protecting his friend was an important role he played in the relationship.

Initially, it would seem that Ricketts and Shannon admired French art to the exclusion of all recent English work. Lucien Pissarro wrote to his father that they only came to value the art of their own country when they saw an exhibition in London of the Pre-Raphaelites; possibly this was the Rossetti Memorial Exhibition held at the Royal Academy in 1883, which so moved Ricketts that he burst into tears.

'Nothing that Ricketts ever did', said Sargent, 'could have been done without Rossetti.'[40] Yeats, too, found that the two artists had much in common; not only their Latin blood but also 'the same apparent lack of philosophy with the same occasional philosophic insight and ... the same occasional over-generosity of praise, but there is something beyond all that which I cannot get at'.[41] The comparison, however, was one that Ricketts himself did not accept. When a friend likened him to the earlier master, he replied, 'You are a dear goose to compare my possibilities to those of Rossetti. He is one of the most singular & original men in art; even his latest and most undelectable works done when he was half-blind & mad with chloral, are unlike anything else.'[42] Similarly, when another friend wrote that one of Ricketts's drawings had surpassed Rossetti, he retorted, 'Please don['t] say anyone surpasses Rossetti. It always makes me angry or cross.'[43] Rossetti fascinated Ricketts, who relished a host of stories about him picked up from books and from friends who had known him. Whilst he was critical of Rossetti's personal life, admitting he was a 'lousy beast', he was particularly influenced by his graphic work and by his poetry, some of which he greatly admired.[44] Rossetti's *Hand and Soul*,

39 CR to TEL (TS), [Apr. 1918].
40 Rothenstein, *Artists*, p. 186.
41 *Yeats and Moore Correspondence*, p. 40.
42 CR to Mary Davis, (4 p.c.s) p.m. (envelope), 3 Oct. 1926.
43 CR to Gordon Bottomley, p.m., 3 Oct. 1918 (on back of envelope) = *SP*, p. 304.
44 *Yeats and Moore Correspondence*, p. 42.

which Ricketts republished in 1899, was a particular favourite: 'it has been an influence upon my life', he confessed, 'it taught me [that] to paint "one's self", one's own limited vision if you will, and not to strive after current ideals, is the best thing to do, since every minute of your life you are urged by friends and critics to paint something else'.[45] Ricketts was particularly deeply immersed in Rossetti around 1890 and 1891, and this influence shows in the work he did at this period; before long, French influences reasserted themselves, particularly those of Gustave Moreau and Puvis de Chavannes. Nevertheless, Ricketts's estimate of Rossetti's stature increased with the years.

Yet when they were later described as Post-Pre-Raphaelites by American critics, Ricketts noted that it 'does not pain us at all but plunges us into surprise for, greatly as we admire Preraphaelitism and its aftermath, we thought we represented a reaction against it'.[46] Whilst they continued the Pre-Raphaelite tradition of imaginative painting, they drew their subjects from the classical tradition in Italian art, their main inspiration being Titian rather than the earlier Italian school with its mediaeval subjects; or, indeed, the more modern subjects painted by some Pre-Raphaelites. Moreover, they abandoned the minute realism practised by Millais and Holman Hunt, preferring more generalized and imaginary settings. Indeed, Shannon turned for instruction to the work of the very authorities against whom Rossetti had rebelled, Sir Joshua Reynolds's *Discourses* and Sir Charles Eastlake's *Materials for a History of Oil-Painting*. The fact that Shannon was largely self-taught was to hamper both him and Ricketts, to whom he later gave lessons. In turning to the Italian classical masters, in preferring to paint in the studio rather than outdoors, in painting in glazes rather than by direct application, they were at variance with everything that was then, and still is, regarded as most progressive in the art of the time.

By the mid 1880s Ricketts and Shannon were occasionally represented at exhibitions at the Royal Academy, the Grafton Gallery, the Grosvenor, the Society of British Artists, and the Society of Painters in Water-Colour. Most of their subjects were biblical, and the medium was usually pastel or water-colour. However, Shannon was doing portraits in oil, one of which, his first exhibited picture, was a self-portrait of the artist 'at the age of 19 with mutton chop whiskers in a

[45] CR to TEL [Aug.–Sept. 1918].

[46] CR Diary, 2 Apr. 1914.

room with an over mantel, penny fans, and photos on the wall, one of them of a football team'.[47] However, they soon ceased exhibiting altogether. They had made a pact. Charles Holmes explained how the two friends, now aged twenty-two and twenty-five, had worked out a programme for their lives together:

> Shannon was not to exhibit again until he appeared as the complete and undeniable master, upon whose princely income Ricketts proposed to live in ease for the rest of his life. Until that great day came Shannon's painting would be done in private, watched and criticised by Ricketts alone, and the painter would work for their common welfare only when collaboration was required, or necessity compelled. It was upon Ricketts that the main burden was laid of providing for immediate wants, by drawing illustrations, advertisements or anything else which would bring in a little money.[48]

Therefore, the two young artists withdrew from the art world. While Shannon devoted himself to these high ideals, Ricketts set about making a career as an illustrator.

At this stage Shannon was the one who was tipped to be the great artist. He was to be *the painter*, while Ricketts was to devote himself to the decorative arts. It was Shannon who took the initiative and visited the older artists they admired, like Watts, Whistler, and Ford Madox Brown. Ricketts in fact always maintained that whilst Shannon would one day become Sir Charles Shannon, PRA, he would remain just plain Charles Ricketts.[49] For the next ten years they put their plan into operation.

[47] CR to Mary Davis [Apr. 1915]. [48] Holmes, p. 164.
[49] A variant of this is Holmes, p. 178.

3 *The aesthetes of the Vale, Chelsea*

RICKETTS'S and Shannon's house in the Vale, Chelsea, has so caught the imagination of writers on the Nineties that it is always mentioned in connection with them, even though they lived there for only six years. A broken-down wooden fence in the middle of a row of shops shielded the Vale from the noise and bustle of the King's Road. On entering, you suddenly found yourself in a small enclave of countryside which had survived from an earlier age. A cul-de-sac, it consisted of only three dilapidated houses set in large, half-wild gardens, and was full of tall poplars, mulberry trees from an earlier orchard, and huge overgrown snowball-trees. On the left stood a large rambling house inhabited by the author and artist William de Morgan. Opposite was a pair of semi-detached houses, in the first of which lived Ricketts and Shannon. Theirs was a small, square, Regency house with a graceful veranda and balcony (Fig. 5). Attached to it was a smaller house inhabited by their friend Reginald Savage.

Whistler, who had met Ricketts and Shannon at the Royal Society of Painters in Water-Colour at Suffolk Street, had been the previous occupant. He had lived in the Vale with his mistress Maud Franklin, who styled herself Mrs Whistler even though the real wife lived near by. The cost of maintaining two establishments finally proved too much for him, so that by August 1888 he had left the Vale and, for fear that its derelict state would dissuade potential takers, had had some repairs carried out. As soon as Ricketts and Shannon saw it they fell in love with it, but as they could not afford the rent they persuaded Sturge Moore and a friend known as 'the dark Clark' (whose place was soon taken by the sculptor T. Sterling Lee) to join them and share expenses, occupying two rooms at the top of the house. Their financial difficulties cast a shadow over the Vale, 'where mere rent-money deprived us of ease and self confidence', Ricketts later recalled.[1]

The main room, which was on the first floor, ran the width of the house, and had french windows on to the balcony. It had a low ceiling,

[1] CR Diary, end of 1901; the description of the Vale is from TSM to Mr Radin (n.d.), Univ. of London, Box 26.

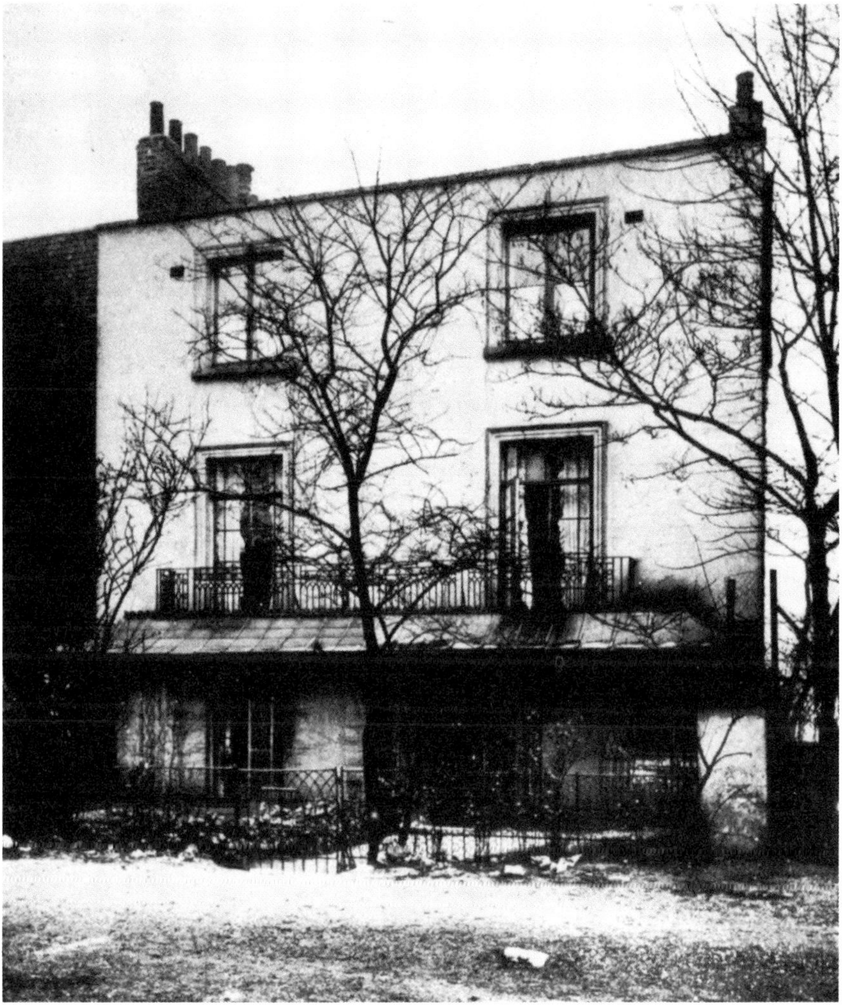

Fig. 5. The Vale, Chelsea (*c.*1888)

an open fire, and was brightened by coloured curtains and a few Japanese prints by Hiroshige, then very cheap, pinned to the wall. Furnished simply with scrubbed wood furniture and a table later usually covered with wood-blocks and burins, this was their work-room as well as the place where they received their guests (Fig. 24). Later, lithographs by Shannon and a fan by Whistler added an 'aesthetic' touch. Whistler's decoration of the house was retained: the panelled dado in some of the rooms and up the stairs had been painted a strong apple-green and the walls above a daffodil yellow. On the top floor Shannon also had his own work-room, where he could retire when he wanted to work alone. On the ground floor was the kitchen. The large garden behind and beside the house was left to run wild, but Shannon sowed several packets of marigold seeds, which brightened it up in summer.

The house's associations with Whistler pleased its new occupants, who admired his early work, but Ricketts later claimed he never liked Whistler as a person, even though the older man had always been charming to them, had liked and gone out of his way to praise Shannon's pictures, and once said to Rothenstein, 'but Ricketts and Shannon are gentlemen', meaning this in the professional sense (though Shannon was at the time appearing against him in the Sickert–Pennell trial of 1897). Recognizing their respect for him, he in turn respected their different views, only making feeble attacks on Wilde to them, since he knew they were his friends. Standing before one of Shannon's paintings, *Les Marmitons*, he is reported to have exclaimed repeatedly 'How is it done? How is it done?' However, the American artist had played an important role in introducing the ideals of Impressionism and undermining the classical tradition in Britain. Ricketts noted, in a deleted passage of his memoir of Wilde, that 'W. E. Henley, a force for harm, helped to destroy what remained from the past; directly or indirectly he supported this French tendency, made more active still by the growing influence of Whistler, whose laughter was against everyone and above all against all things English.'[2] Ricketts also disapproved of Whistler's love of controversy and of being in the public eye, qualities he considered unworthy of an artist. Ricketts noted that 'like Manet, Whistler believed in the press, both kept

[2] CR, 'Roses of Shadow' (BL Add. MS 58092), an early version of Jean Paul Raymond [i.e. CR] and CR, *Wilde*; CR to G. Bottomley, p.m. 10 Oct. 1922; CR Diary, 4 June 1900 in *SP*, p. 38.

scraps of tumbled newspapers in their pockets to read to embarrassed listeners'.[3] Whistler could never resist the chance to exercise his wit at another's expense, and would then gleefully publish the relevant articles or correspondence, while Ricketts and Shannon shunned the sordid world of public controversy. This was partly defensive, as Ricketts could be as touchy and aggressive as Whistler but he never enjoyed rows, which upset him for days afterwards. He preferred to concentrate on the work in hand. The Vale, therefore, provided a haven of tranquillity that was, at the same time, at the centre of artistic London; for many young artists in the 1880s were attracted to Chelsea by its cheap rents and convenience.

No sooner had they moved into their new house than they had to catch up on work. They had promised illustrations for a new periodical, the *Universal Review*, founded by Harry Quilter in 1888. Quilter, a barrister and art critic who had earned Whistler's implacable enmity by appearing for Ruskin in the trial of 1878, eventually fell foul of Ricketts too. During its two years of publication his magazine contained a large number of illustrations, mostly by Shannon, who was paid six guineas for one set of seven drawings, including some decorations. Except for an interest in the nude, Shannon's work shows little of his later quality, but one of Ricketts's designs anticipates clearly what was to come. It depicts the biblical Jezebel—a subject that would have suited Gustave Moreau—as a female type in the style of Rossetti, her dress decorated with figurative roundels taken ultimately from Japanese prints (Fig. 6). Such roundels, also employed by Rossetti both in his paintings and in his book-bindings, are used by Ricketts not only in his book-decorations, but also much later in his costume designs.This suggests how he absorbed Rossetti's influence very early and developed it consistently throughout his life, in different media.

However, they were also getting on with their own work, and began preparing their own 'occasional' magazine, the *Dial*. This was essentially a young man's enterprise, a clarion call that a new generation, a new set of values, had arrived: the first number boldly announced that its 'sole aim . . . was to gain sympathy with its views'.[4] According to Ricketts, its precursor was the *Germ*, the magazine founded by the Pre-Raphaelites in 1860.[5] This had consisted of poems,

[3] 'Marginal Recollections'.

[4] *Dial*, 1, p. 36.

[5] *Bookselling*, p. 508.

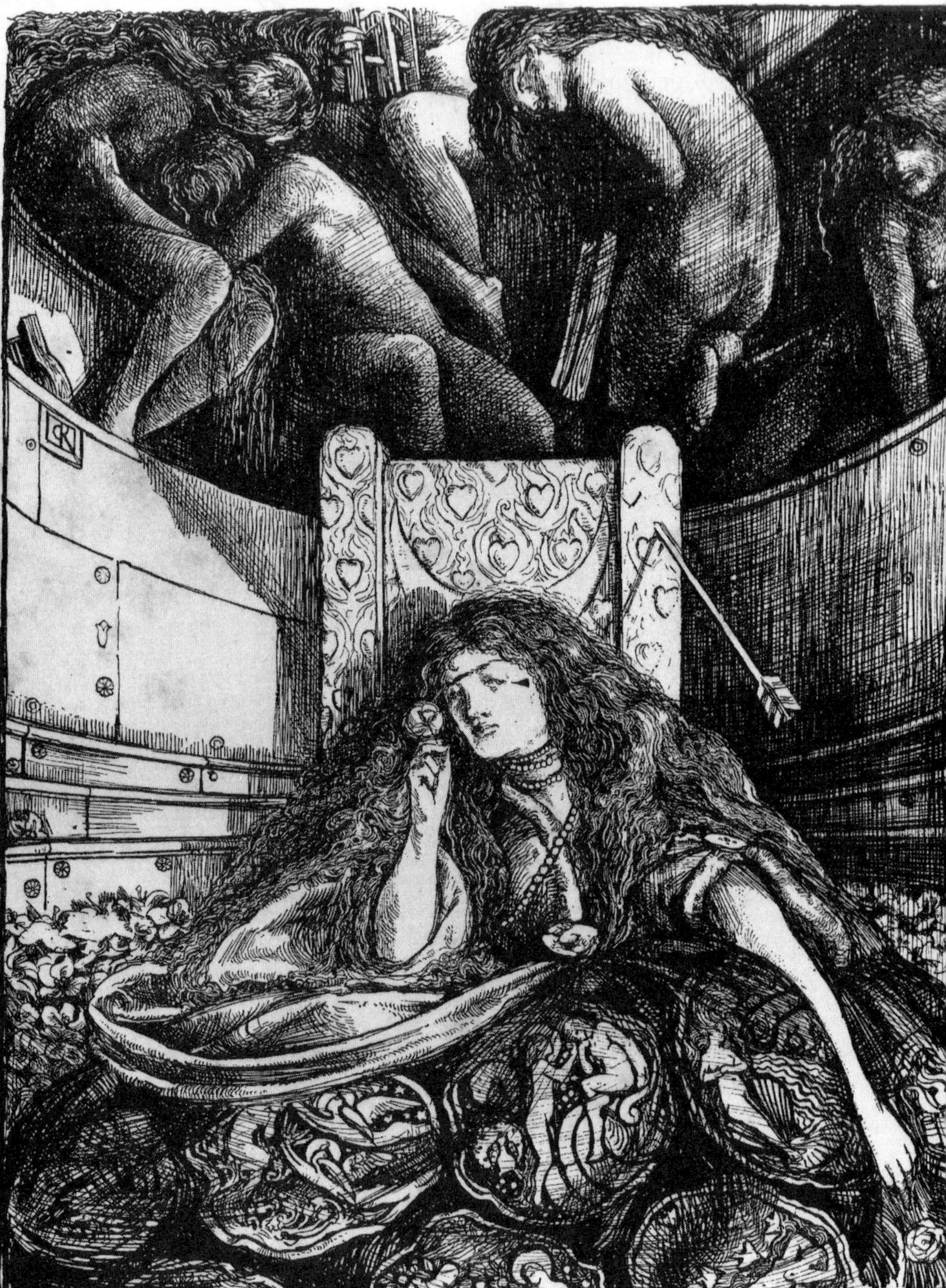

FIG. 6. *Had Zimri Peace Who Slew his Master?* After a Pen illustration by Charles Ricketts for 'Jezebel', in the *Universal Review* (1888)

stories, and essays on art and literature, and each issue contained an original etching. It had inspired another magazine, the *Century Guild Hobby-Horse*, founded in 1884, which was still being published and indeed issued some of Shannon's first lithographs. Two of its editors, Herbert Horne and Selwyn Image, were among Ricketts's and Shannon's earliest friends. Though Ricketts disingenuously omits to mention the fact, it came before the *Dial* and no doubt provided another model. Each magazine represents a somewhat different stream of the Arts and Crafts Movement, with its emphasis on good design and fine craftsmanship, on the handmade as opposed to the commercial and machine made, and on 'fine' artists applying themselves to the humbler objects of everyday life. It was to be some time before Ricketts in particular could realize his own ideals and make his own mark within this tradition.

The *Dial* did, however, add something new. From the outset it reflected a sympathy with French Symbolist art and literature. The first number appropriately contained an article on the Goncourt brothers by John Gray, and one on Puvis de Chavannes by Ricketts, who also contributed a lithograph, *The Worm*, which is much indebted to Gustave Moreau, and two stories, one of which shows that love of gorgeous description epitomized in Huysmans's *À Rebours*. It is full of lines like 'a flower hung near, shaped like an amorous mouth, a flower with lips'.[6] Such exquisitely morbid sentiments had already been voiced in English by Wilde and Arthur Symons. The artistic contribution was more original and represents one of the earliest examples, and appreciations, of Symbolist painting in Britain. Yet when this French influence was widely remarked on by critics Ricketts stoutly—and again disingenuously—maintained that the *Dial* 'had, and has, nothing to do with any form of Continental art'.[7]

It was the product of a tiny coterie: Ricketts, Shannon, Savage, and their new friend John Gray, but was published in the name of Shannon, then its most prominent member. Gray was a poet, and later enjoyed a glamorous notoriety as the supposed model for Dorian Gray. In the Nineties Gray loved clothes, dined out a great deal, and played the role of the beautiful young poet. Ricketts mischievously related how Gray had been flattered when a woman said, 'You are d—d handsome,

[6] *Dial*, 2, p. 30. [7] *Bookselling*, p. 508.

but you're deadly dull'.[8] Of humble background, he had left school at thirteen and worked his way up to being librarian at the Foreign Office, having meanwhile taught himself to read and speak French fluently.

It must have been Gray's interest in French poetry that brought him into contact with Ricketts, who met him in 1889, before Wilde did. Through his friendship with a wealthy young Russian Jew, Marc André Raffalovich, who had been educated in Paris, Gray met Mallarmé and other poets whose work he admired and was to translate into English. Ricketts encouraged the young man to write and translate poetry. Gray looked after the literary side of the *Dial* until Wilde's fall, after which his life changed, and he eventually became a Catholic priest.

Critical reactions to the *Dial* were mixed. The *Pall Mall Gazette* (24 Aug. 1889) found it 'full of quaint and interesting things', and the *Century Guild Hobby-Horse* (Oct. 1889) generously commented that 'no attempt of this kind had appeared in England possessing so much interest and originality or full of more promise of things to come'. Many reviewers, however, did not know what to make of the Symbolist (or what they called mystical) element. The *Academy* (31 Aug. 1889) observed that the 'cover, the initial letters and the plates alike make a fresh appeal to feelings and to methods that have yet to win popularity in this country'. The *Magazine of Art* (Dec. 1889) dismissed the *Dial* as 'nudity and nonsense', and the *Athenaeum* (23 Nov. 1889) went so far as to predict that the first number was 'likely to be the last of the series'. It is surprising that a new magazine by two obscure young artists should have been reviewed so widely. They cheekily appended all of the comments quoted above to the advertisement for the next issue in 1892. Most of the journals that had been critical, however, evidently were not sent copies of the second *Dial*. The touchy Ricketts became very selective about who was to receive review copies.

A complimentary copy of the first *Dial* was sent to Oscar Wilde, whom Ricketts had certainly already met. He had sent drawings to Wilde when the latter was editor of *Woman's World* and had been offered work illustrating and decorating the magazine; his contributions start in 1888. Wilde visited them at the Vale, not far from his house in Tite Street, in May 1889, but their friendship really began in August when the gift of the *Dial* brought him to the Vale again 'to praise generously, rapidly, and not too sincerely our boyish venture', Ricketts recalled. 'It is quite delightful', Wilde said, 'but, do not bring out a second

[8] MF Journal, 1899, fo. 81.

number, all perfect things should be unique.'[9] In fact, this first number proved a failure and a dead end, and the second number was really a fresh start two years later. In spite of all his praise, Wilde himself never contributed to the *Dial*.

The Vale, however, enchanted him. 'What a charming old house you have', he said, 'and what delightful Japanese prints.'[10] The next week he came again; the following week, twice; he ended by coming to 'jaw' three times a week. Their talk, according to Ricketts, was 'inspired "rot"'. For a year Wilde became 'everything to them'. For the next three years his visits were constant, and prolonged well into the night.[11]

It was, no doubt, the deliberate aestheticism of the *Dial* that appealed to Wilde. He was also personally drawn to Ricketts, whom he now described as 'cultivated and interesting', though Ricketts was only twenty-two compared to his own thirty-five.[12] Wilde was already established as the apostle of aestheticism, though not yet at the height of his fame. Only his *Poems*, *The Happy Prince*, and a number of articles had been published. On his first visit he commissioned Ricketts to do a portrait of Willie Hughes, the young actor whom he identified as the 'WH' of Shakespeare's Sonnets. Wilde's 'The Portrait of Mr WH' was being serialized in *Blackwood's Magazine*, and a few days after his visit Ricketts went to Tite Street to hear him read the manuscript. Within a fortnight Ricketts had painted the portrait upon a piece of decaying oak for which Shannon had devised a frame made of worm-eaten moulding, which so pleased Wilde that he wrote to Ricketts: 'it is not a forgery at all; it is an authentic Clouet of the highest authentic value.' Though hard up, Ricketts, with his usual generosity, made it a gift to Wilde, who replied, 'I hope we shall always be friends and see each other often.'[13] The friendship with Wilde not only opened a new era of prosperity in the Vale but was also one of the most important relationships in Ricketts's life. Wilde gave him his first 'break' as a book designer. Ricketts and Shannon between them were to design nearly all of his subsequent books. Curiously, it was rumoured that Shannon was the original of Basil Hallward in *Dorian Gray*.[14]

[9] Ibid., 1900, fo. 165^{v}; Richard Ellmann, *Oscar Wilde* (1987), 281: CR, *Wilde*, p. 28.

[10] CR, *Wilde*, p. 33.

[11] MF Journal, 1904, fo. 24; 1900, fo. 165^{v}.

[12] Wilde, *Letters*, p. 249.

[13] Ibid., p. 250, and in CR, *Wilde*, p. 36.

[14] *The Letters of Ernest Dowson*, ed. D. Flower and H. Maas (1967), 169.

During these months Ricketts was still working for periodicals, and long remembered the depressing time spent in waiting-rooms. By the end of 1889 he was publishing competent, if conventional, designs for *Atalanta*, providing decorative capitals and devices, and writing out some of the poetry in his own script. In one illustration he added a revealing personal touch. In the bark of a large tree his initials are carved next to a heart containing Shannon's. Ricketts was much given to teasing his friend.[15] In 1890 he began contributing to the *Magazine of Art* and in 1891 to a new magazine, *Black and White*. Though his illustrations remain unexceptional, his decorations—initial letters and head and tailpieces—are in a more fluent and symbolic mode that shows more up-to-date Art Nouveau influences and marks the beginning of his own style. Soon, however, Ricketts refused to discuss these early drawings, which gave him little opportunity to express his own personality in his work. Dismissing them as mere hack work, he sought to buy them back and destroy them, thus eliminating collections made by Sturge Moore and others. When he was able to work in more individual style, he was careful to preserve his sketches as well as finished drawings.

Some of these earlier designs, however, caught the attention of one of the most prominent figures in the art world, Lord Frederic Leighton. Wanting to encourage the work of a promising young artist, Leighton commissioned a drawing from Ricketts 'of the same imaginative and fanciful character as those I have seen' and left 'the subject or motive absolutely in your own hands that you may be wholly untrammelled and able to give me an entirely personal and characteristic specimen of your charming gift'.[16] Ricketts called on Leighton at his house in Holland Park and was offered £5 for the commission, a standard amount. When the President of the Royal Academy showed the young artist his paintings, he modestly—and quite accurately—observed, 'I'm afraid you won't care for my work, but I am interested in yours.'[17] Ricketts had little respect for Leighton's work and, recalling that several sketches by Delacroix (which, incidentally, Ricketts and Shannon later acquired) hung together near Leighton's fireside chair, commented that the PRA 'looked at them as though he believed he was doing such things himself'.[18]

[15] *Atalanta*, 3 (Dec. 1889) [189].

[16] Sir Frederick Leighton to CR, 16 Feb. 1891; *SP*, p. 19.

[17] Holmes, p. 164.

[18] G. Bottomley to Kerrison Preston, 28 Oct. 1943.

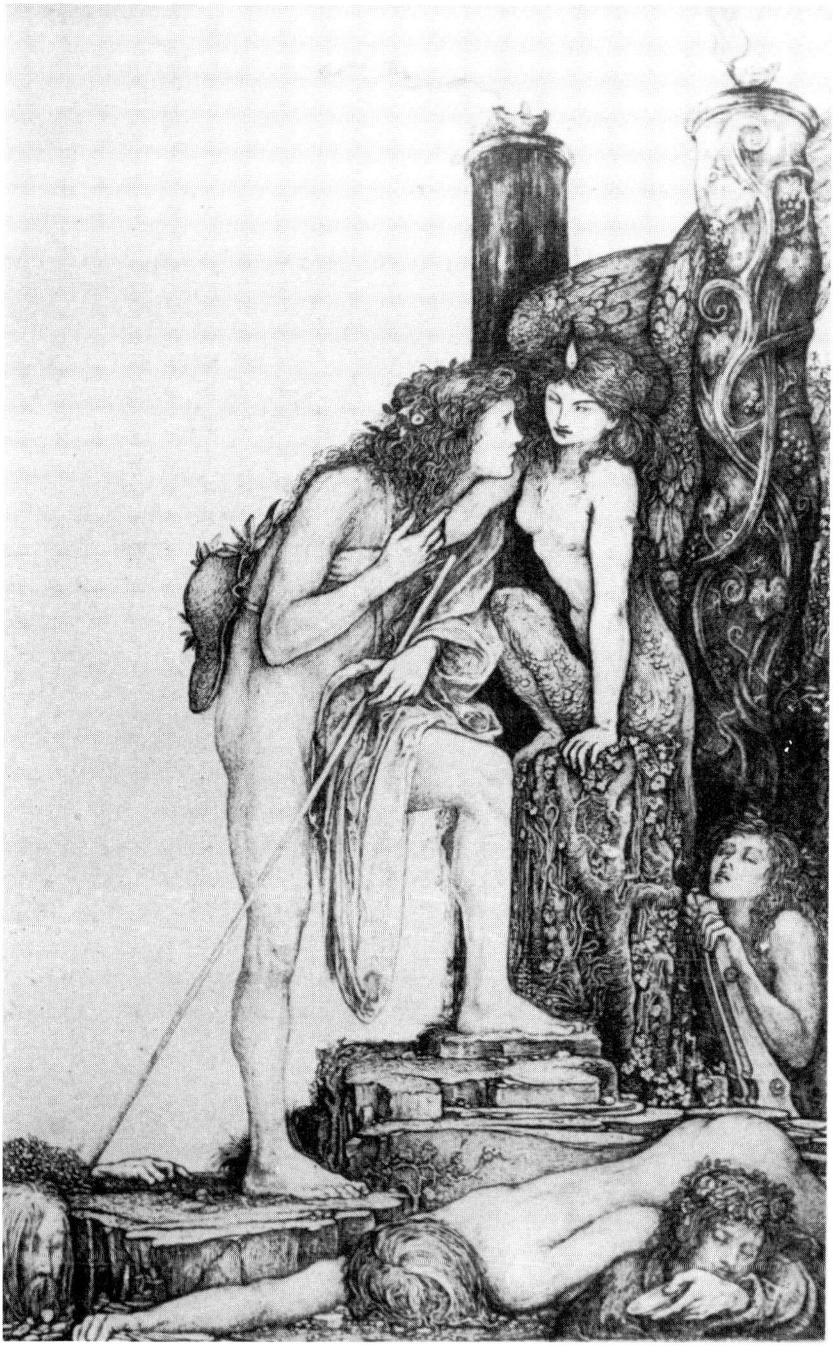

FIG. 7. Oedipus and the Sphinx (1891). Pen drawing by Charles Ricketts (Carlisle Museum and Art Gallery)

Freed from the need to illustrate a given theme, as in his magazine work, Ricketts was able to produce for the first time a drawing that could not be by anyone else. It shows a unique blending of diverse influences. The subject was *Oedipus and the Sphinx* (Fig. 7). The young Greek is staring into the eyes of the female Sphinx, who crouches on an ivy-covered column. Her arm is pushing him away. At his feet lie the bodies of her victims. It is an ancient Greek subject which had inspired Ingres and Gustave Moreau, to both of whom Ricketts owes a great deal. It is not so much because of this influence that the drawing can be rightly be described as 'Symbolist' but because it is unashamedly 'literary' and expresses something of Ricketts's own beliefs and feelings. What fascinated him in mythology, history, and life itself was the struggle between man and his fate. He loved to depict those key moments which illuminate a man's destiny. Here the forces of life and of destruction are pitted against each other; a moment later life will triumph. At the age of twenty-five, Ricketts produced not only one of the first drawings in the Symbolist vein in England, but also expressed for the first time that fascination with man's human position that was to hold him all his life. Yet it is not entirely French in its inspiration. The style, particularly the precise and elaborate draughtsmanship, reveals a strong Pre-Raphaelite influence. It also reflects Ricketts's particular quality as a draughtsman. Though he drew regularly from the life, his real gift was in linear arabesque rather than in three-dimensional drawing, in decorative rather than in realistic interpretation. The result is often highly stylized or mannered, but none the less vital. In many of these early drawings, however, he achieved this by an arduous process of trial and error. He would eliminate any unsatisfactory passages by using Chinese white or by sticking over a piece of paper, and then redraw over the top: the result was rather a mosaic. Later he acquired more fluency.

While Ricketts was establishing himself as an illustrator, Shannon was moving in a different direction. After briefly trying his hand at etching he had taken up lithography, an excellent choice, for it was a medium whose qualities suited his talents. He had first thought of using it for the *Dial*, but had not progressed sufficiently to produce any examples in the 1889 number. In 1888 Shannon produced his first print. Entirely self-taught, he even acquired his own press and printed his own work (which was unique at the time, for artists usually employed lithographic printers), and he soon became a prime mover and leading artist of the revival of lithography in England.

While both Ricketts and Shannon were exploring their artistic powers, they were widening their circle of friends. In their early days at the Vale they lived in seclusion. Their few visitors often found them alone: 'they knew few people, and prided themselves on going nowhere', reported William Rothenstein.[19] As their work became better known, the legend of Ricketts and Shannon, of two young artists completely devoted to art and each other, began to attract the curiosity of writers and artists. As Wilde became a regular visitor, the standard of dress began to improve and the circle to widen. Visitors to the Vale's 'Friday evenings' included Walter Sickert, who for a time ran a school in the de Morgans' studio opposite; with his mordant wit, he could hold his own with anyone. Wilson Steer, however, usually sat in ponderous silence. Max Beerbohm, of whom Shannon did a lithographic portrait, was another early friend and did a number of brilliantly perceptive and funny caricatures of Ricketts and Shannon. *Mr Ricketts and Mr Shannon in the Enjoyment of Popular Success* (1907) catches beautifully Ricketts's sprightly enthusiasm and Shannon's bashful silence before a coarse and unresponsive John Bull (Fig. 8); the contrast between their exquisite aestheticism and the more mundane values of the British public was all too apposite. Ricketts did not always admire Beerbohm's caricatures, but roared with laughter at *Rossetti and his Circle*; he himself did a caricature of Beerbohm scratching his name on the tombs of the great in a book belonging to Robbie Ross, opposite a caricature of Wilde by Beerbohm.[20] Among the visitors from abroad was R. N. Roland Holst, a Dutch artist they met through John Gray. His art was influenced by both Ricketts and Shannon, whose work he helped publicize in the Netherlands. Early in 1891 Wilde brought the young William Rothenstein, fresh from the Académie Julian, to meet the two artists 'of whom Wilde often spoke so admiringly' and 'of whom we knew nothing in Paris'.[21] Sickert and Rothenstein were, according to Ricketts, the only two intelligent men he had met among British artists; the former 'with his limitations [was] intelligent, worldly, shrude & witty', whilst the latter, who had 'the better head-piece', could bore Ricketts with his 'pushing, pulling, and intrcating

[19] *Men and Memories*, i. 174.

[20] CR to TEL, 3 Dec. 1917 and [Dec. 1917–Jan. 1918].

[21] *Men and Memories*, i. 133.

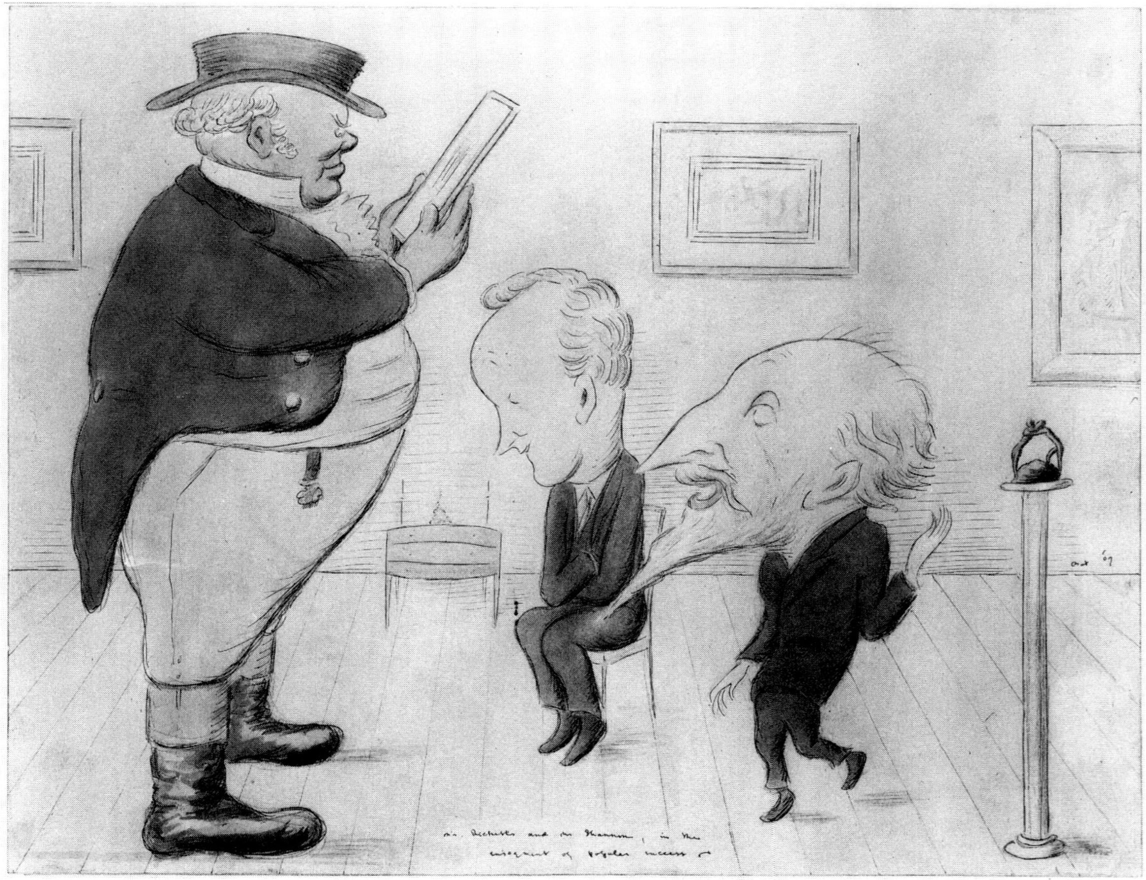

FIG. 8. *Mr Ricketts and Mr Shannon in the Enjoyment of Popular Success* (1907), by Max Beerbohm—the object of popular success can be identified as Fig. 44

egoism' and his 'wish to make all art in his own image'.[22] Rothenstein, moreover, was the only man Ricketts knew, besides Sturge Moore, who would 'face a serious subject'.[23] He produced numerous portraits in pastel and lithography of Ricketts and Shannon, both singly and together. It was also through Wilde that they met Laurence Housman, who had been junior to them at the City and Guilds Art School. Wilde had praised a story by him, 'The Green Gaffer' in the *Universal Review*, and this brought an invitation to the Vale. Even this small group shows a remarkable diversity. Yet at that time the art world was still small and friendly and the younger generation of artists remained free from the type of division by which it was later to be split.

An older artist who became a friend was Alphonse Legros, a Frenchman who had been living in England since 1863 and had succeeded Poynter as Slade Professor of Fine Arts in 1876. Though Ricketts and Shannon treated him with marked consideration, Ricketts, on re-reading his diaries, later decided he had both 'liked & almost hated him'.[24] Legros had known Rossetti and his circle and had earlier been 'a friend of Baudelaire and his contemporaries', Ricketts recalled, and 'spoke of Ingres, Delacroix, and Millet as if they were yet living, and of Manet and Rodin as if they were still promising young men', so his talk could be fascinating.[25] 'Old Legros called, very vivacious, amusing, at times profound, incredibly naïve, and selfish', is one entry in Ricketts's diary, but more typical is Shannon's 'Old Legros turned up and bored us both for about nine hours.'[26] Legros usually visited for the day, and as he spoke no English, in spite of thirty years in England, Ricketts would take him out for visits to galleries or exhibitions, so that Shannon at least could work. They both admired his work, however, and Ricketts confessed that Legros had taught him to appreciate design: 'My personal indifference to Poussin has been scolded into reason', he acknowledged, 'and this great master's constructive gifts held up to admiration.'[27] Legros also did superb portraits of Ricketts and Shannon. Ricketts, when sitting to him, began to expatiate on his own physiognomy, 'at a distance, a holy thing, saintlike ... quite close,

[22] CR Diary, 8 Dec. 1904.

[23] Ibid., 22 Jan. 1903.

[24] CR to CHS [early Oct. 1927].

[25] *Bliss Catalogue*, p. [iv].

[26] CR Diary, 6 July 1901; CHS Diary, 21 May 1902.

[27] *Bliss Catalogue*, p. [iv].

a sensualist's' and was silenced by the terse comment 'Nous allons voir'.[28] Ricketts and Shannon helped to arrange for a painting by Legros to be bought for the national collection, and Ricketts wrote an article on him (*Saturday Review*, 17 April 1897) and the introduction to the catalogue of his prints.

Not all the visitors to the Vale were regarded as friends, however; many were what Ricketts always referred to as 'enemies', even though they continued to come regularly, and probably did not suspect for some time the secretive Ricketts's real feelings about them. To Ricketts, 'enemies' included anyone who either did not share their artistic values or did not appreciate his or Shannon's gifts. It is noticeable that Ricketts's inner circle of friends included artists and writers whom he could help and influence, like Shannon, Gray, Sturge Moore, and, later, the Michael Fields and Bottomley, patrons like Edmund and Mary Davis, and writers who admired and made use of his great abilities, like Wilde and, later, Yeats and Bernard Shaw. These last three show that friends did not need to be subservient to Ricketts, only appreciative of him. Anyone who promoted Impressionism or Post-Impressionism, like W. E. Henley, Sidney Colvin, D. S. MacColl, and Roger Fry, was always under suspicion and later completely beyond the pale. Whilst the last two continued to frequent Ricketts's and Shannon's houses for many years, relations were often strained. With MacColl, Ricketts later engaged in an exchange of letters over some pictures by Camille Pissarro that—on Ricketts's part—bristle with malice and veiled insult.[29] He was trying to hold himself back, but his character had an aggressive side which strained at the leash in the presence of 'enemies'.

His first instinct was to attack, to savage them verbally, as, for example, in an interchange with Wilde. The latter had brought one of his admirers to the Vale, which he felt was too self-contained. The man proved to have no feeling for the graphic arts and as soon as he had left Ricketts, who never suffered fools or philistines gladly, turned angrily to Wilde:

R. Why did you bring this man here?
W. An obscure worshipper who bowed in the outer court, so I, the god, beckoned him in.
R. But, Wilde, you said that gods should never leave their temples.

[28] MF Journal, 1900, fo. 41^{v}. The drawings are in the Fitzwilliam Museum, Cambridge.
[29] The letters are in the Library, Glasgow Univ.

W. You, my dear Ricketts, belong to an older type of deity, who mistake worshippers for food, and when they see one they tear him to pieces.[30]

Ricketts, who admired Wilde's kindness and geniality, could never match them in such situations. Over the years he did learn to control his strong feelings, and, trembling and white with anger, would rush from the room 'in order . . . not [to] be rude to an unsuspecting fool'.[31] When angry, he always turned deadly pale. The fact that he could later tolerate a roomful of 'enemies' at their Friday evenings is further evidence of his increasing self-control.

In one of his sombre, rather malicious, and certainly unfair moods, Ricketts described D. S. MacColl as a 'dangerous specimen of the modern spread of education & false science', which was sufficient to develop 'the smaller attorney faculties' but not that 'larger, more humain view of things & their various conditions & exigencies that characterized the really cultured type of the past'. On the 'rebelious nature of the Celt . . . moulded by centuries of protestantism' was grafted the 'equally dull opportunism of Oxford' and the 'feverish, fantastic, & perverse culture of over-civilize[d] Paris' to produce a nature 'at once pretentious & crude, jealous of all superiorities, combining the snobism of modern culture . . . with journalistic disinterest & disrespect'. However, he would more often relieve his frustrated anger by turning his enemies into figures of fun. MacColl, his favourite target, was pilloried repeatedly in letters to friends. Writing to Robbie Ross about one of his own paintings of Don Juan, for instance, Ricketts complained:

Sturge Moore called to tell me that you had cast suspicion on my expertise in giving the Eunuch a beard. Oh abominable and unobservant man! With Colvin, D. S. M., and other friends of yours, how could you doubt? Learn, . . . as with women in senility, that hair streaks the face in old age, that the date of castration regulates the aspect of an Eunuch who can have a rich beard and other signs of virility. In the Turkish government as in most countries of the world, Eunuchs have often important posts. In this they are most like England, the Turkey of the West. A magnificent and proud looking Turk with a black fan shaped beard was asked at the house of the French Ambassador if it was true that he was a total Eunuch. He answered with quiet pride that such indeed was the case. As they were among men, he was persuaded to show himself, and as a matter of fact there was nothing, nothing

[30] Rothenstein, *Artists*, p. 183. [31] *65 Ill.*, p. [iii].

whatever save the hair on his belly. We will ask D. S. M. to show himself on Monday.[32]

But Ricketts, even at his most malicious, admitted that MacColl could write well.

This tendency to divide up his contemporaries into friends and enemies had unfortunate results. Ricketts saw life in terms of 'us and them'. You were either for him or against him, so that people, like paintings, were divided up into the good and the bad—a rather combative attitude which did not make for peace of mind. This is why, moreover, Ricketts established his own coterie, and in the early days seems to have wanted to found, or at least lead, a school of artists. It is significant that a series of articles on him, Shannon, Savage, and Pissarro published in 1895 was entitled 'The Vale Artists'. Ricketts later complained that at the beginning of their careers, unlike the second generation of Pre-Raphaelites, they had entered no 'set' that had prepared a place and public for them, so that he had had to make his own set based on his own artistic ideals. The *Dial* was part of this strategy.[33]

By no means all their visitors were artists. Though Ricketts claimed there were some contemporary writers that he had met 'but refused to know', like Arthur Symons and Lionel Johnson, this must not be taken too literally.[34] In another place he says he saw a good deal of Johnson, who struck him as a 'typical *Fruit Sec* of his class, time, and training' and who used to 'drink our dull local whiskey in silence whilst the air was fantastic with the sane or insane talk of the moment'. Ricketts doubted Johnson's scholarship, and once caught him 'making a lamentable howler' in a translation of Baudelaire. Dowson came to the Vale twice, brought by Laurence Binyon or perhaps John Gray, when the artists were printing lithographs and 'the atmosphere was probably too strenuous & too technical' for visitors.[35] Binyon, both a poet and an art historian, was one of the most lasting of these early friends. His praise of Ricketts's and Shannon's work (and values) was

[32] CR Diary, 13 Nov. 1900; CR to Robbie Ross, [June 1915], The painting may be *Don Juan in Hell*, *c.*1907.

[33] *Sketch*, on Shannon (25 Jan. 1895); on Ricketts (13 Mar. 1895); on Pissarro (17 Apr. 1895); on Savage (24 Apr. 1895); MF Journal, 1902, fo. 41^{v}.

[34] CR to G. Bottomley, p.m. 16 Sept. 1922.

[35] CR to WBY (late summer 1922); given in *Letters to W. B. Yeats*, ed. R. J. Finneran, G. M. Harper, and W. M. Murphy, 1977, ii. 429.

always whole-hearted and unblemished by the hesitations, reservations, or qualifications that often marred the praise given by other critics. This and the fact that he was sensitive, intelligent, and a good listener endeared him to Ricketts.

Ricketts himself was clearly the main attraction of the Vale, even though Shannon was at the time the better-known artist. When Wilde once took Graham Robertson there and found only Shannon in, he later commented 'that was the wrong one'.[36] A brilliant talker, Ricketts had a gift for evocative phrases and for quick repartee and, although erudite for his age, he wore his learning lightly. John Gray, a regular visitor, recalled his particular charm:

> It has struck me often that Ricketts is able to show things as wonderful now as things were in childhood. How many times did I walk home to the Temple in the small hours dreaming sunnily all the way from the palace of enchantment which the old hygienic Vale really was.[37]

There was something of the child in Ricketts. Indeed, he felt the artist always retained the essentially childlike nature, which he considered one of Wilde's particular charms. Ricketts showed a youthful freshness and vitality until almost overwhelmed by sorrow in the last years of his life. Of this quality he was aware and not ashamed. Once when Robbie Ross was selecting chocolates to send to his godchildren for Christmas, he asked himself whether there were any other children he should remember. He thought of Ricketts, 'the most charming child I know'. When Ross explained his action Ricketts was enchanted.[38]

Holmes described him at this time as 'frail-looking, with a pointed auburn beard and wavy hair:—a Swinburne turned Little Minister one might think, seeing the top-hat, black Inverness cape and grave demeanour'.[39] He was described as 'Shakespeare in a fit' but it was the Christlike in his features that was most often remarked upon. Quilter called him 'a little tin Jesus'. On one occasion, while in a third-class railway carriage, with an amiable drunken workman and a 'harlot' at his side, the workman, pointing at Ricketts, said 'My God, there's bloody Jesus'. Handing him a bottle of beer, the drunkard said, ''Ave a drink, bloody Jesus', but the harlot, with sober kindliness, remarked,

[36] *Letters from Graham Robertson*, ed. Kerrison Preston (1953), 431.

[37] John Gray to MF, 24 July 1907 (Roger Lancelyn Green).

[38] CR Diary, 28 Dec. 1904.

[39] Holmes, p. 141.

'Leave the poor bugger alone'.[40] This resemblance led to his being stopped in the Bayswater Road by the veteran Pre-Raphaelite Frederick Shields, and asked if he would pose for the figure of Christ for the Chapel of the Ascension near Marble Arch, designed by Herbert Horne (which was bombed in the Second World War). Ricketts later felt he was 'just recognizable in Christ washing Peters feet' but confessed to a friend that 'that is not a happy situation, as I can not imagine my washing any persons feet. The other subject for which I sat is Christ and the Woman of Samaria. This I rather like. I always thought he behaved so nicely and sensibly on that occasion.'[41] Shields painted all the afternoon before finally suggesting a rest and some tea. The older artist produced the necessary materials, boiled the water, filled up the tea-pot. When he poured, out came hot water. 'Oh dear, Oh dear', he said. 'I have forgotten the tea. Wait, I'll get it.' Ricketts laid a hand on his arm as he rose and said, 'No, my miracle!'[42] Ricketts loved to recount such stories; he is the source for many jokes against himself.

Conversation is the most ephemeral of the arts and the most difficult to capture in any record. When those fortunate enough to have been present on these occasions describe the experience, we can only share in a pale reflection of it. Wilde, possibly the best conversationalist of the age, would talk of his writing, recite his prose-poems, or recount the plots of work then in progress. Ricketts recalled that he had two sides as a talker: as a reciter of set pieces already polished to perfection, and as an explorer of the possibilities of work that he had in hand. In Wilde's monologues Ricketts noted the absence of the Church-of-England intonation that he so hated in other poets.[43] Wilde's manner to the younger man, according to Holmes, 'tempered laughing assurance with so much friendship and respect as to be immediately attractive'.[44] They had a common interest in the Pre-Raphaelites and in French literature. Ricketts recalled his early impression of Wilde: 'I viewed him as a man who had known Swinburne, Burne-Jones, and who might have met Rossetti. I think he made me read Pater and I

[40] CR to Cecil Lewis, 28 May 1928 = *SP*, pp. 397–8; MF Journal, 18 Oct. 1899, fo. III^{v}. A later but similar incident is *SP*, p. 99.

[41] CR to MF, p.m. (envelope) 9 May 1902.

[42] *Since Fifty*, p. 161; MF Journal, 1899, fo. III^{v}.

[43] CR, *Wilde*, p. 30.

[44] Holmes, p. 166.

made him read Villiers & Verlaine; we had a common meeting ground in Baudelaire and Flaubert.'[45] To begin with, Ricketts lent Wilde Verlaine's *Fêtes Galantes* (1870) and then introduced him to the poet's followers. Ricketts's knowledge of French literature was in some ways greater than Wilde's. In addition to this common ground they had a love of conversation, of matching wits. A vivid picture of Ricketts's manner is given by Holmes:

> Ricketts, perched on the edge of the table, engaged Wilde in a long verbal combat. So swiftly came parry and *riposte*, that my slow brain could only follow the tongue-play several sentences behind, and cannot remember a word of what passed, except 'Oh! nonsense, Oscar' from Ricketts, although it lives in memory as the most dazzling dialogue which I was ever privileged to hear.[46]

In Ricketts's company Wilde was at his best.

Ricketts may in fact have owed something of his manner to Wilde. This was maintained by W. Graham Robertson, who met Ricketts through Wilde, and who told Gordon Bottomley that 'Ricketts had modelled himself so much upon Oscar—voice and speech sometimes [were] identical . . . I suppose Oscar had made such a tremendous impression upon Ricketts when they first met and the imitation followed unconsciously.'[47] Bottomley admitted that, as Ricketts was at a young and impressionable age when they first met, this was possible. 'In his natural delivery—unrhetorical "rhetoric", one might call it—there were', Bottomley felt, 'characteristic cadences and a *portamento* on certain words ("extra*OR*dinary" was one) that seemed part of him' but might have been picked up from Wilde.[48]

Over these meetings there was always an air of somewhat precious discrimination, of aesthetic high-mindedness. One friend commented after a first visit: 'One has to be on one's rarest behaviour—for nothing ordinary is expected; a false tone might be an outrage.'[49] Another difficulty in discussions with Ricketts was his tendency to juggle with ideas, to treat his own and others' opinions as if they were *sub judice*. This playing with opinions often baffled those who did not know him, worried many of those who did (like Sturge Moore), and irritated

45 CR to G. Bottomley, 20 July 1918.

46 Holmes, p. 168.

47 W. G. Robertson to G. Bottomley, [*c.* Jan. 1940].

48 C. Bottomley to W. G. Robertson, 10 Jan. 1940.

49 MF Journal, 1894, fo. 13.

Shannon, especially when he was out of sorts: 'What's the use of saying that?' Shannon would ask, 'You know you don't think it.' As Sturge Moore pointed out, 'Shannon's mental world was as well ordered as his painting materials, and he found it difficult to allow that conclusions he had pigeon-holed were worth recanvassing.' Ricketts, however, enjoyed seeing 'his beliefs set dancing for a while' and knew he would never abandon them if they withstood the test of argument.[50] This way of treating ideas was more French than English, and led Ricketts to conclude that the French in this regard were superior to the English. Commenting on the French painter Jacques-Émile Blanche, he noted 'one can agree or disagree without a hopeless sense of things being all wrong'.[51] Together with his touchiness, his strong feelings about art, and his tendency to change his views suddenly, this playfulness must at times have been exasperating.

Yet most of the time Ricketts and Shannon spent with their intimate friends was characterized by high spirits and a lively sense of fun. Together the two were 'as merry as boys' and there was a great deal of teasing and pantomime between them.[52] They relished good jokes and anecdotes and Ricketts especially delighted in improper or outrageous stories. Their conversations went on late into the night, for Ricketts and Shannon often stayed up till 3 a.m. in winter and slept long in the morning. After a visit to the Vale, an elderly student, North Hall, remarked, 'We laughed the whole time, but I am sure I never knew why'.

These convivial gatherings often included something to eat, but the two young artists were hard up. They often had no more than one square meal a week, so meals at the Vale were usually no more than snacks. Even the bon vivant Wilde would, when the hour grew late, 'share dishes of London eggs improvised by Shannon and washed down with the cheapest drinks'.[53] When, some years later, after Wilde's release from prison, he heard that his two young friends were better off than before, and even making a little money, he commented: 'Ah, I suppose when you go there to supper they give you *fresh* eggs now.'[54] Another friend recalled snacks of 'rolled tongue, bread and butter,

[50] Moore, Original Introd., pp. 60, 68–9.

[51] CR Diary, 2 July 1904.

[52] TSM *Exhibition of the Lithographs of the late Charles Shannon* (Colnaghi, 1938), p. [ii].

[53] *65 Ill.*, p. [ii].

[54] *SP*, p. 17.

quince jam (a speciality of these meals) and great cups of steaming cocoa'.[55] Occasionally there were more ambitious meals, like a 'langouste supper' towards which the guests paid half a crown.[56] These probably came later, and marked a slight improvement in their fortunes. In their first months at the Vale Shannon wrote to the collector, book designer, and first editor of the *Studio*, Gleeson White, for whom Ricketts designed an early bookplate, 'Come whenever you want to, if you do not mind being starved—we can promise you a salad at least'.[57]

Their greatest benefactor, however, was still Wilde. A pastel by Shannon hung in his house at Tite Street, but it was to Ricketts that he gave the greatest help, by introducing him to his publishers. The first commission was *The Picture of Dorian Gray* (1891). The cover's floral design is effective and distinguished, but the spine is not: the title and name of the author are placed where one would least expect to find them (at the base) and the gold lettering on the cream cloth is rather faint and crowded. This experiment with book design is novel but unsuccessful, reflecting not only an understandable desire in a young artist to do something striking and original but also a rather perverse streak in his nature. Especially when young, he enjoyed making startling statements, both in his designs and in his conversation. However, he learned from his mistakes, and never used this formula again.

After *Dorian Gray* Wilde turned to a new publisher, a house recently set up by a pair of Americans, James Ripley Osgood and Clarence McIlvaine. In spite of its very short life in publishing, this firm, largely through Ricketts's contributions, was to affect considerably the way books were designed in Britain. *Intentions*, published in May 1891, was the firm's first book for Wilde. The cover combines a floral emblem with a stylish script derived from Rossetti, and ultimately Blake, while the spare, asymmetrical title-page owes a lot to Whistler, who, after Rossetti, was the most fertile influence on Ricketts's book design. Wilde characteristically inscribed a copy 'to C. Ricketts who made this lovely book. From Oscar Wilde who wrote this lovely book.'[58] *Lord Arthur Savile's Crime*, published a few months later, is less prepos-

[55] Holmes, p. 165.

[56] *65 Ill.*, p. [ii].

[57] CHS to Gleeson White *c.*1889. (Private collection.)

[58] Christie's sale of valuable books and MSS, 4 Dec. 1933, no. 351.

sessing. The most important of these early books was *A House of Pomegranates*, another collection of Wilde's stories. It is the fullest realization of Ricketts's Pre-Raphaelite style and, at the same time, it shows the first signs of what was to come. Ricketts was gradually widening his control over the books he designed. This time the binding, the title-page, and the decorations in the text were all his, whilst Shannon supplied four illustrations. Printed in Paris by a new process, these lovely designs were found to be covered by a dusty deposit caused by impurity in the printer's ink or the chalky paper. When this was rubbed off with a soft flannel, most of the image went with it, so that the illustrations are so faint as to be indecipherable.

No such ill luck affected Ricketts's decorations, which are among the most charming he ever produced. On the title-page a Pre-Raphaelite maiden sits embroidering roundels whose subjects take up the themes of the four stories in the book. Behind is a hedge of pomegranate which combines rich, floral detail with tall, perpendicular trees, a successful blending of Pre-Raphaelite and Art Nouveau elements. A satyr sits by the doorway in the hedge playing the pipe. The design is emblematic and meant to be read: the girl and the satyr are images of the artist at work. Throughout the book there is a repeated motif of a girl gathering pomegranates: as the stories progress her basket fills up, until it is full at the end.

The cover is in a new, more lavish vein. A peacock, a fountain, and a basket of split pomegranates, printed in gilt, stand before swirling branches of pomegranate and waving crocuses in orange-red. All of these images derive from the theme of the book, but the overall effect is Ricketts's own. He was at his best in weaving such rich designs, and the relation of part to part and of part to whole is perfectly controlled and balanced. The curving lines of the design can only be described as Art Nouveau, though Ricketts himself would almost certainly have rejected this description with contempt. However, whilst the design is striking its meaning is not clear. A review in the *Speaker* (28 November 1891) praised the internal decoration, but noted:

> We do not like the outside of the cover ... The Indian club with a house-painter's brush on the top which passes muster for a peacock, and the chimney pot hat with a sponge in it, which is meant to represent a basket containing a pomegranate, or a fountain, or something of that kind, are grotesque, but ideally so. (p. 645)

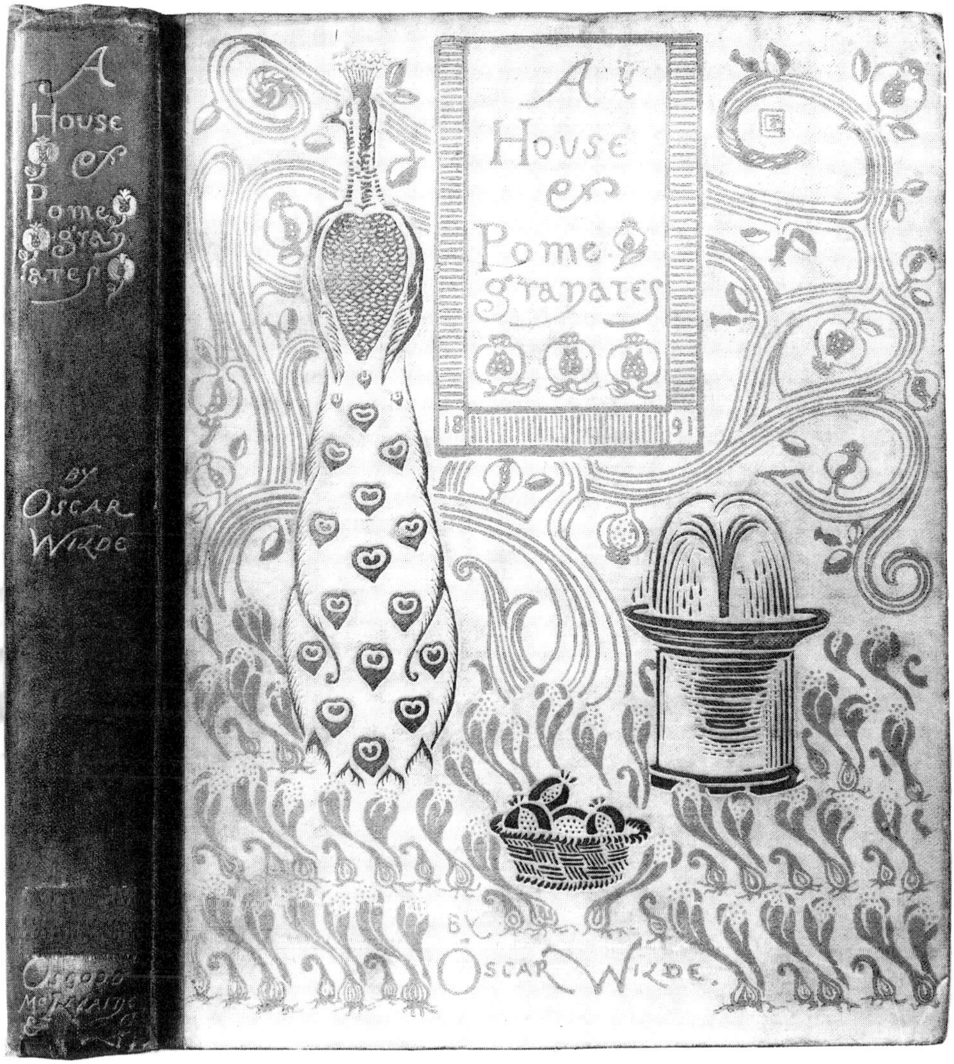

Fig. 9. Cover design by Charles Ricketts for *A House of Pomegranates* (1891), by Oscar Wilde

This was throwing down the gauntlet to Wilde, who rushed to the defence not only of his friend but also of his aesthetic principles and declared that 'there are only two people in the world whom it is absolutely necessary that the cover should please. One is Mr. Ricketts, who designed it, the other is myself, whose book it binds. We both admire it immensely.'[59] Wilde was in turn answered by 'an artist' who, writing in the *Magazine of Art* (Jan. 1892), commented that, in spite of Ricketts's intentions and Wilde's defence, the objects on the cover were indecipherable and 'clever as Mr. Wilde may be, it will puzzle him to discover a single beautiful line of form about this disputed portion of the design' (p. 135). This last must have stung Ricketts, who always regarded the *Magazine of Art*, like the *Studio*, as hostile to his work, in spite of the fact that the balance of their comments in reviews is much more positive than negative. Ricketts was excessively prickly about his work.

He designed two books by Thomas Hardy at this time. Both covers and layout are distinctively up-to-date. The cover of *A Group of Noble Dames* reflects oriental influence in its three stylized flowers in the central of three panels. *Tess of the D'Urbervilles* was decorated with two stems of honeysuckle with stylized blossom that wind across the cover—an elegant sweep of the artist's hand across the space. It anticipates by some five years Beardsley's similar, though more austere, cover for Dowson's *Poems*, and it heralds a new element in Ricketts's book work: the linear aspect of Art Nouveau as opposed to the three-dimensional style of *A House of Pomegranates*. *Tess*, however, was the last book Ricketts designed for Osgood and McIlvaine. In May 1892 Osgood suddenly died, at the age of fifty-eight, McIlvaine rejoined the parent firm of Harper's, and the firm was no more. It had not only given Ricketts his first opening as a book designer, but been responsible for six of his seven designs published in one year, 1891—a brilliant beginning to his career.

Yet Ricketts looked back at this achievement with mixed feelings. He found these books 'crude and hesitating' and regretted some of his excesses. He also pointed out that, although 'not always strictly shaped according to my wishes, [these books] were unlike the ordinary books in the matter of title-page, proportion of margin, and in the designs upon the boards'. Their influence on contemporary book design was great, so that by 1904 the style they had initiated had become so

[59] Wilde, *Letters*, p. 300.

common that it had been adopted by the publishers of cheap editions of popular works: 'today', Ricketts said, 'they might almost pass for the work of Messrs. Bell or Dent'.

'I regret', Ricketts wrote, 'that I had not then seen the "House of the Wolfings" or "The Roots of the Mountains" printed for Mr. Morris as early as 1888; these might have initiated me at the time to a better and more severe style.'[60] His claim not to have seen Morris's work when he first took up book design raises the question of the nature of his debt to the older man. Undoubtedly it was the Arts and Crafts Movement that had led Ricketts to want to apply aesthetic decoration to books. From an early age he had loved the work of the illustrators of the Sixties and (even before taking up book design in 1890) he had turned his hand to wood-engraving in 1882, to caricature in 1886, and to illustration and decorative lettering in 1888. This was before Morris himself had turned his versatile genius to the private-press movement. Ricketts had not yet seen Morris's first commercial designs in December 1890, when, working on his first designs for Osgood and McIlvaine, he told a friend that he intended to buy a press and print small editions of rare books, illustrated and bound artistically, and that he wanted to be a sort of 'Morris' of the book and illustration.[61] By this time, apparently unknown to him, Morris's prodigious energy had already preceded him and, as Ricketts later acknowledged, 'the finer conditions of book building were realised for the first time in modern printing by William Morris in the Kelmscott Press', which was founded in 1890, though Morris had started designing the type the previous year.[62] However, it is certain that Ricketts would have gone on to commercial book design and to founding his own press even if Morris had not anticipated him. His debt to Morris was of a more general nature. Ricketts attributed the revival of interest in book design to two or three calligraphic manuscripts made by Morris and Burne-Jones in an abortive attempt to illustrate *The Aeneid*.[63] It was through the impetus given by Morris not only to book design but to all forms of design that Ricketts was mainly indebted to his great predecessor.

[60] *Defence*, pp. 18–19.

[61] Lucien Pissarro to Camille Pissarro [Dec. 1890]. See also Camille to Lucien [19 Apr. 1891].

[62] *Bibliography of Vale Press*, p. vi.

[63] CR and Lucien Pissarro, *De la Typographie et de l'harmonie de la page imprimée* (1898), 7.

The friend to whom Ricketts revealed his publishing ambitions was Lucien Pissarro, the son of the Impressionist. A simple, black-bearded man, he had come to England in the hope of furthering his interest in wood-engraving. Almost immediately he met Ricketts and Shannon through John Gray, who brought him to the Vale in November 1890. He described Ricketts to his father as being about thirty, rather sickly but highly intelligent; he was in fact only twenty-three, but his premature balding and the authority of his conversation made him seem older. Pissarro added that 'Riquette' and his friends were great admirers of Rossetti, knew little of the Impressionists and, most surprising of all, were all wood-engravers.[64] A return visit was quickly arranged, and they admired Lucien's own wood-engravings. Ricketts invited him to visit them if ever he were bored on a Sunday, assuring him that he was not at all 'comme un Anglais'.[65]

Ricketts and Shannon set about introducing their new friend to English art, showing him drawings by Rossetti, Burne-Jones, and Millais. They put him in touch with the London art world, particularly the Arts and Crafts Movement. Through them he met Herbert Horne, editor of the *Century Guild Hobby-Horse* and another wood-engraver, who took him and Shannon to meet the artist and Rossetti collector, Fairfax Murray. Along with Selwyn Image, they all went to a 'public house' to see Rossetti's mistress and model Fanny Cornforth, who had a large collection of the master's work. They even promised to introduce him to Whistler, who, they assured him, was not so terrible as all that.

Even more important, they helped and encouraged him over his woodcuts. Unlike them, he had had no training in the art; indeed, like his father, he believed that one only achieved a style of one's own by slowly perfecting oneself, rather than by being moulded by a professional artist who taught technical tricks but inhibited the growth of one's artistic personality. This was contrary to the practice of Ricketts and Shannon, who not only had been professionally trained but also deliberately tried to imitate the old masters. Later, Pissarro would find his new friends too dependent on the past. For the moment, however, he was fascinated by them and entered readily into their interests. He was intrigued by their enthusiasm for all the graphic arts. 'Black and white' he wrote to his father, 'is a passion with them, because they do

[64] Rewald, p. 368. Some of Lucien's replies are also given.

[65] Lucien to Camille Pissarro [Nov. 1890]. Quoted in trans. Rewald, p. 369.

almost nothing else but that. Shannon sometimes does pastels but Ricketts and Savage do only drawing and engraving and sometimes watercolour.'[66] Together they pored over reproductions of old master woodcuts and Goya's magnificent etchings of bull-fights. At the Vale, Pissarro was shown how to print lithographs and even took a proof of a stone drawn by Shannon. He was also shown the *Dial* and invited to contribute a woodcut to it. All this could not have been more fortunate for Pissarro, being just what he was seeking in England.

Along with Pissarro, Ricketts and Shannon embarked on a plan to set up their own guild and exhibiting society. The guild was to cover furniture, books, and the decorative crafts, and to include Herbert Horne and Selwyn Image, who already had an established clientele. The French art critic Georges Lecomte, on a visit to England, saw tables, beds, and bookcases constructed and decorated in an original style by Ricketts, Shannon, Savage, and Pissarro, and spoke early in 1892 of an imminent exhibition of such furniture in the Rue Lafitte, Paris. Only financial difficulties had prevented them from carrying out such plans earlier.[67] In fact, with the exception of the books, neither the plans to revitalize interior decoration and continue the Arts and Crafts movement nor the Paris exhibition were realized.

More progress was made with the idea of an exhibiting society, which had been in Ricketts's mind since the age of fifteen. It was to be called 'The Panel Society', from the practice of allowing each member to hang his work on a panel of uniform size. With their work, exhibitors were to submit a wall plan showing how they wanted it hung. The society was to exhibit all forms of art done on paper and to be very liberal, with no jury of selection. Like the *Dial*, this was an attempt to bring their work before the public. A committee was formed and meetings held. Shannon, on behalf of the committee, invited Lucien and his father to join. Other members included Selwyn Image, William Strang, and Wilson Steer, but the founders envisaged a very catholic membership including Impressionists and even a few RAs. It is extraordinary, and a mark of the times, that such a heterogeneous group of artists should have thought of forming a society and exhibiting together. In addition to Camille Pissarro, they hoped to attract other exhibitors from France. They dared to hope they might even

[66] Lucien to Camille Pissarro [*c.*13 Feb 1891].

[67] 'Lucien Pissarro et ses amis', *Art et Critique* (12 Mar. 1892), 141.

get Degas, but they were thinking in particular of the two women Impressionists, Berthe Morisot and Mary Cassatt, who both did a lot of graphic work and were approached through Pissarro. Miss Morisot replied that she would be delighted; Miss Cassatt, however, felt there was no market for graphic work in London; her work was being exhibited there at the time with little response.[68] Nothing came of the Panel Society, but it shows clearly how Lucien Pissarro repaid all the help and encouragement from his friends by widening their European contacts. In fact, the young Pissarro saw himself as a sort of unofficial ambassador of Impressionism, which he felt was not understood in England, even by those artists who had studied in Paris at the Académie Julian. He set about increasing his new friends' understanding and introducing them to many of the artists whom he had met through his father, and who were beginning, after long effort, to make their mark. Ricketts's and Shannon's first reaction was interest, though, it must be admitted, this included a fair degree of self-interest too. They would, understandably, give more credit to Impressionism if its practitioners helped to further their own careers.

One of the first fruits of these contacts was that they were invited to exhibit on the Continent, but the inevitable difficulties of arranging things through intermediaries arose. Lucien asked Georges Lecomte to approach Paul Signac about the possibility of Shannon exhibiting at the Société des Vingt, a group of twenty artists living in Belgium who exhibited with an equal number of invited foreign artists. Founded in 1884, its aim was to exhibit new and unconventional art, and among the English contributors had been Sickert, Whistler, and Wilson Steer. The invitation, however, appears to have been extended only to Ricketts, and when Pissarro tried to sort it out Shannon refused to exhibit. A year or so later there were further difficulties, for Ricketts had not received his invitation until some months after Shannon's. They replied that they would exhibit together the next year, but the Belgians misunderstood and there were further complications. The upshot was that neither of them ever exhibited with Les Vingt before the Society's demise in 1894. However, later they did manage to follow up the contacts made at this time.

It was also through Pissarro that Ricketts and Shannon were invited by the Belgian Art Nouveau designer Henri Van de Velde to join a new international review that he was founding. Lucien at first feared

[68] Rewald, p. 178.

that they might not accept, because of the presence of Beardsley, but he was able to write accepting on their behalf. Their names appear along with other prospective collaborators—Toulouse Lautrec, Signac, Felix Valloton, Max Klinger, Jan Toorop, and four members of the Pissarro family. As with so many of these projects, this never seems to have got off the ground. More fruitful was Pissarro's introduction to Dutch, German, and French dealers, through whom Shannon's work, particularly, began to become known on the Continent. These contacts with avant-garde art, with both Impressionism and Post-Impressionism, show that Ricketts was still open-minded about these movements. They had not yet become the enemy of almost everything that he valued in art, or acquired an almost exclusive dominance over the art market.

However, while benefiting from these wider contacts Ricketts was busy exerting his own influence on the young Pissarro. When he tried to push Lucien towards his own style of wood-engraving Camille became alarmed, and warned his son not to put his trust in the 'préraphas'. Camille began to worry that Ricketts's influence on his son was taking him away from his own Impressionist values. Soon Lucien was obliged to choose between the two.

Ricketts loved to influence others. He gave himself most generously to helping his friends, and his qualities of intellect and imagination suited him to the task. With his quick sympathetic mind, he saw where people had gone wrong, and could show them how not only to improve their work but also to realize more fully their own intentions. He threw himself boldly into analysis and advice, sometimes causing hurt feelings but more often impressing with his critical authority and intelligent sincerity. Sturge Moore felt that Ricketts had the finest critical mind he had ever come across, and he benefited greatly from it.[69] So, undoubtedly, did Shannon. Many friends owed a great deal to Ricketts's interest, suggestions, and practical help, and he left a definite and personal mark on them.

Nevertheless, this domineering trait had serious repercussions. He admitted that his influence was either very blighting or very fruitful. 'Mostly I wither up. People can't stand it', he said.[70] Too much desire for influence can bedevil a friendship; too much help from a powerful personality like Ricketts's can be stifling and even destructive. Both

[69] TSM to Owen Lewis [1933] (TSM Papers, Box 20).

[70] MF Journal, 1902, fo. 175.

Shannon and Sturge Moore suffered from this, and other—less loyal or flexible—friends felt that they had to break with Ricketts in order to follow their own road.

Many have left grateful testimony to his help. John Gray went so far as to write:

> I am able to say ... that with my little talent I was an invention of Ricketts. He used to set me tasks to perform; he was infinitely patient with the performer and very hard on the performance ... He was to me the kind of tyrant I have always approved of.[71]

This was written shortly after Ricketts's death and, even allowing for the exaggeration of grief or modesty, it shows what a direct and constructive part Ricketts took in furthering Gray's career. The tasks set probably involved translations from French poetry. Another friend, Laurence Housman, recalled that within a week of meeting Ricketts he 'had dragged me away from my timid preference for fuzzy chalk drawing, as a means of concealing my bad draughtsmanship, and had set me to pen-work, with Rossetti and the other Pre-Raphaelites as my main guides both in composition and technique'. Chalk is a flattering medium, whose qualities of colour and texture can conceal faulty drawing, and Ricketts felt that sound drawing was the first requirement in an artist. 'The few years during which I was frequently in his company, and very much under his influence', Housman continued, 'taught me far more of what I wanted to know than all the preceding years of dull school-work.' It was thus that he developed his precise and elegant book illustrations and his exquisite bindings, both revealing their debt to Ricketts. However, he found it necessary eventually to break with Ricketts:

> parting from Ricketts [he wrote] I had also to part from Shannon, whose gentle character with its quiet blend of modesty and confidence presented so curious a contrast to the strained and restless brilliance of Ricketts, ever laying down the law with an arbitrariness which the quick change of some of his admirations did nothing to modify.[72]

Many of the friends that Ricketts helped managed to cope with his volatile nature and remained in his orbit for life. Charles Holmes acknowledged that it was Ricketts, his friend and master, who set him on the path to becoming the Slade Professor of Art at Oxford Uni-

[71] John Gray to Gordon Bottomley, 1 Feb 1933.

[72] *Unexpected Years*, pp. 114–15.

versity and then director of the National Gallery. Holmes's early interest in art took the form of painting in water-colour, in which he was working when he first met Ricketts in 1891. On being shown some of these Ricketts pronounced, 'your things are quite accomplished, but you don't know what you're doing. This, for example, might have been a scrap by Whistler: that pile is worthless. You must learn what drawing means, and take up etching.' Here again, then, Ricketts encouraged a young artist to strengthen his foundation in good drawing. From Ricketts Holmes also learnt a great deal about the technique and appreciation of paintings: 'to work under their critical eyes was in itself an education.'[73]

Charles Holmes had first made contact with Ricketts and Shannon over the *Dial*. A second number was being prepared, in spite of their losses on the first one, which they had been unable to get anyone to distribute, even though all the costs had already been covered. Only a few bookshops had taken any, which had then been sold well above the ordinary price, as they could not be obtained elsewhere.[74] Ricketts and Shannon were better artists than business men.

Their first plan was to print the second number themselves, but they soon set about looking for a new printer. A neighbour in the Vale, Henry James Riley, a 'well-dressed little man' who shared a house with Savage, went on their behalf to the Ballantyne Press to get an estimate.[75] Charles Holmes, then working at the press, dealt with the inquiry and sent on the estimate, which prompted a visit from Ricketts. After the details of production had been settled Holmes was invited to the Vale, and soon regularly frequented the Friday evenings there, where he listened to the most entrancing conversation he had ever known.

Mr Riley and Sturge Moore appear to have played an important part in the second *Dial*, for we are told that 'their timely aid made possible the publication of this number'.[76] This suggests that the second *Dial* had run into trouble or proved too much for Ricketts and Shannon. Moore's part is evident, for he was the main contributor, with a story, ten poems, and an unsigned article on Maurice de Guérin. John Gray also supplied two poems of French inspiration and Ricketts produced 'decadent' short stories. Apart from its strong French sympathies, the

[73] Holmes, pp. 146, 176.

[74] Lucien to Camille Pissarro [Dec. 1890].

[75] Holmes, p. 141.

[76] *Dial*, 2 (1891), 33.

main feature of the second *Dial* is its publication of original prints and wood-engravings. Shannon contributed two lithographs drawn on the stone and 'bitten' by him. Pissarro, Savage, and Ricketts all contributed woodcuts designed and engraved on the wood by the artists themselves, making use at last of their earlier training. This was an important stage in restoring these arts to an original, rather than a reproductive, form; Ricketts could claim with some justice that the *Dial* had 'reinvented wood-engraving, and revived lithography', though he ignored the fact that Shannon's lithographs were first printed in the *Century Guild Hobby-Horse*.[77] Thus the second *Dial* really initiated their gradual absorption into the Arts and Crafts movement over the next few years. Not only did they cut and print their own woodcuts, but Shannon also printed his own lithographs—all of these being steps forward in craftsmanship. Publication took place in February 1892.

Complimentary copies were sent to Lord Frederic Leighton, to Camille Pissarro, who praised Ricketts's woodcuts and Shannon's lithographs, and to William Morris, who arranged for them to visit him in Merton Abbey. Of the magazine, he wrote: 'I confess that I looked at the art portion of it with somewhat mixed feelings, as the talent and the aberration of talent seemed to me to be in about equal proportions.'[78] Most of the critics felt the same.

'Please let us know', Ricketts wrote to Pissarro early in 1892, 'when your father arrives from France, as Shannon and I would like to pay our respects to him.'[79] For over thirty years the great Impressionist had struggled doggedly to paint his view of the world, in spite of poverty, of doubts and uncertainty, of public indifference or hostility, and of the nagging of his long-suffering wife. His motto had been 'Only painting counts.' A successful exhibition in Paris in 1891 had been the turning point in his career. Before such integrity and dedication to art, so much like their own, the two young English artists could only feel respect.

There can be little doubt, moreover, that this extended to the Impressionist's work too. When Lucien showed them his father's graphic work, they found his fans and gouaches very beautiful and were even more impressed by his etchings: 'They have the highest

[77] *Bookselling*, p. 508.
[78] William Morris to CR and CHS [1893] = *SP*, p. 20.
[79] CR to Lucien Pissarro [early 1892].

opinion of them', Lucien told his father. After they had particularly admired a sunset with haystacks, Lucien gave them a copy of it 'in exchange for their many kindnesses towards me'.[80] Camille was delighted, as 'nobody noticed it' in France.[81] Two months later, Ricketts and Shannon went to see a collection of Camille's etchings which had just arrived from France, including one *Setting Sun* as a gift for them. Lucien reported to his father

> They were delighted and here is, word for word, what Ricketts told me about *La Place de la République:* 'You know, I definitely prefer certain of your father's etchings to Whistler's, and to find another aquafortist you have to go back to Rembrandt, because Goya's etchings could be done with other techniques'. Isn't that an exqisite compliment? He meant it sincerely . . . Finally, they told me that they wanted to buy two proofs, *La Place de la République* (copper) and *L'île Lacroix*. They chose those two after having hesitated at some length over others.—I was very embarrassed selling them something; it is rather awkward when I'm so much in their debt! But they added: 'Don't tell your father that it's for us'.[82]

Camille would not hear of their paying for the two prints, and suggested an exchange. Some three years later he sent some lithographs of women bathing to them; Shannon in turn sent a portfolio of his own lithographs. Shannon's nudes, Camille thought, were 'treated with a master's hand', though he felt them 'the least bit too studied, too deliberate'.[83] When they were absolutely determined to buy another etching, *Rue de l'Épicerie*, another exchange was arranged.

Ricketts never praised lightly though, as Yeats noted, he was apt to overpraise, and his praise of Pissarro was constant over several years. Nor was he likely, with his very limited means, to want to buy or collect anything but the best. His buying Pissarro's work shows clearly that at this stage he was still open-minded about Impressionism and ready to respond to at least some aspects of it. Unfortunately, the shifting fashions and rivalries of the art world were not only to make Ricketts close his mind to much of the achievement of Impressionism and successive movements, but also to lead to Ricketts and Lucien Pissarro falling out and going their separate ways. Later, Ricketts commented rather sourly that his friendship with the Pissarros was based on the help given to the son and 'an interesting acquaintance

[80] Lucien to Camille Pissarro [Jan. 1891] = Rewald, p. 371.
[81] Rewald, 152. [82] Ibid. 375. [83] Ibid. 253.

with the veteran Impressionist based largely on his genuine praise of Shannon's work'.[84] This, said after his views had hardened, is clearly a distortion of the truth. However, when the divisions of the art world had put them into opposite camps Ricketts decided to get rid of a water-colour and some lithographs by Camille Pissarro by presenting them to the British Museum.[85]

All this lay in the future. In 1892 Lucien married Esther Bensusan, in spite of her father's prohibition. They got the date wrong and turned up a day early, accompanied by their guests, including Shannon and Ricketts carrying a bunch of yellow roses. The whole party nevertheless sat down to the wedding breakfast which had been ordered and returned the following day, 11 August 1892, for the ceremony at the Richmond Registry Office. Ricketts and Shannon, along with Camille, signed as the witnesses. When, after a short stay in France, the young couple set up house in Epping, on the other side of London, Lucien went to London once a week to spend an evening at the Vale.

This summer also brought Ricketts one of the great musical experiences of his life. He saw Wagner's *Ring*, which he had been forbidden to do a decade before. As Reginald Savage could read music at sight, he would often go over a score with Ricketts in preparation for a performance, so as to increase their appreciation. For this occasion in 1892 the soprano Rosa Sucher, who had sung at Bayreuth, was to sing Brünnhilde. Ricketts eagerly secured one of the cheapest seats:

> I was in that enthusiastic and fervent side gallery of Covent Garden, hanging over an iron bar, by which practice something more than the foot-lights could be seen: a magnetic singer, Alvary, or, I should say, actor, had done wonders as Siegfried, and the smoke (called fire) had cleared away, when, from beneath the glittering shield and aegis on the couch, I expected the deliberate and academic apparition of Rosa Sucher to arise, with her eyes firmly fixed on a respectful audience: the harmonica-like chords of the harps, and agonizing violin trills had done their utmost, when, instead of Rosa, an entranced woman appeared beneath her slowly uplifted cloak; whilst her expressive and exquisite hand seemed to touch the encircling harmonies, her eyes smiled to the light in an ecstasy of adoration, they remembered, then saw, and understood the rapture of the return to life and the terrible dawn of passion. The singer was Frau Klafsky who was to . . . leave an unforgettable

[84] CR to D. S. MacColl [8 Nov. 1904].

[85] The gift (BM Acc. No. 1920-10-18-1/6) was made on 18 Oct. 1920.

memory with some elderly men like myself. . . . I doubt if any singer of the past or to come will surpass her rendering of Brünnhilde.[86]

Sucher, unable to keep her London engagement, had been replaced at the eleventh hour by Katharina Klafsky, who was then singing at Drury Lane. It was probably on this occasion that Ricketts broke an olive-wood stick that Shannon had given him in thumping the rail of the gallery in his admiration for Frau Klafsky, who, as well as having a noble voice and beautiful face and figure, had exceptional acting ability. Ricketts often used her performance in later years as a standard of excellence in opera.

At this time Ricketts's and Shannon's publishing venture was also making headway. The first publications from their own press (which they named after the Vale) were albums of prints by Lucien Pissarro and Shannon. An album of Lucien's woodcuts was ready as early as January 1892. The following year *Les Travaux des Champs* (an album of woodcuts by Lucien after his father's designs) and a portfolio of Shannon's lithographs were issued, to be followed by two more such sets of six lithographs and *La Reine des Poissons*, a book of woodcuts by Lucien. These show that Ricketts had first envisaged the Vale Press as an art, as well as a book, publisher. Lack of funds and equipment (for their press at the Vale was not suitable for books) necessitated a slow and modest beginning.

In April 1892 Ricketts and Shannon had begun a project for their first book which was to set the direction of their lives for the next decade. Shannon by chance had come across an Elizabethan translation of Longus's pastoral romance *Daphnis and Chloe* in the British Museum (Fig. 10). So charmed were he and Ricketts that they decided to illustrate and publish it. This was intended to be the first Vale book, but later developments changed the direction of the press. It was an ambitious undertaking and, in its layout and illustrations, set out to imitate one of the masterpieces of Renaissance book production, the *Hypnerotomachia* of Poliphilus, published by the Venetian Aldus Manutius in 1499. The prospectus announced:

An attempt will be made throughout to emulate the build and style of the finest Italian books. The quaint and hearty translation by George Thornley

[86] CR, 'Musical Recollections', pp. 88–9. I have corrected the spelling of names.

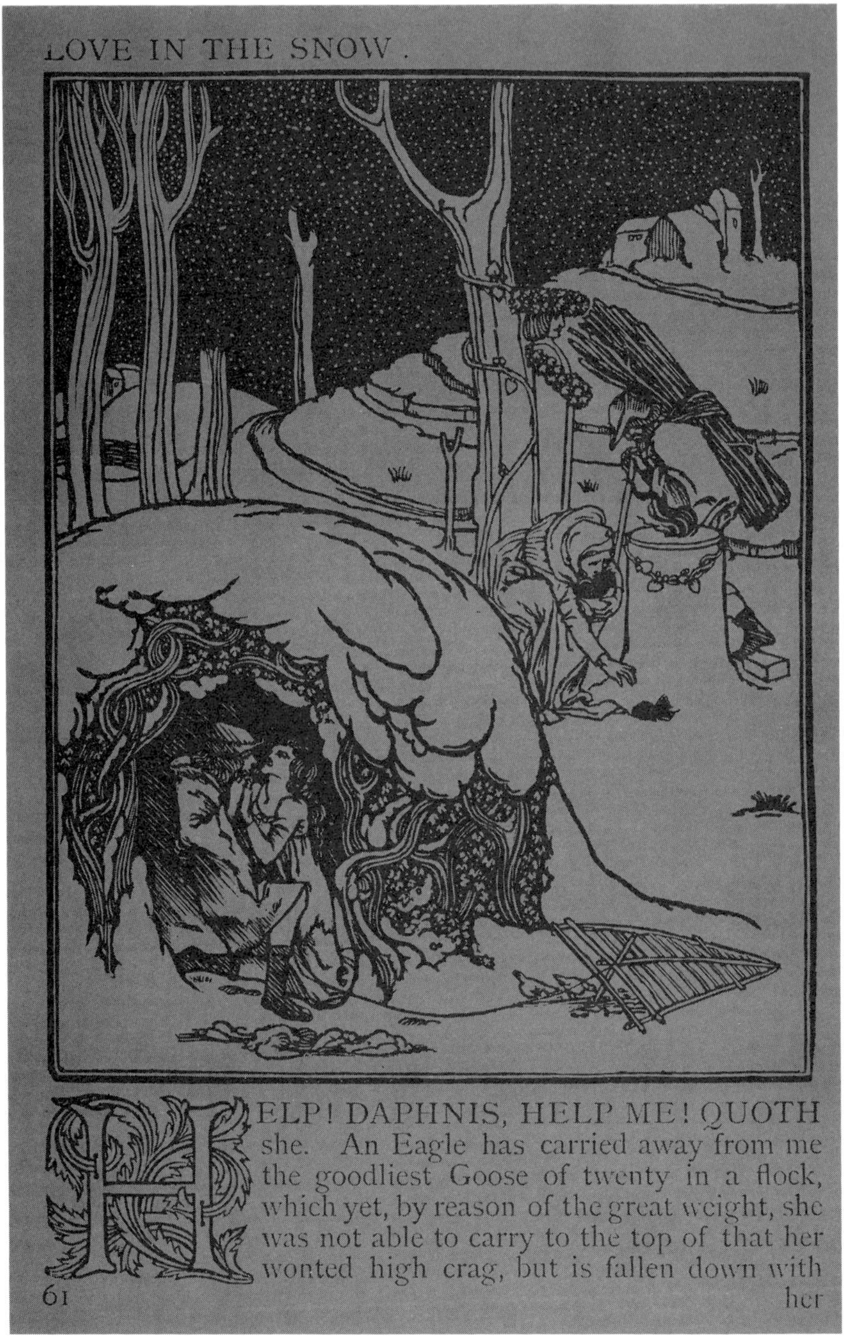

LOVE IN THE SNOW.

HELP! DAPHNIS, HELP ME! QUOTH she. An Eagle has carried away from me the goodliest Goose of twenty in a flock, which yet, by reason of the great weight, she was not able to carry to the top of that her wonted high crag, but is fallen down with

61 her

FIG. 10. *Love in the Snow*. Wood-engraving by Charles Ricketts for *Daphnis and Chloe* (1893)

not hitherto reprinted will be unexpurgated. It is intended that this book shall be the first of a series built under the direct supervision of the artists.[87]

They attempted to finance the undertaking by securing subscriptions. The engraving of the thirty-nine woodcuts was to take them much longer than they had envisaged, for they announced publication for November and did not finally finish until April 1893, the engraving having taken them eleven months. The task of designing the woodcuts was divided about equally between the two artists, but they were all revised and drawn on the wood by Ricketts, so as to ensure uniformity of style, and then engraved by Shannon. As a result of this care it is very difficult to know who did which, and only the signatures on some of the proofs enable us to attribute them confidently. In addition, Ricketts cut numerous decorative initial letters. Aesthetic design and craftsmanship were the book's hallmarks, but it was only a step, though an important one, towards the full realization of the ideals inspired in them by the Arts and Crafts movement.

When Ricketts became more experienced as a book producer, he came to realize that the work was 'poorly printed and overcrowded by initials and lines in capitals, that overset at times the proper balance between illustrations and the text in a manner not foreseen by us', though he always regarded the work with affection and hoped to reprint.[88] Still, it was a considerable achievement. It has been rightly described as 'the first book of modern times with woodcuts by the artist in a page decorated by himself'.[89]

One of the illustrations, a double-page spread of *The Wedding Feast of Daphnis and Chloe*, contained portraits of the Vale group as a private joke. C. J. Holmes gave the following description of the portrait group:

On the extreme right the artists are shown, Ricketts, Shannon, Sturge Moore, Lucien Pissarro, Reginald Savage; the figure standing behind them is Mr. Riley, and on the opposite side of the table to the artists (the opposition was deliberate) sits an unfamiliar Holmes. As Ricketts sat at work among us one evening, some remark made him rock with laughter, and I heard him say, 'Oh damn! I've run the graver through Holmes's moustache. It must come out'. And out it came.[90]

87 *Daphnis and Chloe* Prospectus (1892).

88 *Defence*, p. 20.

89 Denys Sutton, 'A Neglected Virtuoso: Charles Ricketts and his Achievement', *Apollo*, 83 (1966), 141

90 Holmes, p. 169.

Ricketts explained the poor likeness of Holmes very differently. Holmes's hair and appearance had been altered because he was 'silly at the time & troubled to be shown so entirely naked. Now he tells people it is his portrait & when they cannot see the likeness he is disappointed.'[91] This was the sort of story about his friends, or even himself, that Ricketts loved to tell, and perhaps to improve in the telling.

While reviving the use of original wood-engraving in book design, Ricketts was also producing commercial book designs that showed the most up-to-date influences. When a new edition (or more properly impression) of Oscar Wilde's *Poems* was published in May 1892, a new title-page and a splendid cover by Ricketts, of seven highly stylized trees amid gracefully curving vegetation blocked in gold on a pale lilac cloth, gave an extremely rich effect. A similar design was used some nine months later for J. A. Symonds's *In the Key of Blue*, which had a rich brocade-like pattern of hyacinths and laurels (Fig. 11). Originally, the publishers had decided to block the design in gold on an appropriately blue cloth, and some early proofs were so bound. Ricketts, however, objected to this, made a joke about Reckitts blue, a well-known type of whitener, and changed the cloth to cream. The designs for these two books were strikingly original, very early examples of the Art Nouveau style that was soon to sweep through not only book design but every branch of the decorative arts.

An even more striking design was done for John Gray's *Silverpoints*, a collection of poems in a Symbolist vein and of translations of French 'decadent' poets. It is much in keeping with the Vale's French sympathies. Originally, the book was to be published at Wilde's expense, but the final contract shows that it was published on a purely commercial basis. Gray was to settle financially with Ricketts, who was also given one copy of the special edition and three of the ordinary in recognition of his 'aid'. Gray later recorded that he never saw the designs before the publication of the book, and Ricketts seems not only to have kept his work to himself but also to have caused delays in the publication, practices that were to get him into difficulty with authors and publishers in the future. When the book finally appeared in March 1893 it was, in appearance, like nothing that had ever before been published in England, with its use of italics, its tall, slim, narrow format, derived from Aldus again, and its green cover of gilt willow leaves. It had an enormous influence on contemporary design in Britain

[91] MF Journal, 1898, fo. 95.

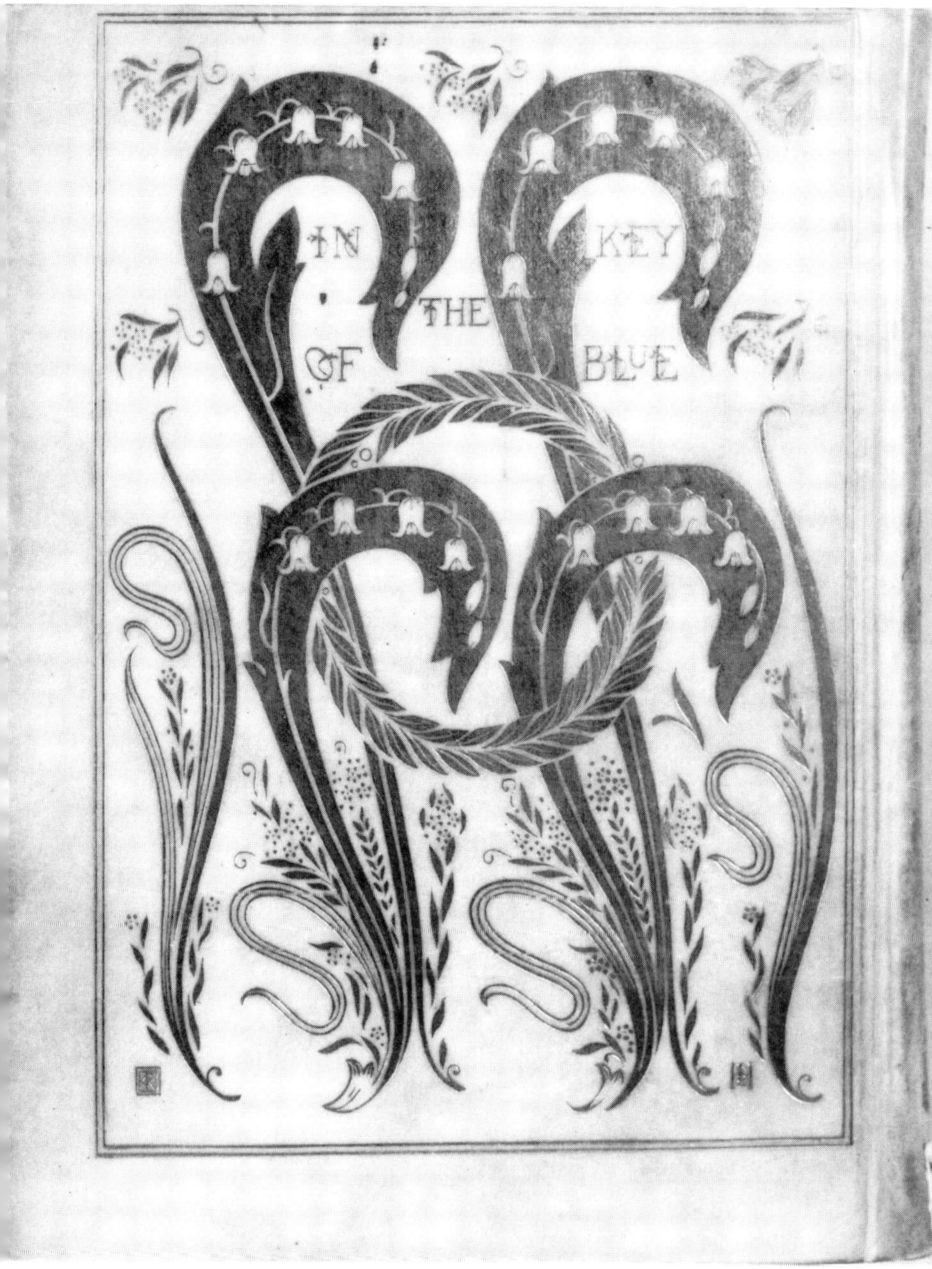

FIG. 11. Cover design by Charles Ricketts for *In the Key of Blue* (1893), by John Addington Symonds

and abroad: Ricketts noted that the book was 'copied outright in America' and then so much further exploited that he wished they had not followed 'those eccentricities and departures from the original that I had thought necessary at the time, or that are there by a final printer's oversight in the carrying out of those last revises allowed by the publishers'. Its influence was not limited to book design, for Ricketts found that it often drifted back to him 'from places where my name is quite unknown; on bindings, end-papers, wall-papers and dress cretonnes'.[92]

Ricketts's next book, Lord de Tabley's *Poems Dramatic and Lyrical*, had an even more chequered history. Lord de Tabley, a sensitive and rather morbid man, had become a recluse in the seventies because of the hostile reception given to one of his books. In the nineties interest in him suddenly revived, and when the Bodley Head offered to publish a collection of his poetry he made his own selection, by dividing them up into two groups labelled 'Dustbin I' and 'Dustbin II'. The apparent modesty of this action did not mean that he was not determined from the first to get his own way over almost every aspect of the book. Such was the man who was to collaborate with Ricketts, who was himself so prone to be secretive, mercurial, and domineering.[93]

By October 1892 Ricketts had been chosen as designer and illustrator. The cover design seems to have been shown to de Tabley shortly after this, and he liked it well enough. It had a semé pattern of devices shaped like parachutes, and a male figure in a heart in the top right-hand corner. When de Tabley's sister identified the shapes on the cover as rose petals he asked Ricketts, who had intended them to be dew-drops, to adopt this description. When asked to do a book-plate for the poet Ricketts produced one including a nude male figure similar to the one on the cover, which rather worried de Tabley, who thought it inappropriate for a private book-plate. He asked Ricketts if he could see his way to 'softening or rendering more vague certain details', which the artist was unwilling to do, replying rather flatteringly that the figure was the personification of Praise. Moreover, Ricketts had selected the poems he wished to illustrate without consulting the author and thus illustrated two poems which had been left out of the

[92] *Defence*, pp. 20–1.

[93] See J. R. Tye, 'The Letters of Lord de Tabley', *Turnbull Library Record*, NS, 8 (Oct. 1972), 17–24, and James G. Nelson, *The Early Nineties; A View from the Bodley Head* (Cambridge, Mass., 1971), 132–49.

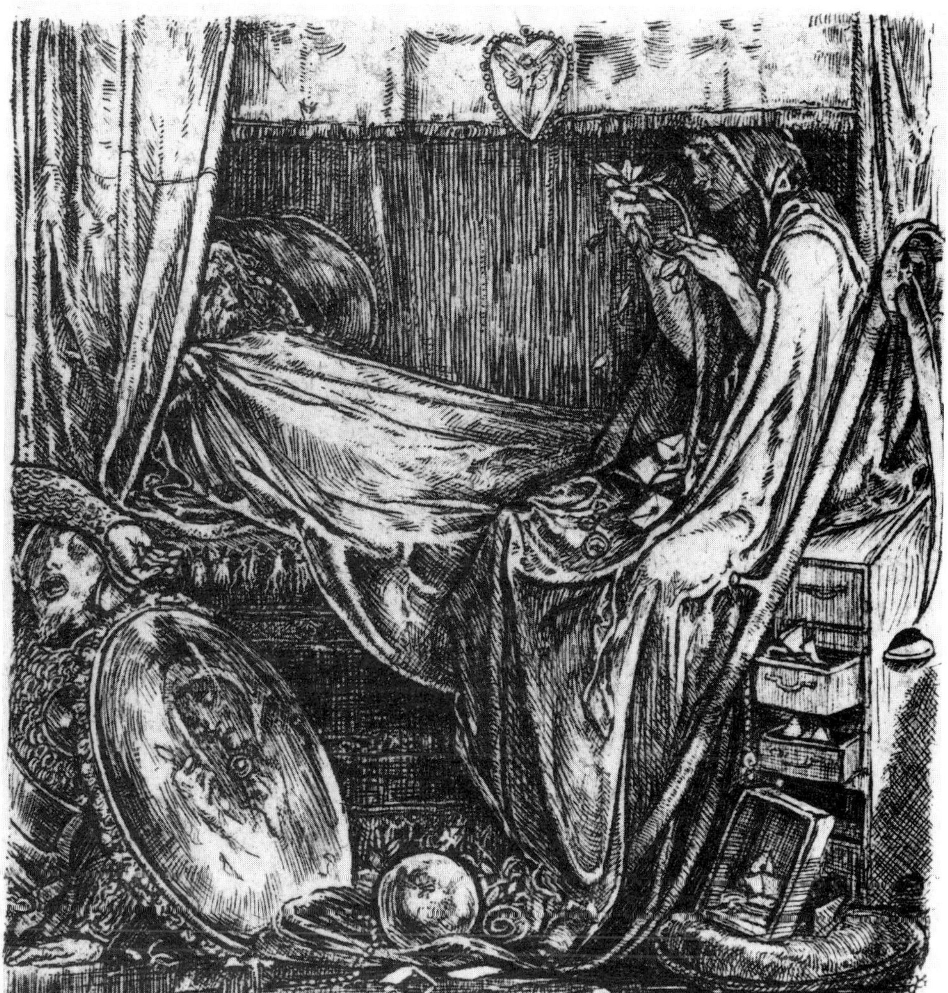

FIG. 12. *The Two Old Kings.* After a Pen illustration for *Poems Dramatic and Lyrical* (1893), by Lord de Tabley

volume, so de Tabley had to write a new poem for 'The Prodigal' and for 'The Two Old Kings', which had inspired one of Ricketts's finest illustrations (Fig. 12). As a frontispiece de Tabley had wanted an illustration of the first poem in the volume, 'A Hymn to Astarte', but Ricketts had not illustrated this, and the splendid 'The Defeat of Glory' was therefore chosen instead. Finally, it turned out that the distribution of the illustrations was very unbalanced, with three of the five coming close together at the end of the book. In a fury, de Tabley wrote to the Bodley Head threatening legal action to halt publication if the male figure went unaltered. Later he decided to have nothing more directly to do with Ricketts, who had taken a month to put a patch of colour over a certain part of the figure, which still remained 'unhealthy, morbid and generally . . . unpleasant'. He had had enough of Mr Ricketts, who was 'self-opinionated' and had given 'so much and such unnecessary trouble', and he was unable to accept any more of Ricketts's 'alterations or interference'.

In desperation, John Lane suggested that they might ask Walter Crane to do the illustrations, but that would delay publication indefinitely. When his partner Elkin Matthews agreed to exclude Ricketts's book-plate and replace it with one by de Tabley's friend William Bell Scott, de Tabley suggested that Bell Scott's name should be given equal billing with Ricketts on the title-page; an arrangement which was, not surprisingly, unacceptable to Ricketts. That Ricketts was rather secretive and dilatory is indicated by a letter from the printer, who wrote on 7 November that he would have preferred to see the illustrations; and although the sheets were printed by the end of January, there was still no sign of the designs. With such pressures being put on him, Ricketts seems to have become recalcitrant, and must also have been in a state of fury. At last, after having kept both author and printer in the dark, Ricketts produced the designs in February, just in time for insertion into the book. No reason is recorded for his delay. He may have been busy with his *Daphnis and Chloe* illustrations, but it is likely that he was hesitating over his designs, or that this was really a show of pique, for scolding or criticism never brought out the best in his nature.

When the book was finally published at the end of March, the title-page design by Ricketts made no mention of the Bell Scott book-plate. The male figure on the cover has sprouted a third wing, which preserves its modesty, but the nude male figure is still represented above the bed in 'The Defeat of Glory'. With all the vexations and difficulties the book had caused him de Tabley was convinced it would be a failure, but it was not. The critics were highly favourable about both its poetic and its artistic aspects, and the edition, which sold out in a few days, went into a second and third reprint. Ricketts's designs show him at his most Pre-Raphaelite, with their intensity, their intricate drawing, and their rather crowded composition, whilst the cover stands as one of the most successful he ever did.

These altercations explain something of the querulous tone of a letter that Ricketts wrote to Lane at this time. Lane, too, was given to delays, especially in paying his artists, so that a financial crisis prompted Ricketts to write:

> I am in absolute need of cash. You promised two weeks ago that you would let me have a cheque for the remainder of the debt as soon as certain moneys had been paid you.
>
> My banking account is overdrawn by some pounds. I fear that through my lack of business habits you have lost all sense of right and dates. You promised to speak to de Tabley about the cover & '2 old Kings' that are still owing to me. I cannot do with a portion of the sum as, thanks to this way of arranging matters, I am disagreeably in debt. I enclose account of the payments made to me, the dates representing the days the cheques were paid into our bank, the actual date of the cheque being a day or two before.
>
> Please forgive the disagreeable tone of my letter. I have no doubt you overlook my affairs out of sheer lightness of heart.[94]

This letter shows not only their lack of business sense and the precariousness of their income, but also the waspishness that Ricketts could show when crossed or provoked. He had a good line in finely-turned insults like the last sentence, which are all the more striking in view of the noble and generous behaviour that also characterized him.

Most of the books that Ricketts had designed up to this point were exhibited to great acclaim at the Arts and Crafts Exhibition in 1893. This was Ricketts's first exhibition in England as a book designer. The critic of the *Studio*, probably Gleeson White, wrote that 'the exquisite *Silverpoints*, Oscar Wilde's *Poems* and Lord de Tabley's *Poems* after C. S. Ricketts's designs are a revelation to those who see them for the first time'.[95]

[94] CR to John Lane [1893] (Houghton Library, Harvard Univ.). The letter is in Shannon's hand.

[95] *Studio*, 2 (1893), 57.

The third number of the *Dial* appeared in October, much along the lines initiated by the second. The only new contributor, Charles R. Sturt, was actually Ricketts, and his article on Gustave Moreau, later reprinted under his own name in *Pages on Art* (1913), was one of the first appreciations of this artist to appear in England. Ricketts may have used this name because he felt that the magazine's association with a small coterie limited its value or appeal, but his rather secretive nature made him resort to pseudonyms all his life. This reserved side made him hide details of his occasional extravagances from Shannon, partly explains the disingenuousness that sometimes marked his need for self-justification, and made him hide his real feeling, whether of unreciprocated love for a young man or of hostility for an enemy, though these sometimes broke through his restraint. He was also given to covert or indirect attacks on 'enemies' in his comments and writings.

Before this, Ricketts had begun work on his most celebrated commercial book, Wilde's *The Sphinx* (Fig. 13). Wilde himself had come to the Vale to read the manuscript, and Ricketts began work in June 1892. The contract stipulated that the designs should be produced before the first of October, but the book was not published until two years later, on 8 June 1894. This delay was attributed by Ricketts to the notoriety surrounding the publication of Beardsley's edition of *Salome*, but this leaves some unanswered questions. Ricketts claimed it was postponed for a year, but barely four months separate the two publication dates. The fact that Ricketts, who commenced work more than a year before Beardsley, should have been published after him may have been due to Wilde, who did not complete *The Sphinx* until 1893 when he was in Paris, but Ricketts too was given to delays, and there was a great deal to see to in this book. Whatever the cause, *The Sphinx* was eclipsed by *Salome*. The novelty of Ricketts's brilliant and original designs was lost after the shock and scandal of Beardsley's work, which not only contained erotic elements but also openly caricatured Wilde. As a result *The Sphinx*, like most of Wilde's books, was, according to Ricketts, a financial failure.[96]

C. J. Holmes records that when Wilde brought Beardsley's newly completed designs to the Vale 'Ricketts was enthusiastic about their accomplishment, praising the more generously perhaps because he would like to have illustrated the play himself'. Holmes felt that Ricketts 'certainly understood Wilde's intentions far better than

[96] CR, *Wilde*, pp. 38, 41.

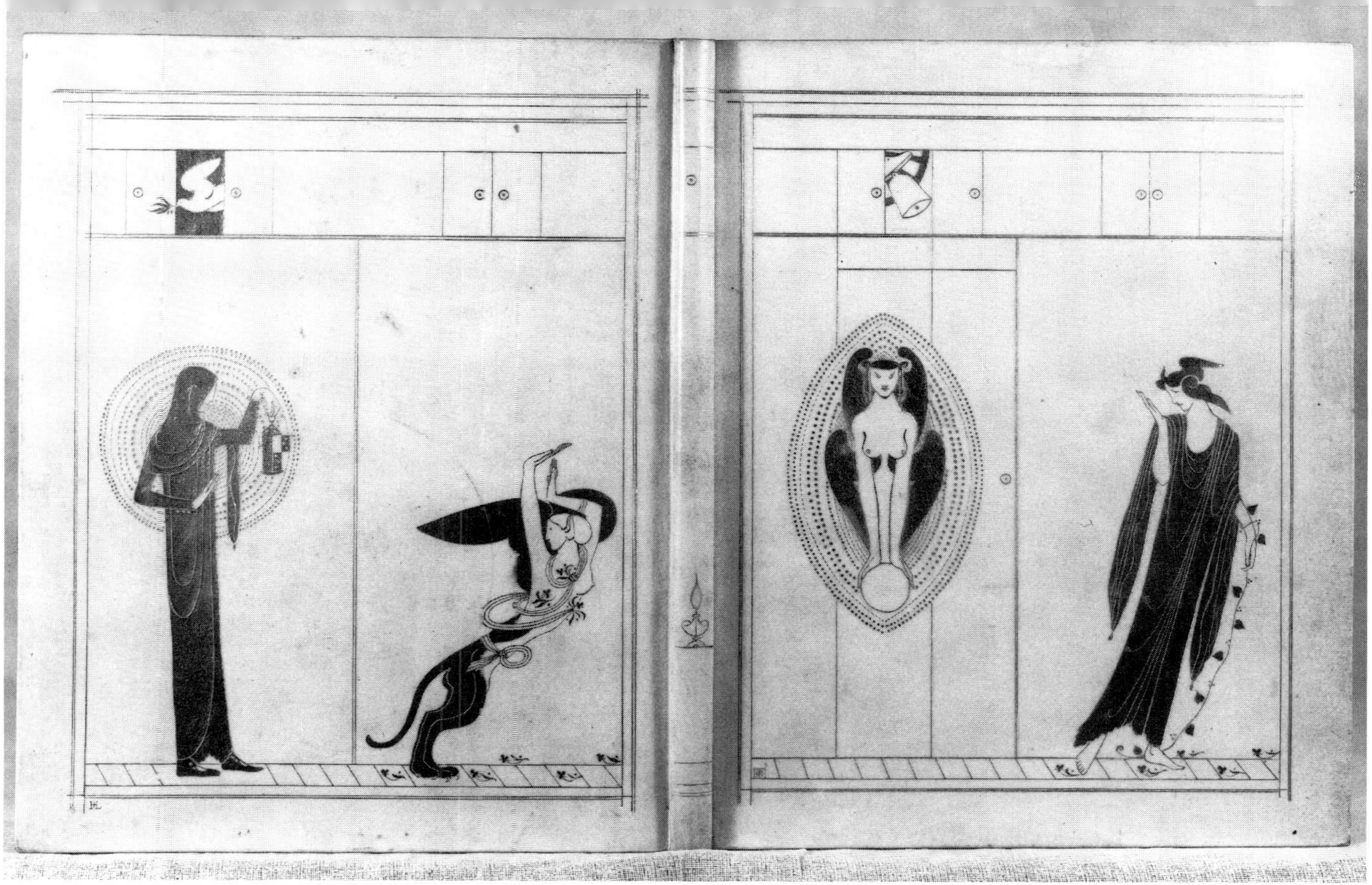

FIG. 13. Cover design by Charles Ricketts for *The Sphinx* (1894), by Oscar Wilde

Beardsley'.[97] The fact that Ricketts saw Beardsley's 'newly-completed' design raises the question of the possible artistic debts of the one to the other. Sir John Rothenstein mentions that Beardsley occasionally visited the Vale, where he saw Ricketts's *Sphinx* designs and was influenced by them.[98] Ricketts himself, who felt that his book-illustration work had been 'swamped forever' by Beardsley's success, later stated that he never knew Beardsley, though (as we saw with Lionel Johnson) he was inclined to be unreliable in such sweeping statements about 'enemies'.[99] That one of Ricketts's designs for *The Sphinx* was published in 1893 in a Dutch avant-garde magazine, *Van Nu en Straks*, supports his contention that publication of *The Sphinx* was delayed by *Salome*. Ricketts, moreover, had certainly developed his own style before Beardsley's appearance. The origin of Ricketts's illustrations and his gorgeous binding for *The Sphinx* can be traced through his work, from *Oedipus and the Sphinx (1891)* through the book covers and illustrations that followed, where Pre-Raphaelite, Art Nouveau, Symbolist, and Italian Renaissance elements mix with those from Japanese prints and the line designs of Greek pottery to produce a new synthesis that is entirely Ricketts's own. If any debt is owed, it is probably Beardsley's to Ricketts.[100] However, Beardsley's brilliant and striking drawings make a more immediate impact than Ricketts's more refined and scholarly ones, and so will probably always be better known.

Another reason for this is that Beardsley was a good propagandist for his own work and courted that contemporary notoriety that so often is the foundation of later fame. 'Beardsley, like Wilde', Ricketts wrote, 'is typical of that decade which clothed its hedonism with brilliance, but also with the wish to astonish and "arrive". In this tendency Whistler had shown the way.' He added that, unlike Wilde, 'Beardsley, intoxicated by success, was not always pleasant to his friends or appreciative of those who helped him to succeed'; indeed, Holmes recalled being offended by Beardsley's 'affected and casual' attitude towards Ricketts.[101] Ricketts and Shannon also showed their

[97] Holmes, p. 167.

[98] Rothenstein, *Artists*, pp. 182–3.

[99] CR to G. Bottomley, p.m. 12 Jan 1915 = *SP*, p. 227.

[100] See Cecil French, 'The Wood-Engravings of Charles Ricketts', *Print Collector's Quarterly*, 14 (127), 194–217.

[101] CR, 'Marginal Recollections'; *Holmes*, p. 168.

disapproval, and perhaps some resentment, of Beardsley; Lucien Pissarro's amusing description of Beardsley as 'un espèce de symbolo-rigolo à la mode ici' must also owe something to them.[102] The Vale artists had refused to contribute to *The Yellow Book*, edited and decorated by Beardsley, claiming that it 'might lead to complications over the fourth *Dial*'; though they found it looked 'extremely well and bright',[103] and Lucien Pissarro wrongly feared they would refuse to contribute to Van de Velde's international art review because of Beardsley's presence. However much Ricketts may have disliked Beardsley, it did not affect his attitude to Beardsley's sister Mabel, whom he regularly visited, or to Beardsley's mother, whom he assisted financially for many years.

Wilde, who since the great success of *Lady Windermere's Fan* in 1892 was a less frequent visitor to the Vale, liked neither Beardsley's nor Rickett's contribution. His aversion to the illustrations for *Salome* was perhaps due to the obvious and unflattering caricatures they contained, and he told Ricketts, 'your drawings are not of your best. You have seen them through your intellect, not your temperament.'[104] Ricketts himself considered them his best work as an illustrator; 'the most typical thing I have done so far, or the work which expresses the larger part of my temper', he wrote in 1911.[105]

The publisher of Wilde's work since 1892 had been the Bodley Head, for whom Ricketts was now regularly producing designs, while at the same time attempting to move into publishing himself. After *Daphnis and Chloe* he had intended to publish 'a less important [book] in bulk and price', another early translation, this time of the Greek bucolic poets Bion and Moschus, with the 'Love Crucified' of the Latin poet Ausonius, translated by Thomas Stanley, a contemporary of Herrick's.[106] However, this project was dropped in favour of another classic, *Hero and Leander*, in the Elizabethan version by Marlowe and Chapman, to commemorate the Aldine edition of the Greek text in 1494. In financing *Daphnis* Ricketts had drawn up a list of subscribers, most of whom were booksellers. Such financial preparations added to

[102] Lucien to Camille Pissarro [*c.*17 Dec. 1894].

[103] CHS to John Lane, quoted in *Books from the Library of John Lane, Publisher* (Dulau and Co. [*c.*1930], item 932.

[104] CR, *Wilde*, p. 38.

[105] Egyptian Diary (1911) (BL Add. MS 58097), p. [10].

[106] CR to John Lane [week of 10 June 1893].

their work and made it dependent upon public interest based on promise rather than performance, a rather risky investment. So the two young artists welcomed the advent of someone prepared to take over the distribution of their venture. Even before completing the engraving of their first book, Shannon had written to Lucien Pissarro:

> You will be pleased to hear that there is every possibility (I may say probability) of our selling the whole book to a bookseller at once for little more than six thousand francs. Printing expenses & binding paid, we will get about four thousand francs. The bookseller will probably make plenty out of it himself but then he has got to sell them. We are very satisfied with the suggestion (I cannot yet call it a bargain).[107]

It was thus that the Bodley Head took over the publication and distribution of Ricketts's and Shannon's books. They were very pleased with 'our new and intelligent bookseller, John Lane ... [He] is publishing everything over here that counts at all'.[108] Lane bought the whole edition of *Hero and Leander*, paying the artists two-thirds of the published price.

A less ambitious and more successful work than their previous one, it gives the first indications of Ricketts's strongest points as a designer. The title-page border is elaborate yet clear, and full of lively beauty. The vellum binding with its blind tooling and touches of gold shows the same ability applied to a more abstract form. The illustrations, which were done jointly with Shannon, are less derivative and more spare and classical. It is clear that Ricketts and Shannon were thinking of setting up their own firm even at this early date, for they had had the paper specially watermarked with a Vale Press monogram for the first time, and a similar device, with a P emerging from inside a large V, combined with an S and an R, appears at the end of the book. Publishing was still envisaged as a joint venture.

Another of John Lane's authors at this time was 'Michael Field', the pseudonym of two women, an aunt and a niece, Katherine Bradley and Edith Cooper, who formed a sort of female counterpart to Ricketts and Shannon, and also followed an aesthetic creed in high-minded seclusion. So close was the collaboration in their writing that they claimed—rather optimistically—that no one could detect the seams where their different contributions met. True aesthetes, they were

107 CHS to Lucien Pissarro [late 1892].

108 CR to Lucien Pissarro [early 1893].

interested in painting and deeply concerned over the appearance of their books. When Ricketts and Shannon had come across Michael Field's *The Tragic Mary*, with its striking cover by Selwyn Image, in the 1890 Arts and Crafts Exhibition they were so impressed by it that they bought a copy and sent a fan letter and a complimentary copy of the *Dial* to its author, whose real identity was still unknown to them.[109] The Fields, as they were known, were not impressed by the *Dial*, describing it as 'that mad journal . . . with some good line work in it and rubbish for prose—ghastly translations of Verlaine and Rimbaud'.[110] Nevertheless, they were curious to meet these two young artists whose life together seemed so similar to their own. Rothenstein, who had met them at the Rossetti Exhibition at the New Gallery on New Year's Day 1894, by their 'express entreaty' took them to the Vale within a few days. They were fascinated. Shannon was 'exactly like one of the comely angels of Della Francesca—the face round, the features round, a smiling sobriety of expression'; and Ricketts, 'an unaurioled decadent Christ'.[111] This recalls Wilde's description of Shannon as a marigold and Ricketts as an orchid.[112]

On this first visit to the young artists the Fields were well aware of their own status as established authors who had already published some ten books, but they soon fell completely under Ricketts's spell. The aunt was fourteen years older than the niece, who was much the same age as the two artists. The older woman was 'small, ruddy, gay, and buoyant and quick in word and temper', Ricketts recorded, while the younger was 'tall, pallid, singularly beautiful in a way not always acknowledged as beauty, that is, white with grey eyes, thin in shoulders, with delicate hands, touched through imperceptibly with grey long before the greying of the hair'. In her face Ricketts saw a 'Tuscan quality of refinement, the outward expression of an inward beauty of thought', and after her death he described her as having had 'the rarest & most delicate nature of any woman—perhaps human being—I have met'.[113]

Along with many noble qualities they had others which were silly and

109 See *Field*, and Paul Delaney 'Book Design: A Nineteenth-Century Revival', *Connoisseur*, 198 (Aug. 1978), 282–9.

110 MF Journal, 1892, fo. 61.

111 Ibid., 1894, fo. 13.

112 *Men and Memories*, i. 173.

113 *Field*, p. 2; CR to Bottomley, p.m. 1 Dec. 1914.

infuriating. They indulged in the contemporary fashion for nicknames. Their friend Bernard Berenson was always referred to as 'The Doctrine'; Rothenstein, as 'The Heavenly Dog'; and Ricketts was given a succession of such names: 'Lizard', 'Basilisk', 'Pan', 'Fairyman', 'Fay', 'The Painter'. They never seemed able to pin down his brilliant, mercurial quality. Miss Bradley (Fig. 15) was known as 'Michael'; Miss Cooper (Fig. 14) as 'Field' or 'Henry'; thus they addressed and referred to each other. Their conversation, nurtured in isolation like some delicate household plant, had become so idiosyncratic as to be almost incomprehensible to the uninitiated. Katherine Bradley, moreover, was emotional and could be overbearing, and often provoked Ricketts, who after one visit commented in his diary:

> To the Michaels. The older one spread about her at times a stifling exhalation of sentimentality, cant, and nagging femininity. One is astonished how the younger one has kept her wits and a great measure of perception and delicate response.[114]

Their seclusion and preoccupation with each other and their work had produced an atmosphere charged with emotion that sometimes spilt over into absurdity. They often sparked off a perverse and provocative element in Ricketts's nature: 'I have no affections only tastes', he told them,[115] and they recorded his every word and gesture in their diaries. Over the twenty years of their friendship they brought out the best and the worst in his character.

After the ladies' first visit to the Vale a return visit was soon arranged to their house in Reigate, a typical comfortable villa with aesthetic modifications, such as William Morris wallpaper and photos of Italian paintings on the walls, which, Ricketts observed, showed the influence of Berenson. The women were thrilled by the young men's knowledge of their writings, and they discussed art and literature. They noted that Ricketts 'expects Keats is very popular in Paradise' and defined Marlowe as 'Keats in pre-existence'. When the conversation turned to gossip, Ricketts informed them that Beardsley, 'who gesticulates now and leads conversations, is the only man who sits on Rothenstein with success'; Ricketts was to become one of their chief sources of information about the outside world. In the garden, every herb was examined by 'eyes that discover quality in the leaves' and Ricketts

[114] CR Diary, 21 Nov. 1901.

[115] MF Journal, 1900, fo. 60.

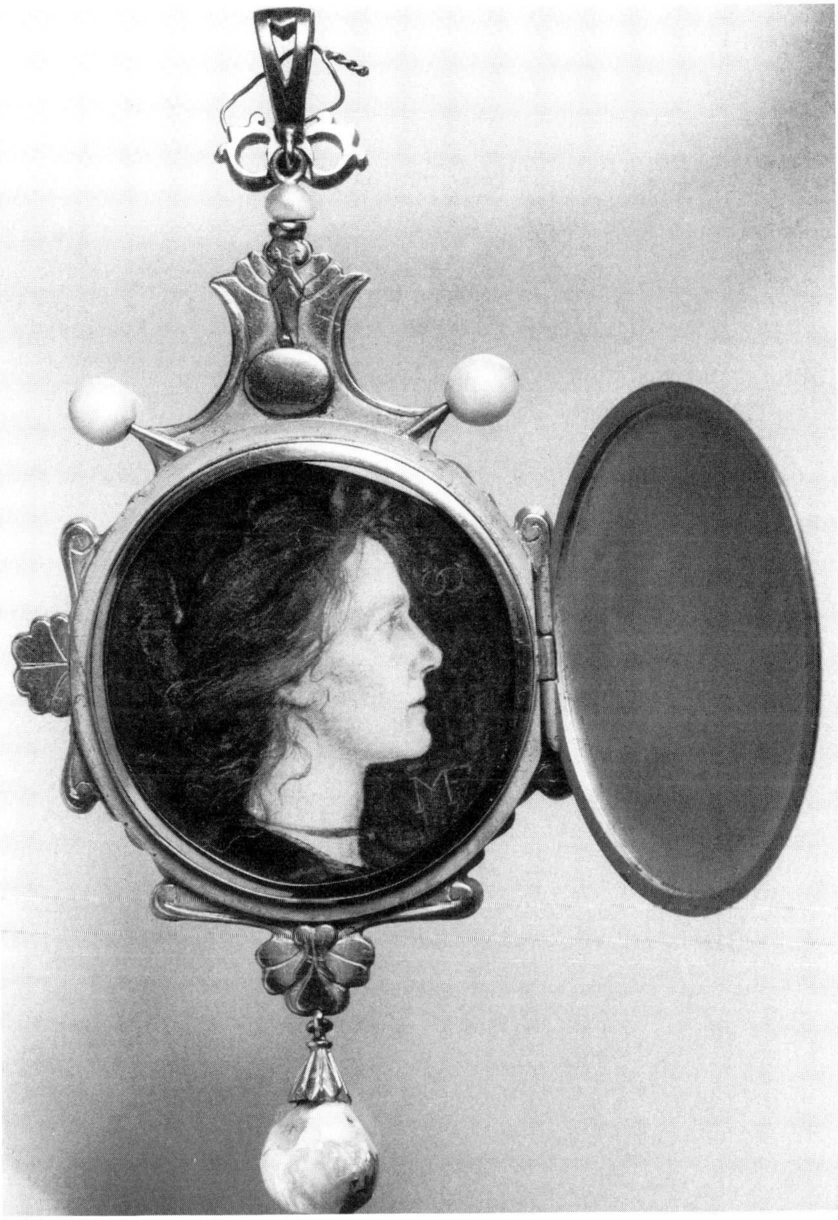

FIG. 14. Portrait miniature of Edith Cooper (1901) by Charles Ricketts (Fitzwilliam Museum)

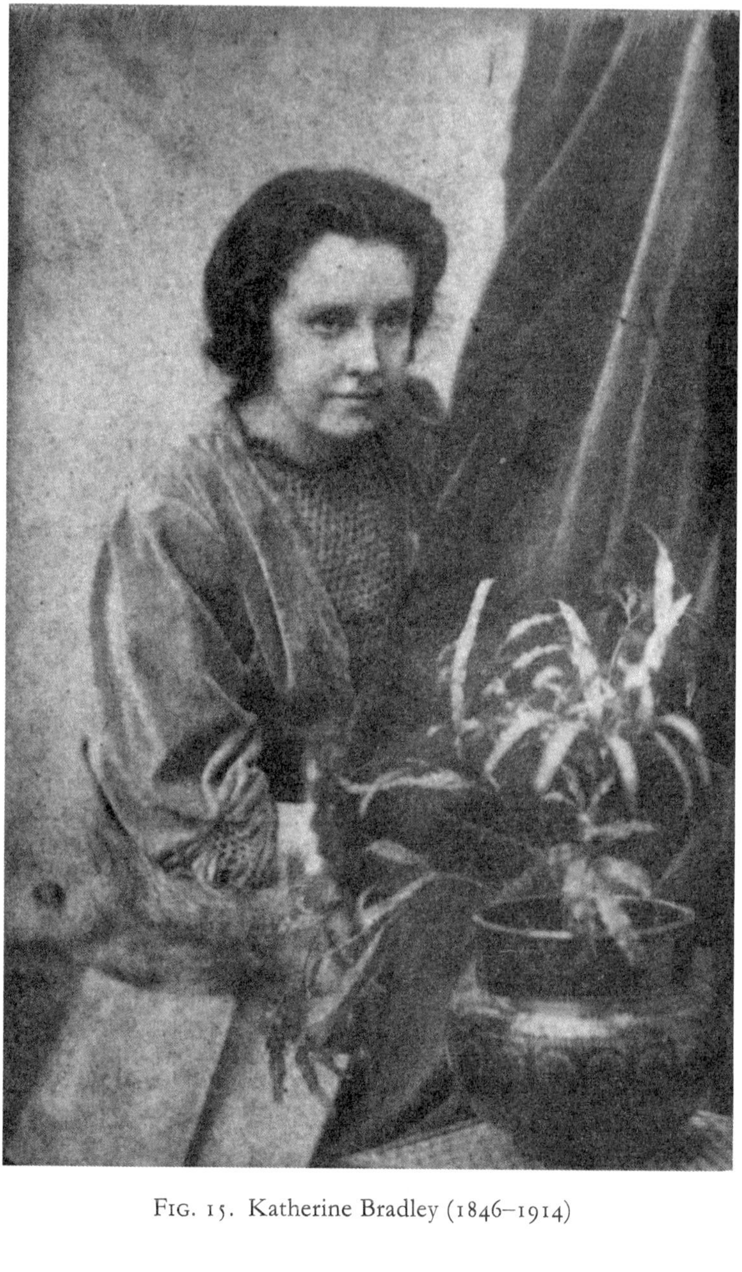

FIG. 15. Katherine Bradley (1846–1914)

waxed lyrical over the lettuces, 'those green roses'. Though knowledgeable about flowers himself, he commented that Shannon asked the name of the buttercup every spring. Ricketts, who wanted to have a carbuncle carved in the shape of a heart to give to Shannon, teased him because he worked in a different room: 'I call him sulky.' The two women were charmed by their affectionate banter, by Ricketts's 'nervous hilarity' and Shannon's 'boyish self-inculpating laugh'.[116] A deep love and knowledge of books, pictures, and flowers, enlivened by wit, teasing, gossip, and an engaging preciosity, was the basis of their friendship.

Apart from Wilde, Lucien Pissarro was their most helpful friend at this period, and his European contacts were now bearing fruit for them. As well as *The Sphinx* drawing published in *Van Nu en Straks*, Ricketts's and Shannon's work appeared in other journals in the forefront of the modern art movement. In 1894 a woodcut by Ricketts, *The Deluge*, and a lithograph by Shannon, 'Biondina' (called *La Femme aux Chats*) appeared in André Marty's *L'Estampe Originale* (1894), and a Shannon lithograph, *The Nursery Floor*, was published in Ambroise Vollard's *L'Album D'Estampes Originales* (1897). Their work was also exhibited on the Continent at the Nederland Etsclub (the successor to Van de Velde's Society) in 1894, at the 1894 and 1895 exhibitions of *Le Livre Aesthetique* in Brussels, and at an exhibition in The Hague in 1894, where a case was reserved for Vale work, which, along with the Kelmscott Press, was chosen to represent England. It is not surprising, therefore, that much of their early work, both prints and books, went abroad, especially to Holland and Germany. Eight copies of *La Reine des Poissons*, for instance, were sold by *Van Nu en Straks*. A large exhibition devoted entirely to Ricketts's and Shannon's books and woodcuts and Shannon's lithographs was held in Van Wisselingh's gallery in Amsterdam at the beginning of 1895. Ricketts's reputation led to his being invited by Samuel Bing not only to exhibit, along with Morris, in his great exhibition of books in Paris at L'Art Nouveau Bing (the shop that gave the movement its name) in May 1896, but also (earlier) to supervise the printing of a catalogue for Bing at the Ballantyne Press. This was not the sort of work he envisaged for himself, so he refused; nor did he approve of the exhibition—Lucien Pissarro noted that Ricketts distrusted 'the Americanism of such people and the Champs de Mars atmosphere'; that is, the commercialism and

[116] MF Journal, 1894, fos. 45–7.

vulgarity.[117] Ricketts's work, though not Shannon's, found less favour in France (where Durand-Ruel sold some of Shannon's lithographs) than in Germany and the Low Countries.

For their part, Ricketts and Shannon attempted to promote the work of Camille Pissarro as well as Lucien. The dealer Van Wisselingh, an intelligent and cultured man who had The Dutch Gallery in London as well as the one in Amsterdam, was much under Ricketts's influence, Lucien assured his father, and they could exhibit in his gallery. The Dutch dealer, however, was rather timid and not much of a business-man, according to Ricketts, who in handling him showed a good deal of resource and imagination. A plot was hatched to stimulate his interest in the elder Pissarro's work. Having invited the dealer to dinner at the Vale, they first showed him a number of Camille's lithographs. Another of his works, a fan of geese in a flood which had been borrowed from Lucien for the occasion, was hung just opposite where Van Wisselingh was sitting at the table. In the middle of the meal he 'discovered' it himself, and got up to examine it. After dinner, when shown some of Camille's etchings along with some by Whistler, the Dutchman pronounced that the Frenchman was the 'only etcher'.[118] A selection of Pissarro's work was soon to be found in The Dutch Gallery.

Another plan of Shannon's was to go to Eragny, after a trip to Paris, to select some of Camille's gouaches, paintings, and water-colours, which could be brought back to England and hung in a large room in their house. A private view would be organized to which all artists, dealers, and collectors interested in current art movements would be invited. Among these would be Van Wisselingh, who would then be asked to move the whole exhibition to his gallery.[119] Though this plan seems never to have come to anything, it shows not only the machinations of young artists to secure recognition but also the surprising spectacle of Ricketts and Shannon exhibiting and promoting Impressionist art in England.

This exhibition had not been planned for the Vale, for in October 1894 the artists had left it for a house nearby at 31 Beaufort Street, which Charles Conder later shared with them for a time. While the Vale was being redecorated for its next occupant they went off for a

[117] Lucien to Camille Pissarro, 14 Mar. and 28 Apr. 1896.

[118] Lucien to Camille Pissarro, Sun., p.m. [*c.*15 Nov. 1894] and 8 Dec. 1894.

[119] Lucien to Camille Pissarro [Mar.–Apr. 1895].

holiday, probably in Broadstairs, where they went each year: 'Ricketts has come back', commented Shannon, 'looking like a tomato covered with freckles, while I make the lobsters look greenish.'[120] Sturge Moore spent the months of September and October helping them to move. It is likely that they could no longer afford the rent at the Vale, for they were about to launch on their most costly and ambitious venture so far.

[120] CHS to Lucien Pissarro, 2 Sept. 1894.

4 *The 'Book Beautiful'*

'My dear Ricketts, you look like a conspirator with a mask. May I come in. . . .' Ricketts was alone in Beaufort Street, working at his wood-engravings. When he answered Oscar Wilde's loud knocking at the front door on that misty day, 28 February 1895, he was still wearing the visor that engravers use to protect their eyes from the strong light. On entering the low, gloomy room in which he had been working, Wilde saw in the dim interior the single lamp and the glass reflector that intensified the light on the wood-block, and went on 'or are you an Alchemist turning wood blocks into gold? I can see your glass or crucible. Tell me now of your magic practices?'[1]

They laughed. Wilde had often referred to Ricketts's 'magic' gift in book designing: 'making beauty out of a little coloured paper smeared with gold.' In spite of this jovial beginning, there was a brooding sense of silence and estrangement over their meeting, the awkwardness of old friends who have grown apart. It was some weeks since they had met each other, probably not since the first night of *An Ideal Husband*, which Wilde had insisted they see on 3 January. Ricketts had heard rumours of acts of imprudence on Wilde's part, and knew of his love of luxury, which made him live in expensive hotels (which Ricketts considered 'paradises for the vulgar') and sometimes in strange company. He, moreover, had been working in solitude for many months, and was rather disorientated by this sudden intrusion from his past and from a wider social world. 'But why this darkness? This place is like a tomb!', Wilde exclaimed. Ricketts lit two candles. 'My dear fellow', Wilde went on, 'how could you leave the Vale for this dull, dark house? Tell me about yourself, why this change?'

Ricketts explained that he was about to become a printer and publisher with a small shop of his own. The hope he had expressed more than four years before had suddenly been made possible. William Rothenstein, always eager to help his friends, had introduced Ricketts to William Llewellyn Hacon, who had then become a partner in the Vale Press. Hacon had originally offered to finance Rothenstein, who

[1] CR, *Wilde*, pp. 38–43, is the source of this account, unless otherwise indicated.

had painted his portrait, but the young artist suggested that Hacon might help Ricketts to found a press. By October 1894 the papers were being signed, and the firm of Hacon and Ricketts came into being.

A jovial club-man, a keen golfer and yachtsman, and a patron of artists like Conder and Rothenstein, Hacon was a barrister by profession, though he had never actually practised at the bar. He put up £1,000 towards the press. A few months later Ricketts was able to contribute some money inherited from his grandfather, who died on 20 April 1895 and left each of his eight grandchildren an eighth of his estate. This amounted to £235, though Sturge Moore puts the figure at £500. Originally, there had been a vague understanding about sharing the profits equally between Hacon and Ricketts but (as Hacon was a sleeping partner and Ricketts responsible for all the artistic and practical side, the design and decoration of the books, and the supervision of the printing at the Ballantyne Press) the arrangement came to seem very unequal, and led to some hard feelings on Ricketts's part. Still, Hacon had made the press possible and freed Ricketts from dependence on other organizations, like Lane's Bodley Head, to produce his work.

The lease of the Vale had been taken over by Hacon, who had married a beautiful model, called Edith Broadbent but always known as Amaryllis or 'Ryllis. She had sat more than once for Shannon; as Delia, for instance, in *In the House of Delia*. Rothenstein, who had first seen her at the Vale, did a pastel of her, which was exhibited at his joint show with Shannon in May 1894, and purchased by Hacon, who before long had secured an introduction to the original. She had the rich golden hair and voluptuous figure associated with Titian and the later work of Rossetti. Having come up to London to make her fortune, she soon found herself in a whirl of poets, painters, and men about town. She was eventually set up in a flat in Regent's Park by Herbert Horne and Arthur Symons, who had based some of his stories in the *Savoy* on her life. After their marriage, the Hacons began to keep open house at the Vale to their literary and artistic friends.

Through the last months of 1894 and all of 1895 Ricketts was at work preparing for his press. His original plan had been to use Jenson's type, but before long he decided like Morris to design his own. After careful study of the great masters of typography he produced his fount, though the six months he had set aside for this lengthened into a year. In addition there were initial letters, borders, and illustrations to do, all of which he designed and cut himself, an arduous and time-consuming task when a border alone could take up to a month to engrave.

1895, in spite of these promising beginnings, was not without its set-backs. After Wilde had left Ricketts that night he had met Shannon in the street (who found him looking tired and preoccupied) and then hailed a cab to take him to a club near Piccadilly. It was there, on that very evening, that Wilde had found the insulting card from Lord Queensberry 'to Oscar Wilde posing Somdomite [sic]' which led to his ill-advised prosecution of the peer, and ultimately to his own downfall. During April and May, the months of the Wilde trials, Ricketts and Shannon had no news from him, but they followed every development. As personal friends and as the designers of Wilde's books—on which Ricketts's reputation was still largely based—they were also caught up in this tragedy. After Wilde was sentenced to imprisonment, when his name became taboo and was struck off the billboards advertising one of his plays, and when John Lane withdrew all Wilde's books from the booksellers, they had every reason to fear for their own future. It was a time of great anxiety and dread, and Ricketts's health seems to have given way under the strain.

To me, the shock and stupour [he wrote years later] were slow to pass away. I was lucky in an inborn fund of 'self-concentration' and a blending of irony & patience which is also inborn; but, something happened from which I have never quite recovered, a mistrust of the British conscience, a mistrust of modern civilization.... I have never forgiven the emminent men of the time, Meredith, Morris, Hardy, who were silent, though I know this is unreasonable.[2]

In his diary, he was more explicit:

I believe the crash of his scandal affected his contemporaries to a degree unknown to younger people. Not only did eminent men, his seniors, do nothing, but several friends behaved badly, Sarah Bernhardt, for instance, Degas, Gosse, and I believe Burne-Jones. Meredith was brutal and Morris indifferent. To the young of today, there is no kind of moral prejudice, it is only with men of my age, whose career was actually affected by the crash, that the story shows itself as a tragic event, in which the fiber of the nation seemed of poor quality. Personal friends of Oscar's who had positions to maintain may have passed unpleasant moments, and still remember this.[3]

[2] CR to G. Bottomley, p.m. 27 July 1918; *SP*, p. 300, gives only part of this.

[3] CR Diary, 9 Feb. 1916.

At the time, however, they were so near 'its tragedy & horror' that they 'did not focus the fact that it had hurt anyone else', that Beardsley, whom they viewed as an exploiter of the disaster and 'a purveyor of the stuff the public wanted at the time or has wanted since', 'suffered persecution' too and was dismissed from the *Yellow Book*.[4]

Ricketts came to regard Wilde's trial as a social vendetta inspired by a deep hatred of the artist in the British mentality. This hatred was fostered by the public-school spirit, which was fundamentally hostile to ideas and emphasized athletics rather than art, and the team spirit rather than the value of the gifted individual. (He also deplored the lack of idealism which made the pragmatic 'Britisher' prefer landscapes and portraits to classical or historical paintings, such as Ricketts was to paint himself.) When, as in the case of Wilde, the artist got 'outside his field into publicity of an undesirable kind' the latent hostility to art was called into activity.[5] (Incidentally Ricketts's readiness to support any artist seen to be misunderstood and persecuted by a philistine society suggests that he often saw himself in that role.)

What Ricketts nowhere mentions is what might have been thought to have concerned him most about Wilde's trial and imprisonment, the sexual crime for which he was condemned. Both as a friend and as a fellow homosexual, Ricketts must have felt keenly the humiliation and vindictiveness that Wilde had suffered. But as Ricketts and Shannon, whatever their private relationship, belonged to no homosexual circle, they had no real need for fear. They never frequented the *louche* people or establishments that Wilde had known and, indeed, disapproved of such associations; it is most unlikely that they ever discussed sexual matters with him. Ricketts and Shannon, consequently, were not among those who took the opportunity that Wilde had spurned—to catch the boat to France.

Because of the Wilde trial, their last project in collaboration with John Lane got into difficulties. More than half the copies of *La Reine des Poissons*, Lucien Pissarro's first book and originally a Vale publication, had been transferred to Lane. As Ricketts knew, Lane was slow in paying, but Pissarro was desperate for the money, and he and Shannon were too badly off themselves to help him out. Lane, having been dragged into the Wilde scandal, had made himself unobtainable. Shannon advised Pissarro that if a stiff letter had no effect he would

[4] CR to G. Bottomley, p.m. 16 Sept. 1922.

[5] CR Diary, 29 May 1914 = *SP*, pp. 195–6.

call on Lane himself. Then Lane suddenly demanded a prospectus which had not been included in the original agreement. A compromise was finally worked out by Ricketts, and Shannon commented: 'the sooner the affair is settled, the better for everyone. We can then bid a final farewell to John Lane.'[6]

An additional hardship during this period was their own lack of money. In order to prepare for the press, Ricketts gave up his commercial book-designing, which Wilde's disgrace would probably have affected adversely anyway. He was fortunate to have already changed direction before the crash came. With such work, however, went Ricketts's sole income, and the money Hacon had promised was tied up in the firm. Even when they had first left the Vale, they had been anxious about money. Visiting them in Beaufort Street on 13 December 1894, the Fields found them 'starvy and sad', and Ricketts confessed that there was one tragic moment before they left their old house—'it was easy to see it was a moment when Famine pointed his skinny finger at them'.[7] This must have been the start of the period of 'trouble & starvation, of work made outré by harass of illness—of two years of life taken away by sickness and dread' which Ricketts spoke of later.[8] During this period, he did some pen-and-ink drawings to illustrate a projected edition of *The Kingis Quair*. They are in his best Art Nouveau style and have that marvellous sense of design and decorative effect at which he excelled. Though he could be weak as a draughtsman from life, he is unsurpassed in designs such as these, where imagination can cast realism aside. These splendid designs, however, were associated in Ricketts's mind with the period of poverty, despair, and fear brought on by their change of direction and Wilde's fall. The collector Laurence Hodson purchased two of them (Fig. 16), which were reproduced in the *Dial* (1896). Later, Ricketts asked for them back, either for the price Hodson had originally paid or in exchange for something else. When Hodson asked why Ricketts wanted them, he was told it did not matter, but when he insisted Ricketts said, 'I was starving while I worked on that drawing in dejection and misery. I would destroy all I did in those months if I could.'[9] So Hodson refused to part with them.

Throughout this difficult summer Shannon was gathering material

6 CHS to Lucien Pissarro [Feb.–Mar. 1895].

7 MF Journal, 1894, fo. 139^{v}.

8 Ibid., 1901, fo. 11.

9 G. Bottomley to TSM, 26 Oct. 1940.

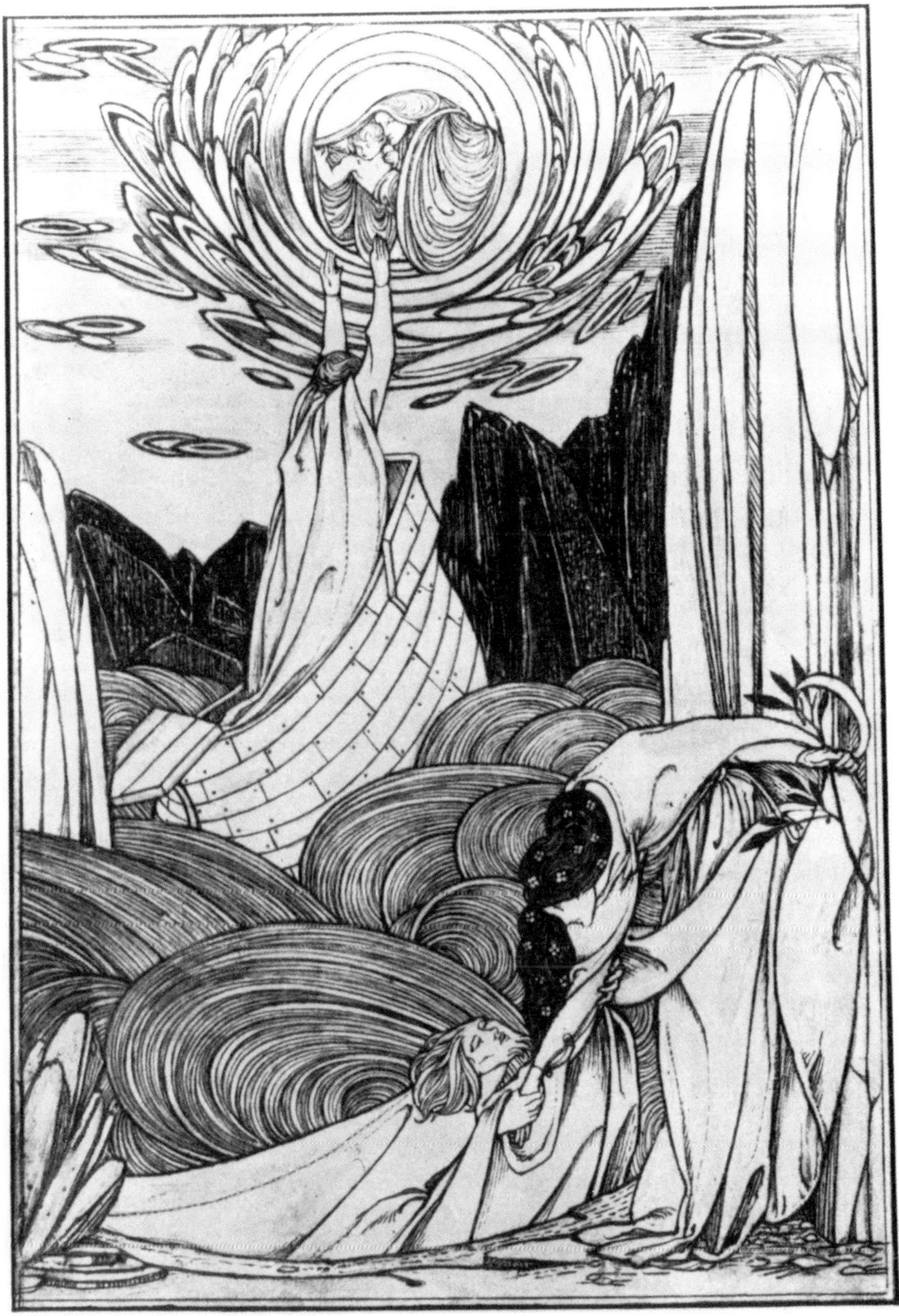

FIG. 16. Pen illustration for *The King's Quair* (1895; collection of Frederick Koch)

for the *Pageant*, an annual magazine which he was editing with Gleeson White. A Christmas-gift book, it was, like the *Dial* and the *Yellow Book*, to have both artistic and literary contents, but unlike the latter it was not to achieve notoriety. Ricketts and Shannon believed that the Arts and Crafts movement in Britain was beginning to flounder in the hands of inferior artists who did not understand the great tradition established by the Pre-Raphaelites, so the first issue was devoted to these latter in the hope that this would have a beneficial influence on the movement. Apart from an original lithograph by Whistler, the art contents were mostly reproductions of Millais, Burne-Jones, and Rossetti, whose *Mary Magdalene at the Door of Simon the Pharisee*, recently discovered by Ricketts and Shannon, was used as the frontispiece. Paul Verlaine contributed a poem on another Rossetti picture, *Monna Vanna*. The contributors ranged more widely than the usual coterie, and included Robert Bridges, Cunninghame Graham, Max Beerbohm, and Yeats. Ricketts designed the cover and provided two illustrations; he was the subject of an article by Gleeson White which was considered 'thoughtful' by the *Academy* (14 Dec. 1895) and 'injudicious' by the *Athenaeum* (14 Dec. 1895). As with the *Dial*, the artistic contents were more generally liked than the literary.

Ricketts had by this time finished his preparations for the Vale Press. One of his first priorities in the last few months had been to find a shop that was not expensive but was in the right sort of locality. By the end of the year they had found 52 Warwick Street, which was reached through an archway from Regent Street (but disappeared later in the rebuilding of the Quadrant). It had two rooms, a front one with a counter and another behind with eighteenth-century panelling where small exhibitions were planned. Inside, the shop was painted green, 'the colour of hope', Ricketts noted.[10] Shannon painted a signboard which hung outside, with the design of a sundial on one side and on the other the winged horse Pegasus, who represented the inspiration of poetry, and the legend 'at the Sign of the Dial'. Ricketts was always there on Saturday afternoons, when he saw friends, but most of the time he worked at home or at the Ballantyne Press, where the Vale books were printed under his supervision. Mr. E. LeBreton Martin was the office manager and there was a pleasant efficient office-boy named MacGregor.

In addition to acting as a meeting-place for their friends and patrons,

[10] CR to MF [Apr. 1896].

the shop sold prints by selected artists, woodcuts by Lucien Pissarro and Sturge Moore, etchings by Legros, and lithographs by Shannon and Rothenstein. The Vale Press also acted as agents for most of these artists, although when Whistler had asked them to act for him they had tactfully declined, knowing how difficult he could be. The press printed catalogues for their own exhibitions and for many held in Van Wisselingh's gallery, and offered to print 'small editions for Societies or Private Persons', to provide bindings in pigskin and morocco designed by Ricketts and carried out by Rivière, and to supply book-plates designed by Ricketts 'on the condition that they are engraved on wood under his supervision'.[11] Special copies of Vale books, moreover, could be ordered in vellum.

With their establishment in a shop, they were able to put the finishing touches to the fourth number of the *Dial*, which had been ready, waiting for the new address, since the middle of December. The artistic contents included original woodcuts by Pissarro and Sturge Moore and lithographs by Shannon, as well as reproductions after Shannon, Savage, and Ricketts's illustration to *The Kingis Quair*. A leaf of Vale type was also inserted.

Though they were still in the middle of the painting and carpentry work in the little shop, they felt they should issue a prospectus of their publications. An earlier anouncement had put forward the general aims of the press:

> In offering the following books to the attention of book-lovers, to collectors of shapely books and sound editions, we would urge that all of them have been carefully edited and printed in the spelling in which they were written; in many instances for the first time since they originally appeared.
>
> The decorations & woodcuts they contain have been engraved throughout by the designer; the type in which they are printed has also been designed by the maker of the decorations & the builder of these pages.
>
> We would suggest such conditions have not been realised by any printer or publisher in the past.[12]

This represented not only the acme of Ricketts's involvement in the Arts and Crafts movement but the greatest control he was to exercise over book production, so that the Vale books are really the fullest expression of a single artistic personality. Next, however, a list of titles with publishing dates was necessary. Six books were already in press,

[11] *Vale Press Prospectus*, [1897–8], and [late 1898].

[12] Ibid. [after Apr. 1895].

and Ricketts wrote 'I am praying to the gods whose names must never be mentioned that [they] ... may appear before the summer'.[13] The first book, *The Early Poems of John Milton*, was announced for March, with two more to follow in April and two more in May. The Milton in fact did not appear till 14 April. Though the decorations are attributed to Ricketts, his old pseudonym (Charles Sturt) was used—not for the last time—for the editor who had seen the book through the press.

From the beginning Ricketts had conceived the press as publishing 'limited editions of chosen and carefully edited reprints', but they were still feeling their way. Among the books they planned but never published were Lamb's *Essays of Elia*, *The Tragedies of John Webster*, *The Voiage and Travaille of Sir John Maundeville*, *The Song of Songs* (illustrated by Ricketts), Crashaw's *Poems* (which together with an edition of Suckling was envisaged in italics, with the tall narrow shape of *Silverpoints*), Greek and Latin classics, including editions of Catullus and Plato (for whose *Symposium* Ricketts did an illustration), along with a Propertius announced in an early prospectus.[14] Soon, however, they had settled on a line of English poetic classics, mainly from the sixteenth and seventeenth centuries, like Suckling, Campion, Constable, Sidney, Shakespeare: but they were also to include Romantic poets, like Keats, Shelley, and Wordsworth, and Victorians, like Browning and Rossetti, as well as a few prose pieces and one Latin work.

In spite of his intention to print only the classics, Ricketts did issue some contemporary work. On his visit to Beaufort Street Wilde had suggested that Ricketts might publish his 'Portrait of Mr WH', using as a frontispiece the portrait Ricketts had done for him of Willie Hughes, and an embarrassed Ricketts had reluctantly agreed to publish it if the press was successful. He felt modern books were too risky, and remembered that in spite of Wilde's success his books tended to be financial failures.[15] It seems rather surprising, therefore, that as early as 1894 Ricketts had approached the Fields over the publication of one of their plays, and eventually published four of them. The fact that these and a few other modern books did not sell well justified his initial caution, however, and he wrote later, 'When a publisher, I found well-

[13] CR to MF [*c*. Feb. 1896].

[14] CR to Mr Day [*c*. Oct. 1894].

[15] CR, *Wilde*, p. 41.

known classics pay, when other books (for which there was a crying need) did not go down'.[16]

In June, having delivered the first productions of the press to the world, Ricketts and Shannon fled to Paris: 'We could not stand the sights of London'. It was terrifically hot, but 'the roses alone made up for the heat & glare'.[17] The dealer Ambroise Vollard was organizing an exhibition, and Shannon had been invited to show his work. Lucien Pissarro joined them in Paris, and when they dined together he and Ricketts planned to write a little book on engraving. On their return they took in Amiens in order to see the murals by Puvis de Chavannes in the Musée de Picardie there.

Back in London, Shannon began collecting material for the second volume of the *Pageant*, which had to be ready for the printers by the beginning of July. In this issue an effort was made to broaden the scope and give more space to the French artists and writers of whom Ricketts approved. In addition to an article on Gustave Moreau by Gleeson White and two reproductions after Puvis de Chavannes, there were contributions on French themes by Edmund Gosse, Lionel Johnson, and Dowson, to supplement the material on the Pre-Raphaelites and such contemporaries as Walter Crane, William Strang, and Rothenstein. Ricketts contributed a reproduction of his *Autumn Muse* and a brief history of original wood-engraving, while Shannon reproduced a drawing and, more significantly, an oil-painting entitled *A Wounded Amazon*, a version of the lithograph *Atalanta*. When the *Pageant* appeared, it was pointed out in the *Academy* (5 Dec. 1896) that the artist had 'too obviously forgotten the surgical ceremony which the Amazons are reputed to have observed!' (So as not to interfere with the bow arm, girls had their right breasts amputated, and to this part of the anatomy Shannon's composition lovingly draws attention.) This was the painting described by Wilson Steer as 'the Wounded Watts', which aptly pin-pointed one of its principal influences. When this remark was repeated to Shannon *he* was 'wounded', and there was a coolness for a time with Steer. By the time the periodical appeared at the end of 1897 Shannon, with great éclat, had made his reappearance as an oil-painter.

Ricketts too was making headway. At about the same time two influential periodicals carried appreciative notices of his work. The

[16] CR to Laurence Binyon [early 1906].

[17] CHS to MF [*c*. June 1896]. CHS to Lucien Pissarro [*c*. June 1896].

Saturday Review (4 July 1896), in a long article on the first three Vale volumes, praised the wood-cut frontispiece to the Milton for its 'marvellous breadth and exquisite skill', described Landor's *Epicurus, Leontion, and Ternissa* (1896) as 'an exquisite piece of bookmaking', and singled out *The Poems of Sir John Suckling* (1896) 'with its border of honeysuckle, and its binding of flowered paper' as 'perhaps the most attractive of the first three of Messrs. Hacon and Ricketts's Vale books'. The article concluded that Ricketts's work was not of the sort that would achieve popular success: 'He is too great a scholar, too fine a craftsman, to appeal to any considerable audience.' The long note in the *Studio*, commenting on the same three books, praised this 'worthy and accomplished effort to produce an edition with its own individual distinction'. It is astonishing that Ricketts always believed that the *Studio* had been hostile to his press.

The work of the press continued, though with vexing delays; their September books likely to be appearing in November and the *Pageant* tarrying 'for no ascertainable reason'.[18] Michael Field's *Fair Rosamund,* being in dramatic form with a great many stage directions, was difficult to print. Arnold's *Empedocles*, to be published at the end of the year, was only arriving in parts very slowly. Over the next four years Ricketts laboured steadily at producing borders, capital letters, frontispieces, and illustrations. Almost every book has some such decorative feature, and the most successful are the ones which stimulated his imagination to make the richest decoration. Among these is the splendid first opening of Drayton's *Nimphidia* (1896), comprising a frontispiece on the left and a border on the right, both based on a motif of honeysuckle (Fig. 17), and the two volumes of Blake's poems, *The Book of Thel: Songs of Innocence and Songs of Experience* (1897) and *Poetical Sketches* (1899), for which Ricketts designed, as well as border and frontispiece, exquisite capital letters which illustrate the poems they decorate. For John Gray's *Spiritual Poems*—translations of Christian poetry originally planned before the rupture with John Lane and the founding of the Vale Press—Ricketts had made designs for a cover with nuns in the convention of *The Sphinx*, but it was issued in pale grey paper boards and one of the nuns was put in a woodcut frontispiece. Though the first Vale book had a white cloth binding and the second, plain paper boards, Ricketts soon began using patterned paper boards: the Suckling had a floral design and the *Nimphidia*, a charming motif, nut and

[18] CR to MF [Sept. 1896].

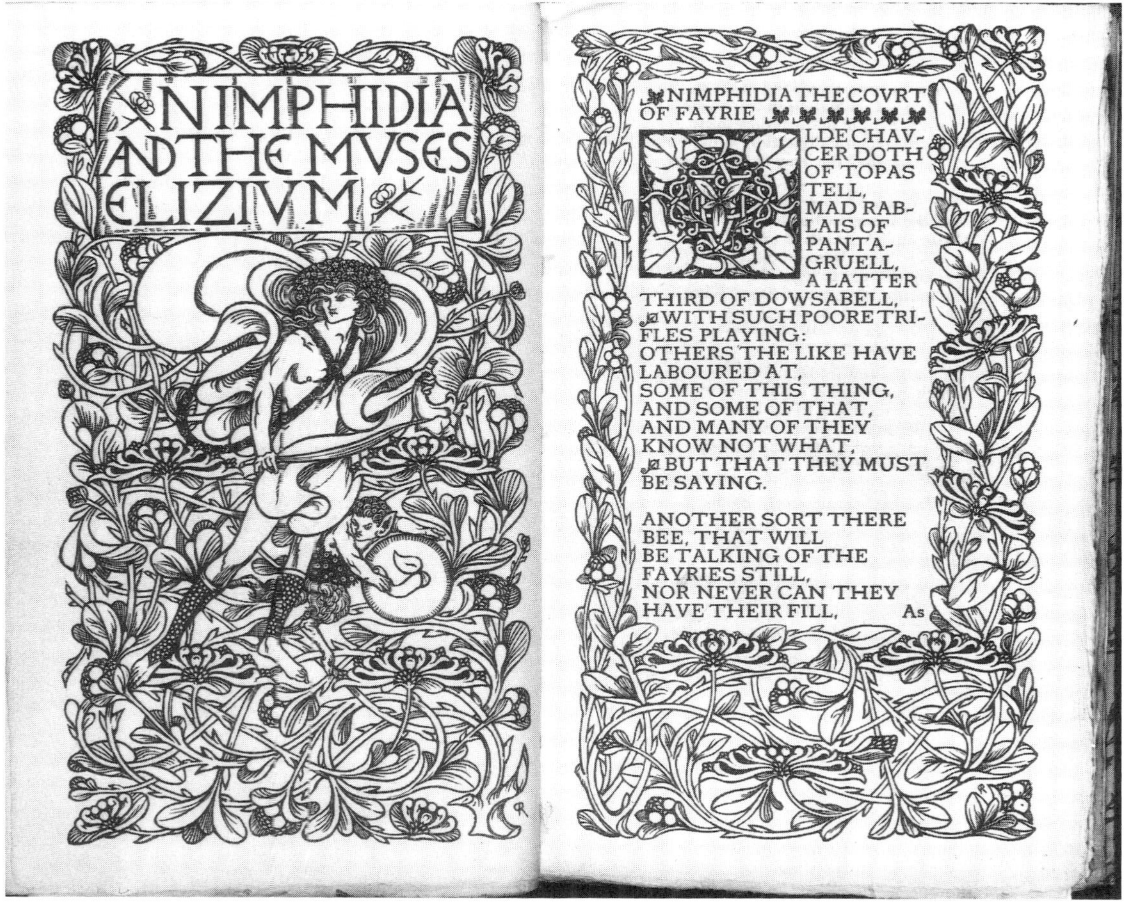

FIG. 17. Border and frontispiece designed and cut in the wood by Charles Ricketts for *Nimphidia*, by John Drayton (Vale Press, 1896)

mouse. Such paper covers had been inspired by those of Lucien Pissarro, and employed motifs appropriate to the books' contents, though expedience sometimes intervened, as when Ricketts upset the Fields by using a paper of fir-cones—designed originally for their *The World at Auction* (1898)—for Sidney's *Sonnets* (1898). In his Vale Press decorations Ricketts's strength is his sense of linear design, of clear yet intricate pattern that is as vital as it is elegant.

Ricketts's weakness as a publisher was his lack of business sense, which he had tried to offset by hiring an office manager, who in the summer of 1896 left for a more remunerative job on one of the daily papers. He was replaced by Charles Holmes, whose seven years of experience in printing and publishing made him particularly well qualified for the job. After Eton and Oxford he had worked as assistant, compositor, proof-reader, and book-keeper at three different publishers, and so had more experience of different aspects of publishing than either Ricketts or Shannon. He was also much more of a businessman. Though he took the position at a salary of £80 a year on a temporary basis, he was to stay with the press until it closed in 1903, and his contribution to its success was considerable. On taking the job he checked through the accounts, and found that all of the original £1,000 had been spent and that he had to get another £50 from Hacon 'merely to keep things going'.[19] He also took over the responsibility of negotiating with booksellers, a task at which both Ricketts and Shannon had proved too sensitive to be successful. Ricketts's attempts to sell his publications to friendly booksellers had failed because, after exasperating him with criticisms of his books, they managed to extract large discounts from him. Holmes told the Fields of one such occasion:

> Ricketts was to sollicit the London trade ... he met the head of a city firm & yielded in every point to the man's demands: business over, the man began criticism of the book offered, Ricketts shrivelled up to the murder-point ... He swore ... he would never take a book round again.

Ricketts, Holmes remarked, 'rages against unfair conditions & prices, etc., only to yield at once when the coarse business begins'. Shannon had no more success in dealing with the provincial trade: his first journey for the firm was described by Holmes:

[19] Holmes, p. 172.

He went to Birmingham, a great prosperous place, & had his hair cut. Then he went to Manchester & had his hair cut & he arrived at a bookseller's. The man was at the top of a ladder. 'Are you Mr. Shannon?' 'Yes.' 'Then I suppose you think yourself very clever?'—No answer! Shannon's heel was beyond the door & those feet never entered a bookseller's again.[20]

The arrival of the more hard-headed Holmes may well have saved the firm from an early disaster. In addition to attending to the accounts and to the booksellers, he saw to the printers and read all the proofs. Later he was also to edit several texts.

At the beginning of May Lucien Pissarro, after dining with Ricketts and Shannon at Beaufort Street, found himself too exhausted to get up: the next day an illness began which was to linger throughout the rest of the year. Relations between him and Ricketts and Shannon had begun to show some strain, and this illness was to produce further difficulties. In September 1896 Pissarro had written to his father that he had not achieved what he should have in his publishing, due to the delays for which Ricketts was responsible, and that Ricketts was trying to drop him ('tâche de me laisser de côté'), because his art was inspired by nature rather than art, and because Ricketts feared the competition that Pissarro was giving him in the rather limited market for beautifully produced books.[21] One of the consequences of Pissarro's illness was that Ricketts had to carry out alone the book on fine printing, *De la Typographie*, which they had planned together in Paris, and which Pissarro had already begun to print. Ricketts proposed generous terms to Mrs Pissarro, offering to take 100 copies of the edition and paying half of the cost of printing, which would leave Pissarro and his French publisher, if he could find one to pay the remaining half, 150 copies. A running account was kept by Holmes which led to confusion and difficulty later and to some bitter letters from the Pissarros.

News confirming Pissarro's illness had reached Ricketts and Shannon at Christchurch in Hampshire, where they had gone for a short stay. Shannon promised to visit their sick friend as soon as he returned to London on Tuesday, 10 May, but Ricketts was unable to do so because he was to go to Reading that day to visit Oscar Wilde in prison. Ricketts had applied unsuccessfully a number of times to visit Wilde, and had been in communication since January with More Adey, who, with Robbie Ross, was managing Wilde's affairs. Ricketts

[20] MF Journal, 1898, fo. 127^{r-v}.

[21] Rewald, p. 383.

had written to Wilde in April expressing his disappointment at not being able to see him, and offering to publish something by him as soon as possible. He offered too to accompany Wilde to a Trappist monastery on his release (his idea being that after six weeks of retreat Wilde should then immediately begin a quiet re-entering of English society), but Wilde, who had had enough of solitude and austerity, declined the offer. Neither Ross nor Adey was particularly anxious to have Ricketts with them, as there were delicate and difficult matters to discuss with Wilde about his imminent release; nor was Wilde himself keen on seeing him, as he was upset about what he considered the mismanagement of his affairs by Ross and Adey. But as Ricketts was an old friend, 'of whom I am very fond', wrote Wilde, and had been hurt by not being included in the last visit, Wilde agreed that he should come, though he would be asked to leave after half an hour so that the others could discuss business.[22]

Ricketts met Ross and Adey at Reading Station.

> Our manner at this meeting was constrained [he recalled], we were almost strangers and our behaviour resembled that of men about to attend a funeral, Ross even developing a touch of hysteria on reaching the Prison gate. Acquaintance with the life of Verlaine, and traces of Latin realism in my upbringing, saved me from such qualms till I found myself alone in a small putty-coloured waiting room, decorated with a chocolate dado and an ebonized clock.[23]

Presently, Ricketts was summoned and found Wilde sitting at a green-baize-covered table, looking radiant with health and good humour and stouter than before in his loose-fitting prison clothes, which made Ricketts wince. Later on, Ricketts learnt that Wilde had been 'difficult, arrogant, and unreasonable, impatient with all arrangements, and utterly rebellious to all proposals for his immediate future, though long since discussed and agreed upon'. Nothing of this was visible, and Wilde's cordial greeting and enquiries 'might have been spoken in Bond Street'. Conversation, however, was difficult, with Wilde protesting that he had had too much of silence and privacy to want to go to a monastery or to Venice. They diverted the discussion to speak of *Salome*, which had been produced recently in Paris, and *Lady Windermere's Fan*, which had been given in Richmond, and Ricketts repeated the recent praise of Wilde by Ellen Terry and other actors.

[22] Wilde, *Letters*, p. 534.

[23] CR, *Wilde*, pp. 44–9, is the source of this account.

Wilde then enquired about his publishing, adding that the Vale must print *The Portrait of Mr WH*. 'Your picture, Ross tells me, has vanished, it was not in the sale, but you will design me another wonderful frontispiece.' Ricketts had promised Wilde to issue this work if his publishing venture was successful. Now, with its homosexual associations, it was not a work whose publication would do the reputation of either Wilde or the firm any good. John Lane, who had actually announced it on his list, had dropped it along with Wilde's other works for fear of losing his other authors. So Ricketts tried tactfully to suggest that the firm might publish *La Sainte Courtisane*. 'Alas', exclaimed Wilde, 'She no longer says marvellous things, the robbers have buried her white body and carried away her jewels ... Yes, perhaps you are right ... *Mr WH* might be imprudent ... the English public would have to read Shakespeare's sonnets.' They all joined in nervous laughter, and Ross suggested Ricketts should publish Wilde's play about the Pharaoh instead.

In his *Recollections of Oscar Wilde* Ricketts did not give a full account of his visit. It had not gone smoothly; his comments upset Wilde, who afterwards wrote to Ross:

> I was wrong to have Ricketts present: he meant to be cheering, but I thought him trivial: everything he said, including his remark that he supposed time went very fast in prison (a singularly unimaginative opinion, and one showing an entirely inartistic lack of sympathetic instinct), annoyed me extremely.[24]

Though Ricketts still hoped to publish *La Sainte Courtisane* at the Vale Press, the manuscript never came, as Wilde lost it in Paris in 1897 or 1898 by leaving it in a cab. The loss of the almost completed play meant that no work of Wilde was to be published by the firm, though Ricketts was later to design the cover for the author's edition of *The Ballad of Reading Gaol* published in March 1898.

Wilde's friends had attempted to raise money for his needs on his release. In this connection, Frank Harris went to Ricketts and talked at length about Wilde's financial difficulties; for most of the money that had been sent to him by his friends—£1,150—had been used up in repaying debts and expenses, and Wilde was extremely annoyed. Ricketts contributed money; a day or two before Wilde's release, he wrote to More Adey:

[24] Wilde, *Letters*, p. 541.

I have seen my lawyers in the matter of raising a small sum & I think I can promise £100 in the space of two or three weeks. This I think you can safely reckon on though the date is somewhat uncertain since mortgaging has to be gone into & this I do not understand.[25]

In a postscript he added: 'Of course, I don't want Oscar to know anything about this matter of mine. It would be very unwise.'[26] Robbie Ross, who knew of this generosity that Ricketts could ill afford, ensured later that the money was repaid out of the Wilde estate. A year later Shannon recorded in his diary: 'Ricketts had a letter from Oscar asking for the loan of £20; awfully unfortunate as we are excessively hard up—Ricketts sent it with a charming letter.'[27] Such actions with Ricketts were spontaneous and without fuss and show the deep compassion and generosity that lay beneath the nervous and aggressive side of his character.

The summer of 1897 brought to fruition their long-term plan for Shannon's career. After his years of apprenticeship to the masters of oil-painting, particularly Titian, he had begun to exhibit his own oils. Along with some of his lithographs he sent two paintings, *A Wounded Amazon* and *The Man with the Yellow Glove* (a portrait of his old friend Thomas Sturge Moore) to the Annual Exhibition of the Fine Arts at the Royal Crystal Palace, Munich. Together with Burne-Jones and Swan, he won a first-class gold medal. Shortly afterwards he began painting his fine self-portrait, *The Man in a Black Shirt* (1897) (Fig. 18), and then its companion portrait of Ricketts, *The Man in the Inverness Cape* (1898) (Fig. 19). (When the latter was exhibited in the New English Art Club Ricketts urged the Fields to see it: 'I am turning away from the 20th century to think only of the 15th.' Later he added: 'I look as if I had written to Ariosto, the book at my side has been sent to me by Aretino with a hint that a silk doublet would be acceptable. We are in A.D. 1515.'[28]) Shannon continued as a painter and a portraitist, so that lithography was never again his principal field of activity.

His career, however, was plagued by the greater contemporary fame

[25] MF Journal, reporting Robbie Ross, says CR paid the money out of a legacy (1904, fo. 35) while Sturge Moore says it came out of the 'meagre' profits of the Vale Press (*SP*, p. 106).

[26] CR to More Adey [18 May 1897].

[27] CHS Diary, 16 Aug. 1898.

[28] CR to MF, p.m. 16 and 30 Nov. 1898.

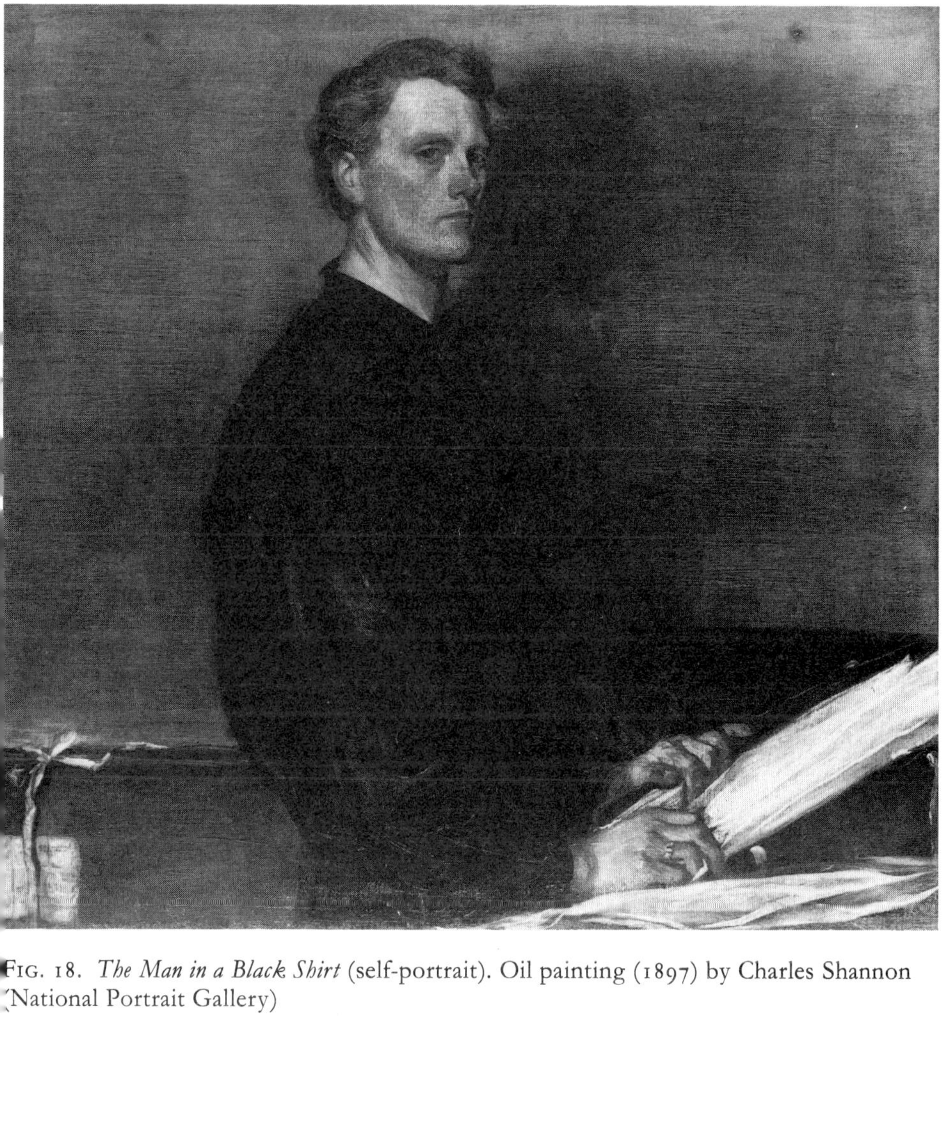

FIG. 18. *The Man in a Black Shirt* (self-portrait). Oil painting (1897) by Charles Shannon (National Portrait Gallery)

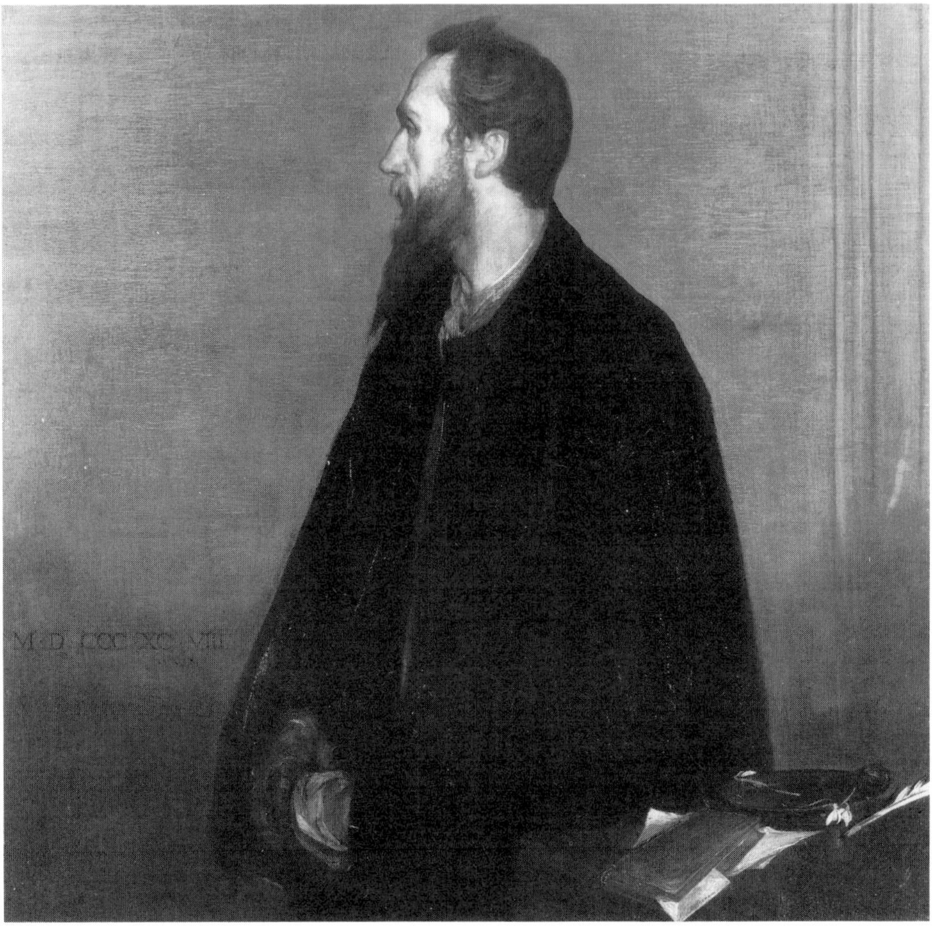

FIG. 19. *The Man in the Inverness Cape* (i.e. Charles Ricketts). Oil-painting (1898) by Charles Shannon (National Portrait Gallery)

of his namesake, the portrait-painter J. J. Shannon. So often was Charles Shannon mistaken for the other that he became tired of having to explain who he was, and that Ricketts observed:

I am sure that when Shannon goes to Heaven he will be told that he is there all ready and have to put up with some other place, as it is hardly likely that Heaven will admit they have made a mistake—I can imagine my arriving in 'tother place' and finding him there, and when I ask him, with some surprise, why he is not with the angels, being told of the mistake in the initials.[29]

Such are the vagaries of fashion that Charles is now far better known than James Jebusa.

At the end of this year, the fifth and last number of the *Dial* was published. Its make-up was similar to earlier numbers: woodcuts by Shannon, Pissarro, and Sturge Moore, lithographs by Shannon. Ricketts supplied two drawings, another of the illustrations for the projected *Kingis Quair*, and one subsequently engraved on the wood for *De Cupidinis et Psyches Amoribus*, published by the Vale Press in 1901. In addition to the usual contributions by John Gray, Sturge Moore, and Michael Field, the literary side included works by Laurence Housman, and Emil Verhaeren, the Belgian poet whom Ricketts and Shannon found so very dull when he visited them at Beaufort Street. Ricketts make his last appearance under his pen-name of Charles Sturt with an article on the Japanese artist Utamaro—but he was to use other pseudonyms in future. Explaining their decision not to issue any more numbers of the *Dial*, the Vale Press announced: 'The early parts were produced when it seemed impossible to launch original work in any other way. This is no longer the case, and the series ceases with the necessity for it.'[30]

[29] CR to Mary Davis [*c*.1913]. [30] *Vale Press Prospectus* [mid 1898].

5 *The 'Seven Devils of the Collector'*

ALL this while, since Ricketts and Shannon had first met in 1882, they had been relatively hard up, but they now entered a more prosperous period. Shannon's lithographs, drawings, and paintings were selling and he was getting commissions for portraits, while the Vale Press under the able management of Holmes brought in a more regular income. They could now permit themselves to collect and travel in a way they had always wanted, but could not have done before.

Even during their poorest period they had been able, by living frugally, to buy an occasional treasure. In November 1892 they acquired some prints by Utamaro to add to those by Hiroshige they had at the Vale, and Ricketts's interest in the later Japanese masters, then generally ignored by collectors, enabled him to increase their collection without any great expense. They frequented the sale-rooms but, their funds being limited, were often outbid. At the J. Andrew Rose sale at Christie's in May 1891, for instance, Shannon bid up to twelve guineas against a dealer for some Whistler drawings, but failed to get them. He felt he would probably have 'landed' them for thirty shillings if the dealer had not been there, for no one else showed any signs of bidding. At the same sale he watched Fairfax Murray buy nearly all the Rossetti drawings and water-colours at 'absurdly low prices' which they, however, could not afford.[1] Murray, by buying up all the Rossettis, was making it impossible for them to get any. With vigilance and patience, they were able to pick up things that sold cheaply from time to time. There were occasional bits of luck, as when, in May 1898, they came across Rossetti's large pen-and-ink drawing of *Mary Magdalene at the Door of Simon the Pharisee* at a furniture dealer's. 'It is a find of extraordinary importance', Shannon noted in his diary, for the thing was rumoured to have been lost. It had just been left for sale at the shop for £150, and they carried it away with them. At first they hoped to exchange it with Murray for a water-colour, possibly *The Christmas Carol* and something else besides, but it remained in their collection. 'This we both feel one of the finds of our lives', Shannon

[1] CHS to TSM [May 1891].

noted excitedly, which made Rothenstein, on hearing all about their good fortune, exclaim, 'Then there is a God!' However, it had 'about soaked up all Ricketts's money for the present'.[2] That they could afford such a price, nevertheless, was a sign of a certain prosperity.

The foundations of their collection were really laid at Lord Frederic Leighton's sale at Christie's in July 1896. They obtained a Tintoretto study, two drawings by Claude Lorraine, sketches by Delacroix, a William Etty copy after Veronese, and prints by Harunobu, Kiyonaga, and Hokusai. In addition they bought back copies of their own *Daphnis and Chloe*, *Hero and Leander*, *The House of Pomegranates*, and *The Sphinx*. Ricketts's *Oedipus* and *The Sphinx*, for which Van Wisselingh had put in a bid on their behalf, was sold to Dunthorne for six guineas, a high price when a Burne-Jones sold for five pounds and a Sir Thomas Lawrence for four pounds fifteen shillings. Though Holmes says that Ricketts repurchased the drawing at this sale, he must have acquired it afterwards from Dunthorne.[3] However, they did not get all they wanted, and were outbid for a signed painting by Lippo Memmi which Leighton had shown Ricketts and which they hoped to buy for fifteen pounds. Soon the bulk of their income must have been spent on their collection, for they still lived very carefully. 1898, the year they found the Rossetti, was a time of great opportunities, but after this purchase their funds were very limited. In July, at the Burne-Jones sale, Ricketts bought two drawings for twelve pounds, and he made a few purchases at the William Morris sale in December. A few months later came another of the 'great hauls' of their lives.

On 18 November there was a sale of the Japanese collection of Captain Frank Brinkley, who had spent much of his life in Japan. For ten guineas they obtained some fifty-three drawings of the personages of the Suikoden by Hokusai, and for a further forty-nine pounds secured six albums of miscellaneous drawings, two hundred in number, attributed to the same master. Ricketts and Shannon accepted this attribution, but subsequent scholarship has demoted many of them to his school or others. To reach their total of nearly sixty pounds they had to borrow from all quarters, Holmes actually lending them twenty pounds in the sale-room. It was not all good luck and success, however, and they made the inevitable mistakes. At the end of 1898 there was a sale of Tanagras, Greek terracotta figurines, at which they obtained

[2] CHS Diary, 27 May 1898; MF Journal, 1898, fo. 92^{v}; *All for Art*, pp. 70–1, pl. 28.

[3] Holmes, p. 164; *All for Art*, p. 51.

three out of the five they considered best. One of them, the weakest of the three was purchased after the sale from the dealer who bought it. It cost them thirty-five pounds, the most they had ever paid for any single figure. 'Both our banking accounts vanished in this sale', Ricketts later noted, 'The Tanagras proved to be forgeries and were given away.'[4] A year later four Tanagras bought at Spink's for sixty pounds also proved to be fakes. The losses were quickly made up by other figures, including one at the Carmichael sale standing with arms akimbo that Ricketts was determined to get at any price. This and another figure were bought for thirty-five pounds, the price of a trip to Italy. This, Ricketts told the Fields, was the true measure of money, its value in terms of other delightful things.[5] Visits to other collections of Tanagras only increased their pride in their own. They would not have exchanged the best six of theirs for the best six in the Ionides Collection, and Shannon felt that only one in the celebrated collection left by Lady Saumarez to Girton College, Cambridge, would have felt at home in their collection.[6] Ten of their Tanagras, along with an archaic Greek statue of a woman and four Greek pots, were exhibited at the Burlington Fine Arts Club in a major exhibition of Greek art in the spring of 1903.

Collecting was a passion with them, and they pursued it assiduously: 'Alas, to be possessed of the seven devils of the collector', Ricketts once acknowledged, 'when in the throes of concupiscence and hesitation.' When Rothenstein mentioned he had seen a drawing by Gainsborough in a shop near Piccadilly, but would not tell them where, they spent an afternoon searching London in vain. After having secured some Japanese prints from Van Wisselingh, among them a mother and child by Utamaro which they regarded as the most beautiful print they had ever seen, they rushed off to the furniture dealer from whom Van Wisselingh had obtained them, and bought some more. They ended up with '3 first rate Harunobus, 10 first rate Outamaros, and one first rate Hokusai'.[7] Later, having secured yet more prints from a German dealer, 'a thief who asked high prices for them', Ricketts made the surprising comment: 'I think at their best that nothing quite touches a first rate Jap print; excepting a good Greek kylix or a first rat[e]

[4] CHS Diary, 18 Nov., 20 Dec. 1898 = *SP*, p. 22.

[5] MF Journal, 1901, fo. 28^{v}.

[6] CHS Diary, 12 May 1902.

[7] CR Diary, 17 July 1900 = *SP*, pp. 41–2.

Tannagra: even the latter hardly compare; only the masterpieces of the greatest masters go beyond: picked Titians or Rembrandts or world famous frescoes.'[8] He was thinking in terms of design and aesthetic success, rather than of rarity or monetary value, but his judgement here seems to be particularly of the period, when the taste for Japanese art was still at its height. It is noteworthy that Ricketts and Shannon did not worry overmuch about the condition of their purchases. They were not so much interested in rarity or investment as in artistic achievement.

They were ever on the alert for new art-forms. A new turn to their collecting was the acquisition of some important Persian drawings at the sale of the collection of Capt. W. J. Myers, who had been killed in the Boer War. Myers showed 'undoubted signs of a fine taste or instinct in the choice of things', Ricketts noted, after viewing the collection. Owing to its reputation, he had feared that they might have little chance at the sale, but they secured five drawings, 'all the plums at a high price for such things, that is, from 11£ to 4£ a drawing'. The total cost was thirty-five pounds. One of their purchases, *The Camel*, Ricketts felt was the best Persian drawing he had ever seen. This and another drawing of a 'young fop bandaging his foot' showed, Ricketts thought, the influence of some stray sketches by Gentile Bellini which had found their way to Persia. Shannon, who had always coveted them when exhibited, noted that they had made an excellent start to a collection.[9] These turned out to be some of the most astute and lucky purchases they ever made and when, years after, tragedy made Ricketts sell *The Camel* and three other Persian drawings they realized a vast sum, £4,000.

Ricketts said that his 'one perception of finance is to spend money intelligently not to save it'.[10] He sought good value in his purchases. He would make every effort to secure an important piece, but he knew his limits. Though as a collector he made mistakes, he had a natural prudence which kept him from disastrous over-expenditure, however desirable an object might be. Sturge Moore records that Ricketts 'adored recklessness in collecting, in giving, in gaiety, in enterprise, in devotion to the work of others'. However, he never indulged this recklessness beyond the point of no return. He admired instinctive generosity and practised it himself; he bought things to give away, like

[8] CR Diary, 30 Oct. 1900 = *SP*, p. 47.
[9] CR Diary, 10, 12 July 1900 = *SP*, pp. 40–1; CHS Diary, 12 July 1900.
[10] CR to Mary Davis [*c.* Mar. 1929].

the Chinese hair ornaments, made of kingfisher feathers, which he would present to selected lady visitors. Special friends were showered with gifts. (Yet he was quick to spot insincerity and when a woman friend had not appreciated a beautiful necklace he had given her he said, 'Henceforward I make a rule never to give away anything I really care for.')[11] Of course, he never gave away any of the major objects in their collection, which kept growing and improving till his last years, when he lost heart.

In these years at the turn of the century, when old master drawings were still available at prices accessible to those on relatively modest incomes, like Ricketts and Shannon, they secured many of their prize pieces. In the days preceding the Robinson sale of old master drawings at Christie's, in May 1902, they lived in a whirl of excitement. Sir Charles Robinson was Surveyor of The Queen's Pictures and had often outbid Ricketts and Shannon at auctions. Among the lots in the sale were '2 first rate Credis, 3 Carpaccios, 1 head by Dürer, ... frame of 3 drawings by Raphael, 2 superb drawings by Michael Angelo, 6 magnificent drawings by Rubens, 2 grand Rembrandts at least, ... exquisite landscape by Van Dyck ... damaged but genuine Perugino, damaged but exquisite cartoon by Bigio', along with other doubtful, misattributed, or less interesting drawings. Ricketts's observation on the sale show that his enthusiasm never clouded his sure, independent judgement; he trusted his instincts, and they rarely let him down.

> I can not understand how it is that people will not bid sometimes, or hesitate at tiny prices for drawings that ultimately fetch a great deal. I was almost scandalized at the fatuous behaviour of Colvin when he declared that the magnificent head by Durer was not by him. Old Lippman behaved with the tranquility and ease that should characterise a man who knows, though I was surprised at his indifference to some things and his interest in others, the Lotto drawing and Philippino drawings, for instance, mere matters of fashion, and his sudden interest (as if seeing it for the first time) in drawings of obvious excellence, like the Van Dyck Landscape, or one of the pseudo Lionardo sketches and the Pollajuolo head, both plausible contemporary scraps about which there should have been no matter of doubt ... I do not understand the almost absence of bidding for the frame of genuine Raphael drawings ... I feel the atmosphere is still clouded by Morelli's mistakes over sketch-book, etc.[12]

[11] *65 Ill.*, pp. [vi, iii].

[12] CR Diary, 10, 12 May 1902. In examining the sketch-book (generally no longer attributed to Raphael) in the Accademia in Venice in 1903, Ricketts had been staggered

Ricketts's and Shannon's limited means left them out of the running for the most important drawings, but they secured one of the marvellous silverpoint heads by Lorenzo di Credi; a Magdalen, and a man in a hat, both by Rubens; a 'fascinating study of costume called Pordenone'; and the damaged drawings by Bigio and Perugino. They also secured, for two pounds ten shillings, a Van Dyck drawing of the Nativity bought by a dealer. They offered a 25 per cent profit to another dealer who had bought the other di Credi head, but withdrew when he asked for more.[13] Other acquisitions came pouring in: a Van Dyck drawing of dogs, a Watteau of a flute player, and Rossetti's *Beatrice denying her Greeting to Dante on the Public Stairs*, which they acquired from Carfax for thirty pounds. They also made an offer of £100 for the ceiling by Alfred Stevens at the Crystal Palace, but this was painted on plaster and could not be removed. It was destroyed in the fire in 1936.

In purchasing work from the past, Ricketts and Shannon were more than mere collectors. Their collecting was not simply greed or a passion for acquisition, though it had something of these, but grew out of their belief in the value of art as an expression of what is noblest in the human spirit, and as the only form of immortality given to man. It was equally important to them to see that the nation respected and preserved outstanding works of art. For this reason Ricketts was a founder-member of the National Art Collections Fund, an organization which sought to secure for the nation key works of art threatened with export. By the turn of the century works of national importance, such as Holbein's portrait of Henry VIII, had left Britain in a growing exodus of works to Germany and America. Ricketts was involved in a scheme to obtain an 'interesting and characteristic' painting by Piero di Cosimo, *The Fight between the Lapiths and the Centaurs* (Fig. 20). After having been refused by the Berlin Gallery, 'unaccountably', as Ricketts thought, it had been accepted by the Louvre, and then 'rejected by some minister on the score of expence'. It turned up in England in 1901. Although its great width (eight feet) and its violent subject (the

to find that, except for some late washes, most drawings were indeed by Raphael, and not by Perugino, Pinturicchio, and other lesser masters to whom scholars like Giovanni Morelli had given them: 'the bulk of the more interesting drawings, wether copies after Pintorichio or Perugino, are by the same hand i.e. Raphael, even some tame & very dry ones are also by him, see hands, feet, toes, thumbs, angle of cross hatch on the shading. Some 28 . . . are however by nobody' (see CR Diary, Italian trip, 1903, pp. [22, 23]). Some of the drawings had been added to and retouched.

[13] CR Diary, 10, 12 May 1902; *All for Art*, pp. 51, 73–4.

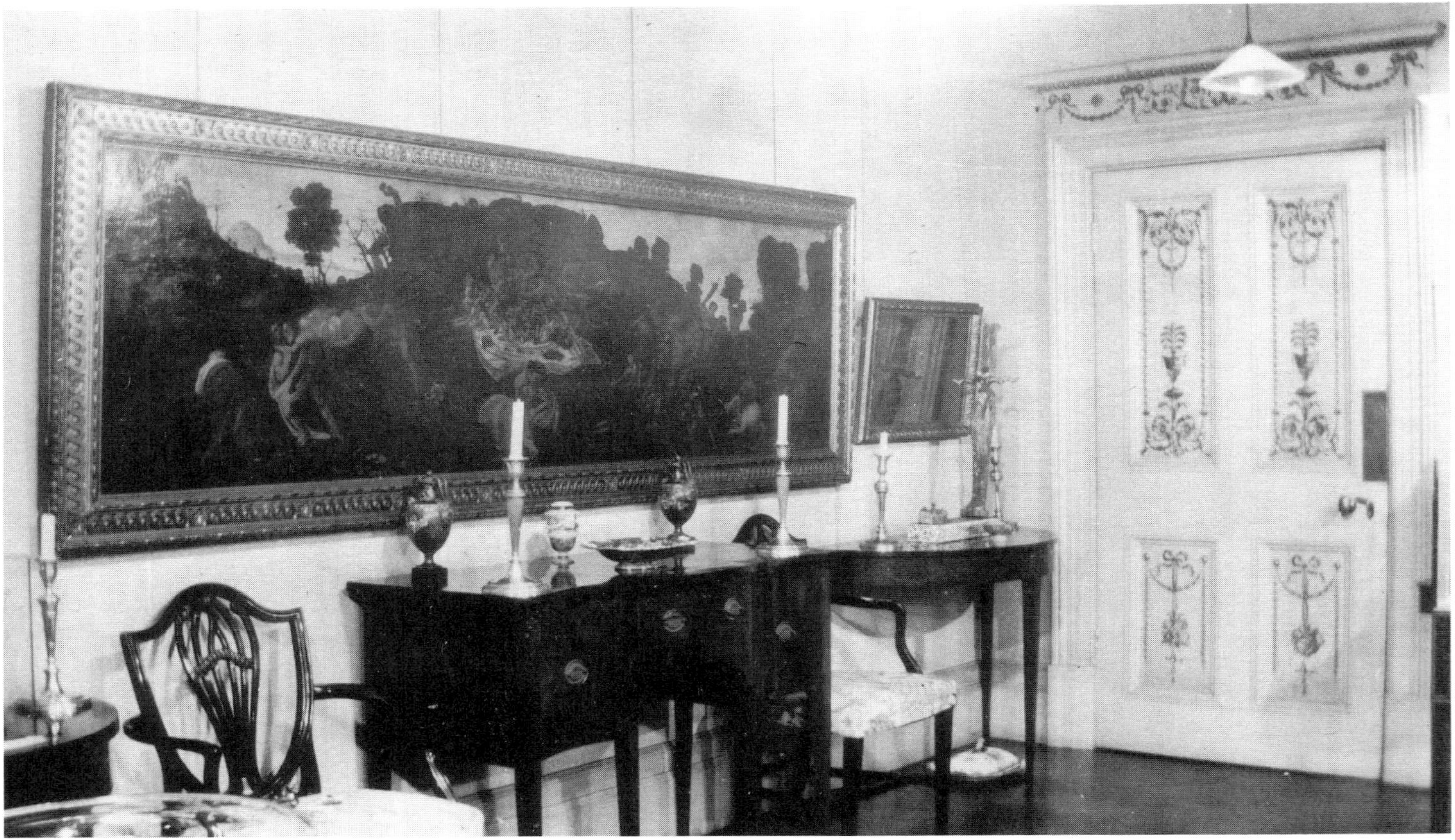

FIG. 20. Lansdowne House, showing *Lapiths and Centaurs* by Piero di Cosimo (Fitzwilliam Museum)

rape of the Lapiths) perhaps made it unattractive for a private collector, Shannon tried to get Edmund Davis to buy at £1,500, the price asked. With Herbert Horne, who brought the picture to their attention, they enlisted the help of Roger Fry and Arthur Strong to devise a strategy for its purchase for the 'ungrateful nation'. These plans went on for a year, but in vain. Meanwhile, the price had doubled; 'so like those thrifty Italians', Ricketts observed. The picture was still at the Carfax Gallery in August 1904 when, unexpectedly, it was offered to Ricketts and Shannon for £700. After hastily checking their finances they decided to go for it, only to learn that someone else, prompted by Fry, was also interested; so they backed out. Robbie Ross hastened round to persuade them to buy it and the same day telegraphed that it was theirs for only £500; their original offer. (The vendor was a rather shady character who asked for a cheque with the name of the recipient unspecified.) To their great joy it arrived the next day, the pride of their collection and probably the most important picture they were ever to acquire, though not the last great treasure of these remarkable years.[14]

'As a test of human nature the sport of collecting is notorious', Holmes commented; but, whilst it brought out greed or meanness in some collectors, this was not so with Ricketts and Shannon. They would never bid against any of the museums, for their devotion to art was such that they could sacrifice their own for the nation's gain. Nor would they compete with some close friends. Holmes once put in a bid of twenty-five pounds, the limit of his funds, on a landscape attributed to Wilson. Shortly afterwards, Ricketts and Shannon saw it but, on learning that Holmes had put so generous a bid on it, decided not to bid against him. Holmes was often sent to the sale-rooms to bid for them. Once, after they had just secured seven Tiepolo drawings, Ricketts discovered that the British Museum was without any, so he wrote to Binyon to say that if this were true then he and Shannon would 'make a pompous but stingy present of some three or four'.[15] When, in 1927, he bought for thirty-five shillings a small roundel of God the Father that he believed to have belonged to the frame of the Pisan altar-piece by Masaccio, the central panel of which is in the National Gallery, he presented it to the gallery through the NACF.[16]

[14] CR Diary, 16 Apr. 1901; 19 Aug., 29 Sept., 5 Oct. 1904; CHS Diary, 16 Apr. 1901.

[15] CR to L. Binyon [*c.*1904].

[16] National Gallery Acc. no. 3627. See *SP*, p. 149. It is no longer attributed to Masaccio.

Their treasures were housed in a new home. By the end of March 1898 they had announced to the Fields that they were leaving their dark house in Beaufort Street and moving to 8 Spring Terrace, Richmond, 'discovered for them by Holmes'. They could live more cheaply there, and it was a marvellous place for parks and gardens:

The place is an ideal one for walks, there is the heart of Richmond Park, mists, water, stags, fallen trees, desolation and rabbits, there is Petersham park, full of sun, planted by Burne-Jones when he painted the garden of Pan, this overlooks the Bute estate, planted by Rubens, with a white forgotten pillared house, all these places are paradises. Hampton Park & Palace are a walk or buss drive. Kew Garden (a world) a tram fare or walk past a lovely river side walk, there are other river paths and, at Isleworth, the Northumberland park hidden in mists.[17]

These places were to become regular haunts of Ricketts, particularly at moments of stress and depression. Another great advantage of Richmond was that living there was cheaper than in Chelsea, and it was only twenty minutes from Waterloo Station by train. It was their friend Sydney Cockerell, who lived there himself, who had persuaded them to move. Cockerell, whom Ricketts described as 'a gruff diamond', had been secretary to William Morris and the Kelmscott Press, and by 1911 had persuaded Ricketts and Shannon to bequeath their collection to the Fitzwilliam Museum, of which he had become Director in 1908.[18]

Preparations for their departure started at the beginning of May. Drawers were turned out, and a lot of rubbish was burnt, including much of their own early work. Shannon went through his drawings, reducing them to about a quarter in bulk. Savage came round to 'grub' (they always used this Rossettian slang) one evening and helped to take the printing press apart. The official date of the move was 5 May.

The new house, a detached, eighteenth-century brick building, was described by the Fields, whose first visit there took place on 23 September:

The little house is delicious. There is behind a small lawn . . . it has a vista of little apple-trees in avenue, the Oxford bust[19] under a canopy of ivy &

[17] CR to MF, p.m. 1 Jan [1899].

[18] CR to R. Ross (n.d.). I am indebted to John Christian for this information on Cockerell.

[19] Presumably bought on a visit to Oxford with the Fields in Sept. 1897.

> Verrochio's Cupid & Dolphin by a tumbled wall of grape-vine with grapes green & hard as jade.
>
> The large room is grey & white, filled with Sheraton furniture, old Sheffield plate, & Greek pottery. On the table a few violets, the wriggling stems supported delicately in a glass with columnar neck & a single dandelion clock beaming from a bit of old China. Lunch in the willow-green parlour (fowl, chocolate & custard pudding & banana fritters ...) Above my head, the original drawing of Rossetti's *Mary Magdalen* ...[20]

Gradually, the house was to fill even more with treasures, with glass cases of Greek pots and Tanagra figurines. Drawings soon became so numerous that they had to be hung on doors or stood on chairs. However, their attitude to their possessions could be rather playful. They introduced the Fields to 'Fatty', 'Swallow', and 'Bullfinch', three new Sheffield jugs, and to a jack-in-the-box named Dr Ibsen. They believed that objects that were dear should be given names, calling a favourite old chair 'Jennie', for instance.[21] At a dinner for the Fields—rather more lavish than those at the Vale—there was a mechanical mouse running round the table.[22]

In the spring of 1898 they had an exhibition in their shop of the woodcut illustrations of Millais. The idea for the exhibition had come to them when they went to the big Millais show at the Academy in January and, horrified to see that the whole field of that artist's book illustration had been untouched, they had decided immediately to have a show at Warwick Street.[23] This, like their earlier shows of the work of A. Boyd Houghton, indicates not only their interest in the illustrators of the 1860s but also their early recognition of the field of illustration as a distinct and valuable part of an artist's work.

This same spirit, together with their long-standing interest in founding a society for the purpose of exhibiting, finally led to the first exhibition of original wood-engraving at The Dutch Gallery, which opened at the beginning of December. To some extent it was still the work of a coterie, but in addition to Ricketts and Shannon, Sturge Moore, Savage, and Lucien Pissarro, there was also work by William Nicholson, whose colour wood-engravings had already attracted a great deal of attention, and the Frenchman, J. F. Millet, who was

20 MF Journal, 1898, fo. 92^{v}.

21 Ibid., 1899, fo. 17^{v}.

22 Ibid., 1900, fo. 145.

23 CHS Diary, 12 Jan. 1898.

probably included to give a European breadth to the show, and whose drawings were admired and collected by Ricketts and Shannon. Perhaps as his was the only work by an artist no longer living, it was hoped that this exhibition would promote his (neglected) reputation.

Many of the wood-engravings submitted by Ricketts and Shannon to the exhibition had in fact been done as book illustrations. Most of Ricketts's work was from *Daphnis and Chloe*, *Hero and Leander*, and *Cupid and Psyche*, but there was also a sombre, colour woodcut portrait of himself, done in Renaissance style at the beginning of the year. Ricketts also included one print after a design by Savage and three other lugubrious ones, all featuring the personification of Death, engraved by him after drawings by Alphonse Legros; Ricketts's interpretations of his work delighted the older artist.[24] Shannon's contributions were more cheerful, a delightful series of twelve roundels on the theme of the labour and activities of rural life; binding sheaves, raking up leaves, gathering fruit. Printed in colour from more than one block, they were, like Ricketts's self-portrait, chiaroscuro woodcuts, but neither artist used this medium again.[25]

Ricketts and Shannon had, earlier in the year, become involved in the formation of a new artistic group, to be known as the International Society. In April 1898 Whistler was elected its first President, with the sculptor Alfred Gilbert as Chairman and the painters James Guthrie, John Lavery, William Strang, and William Rothenstein, as well as Ricketts and Shannon, forming the committee. All progressed smoothly until Whistler wrote from Paris suggesting that Joseph Pennell and Albert Ludovici also be co-opted. Both of these were his faithful followers, though neither of them 'was taken seriously as an artist', as Rothenstein pointed out.[26] Pennell was one of Ricketts's 'enemies'. When the committee opposed the suggestion, Whistler insisted and it gave way. Rothenstein left, to be followed shortly afterwards by Ricketts and Shannon, who later rejoined the society and became two of its most active and influential members.[27] Shannon also resigned from the newly formed Pastel Society, to whose first exhibition he was a contributor, because of Pennell.

[24] See CR, 'Legros', *Saturday Review* (17 Apr. 1897), 406–7.

[25] See Frank Gibson, 'Idylls of Rural Life: A Series of Woodcuts by Charles Shannon, ARA', *Studio*, 58 (15 Sept. 1916), 211–15.

[26] *Men and Memories*, i. 335–6.

[27] See Philip Athill, 'The International Society', *Burlington Magazine*, 127 (1985), 21–6.

Earlier, Ricketts had been rather reluctant to join societies in whose formation he did not play any part. When Sydney Cockerell wrote in 1893 to say that Noel Rooke and George Frampton wished to put him up for membership of the Arts and Crafts Society, he declined to join, though flattered by their good wishes: 'I have perused the rules you sent me but feel myself quite unfitted to comply adequately with them.'[28] An instance of Ricketts's independence, this also contains a hint of disdainful superiority, but in 1897 he was delighted to be elected a member, without in any way seeking the honour.[29]

Not all responses to his work were so favourable, and Ricketts, so touchy and emotional, was apt to over-react to any reproach or criticism. Shannon tried to soothe him or, even better, to protect him from adverse comment. Laurence Housman told the printer and designer C. R. Ashbee about this aspect of Ricketts:

> he's such a touchy fellow, can't bear a word of criticism. And Shannon used to temper the wind to the shorn lamb for him so beautifully; I remember once when —— blurted out some adverse review of his work—Ricketts turned silver with rage (he always does turn silver you know)—& poor Shannon muttered:—'There's my peace of mind gone for a month—What *did* you tell him for? I've been keeping it from him for ever so long.'[30]

Such violent reactions were due to the great sensitivity that made him so responsive to art, the sense of the superiority of his gifts that could make him arrogant and domineering, and above all the deep insecurity that underlay his character. Little things could cause disproportionate reactions, even make his world seem to collapse. He once decided to buy a Tanagra of a Satyr and Bacchus, even though there were doubts about its authenticity. When he wrote out a cheque for £10 he discovered that Shannon had misread the price from the dealer's letter: it was £60. Ricketts felt 'like the young man who thought he had acquired a silver chandelier at Christies for 15 shillings' when it was that price per ounce: 'I blushed crimson, burst into perspiration and was shaken in nerve and temper for the rest of the day and the major part of the night.' Even the next day he was 'still ruffled and hurt'.[31] Adverse criticism of himself or his work made him suffer even worse fretful

[28] CR to Sydney Cockerell [*c.*1893].

[29] TSM to A. H. Fisher, 16 Aug. 1897 (Sturge Moore Papers, Univ. of London). See also *Affectionate Cousins*, p. 101.

[30] C. R. Ashbee Papers, Diary i, Apr. 1900 (King's College, Cambridge).

[31] CR Diary, 28 Aug. 1905.

days and sleepless nights. Much of this was hidden behind the protective shield of his relationship with Shannon, but occasionally it burst forth in a more public way.

A review appeared in the *Athenaeum* (23 July 1898) assessing the achievements of the Vale Press. Inevitably, Ricketts was compared with Morris, his precursor in the private-press movement. With regard to typography, the reviewer observed that neither of Ricketts's founts could claim 'a complete victory'; in decoration, Ricketts had much of Morris's 'judgement and fecundity and a strong sense of line, but often when he has overcome a difficulty he leaves a sense of striving which lessens his effect almost to the point of failure, where the older master would have evaded the difficulty with an air of triumph'. Turning to the wood-engraving in the Vale books, he singled Ricketts out as the first wood-engraver of the day: 'His beauty of line, his mastery over colour on the wood, approach the marvellous.' As Ricketts was still young and had the ability to learn from his failures, the reviewer felt that his 'chief achievements and triumphs' lay before him.

Worse was to come. In August the *Contemporary Review* published an article on 'The Kelmscott Press and the New Printing' by Albert Louis Cotton. After a long and glowing account of Morris's achievement and contribution to the revival of fine printing it turned to Ricketts and the Vale Press, which demanded an honourable mention, though marred by 'needless affectation and excess'. Ricketts's type was compared unfavourably with Morris's, the interrogation mark, the ampersand, and the paragraph signs possessing an 'eccentricity of form which at every turn annoys the reader'. Ricketts's striving for modernity, he went on, led him to overstep the bounds of art by such affectations as the practice of introducing words in the lower case into a page printed in capitals. Such faults, Cotton claimed, prevented Ricketts's efforts from taking the place in artistic typography that would otherwise have been due to them. However, in the borders and decorations for his books Ricketts was more original than Morris and deserved to be considered the first among Morris's successors, whose work could only be compared with Aubrey Beardsley's magnificent borders for *Le Morte Darthur*. The illustrations in such books as *The Marriage of Cupid and Psyche* were of a high order of workmanship.

These two articles were to sting Ricketts into a justification of his aims and practice in *A Defence of the Revival of Printing*, which he published the following year. Both articles had balanced praise and criticism (though in different proportions), but it was only of the criticism that Ricketts took notice. At the beginning of his *Defence* he clearly states that he is answering the objections he has met with, though his essay is, in fact, more concerned with the history and principles of good printing and book decoration. He deals with the eccentricities in his type design. The ugliness of the paragraph marks was a matter of opinion which he would not discuss. The contraction for 'and' combined a mediaeval uncial 'e' with a smaller case 't', thereby producing the 'et' or 'and' that the standard ampersand indicated by a twiddle of the pen, which was unsuitable to type. As for the interrogation mark, Ricketts pleaded extenuating circumstances, as he had found a form similar to the one he had used in Jenson. The comparison of his borders with Beardsley's, though meant in approbation, he found invidious, for these early borders done by Beardsley before he evolved his own personal epigrammatic manner were quite bad; and with this opinion Beardsley himself had agreed. The eccentricity of putting a word in lower case in a whole page of capitals was a trifling fault which he now regretted, though he affirmed that the writer of the *Contemporary Review* article was wrong to suggest that insufficient time was devoted to page layout: 'no amount of pains has been spared in the building of such pages; they have been cast and recast by me in tracing paper over a solid page of type, the initial changed or redesigned, the type filed, the compositor made unhappy.'[32] Ricketts later regretted the personal tone of this pamphlet; 'that is, one or two unnecessary pieces of self-defence which take from the importance of the remarks I then made on the shaping of beautiful books. There was no need for any defence whatever.'[33]

However sensitive he was to criticism himself, he was always ready to criticize the work of his friends. The intense and demanding Fields were drawing closer and encouraged him to criticize their work. They submitted their manuscripts to him to read and he blue-pencilled passages that were prolix, prosaic, or stilted, and made witty comments in the margins which were usually acted upon. They felt Ricketts, unlike Berenson, had a true appreciation of their work and a predilection 'in favour of what is great & poetic & of the chosen mood'.[34] However, about his own work they found him touchy and secretive.

32 *Defence*, pp. 16–17.

33 *Bibliography of Vale Press*, p. v.

34 MF Journal, 1902, fo. 71^v.

The cover for their *Fair Rosamund* (1897) was to be a surprise, he told them, and when Miss Bradley asked if they should like it, he replied 'I shall be *immensely* wounded and *unforgiving* if you do not'.[35] It turned out that they did dislike the paper cover with roses and doves that Ricketts designed for them. The plump dove they considered a 'Valentine symbol' and objected to the 'death dealing' arrow that pierced it, as their play was intended to be an affirmation of love, not death. They finally decided that Ricketts (an artist more prone to attenuate his figures) had fattened the doves deliberately in ironic revenge because Miss Bradley had once irked him by praising 'rounded outlines'. Concerned about the cover of their next book, *The World at Auction* (1898), the first of a trilogy, they made an exploratory visit to the Vale office, where they found Ricketts unusually forthcoming. He gave details of the emblematic border he was going to design for them, though he generally did not 'like describing a design, which should be seen'.[36] This quietened their anxiety, but a letter written a few weeks later showed a fear that the mischievous Ricketts might find a pretext for another spiteful jest. 'You must not let a careless comment of mine on your *Autumn*', Miss Cooper wrote, 'obliterate the great vision of Fortune and the flower-tossing dancer beside her.'[37] This time the cover was, in their eyes, fully successful. The border of the last of the trilogy, *Julia Domna* (1903), nevertheless brought some difficult moments, though the Fields showed a new confidence in Ricketts's ability to interpret their work in imaginative designs. 'I want to talk to you about *Julia Domna*', Miss Cooper began, as she recounted in her diary:

> 'But I dont want to hear: I don't want to draw the temple of Vesta'—he exclaims like a wilful child. 'Then you shan't', I hasten to say. 'I don't want to talk to you about your part in *Julia Domna*—an artist must be left entirely to himself—but about my part. On Thursday, I shall hope to send you the new poetic *Julia* & I want you one evening to take your pencil & mark it where you think it could be improved. Your advice has already saved it'—'It would have been most unwise to publish it as it was', he interpolates—'And I know from your cutting of a finished work—*Anna Ruina*—your pencil may be trusted'.[38]

[35] MF Journal, 1897, fo. 12.

[36] Ibid., 1897, fo. 26.

[37] Edith Cooper to CR, Easter 1897, transcribed in MF Journal, fo. 46.

[38] MF Journal, 1902, fo. 88^{r-v}.

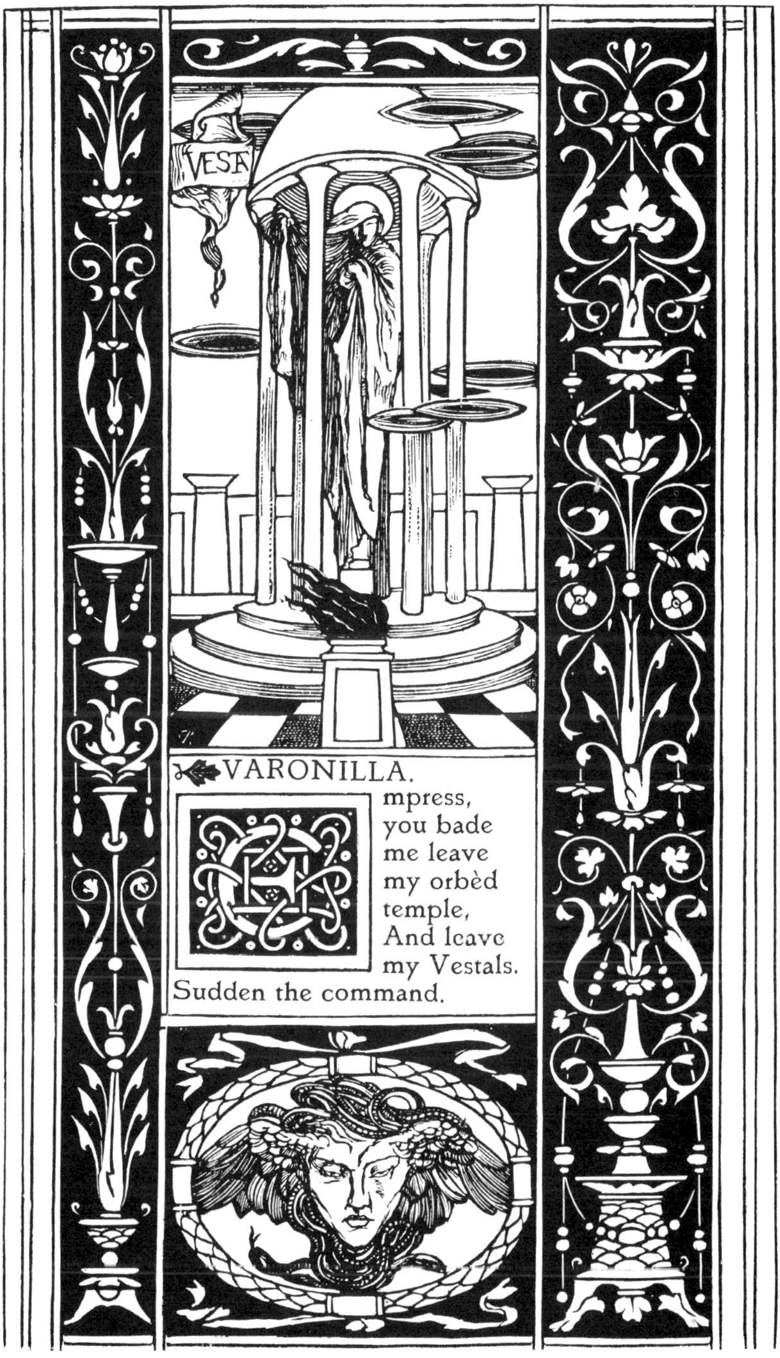

FIG. 21. Renaissance border designed and cut in the wood by Charles Ricketts for *Julia Domna*, by 'Michael Field' (Vale Press, 1903)

They were evidently discovering—or at least Miss Cooper was—how to handle Ricketts, but there were further difficulties ahead, as Ricketts complained a few months later that he had become 'stuck' in the design, because he had been doing borders for seven years and could 'get no freshness into the work'.[39] It was another four months before the border was finally shown to them (Fig. 21). It included the Temple of Vesta that Ricketts had at first refused to do and, with the two other volumes of the trilogy, gave the Fields great pleasure: 'We have not felt so strongly excited for years.'[40]

On the first day of 1899 Ricketts replied to a letter from the Fields, who had announced their intention to move to Richmond. It was he who had initially suggested the move. The decision prompted a flurry of letters and visits to Richmond, and the move dominated the first three months of the year at Spring Terrace. Shannon soon wrote that they had found a house, 'one of the most charming places (as far as outlet is concerned) in Richmond'. It had a small garden running down to the river tow-path, a view of Richmond Bridge, a balcony overlooking the river and (dating from the early nineteenth century) it was meant to 'hold Sheraton (not Chippendale) furniture'.[41] A few days later he sent a room-by-room description. The garden was so steep that their basset-hound might slide into the river if it were not for the garden door: 'Unprecedented opportunities for pushing relatives thro' doorway at high tide.'[42]

At the beginning of February Michael Field, after several visits to London, became the mistress of 1 The Paragon. The question of decoration arose, for the poets were as much concerned as the artists that their surroundings should be aesthetic; and this second household devoted to beauty was set up only ten minutes' walk from the other. For advice on what and where to buy the Fields turned to the artists, who responded with quiet amusement at their excitement: 'You mention more pieces of furniture than were ever designed by Sheraton.' A furniture shop in Fulham Road was recommended, where not only were the prices moderate but 'you can always knock him down considerably. In fact he prefers it'.[43] On another occasion the Fields

[39] MF Journal, 1902, fo. 150.

[40] Ibid., 1903, fo. 47^{v}.

[41] CHS to MF, Thur. [12 Jan. 1899]. See *Some Letters*, p. 16.

[42] CHS to MF, p.m. 13 Jan. 1899.

[43] CR to MF, 26 Feb. 1899 = *Some Letters*, p. 18.

set out with Ricketts to buy rugs, with the two women making many 'conjurations' that they would leave him to settle everything. Ricketts, though, showing 'that fuss of caution & self-importance a man so amusingly makes when he goes to a shop in women's company', was so low-toned and submissive in manner that, when they found a French carpet of roses they liked, 'never a breath of bargaining does he draw'.[44] The artists offered to be their 'slaves' in the matter of meals 'or other small matters of assistance' on the day of their arrival.[45]

'The wretches of artists', noted Miss Cooper in their diary, 'contemplate starting for Italy about 20th April, which will hurry our work.' This journey was probably a flight from the excitement and importunities of the two women, and from the vast preparations for the impending move itself which was to take place in mid May. He and Shannon made their wills (they had got Whitaker's *Almanack* and were their own lawyers) 'to insure our safety', Shannon wrote to Mr Dowdall, 'just as one takes out an umbrella to prevent it raining'.[46] They did not dare go to Italy without ensuring that should anything happen to them their Hokusai collection would go to the British Museum, their bits of Antiquity to Oxford (this was before Cockerell had got hold of them), and their work to Sturge Moore. All this would otherwise have gone to Ricketts's sister, and to Shannon's family in the depths of Lincolnshire, who would wish to have 'all dear Charlie's drawings'.[47]

They left for Italy on 24 April by way of Switzerland, which they found astonishingly fine and 'not a bit as bad as it is painted'.[48] Only two days of rest were spent in Milan, which they disliked as having 'all the vulgarity of a modern town and none of the seriousness. The Duomo rivals the Albert Memorial and looks like an American organ in sugar.' They arrived in Venice on a 'charming silver evening' and were enchanted and surprised. Soon, however, they were getting about in gondolas and fed up with the incessant tipping of those who held a gondola for you, opened a door or sacristy, or merely looked at you: 'one starts in the morning with enough coppers to sink a gondola, &

[44] MF Journal, 1899, fos. 58–9^{v}.

[45] CR to MF, 26 Feb. 1899 = *Some Letters*, p. 18.

[46] CHS to Mr Dowdall [Apr. 1899] (Henry E. Huntington Library and Art Gallery).

[47] MF Journal, 1899, fos. 58–9^{v}.

[48] CR to Lucien Pissarro [June 1899]. The diary of Ricketts's and Shannon's first Italian journey (BL Add. MS 58094), part of which is printed in *SP*, pp. 23–8, is the source of this account, unless otherwise indicated.

in the evening one lacks a penny to buy an orange'. As obvious visitors, they were disturbed by the constant attention they received. As for the language, Ricketts claimed that it posed no problem: 'I find it quite unnecessary to know Italian. If you speak in dictionary Italian, they answer you in French; if you talk French to them, they answer you in English; they seem to talk German to each other.'[49]

For more than two weeks Ricketts and Shannon visited and revisited the churches and museums of the city, examining each painting, comparing it with the impressions they had gained from photographs, and sometimes disagreeing with attributions. In his diary Ricketts carefully noted details of technique, quality, and condition, and of restoration. Bellini and Titian emerged as their favourites and they were charmed by the Carpaccios in San Giorgio degli Schiavone and discovered in the church of Mater Domini the delicate Catena, whom Ricketts decided was 'the best Bellinesque after Cima, though I used to think badly of him in London, perhaps because fools liked him so much'. However, there were also disappointments. They found they hated Tintoretto ('the most beastly artist in the world') and at the Doges' Palace they found Veronese's famous ceiling, which Ruskin advised should be seen again and again, since the traveller would 'not otherwise enter so deeply into the heart of Venice', absolutely appalling. 'I shall live in the Gondolas if Venetian painting is like that', observed Ricketts. Later, in the Accademia, he modified this view, and found a large painting by this master very good: 'Our first impression over his other pictures was one of intense dislike, though on second inspection, there are good pieces. We must look at him more carefully.'

The minute attention he gave to the paintings, his readiness to change his mind, to accept new impressions, his determination to judge a work on its own merit, by the standards of its own time, rather than by current opinion or by the expectation created by his study of photographs or reproductions, all show how Ricketts's mind worked. He was always independent and never doctrinaire; he never worked from a theory, but judged what was in front of him. Ultimately, his standards were those set by the greatest masters in any art, but he could also enter into the spirit of whatever he was looking at, whether it was European, Oriental, or even primitive Amerindian. Together with these standards derived from study and his wide imaginative

[49] CR to MF 9 May 1899 = *SP*, p. 28; CR Diary, 4 Apr. 1899.

sympathy there was also a good deal of intuition, of 'something in the pit of his stomach', in his response.

Soon their first delightful impression of Venice turned to oppression at the 'decay and degradation of the place, the ruined pictures in the churches and the brutal restorations in St. Mark's'. At the church of Santa Maria della Salute, they were delighted with Titian's *St Mark enthroned with four Saints*, which was 'well-preserved, though sunken and dusty'. *The Descent of the Holy Ghost* by the same artist, however, was 'dead, sunken, blighted, without hope of recovery, of course tending towards that side of the later Titianesque I like least'. At the larger Gothic church of the Frari, one of the most beautiful in Venice, the Pesaro Madonna, which they had worried might be a wreck, struck Shannon more favourably than it did Ricketts, who found it 'sunken & dirty to a degree difficult to imagine outside an Italian church'. The paintings by Bartolomeo and his nephew Alvise Vivarini were as bad as usual, but the altar-piece of Madonna and Saints by Giovanni Bellini was as good as the photographs had led them to believe: 'lustrous and wholesome, tender & grave, beautiful & real'. It was, Ricketts felt, his masterpiece, 'and that is almost a dangerous thing to say' (though he later scored out this observation); it should 'be hung in Paradise'. Also in the sacristy was Tiepolo's *Stations of the Cross* (now given to the younger Tiepolo), which Ricketts found unexpectedly fine and serious in invention, and lacking in that unpleasant quality he felt that Tiepolo's religious painting so often had. Of course they were dirty and damaged beyond belief, which 'upset & moved' him, and after seeing them again just before their departure he noticed:

> the Entombment is ruined; the Deposition practically so; the Crucifixion seriously damaged; some rain water was dripping from one. They all of them have pieces of paint scaled off, varying from an inch to three inches in size; some have bulged from their frames, most have broken stretchers loose behind the canvas.

This distressed Ricketts, who felt something should be done about the terrible condition of the paintings in the churches. Both the Bellini and the Tiepolos should, he believed, be in the Accademia.

> It is wicked, idiotic, & sentimental to leave good pictures in churches; why not in public w.c.'s or at the corners of streets? Art has ceased to belong to the world, and should be handed over to the museums, idiotic as they are. Sentimentalists who like to see things 'in their proper place', as they say, say

so because they do not wish to see them at all, but only to talk about them. The sentimentalist hates Art, and all other direct appeals to the intellect; therefore he does not mind the smoke of candles & the breath of the faithful (when they exists).

All the Vivarinis should be sent back to the churches. Copies might replace originals behind the trumpery one finds on the altars, the sentimentalist might then imagine the beautiful effect of the original when it was there, & sentimentalize to his heart's content.

Ricketts also felt that the 'damned' law preventing the exportation of Renaissance Italian paintings contributed to the continued presence of great works in the churches, when they could have been in much better hands in a museum in London. It was his concern for the works themselves that prompted this protest; he would have been content to see them properly cared for anywhere in the world.

Their visits to the churches and museums were interrupted by trips to the doctor, for Shannon was unwell, suffering from a sore throat. In fact he was never a good traveller, and was often ill as soon as they set off for the Continent. This sore throat was to last the whole trip, and the repeated visits to the doctor (in gondolas) eventually made Ricketts loathe gondolas and gondoliers. On their second visit to the Accademia Ricketts was also smarting from two flea-bites and a hatred for modern Venice. But Titian's *Assunta*, which had astonished them on their first inspection, looked even better on a second. 'That shit Holmes' had frightened them with his description of the painting: 'I fancy . . . .that he is unused to monumental painting & missed those blurs and blunderings proper to cabinet pictures, due in his case to a study of Turner, who is too clever, & of other rather vulgar english tricky painters.' Whilst the famous Tintoretto *St Mark rescuing a Slave* was, though inconsistent in quality and pretentious, on a higher level of drawing, execution, and composition than usual for him, Ricketts felt that 'no man has ever existed in art less equipped as a draughtsman and with a more limited brush-handling'. The comparisons made between Tintoretto and Titian infuriated Ricketts ('made me foam at the mouth') like those between Romney and Reynolds. It was Ruskin, 'who is a lady [and] overawed by Tintoretto, just as he might have been silenced by somebody who bawled him down or used slang', who was responsible for the current high estimation. Reynolds, with his usual good sense, had disregarded 'Tinto'. However, in the church of the Gesuiti Tintoretto came off better than expected, some of the work being very fine indeed: 'I am bound to say that much of it was what one had imagined Tinto to be before one had been to Venice'. The following day, though, he noted that Shannon, at the Madonna del Orto, 'very properly refused to look at Tinto whose pictures I had rather liked in a cowardly & backsliding sort of way the day before'. He concluded that 'Tinto & even Veronese lowered one of the most perfect conventions the world has ever seen since the arts of Greece.'

Both artists were suffering from the pace. Shannon was still 'soaked in gargles' and both of them were sleeping badly and felt fretful 'under the glare, squalor, & restorations' of the city. In spite of this, they kept up their visiting, sometimes together, sometimes separately. While Shannon went off to the Accademia, Ricketts went to see more churches, 'to settle their hash'. Then Shannon went to the Modern Exhibition, while Ricketts returned to San Giorgio and San Marco, the latter of which he found 'too complicated & fascinating to write about', except for the naughty subjects in the vestibule, which charmed him: 'Potiphar's wife, Noah on the bust, circumcision of adult, the prick of the patient drawn as a cad draws a prick on a wall.'

One evening they met and fell 'in love with a dear old buggins' who looked like a sea captain and whom they met in the smelly smoking-room of the hotel. 'He was insulted there by a bellied, commercial-looking Briton to whom he had spoken, and being nice, well-bred young men, we instantly handed him matches, a recent English newspaper, and in addressing him used the conventional "Sir".'[50] It turned out to be Samuel Butler, some of whose writings they had read. He had a more than usual 'grip & sense of art' with 'genuine feeling and insight in an arbitrary way', was all sentiment and lacked any documentary knowledge of art. He loved the Italian primitives, and shared their 'wholesome hatred' of Tinto. They met him every evening in the smoking-room and he handed on to them a piece of advice that had once been given to him as a test of real appreciation for a work of art: 'Ask yourself whether you would really like to see it a second time alone.' They were both struck by the soundness of this, and thought of him as 'a gouty angel in carpet slippers'. He called on them after their return to London, but the friendship did not ripen.

They reached Padua in a shower, but it seemed fresh and sweet and water-tight after Venice: 'We like the clean, cloistered streets, the absence of beggars, the large brick churches.' There was a curious

[50] CR to Cecil Lewis [1926] = *SP*, p. 363. See also *SP*, pp. 26–7.

contrast between Padua, so clean and well-to-do, and Venice 'rotting away in dirt'. The Arena, with its frescoes by Giotto, gave more satisfaction than almost anything in Venice. The colour was delightful; the photographs gave no idea of the handling; the work of assistants was curiously unobtrusive, and of restorers so well done that it was difficult to detect at first. Ricketts admired the scenes from the life of the Virgin and found the Vices and Virtues exquisite: 'the clenched fist of Despair, the ears of Malice, the half-shut eye of Infidelity; [Giotto] is able to summarize an episode in one figure and foresee the claws in the hands of Injustice.' Ricketts was also struck by the comparative innocence of the hell portions: 'the absence of the w.c. and humourous side we find in French cathedrals; even my dirty mind detected only one figure, a monk having something unpleasant done to his parts, & I daresay he deserved it.' The Giotto he described as one of the great impressions he had received in his life. After the hard work of visiting churches and museums they relaxed, hunting grasshoppers by the river and strolling in the town at night. Ricketts was rather 'liverish through eating and drinking like a beast'. Their hotel 'like Paradise has [a] snake in its Premises and the bill made no pretence to follow tarif or verbal agreement'.

They did the churches and museums of Vicenza and Verona and, skipping Brescia to avoid the Civico Museum, they returned to Milan and its museums, where the Brera was even better than they had thought before and the Ambrosiana had 'masses of distressing rubish attributed to Lionardo, few only by him'. At Basle they were 'smiled at by all the Hotel people (smiles will probably be put on the bill)'. After a day of rest, they set off for the museum to study the important collections, especially the Holbeins, which were better than the photographs had led them to expect. In Paris, where the weather was cold after the heat of Italy which had given them sleepless nights, they were unable to call on Camille Pissarro as they had hoped, because Shannon was still unwell. They thought that Pissarro's pictures were 'by way of looking the best' in the Impressionist Room at the Luxembourg Gallery. The whole trip was a sad disappointment. 'Most things there are invisible or if visible so damaged that one feels sick to look at them; then they are often repainted (by house painters).'[51] Shannon's sore throat, moreover, had persisted for the whole of the journey and prevented their going out at night.

[51] CR to Lucien Pissarro [June 1899].

On their return Ricketts had to sort out difficulties with Pissarro over his vellum copies of one of his books. In June the question of premises for the firm came up, as their tenancy was due to expire and they were keen to move. Customers and important visitors often called, and they were always embarrassed by the 'unwashing personality' of their Jewish landlord, a tailor, and his formidable wife, as well as by some ladies of another ancient profession on the upper floors of the building.[52] On 10 August they heard that they had obtained premises at 17 Craven Street, Strand, a less convenient but more salubrious address, to which they moved officially on 7 September.

The other great move of the year, that of the Fields, was settling down at last, though Ricketts and Shannon were not to escape so easily, as the Fields were to tease them for not helping them sufficiently in setting up their 'Doll's House'. Ricketts and Shannon became regular visitors, going over for short chats as well as elaborate meals, and joining the two women for walks in Kew Gardens and Richmond Park. When the Fields pressed Ricketts for details about his designs for their Vale Press books he was reluctant to talk, but he was more eager to discuss jewellery, on which he had started to work. He was designing the figures of the Muses in a crystal: 'tragedy sweeping her cloak away from the touch of comedy who mocks her; lyric poetry half asleep by falling water.' Ricketts felt that women should only wear three kinds of adornment: a ring, chains, a pendant. Not bracelets, for though the Greeks wore them they had them clasping the upper arm.[53] For the wedding of Amy, Edith Cooper's sister, to an Irish doctor, he showed them designs for a salt cellar and for a pendant and asked them to choose, and Miss Bradley chose the latter. The result was a gold scallop-shell set with ruby and hung with pearls. A visit was made to Spring Terrace to choose a design for a piece of jewellery from Amy to Katherine, and Ricketts tactlessly suggested a copy of Amy's cockle-shell ornament. The ladies were hurt, and so was the sensitive Ricketts who exclaimed: 'You are cold to my designs.'[54] Soon, however, Ricketts brought round a pendant of thrush-egg turquoise set in gold with two pearls and an amethyst, which he said he was going to paint as hanging on the breast of Icarus in a picture. When Katherine delicately

52 Holmes, p. 173.

53 MF Journal, 1899, fo. 98. See Diana Scarisbrick, 'Charles Ricketts and his Designs for Jewellery', *Apollo*, 116 (Sept. 1982), 163–9.

54 MF Journal, 1899, fo. 113^{r-v}.

asked if her jewel could be like this one, he promised this very one to her, if she was 'very good', and when Icarus had had his pendant safely painted.[55] Miss Cooper was not to be left out of the distribution of jewels, for Miss Bradley commissioned an enamelled gold pendant of a blue-bird picking at red berries for her niece's birthday in the following year (which was in fact delivered before the end of 1899).

A succession of jewels for the Fields followed, either as gifts or commissions: a ring of lapis lazuli and chrysoprase similar to one he did later for Shannon; a pendant called 'Music', after the Fields's basset-hound, depicting a hound hanging from a lyre; and, finally, a ring called the Sabatai ring, representing the one given to the Jewish mystic Sabatai, to make him convert to Islam; the theme of one of the Fields's plays (Fig. 22). The ring symbolized the Mosque of Omar, supported by a shrine and with a star sapphire set in the dome, a loose emerald inside the shrine which could be seen and heard) and ambergris smeared on the doors; it was intended to appeal to all five senses. It inspired a poem by the Fields. Such jewels were received with great joy and a great fuss, and sometimes decorated the tables as well as the persons of the Fields. Ricketts believed that 'there is no good in any gem we do not raise to the mouth as a sweet', and that women liked jewels to be 'twinkly, twinkly' while men liked them in so far as they suggested substances, 'a drop of milk, a dew, or a flame'.[56] Most of his designs were based on early models; Roman, Byzantine, or Renaissance.

His most famous jewel for the Fields was done in 1901, and included a marvellous miniature portrait (Fig. 15). In a locket decorated with a relief of Pegasus at the fountain of Hippocrene in gold and enamel Ricketts had placed a profile of Miss Cooper done from a photograph. It was done in pencil and water-colour, heightened with gold. At first Ricketts had tried to make the face very like, but then had hit upon the idea of showing the face seen under a flash of lightning. 'Then he discovered', recorded the Fields, 'the infinite greyness of gold used as shadow & that gave the pallor and sensitiveness to the face, specially the nose. He liked working on the nose exceedingly. He thinks the flash shows the nose, eyes & eye-brows—the rest is only what surrounds them.'[57] Ricketts was angered by Holmes's criticism that the

[55] MF Journal, 1899, fo. 127. See Mario Praz, *The House of Life*, trans. Angus Davidson (1964), 237.

[56] MF Journal, 1900, fo. 135.

[57] Ibid., 1901, fo. 67^{v}.

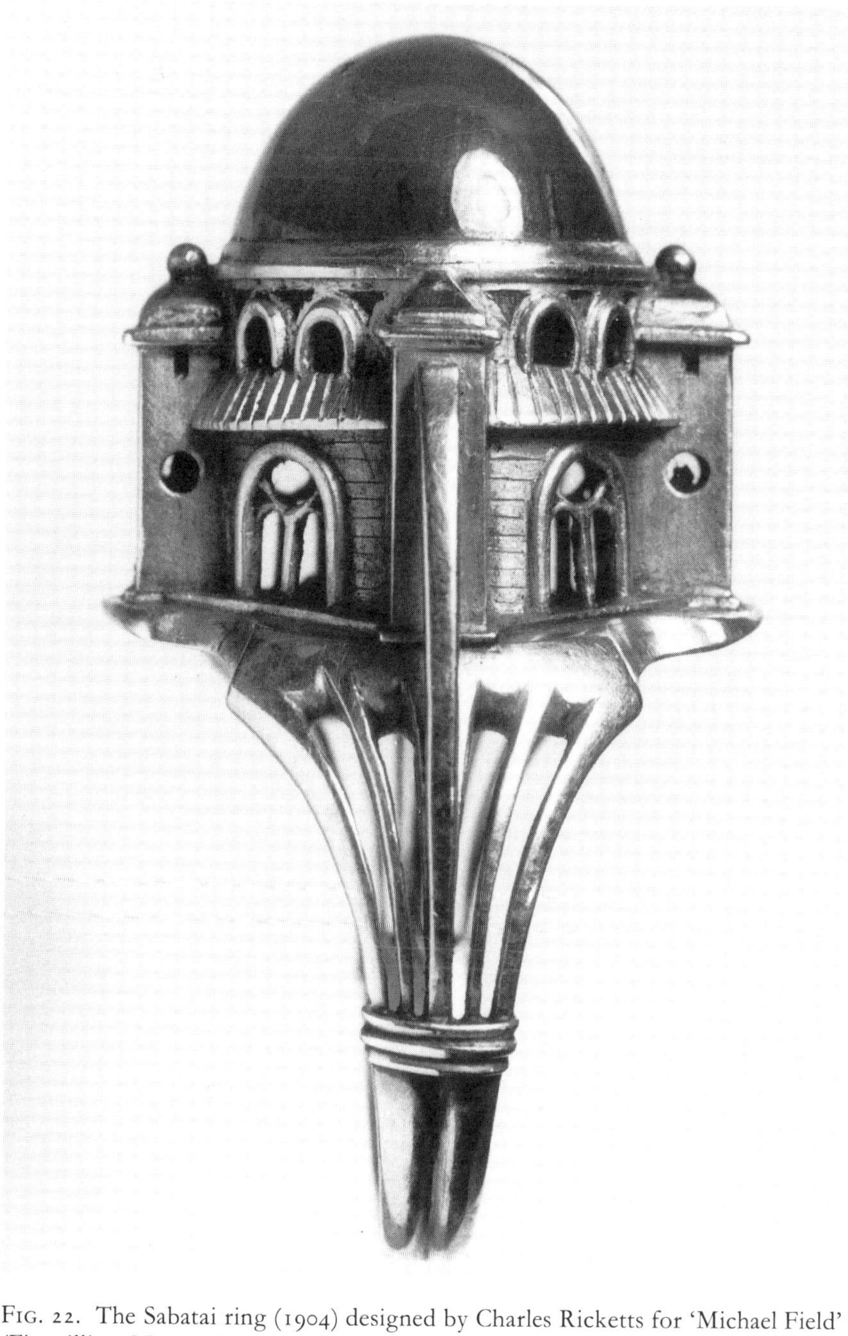

FIG. 22. The Sabatai ring (1904) designed by Charles Ricketts for 'Michael Field' (Fitzwilliam Museum)

eyes were too light. When Sydney Cockerell told Shannon that it was the finest thing that Ricketts had ever done, Shannon replied that Ricketts had done more important things, as he, Cockerell, ought to know, for Cockerell was a great bibliophile.[58] Most reaction was highly favourable. Robbie Ross commented that you would have to go to the Renaissance for such a face. 'Is she really like that?' And Ricketts said, 'Exactly like that'. The Fields were as thrilled by this comment as by the jewel itself.[59]

Ricketts designed more than half of his jewels for the Fields, but other female friends also benefited. One of his earliest designs, if not the earliest (of a shell like that for Edith Cooper's sister), was for the marriage of Mrs Charles Furse. Then in 1900 he modelled a portrait of 'Ryllis Hacon in gold, using a knife and a lead pencil. For May Morris, daughter of William Morris, who they took on boat rides on the river at Richmond, he had designed a Bishop's ring in the form of a buttressed tower with a jewel beacon, which, along with bishop's gloves which she had embroidered to his designs and which she claimed as her own, were exhibited at the Arts and Crafts Exhibition in November 1899. He had been very disappointed that she had only worn the ring for a week, complaining that she was an 'indolent' woman.[60] For both Miss Bradley and Miss Morris, he also designed and painted a fan. He appears also to have done a portrait jewel of Mrs Green, who entertained Ricketts and Shannon at her house in Kensington Square, where they met such adventurous characters as Miss Mary Kingsley, after her return from Central Africa.

They also made new friends. At the end of November Ricketts met W. B. Yeats, Arthur Symons, and Havelock Ellis at a Dolmetsch concert. Ricketts liked Symons because he was kind and 'will serve you as you want to be served by him', and Symons was invited with Yeats to dinner at Spring Terrace so that Yeats could 'be shown to Shannon'.[61] The artists liked Yeats, and found him provincial in a nice way, giving you the name of an artist or writer if you paused for it, and overjoyed if you had read the book that he had just been reading. This friendship was to grow into an important one, and Yeats later attributed his artistic education to the influence of Ricketts and to the

[58] MF Journal, 1901, fo. 160^{v}.

[59] Ibid., 1901, fo. 169.

[60] Ibid., 1899, fo. 129.

[61] Ibid., 1899, fo. 125^{v}. See 'Heirs'.

frequent visits he made to Spring Terrace and, later, Lansdowne House, where he was often to be found at their Friday evenings.

The year 1899 ended in tragedy. At midnight on 8 December a fire at the Ballantyne Press destroyed a great deal of the firm's unsold stock, many of the wood-blocks and ornaments which Ricketts had cut, all the labour of the previous six weeks on the Shakespeare edition, and the proofs of the first volume, *Hamlet*, which was already over-subscribed. The initials alone represented the labour of a year. It was a terrific blow to the small firm, and an important factor in Ricketts's decision to close it down, after the books to which they were committed had been printed. This included the monumental edition of Shakespeare which had to be started all over again from the beginning. When the first volume came out in 1900 it bore the image of a phoenix rising from the flames and the motto: VALEO, SED VALE DICO (I am well but say farewell).

Shortly after the fire, there was a knock one evening after dinner on the Fields's door. The maid announced 'Mr Higgins', the name she persisted in giving Ricketts. He came in, white and impetuous. He had quarrelled with Shannon, who complained that he was going on too much about the fire. To the two sympathetic but weary women he poured out all the agitating sides of the loss. As he slowly emerged from his distress, stories began to appear. The painter Watts had been approached by Mrs Kipling, Burne-Jones's sister-in-law, to paint her son's portrait.

WATTS: Let me see, he writes little Indian tales.
WIFE: You will paint him
WATTS: I haven't looked at him yet.... No, there is nothing to paint.... But you are related, I think, to some of these Academy chaps. You should get them to paint you.

Ricketts laughed with glee at Watts's informal nastiness. He himself could not get on with Rudyard's tales, though he had been struck by the 'cleanness of treatment' in them, the quality of facing the facts which he found in Maupassant and his master Flaubert. After talking of Tennyson and their own work, and suggesting that they should read Boccaccio ('even the very naughty stories'), Ricketts went home at about eleven to his 'quarrelling mate'.[62]

[62] MF Journal, 1899, fos. 133^{v}–5.

6 *An 'undiscovered Master' of the nineteenth century*

'SHANNON', asked Ricketts, 'do you think I shall live into the 20th century if I eat any more cream?' They were at The Paragon sharing the Fields's ample New Year dinner, where two servings had become 'pleasure's law'. They had rightly, though with some misgivings, waited to celebrate the advent of the new century until the last day of 1900. However, Ricketts's instincts were all with the past. Edith Cooper recorded: 'Of course Ricketts protested he was of the Nineteenth Century, & always would be—the wonderful century of Keats, of French literature, of Beethoven & Wagner. He had nothing to do with the 20th Century. He is an undiscovered Master of the 19th'. The Fields wished to unite the two centuries in their work. Shannon maintained that he belonged to the new century and savoured the opportunities it offered him, but Ricketts declared firmly that he remained with Beethoven.[1] His instincts proved to be right. As the new century progressed he was to find it ever more alien and uncongenial, and himself in conflict with many of its leading tendencies.

Two deaths underlined the break with the past and the direction of the future. Oscar Wilde died in Paris on 30 November 1900. Even more than the death of Queen Victoria two months later, it marked for Ricketts the end of a period, the end of youth. Ironically, when he first got a hint of the bad news he was attending the revival of *Patience*, that parody of Wilde and the Aesthetic Movement. Absolutely charmed by the operetta—'its clever plot, its smartness, grace and, even though I hardly am able to believe it myself, "Its good taste"'—he saw it three more times that season. But he overheard someone say, 'How sad about Oscar!' It was not until Sturge Moore brought the news the next day that his friend's death was confirmed. In his diary, some of which he later deleted, he wrote:

[1] MF Journal, 1901, fol 5.

I feel too upset to write about it, and the end of that Comedy that was realy Tragedy. There are days when one vomits ones nationality, when one regrets that one is an Englishman.... I know that I have not realy felt the fact of his death. I am merely wretched, tearful, stupid, vaguely conscious that something has happened that stirs up old resentment and the old sense that one is not sufficiently reconciled to life & death.

No sooner had Sturge Moore given them the news than the doorbell rang and Katherine Bradley arrived sobbing in the hall. The news was broken to Shannon just before a meal; all he could say then and afterwards was 'I can't believe it'. At first Ricketts could not discuss the death. He stopped Holmes and Rothenstein from giving accounts and Miss Bradley from asking questions. His grief was poured into his diary. Most of the entries from 6 to 19 December were later torn out, but one entry that survives, on 17 December, records a dream about Wilde that upset him greatly. At the end of the year he noted that the death, 'at first hardly felt ... affects one at stray moments, when one is off ones guard, at Sundown or at Sunrise, moments with me of introspection, hesitation, or regret.'[2]

However, when Ricketts had first met Wilde he had been too young to appreciate him. With youthful optimism, he imagined he would meet many more of Wilde's calibre. As, over the years, this did not happen he began to realize how remarkable Wilde had been. On 25 September 1905, after seeing *An Ideal Husband*, which Oscar himself had criticized severely, he wrote:

Oscar was always better than he thought he was, and no one in his lifetime was able to see it, including my clairvoyant self. It is astonishing that I viewed him as the most genial, kindly, and civilized of men, but it never entered my head that his personality was the most remarkable one that I should ever meet, that in intellect & humanity he is the largest type I have come across. Other greater men of my time were great in some one thing, not large in their very texture. In this estimate I do not include Puvis de Chavannes, whose simplicity and oneness masked a suffering and clairvoyant nature, a man who had no time to waste outside his work.[3]

Ricketts, nevertheless, did not admit any of Wilde's work into the first class, disapproved of his later life-style, which was unseemly for an artist, and had also been irritated by Wilde's flamboyance; after Wilde had inscribed a copy of *The Sphinx* with a flourish that went across the

[2] CR Diary, 4, 5, 17, 31 Dec. 1900 = *SP*, pp. 49–50; MF Journal, 1900, fos 163–7.
[3] CR Diary, 23 Sept. 1905 = *SP*, pp. 124–5.

page, he had exclaimed 'Vulgar beast' and torn it out.[4] His failure to appreciate Wilde's true worth produced a sense of guilt and regret which, together with the feeling that Wilde had been persecuted because he was an artist, prompted Ricketts some twenty years later to write his *Recollections of Oscar Wilde*.

Another death, earlier in the year, produced a very different reaction: 'Read with some pleasure of R. A. M. Stevenson['s] death. Though toothless and belated, almost harmless, he had lived on the works of others. He was insencere and petty fogging to the last, and against all Art in its higher, & creative manifestations.'[5] As art critic for the *Saturday Review* and then for the *Pall Mall Gazette*, Stevenson was sympathetic to the aims of the Impressionists and of the New English Art Club. His book on Velasquez (1895) soon became a classic and, according to Frank Rutter, a bible for art students in the Nineties, and fostered naturalistic and other modern tendencies in art.[6] Moreover, he fathered a school of critics which encouraged these trends, including D. S. MacColl and Roger Fry. To a man of passionate beliefs like Ricketts it was insincere to mourn for an enemy. He felt there was a 'detestable element of weakness and indifference' in the modern notion of the 'impartial dropping of a tepid tear' at the graveside of one whose values and influence you hated. It was only with the greatest difficulty that he restrained himself from directly attacking his living opponents. When he told D. S. MacColl how glad he was that Stevenson was dead, he recorded, 'I felt wretched in my half hearted attempts not to hurt him over much & not to grow irritated.'[7] Ricketts was gradually learning to keep his emotions in check, and at the end of 1903 noted with satisfaction how he had grown in self-control.

Yet, while he could be so cold and ruthless about an 'enemy', he was extraordinarily sensitive and vulnerable to the crash of larger events. War and its cruelty both fascinated and repelled him. He acknowledged to an acquaintance who had drawn up a character analysis of him on the basis of his horoscope that this fascination with cruelty corresponded to an element in his own nature: 'I even admit ... to the possibility of certain phases of "non-brutal" cruelty, though this element is absent in all its practical aspects, yet the fear and hatred of it and a certain insight into it may argue the presence of an

[4] *65 Ill.*, p.[iii].

[5] CR Diary, 19 Apr. 1900.

[6] Rutter, pp. 16–17.

[7] CR Diary, 8 May 1900.

"atrophied" tendency."[8] Not all his friends or 'enemies' would have agreed that this tendency found no practical outlet, though this was relatively rare. Reports of the Massacres of the Legations in Peking in July 1900, for instance, worried his nights. On one occasion he was conscious of hugging thoughts of pictures by Watteau 'to escape from worry in these half awake moments when horror will sometimes persist & hang about one'.[9] Such incidents haunted his imagination and, together with the deep pessimism and insecurity in his nature, made him fear for the future. Similarly, the Boer War affected him personally: 'Feel dumpy over war news. This has depressed me & my work for over a week.' Reports of the primitive cruelty unleashed by war distressed him. Nineteen wounded men were said to have been shot by the Boers, and the pistol shots in the head 'moved and pained' him as nothing else had in the war. The incident conjured up for him 'those few dark drop[s] in ones blood, those few drops left in us all from pre historic times'.[10] However, he contributed to the war effort by donating his drawings to war charities. A drawing for *The Sphinx* was knocked down for five pounds at Christie's (26 Feb. 1900), two pounds less than he would have given for it himself. Earlier he had had to buy back another drawing, *The Autumn Muse*, when he would have preferred to spend his money on old master drawings. In spite of his dislike of British philistinism, Ricketts readily supported his country in its times of crisis.

Though he still carried on with his publishing, his energies turned increasingly towards painting. Early on, Shannon had given him lessons in this new art. However, as Shannon himself had never been trained as a painter by an experienced master, but had learnt his craft out of books, this meant that Ricketts inherited some of his faulty techniques. One of Ricketts's early paintings, for example, given to Holmes, had so darkened by the early 1930s that it was no longer hung;[11] *Tobias being comforted by the Angel*, a painting done for the Fields (Ashmolean), is now almost invisible. In trying to paint like the old masters, they copied the deep golden tone which had actually been caused by the darkening of the final varnish, and was not intrinsic to

[8] CR to Hamilton Minchin, 29 June 1914. Quoted in Edward D. Nudelman, *Catalogue*, 9 (Seattle, Wash., Dec. 1985), 34.

[9] CR Diary, 5 July 1900.

[10] Ibid., 30 Jan., 8 Mar. 1900.

[11] C. J. Holmes to F. E. Jackson, 2 Sept. 1932.

the paint, a fact which has only come to be understood since Ricketts's day. Thus they artificially aged their paintings at the very start, so that natural ageing has darkened some of them even further. Moreover, Ricketts often envisaged his paintings in a low key, which looks very dark to anyone accustomed to the lightness of the Impressionists. When Durand-Ruel (an 'old ruffian' whom Ricketts decidedly liked) saw his paintings, in 1905, he protested that none of the old masters had painted in so black a tone, and implored Ricketts to change his key.[12] This he eventually did, and only the early pictures have been affected by excessive darkening. Without the training that any young French artist of the time would have been given, he had more than the usual struggle with his work. His diary abounds with descriptions of his difficulties. Much depended on his mood: 'out of form, retouched sketches & generaly spoiled things' is a typical entry, his mood being influenced by the weather as well as the events of the world and of his daily life.[13] There were many false starts, but most of the pictures that were completed involved two figures in relation to one another: Job and his wife, Centaur and Child, Jacob and the Angel, the Good Samaritan. These gave him scope to develop the dramatic element that was to dominate his work. Several versions of each were made. An early version of the Samaritan, which prompted Berenson to say that Ricketts might turn out to be the first intellectual artist that England had ever produced, was later painted over, much to the distress of the Fields. They had asked for, and been given, a copy of *Tobias being comforted by the Angel*, which, he told them, they must charge Berenson sixpence to see.[14] Among other early pictures, almost all of which were biblical or classical, was a curious depiction of a centaur swimming, and another of centaurs quarrying. What drew him to these mythological creatures is uncertain, and to the Fields he insisted that the centaurs meant nothing.[15] They are usually seen as symbols of man's animal nature, and in antiquity and in the Renaissance usually depicted as drunken and lecherous, but in Ricketts's paintings, they are found in idyllic or purposeful activities—except in a later painting, *Faust Riding the Centaur* (1909), where the mythical creature may represent man's baser nature. His interest in them owes nothing to Piero di Cosimo's

[12] CR Diary, 10 July 1906, 14 Nov. 1905.

[13] Ibid., 22 Feb. 1900.

[14] MF Journal, 1903, fo. 35.

[15] Ibid., 1902, fos. 27^{v}, 55.

Lapiths and Centaurs, as it predates by at least a year their first sight of this painting on 16 April 1901. No doubt the general inspiration was the numerous classical and Renaissance depictions he must have seen in the Louvre and the British Museum, and possibly *A Dead Poet and Centaur* by Gustave Moreau, the dominant influence on his painting style. Moreau depicted repeatedly many of the themes to which Ricketts also returned again and again, the Good Samaritan, Salome, the Pietà, the Deposition; and there is, moreover, some stylistic influence in the depiction of rocks and cliffs, for instance, in the attitudes of some of his figures, and in the atmosphere of charged drama he often evokes. His early work also shows, in the loose handling, the blurred outlines, and the restricted colouring, the influence of Daumier, an artist Ricketts admired. He and Shannon went to the large Daumier exhibition in Paris in 1901, and in 1905 bought a painting which Shannon said was of 'a man bathing (very nude)'.[16] Ricketts's *The Argonauts*, with its row of naked male figures, was, he felt, without a market, as it could only be hung in a public gallery or, possibly, a billiard room; but the picture was a failure, and only a magnificent sketch survives.[17]

Increasingly, his painting became Ricketts's major interest. As the first step in his new career, he was 'reborn as a painter' at the great exhibition at Wolverhampton in 1902.[18] Laurence Hodson, who was organizer of the exhibition and an early patron of Ricketts and Shannon, had consulted them over his plans. Ricketts liked the way that Hodson had had a gallery 'panelled & papered by Morris and Co.'[19] In the end, Ricketts showed only a single oil, one of the Good Samaritan pictures, along with his books and woodcuts, while Shannon showed five oils, three lithographs, two woodcuts, and nine drawings; but this was Ricketts's first appearance at a major exhibition of painting. He had chosen his moment well. The emphasis of the show was on the work of a rising generation of young British artists such as John, Tonks, Orpen, Steer, Clausen, Nicholson; artists who were mainly associated with the New English Art Club. The critic Frank Rutter counted this with the first International Exhibition of 1898 as 'the two exhibitions of Modern Art which have most deeply thrilled me'. But whilst the International had brought to his notice the

16 CHS Diary, 13 Feb. 1905.

17 CR Diary, 19 Jan. 1900 = *SP*, p. 43 (where it is misdated).

18 CHS Diary, 2 Feb. 1902.

19 CR Diary, 22 Feb. 1902.

work of French artists, this exhibition had shown that exciting things were also being done in England. Contributions from members of the Royal Academy were limited, and Rutter points out that the exhibition was a severe blow to the prestige of this venerable institution.[20] Of course, Ricketts and Shannon were not at that time associated with it, and still found their place among the rebellious younger generation.

When they visited the exhibition Ricketts found that his and Shannon's pictures 'suffered' from the bright green wall, but he noted, 'I think we might work left-handed for 5 years and still top our contemporaries'. Ricketts was never one to undervalue himself or his abilities, and Shannon shared in this approbation. However, this judgement is not as blind and egotistical as it may at first appear. Given Ricketts's canon of taste, one that was shared by few other artists of his generation, it was inevitable. He believed passionately in a kind of subject and treatment that found little space on the walls of this show. If his contemporaries were to be judged solely by principles derived from a study of the old masters, ignoring the discoveries of Impressionism and their consequences, then most of them would be found wanting. His position is clarified if we juggle with time and look at artistic developments through the eyes of an old master, say Titian. He might find some of the objectives of the Pre-Raphaelites odd, but he would be able to understand their work. Similarly, he would find things to admire in Puvis de Chavannes and Gustave Moreau, and recognize them as belonging to the same tradition that he practised. What would he make of the Impressionists? Whilst he might wonder how they achieved the light and sparkle of their pictures, he would not rank them very highly, for landscape was merely the background to the depiction of human life. Eventually there would come a point in artistic development when Titian would become completely baffled. Could he have made much sense of post-Impressionism and its disregard for perspective and fidelity of representation? What would he have thought of the expressive line of Matisse, of Cézanne's toppling bathers? Something that was fundamental to his understanding of art had disappeared, and this was to lead to abstract art, conceptual art, pop art, forms that he simply would not have envisaged as art at all. Nor would he have understood, in spite of eloquent apologies by some critics, how they could claim descent from the conventions of Western-European art. However, he would, perhaps with relief, have acknow-

[20] Rutter, p. 77.

ledged Ricketts and Shannon as his successors, however imperfect their technique and draughtsmanship. By this standard they judged not only their own work but also that of their contemporaries.

Ricketts deplored the loss of the richness, complexity, and subtle balance found in Renaissance art. In the work of his contemporaries he objected to the tendency to sacrifice the exquisite balance of line, tone, and colour that the old masters had achieved, in order to concentrate on one of them, so as to obtain a greater impact. These three elements came to be separated and used for their own sakes. Of the artists at Wolverhampton he wrote:

> None of our contemporaries seem to realise that, even in realism pure and simple, painting should contain more than a single aspect of a thing, that line, tone, and colour should play in and out of the surface, flash and twinkle like a sinking fire ... Then in the matter of stylistic control, they know nothing; the most elementary continuity is attained by mere raw, rapid painting and the chances of a lucky hit.

The artist should be in control of all the aspects of his craft and make use of them to maximum affect. Ricketts stood for an art that was deliberate rather than spontaneous, balanced rather than striking, subtle and modulated rather than simple and direct.[21]

Such criteria he applied to his own and to Shannon's work. When Ricketts, at about this time, set off alone on a trip abroad, he wrote a letter to Shannon in case anything should happen to him. After details of bequests and what should be done with his unfinished writings and pictures, he went on

> Dear old chap, do not fret if you can help it; that is, think I should like you not to fret. When your work worries you, consult the broad emphatic painters, keep your work broad and vivid, well-contrasted and broken from the first. I think ticking suits you best so far and that burnt siena is the colour you manage least well. I think a light warm yellow should be used more in your flesh, most flesh painting is really whiter than you make it, more creamy in tone.
>
> I hardly know what to say that you do not already know.[22]

What is touching in this letter is his deep concern for Shannon's well-being that lay behind the reticence of the last sentence, a reticence all the more sad and ironic in one who so readily expressed his fears,

[21] CR Diary, 19 Sept. 1902.

[22] CR to CHS, 5 Mar. [1903], TS formerly belonging to F. E. Jackson.

anger, and other negative emotions. Here also is Ricketts the perceptive and constructive critic. What is tragic is that it was this very quality in Ricketts that helped to make Shannon hesitant and unsure over his work. Too much criticism can be undermining, even when it is given with the greatest love and concern. Shannon's diary reveals that, while Ricketts was generally helpful, his criticisms could have an adverse affect on Shannon's painting: Ricketts's 'criticism seems to get me', he noted on 30 March 1906. Over this he sometimes became rather touchy, and Ricketts recorded an 'unpleasant dinner with Shannon, who flared up over the finishing of his unfinished pictures': nor eventually would Shannon allow Ricketts to see a work until it was completed. Holmes thought Ricketts had made Shannon too self-conscious and over-educated, so that his earlier, sunny nature had become overclouded with doubts and tensions.[23] There is, perhaps, no need to say that Ricketts would never knowingly have hurt Shannon, but was merely acting in what he saw (in his clear and definite vision) as Shannon's best interests.

Meanwhile, Shannon's reputation was increasing. His portrait of the Hon. Mrs Challoner Dowdall, *Lady with a Cyclamen*, one of his most enchanting pictures, had been well received and widely exhibited. In August Shannon was awarded one of twenty-six medals at the Paris International Show. Ricketts, as always yielding first place to him, reflected bitterly: 'How wise of me not to show work; mine would have been an honourable mention.' All one got out of these exhibitions, he thought, was damaged frames to one's pictures. The night before he learnt of Shannon's award Ricketts spent in sleeplessness because a note in the newspaper announced a very serious fire at Van Wisselingh's Gallery, where two of Shannon's pictures, both unphotographed, were being stored. A permanent anxiety about fire was a legacy of the conflagration at the Ballantyne Press. The next day Ricketts rushed up to town only to find that the 'not' before 'very serious' had been omitted in the newspaper account.[24]

One of these pictures was a portrait of the railway manager James Staats Forbes, a collector, which was rejected by him as not sufficiently like, 'with all the pomp of a grocer returning his photo to a photographer', recorded Ricketts. It was bought by Van Wisselingh, who

[23] CHS Diary, 30 Mar. 1906; CR Diary, 3 Sept. 1905; MF Journal, 1902, fo. 140v. See Frederick L. Gwynn, *Sturge Moore and the Life of Art* (1952), p. 24.

[24] CR Diary, 16 Aug. 1900.

sold it to the Bremen Kunsthalle, the first work by Shannon to be purchased by a public collection.[25] Forbes, however, had already bought Shannon's *The Shell Gatherers* (1894–7), and he then astonished Ricketts by buying one of his Centaur paintings, which was at Carfax, the gallery run by Robbie Ross, who told Ricketts that Forbes had described him as 'such a nice *modest* fellow'. He had never been called that before. He and Ross roared with laughter, until Ricketts found that Ross was laughing too much: 'I then feebly began to explain that my conceit was often mere childish egotism and childish trust in the friendliness and sympathy of others.'[26]

The Fields often noted Ricketts's 'usual fine estimate of himself!'[27] (though this was redeemed by his engaging ability to laugh at this, and other aspects of his character). He knew that he was highly gifted, but there was also a strident, often defensive, egotism that came out whenever he felt threatened. He liked being a centre of attention, and his wit and ebullience usually ensured that he was. He did not like to give much credit to his rivals, or at least he was very selective about which ones he would acknowledge to be on or near his own level of achievement. On being asked to write an article on the revival of printing for the *Encyclopaedia Britannica* in 1900, for instance, he concentrated on the Kelmscott and Vale Presses and ignored all others (including the Eragny) 'with perfect tact & distinctness', he felt, and was 'not a little ruffled' when this rather narrow view was challenged.[28] Earlier, he had acknowledged the *Dial*'s debt to the *Germ*, but not to the *Century Guild Hobby-Horse*. This ungenerous view was compounded of a sense of superiority and one of defensiveness, so that he would only associate himself with the giants, like Rossetti and Morris, but not the lesser figures. However, when he did not feel threatened he was extraordinarily generous, as his friendships with Sturge Moore, Wilde, the Fields, Bottomley, Shaw, Yeats, and many others show. Nor did lack of generosity bedevil his relationship with Shannon, to whose needs and priority he readily deferred.

When achievement did not encroach too closely upon his own, or when a budding talent needed encouragement or guidance, Ricketts was a fervent and generous ally. At the beginning of 1901 William

[25] Ibid., 17 July 1900 = *SP*, pp. 41–2.
[26] CR Diary, 5 Aug. 1902.
[27] MF Journal, 1901, fo. 38^{v}.
[28] CR Diary, 27 Oct. 1900.

Rothenstein showed Ricketts and Shannon some drawings by a young artist, Augustus John. Ricketts was 'hugely impressed', and bought two. He felt that if John 'could concentrate all this real ability into some sort of medium, he would count; as it is, he is one of the most able draughtsmen we English have had of the Wilkie & Millais order; that is, the clever piece draughtsmen'.[29] Some fourteen or fifteen years later, when a cartoon by John was exhibited which angry friends had assured Ricketts was 'farcical', Ricketts found it 'full of effort and sincerity. It shows that "Fundamental Brainwork" which is necessary and rare.'[30]

Wherever possible, Ricketts encouraged young artists whose work he admired. Of course, what was personal or original in their work had to be expressed within the conventions of the old masters. Legros had brought to his attention the work of a young sculptor, R. F. Wells, a pupil of Lanteri. Ricketts and Shannon undertook to purchase one of his things in terracotta, a mother and child. Wells soon called at Richmond (in February 1902) and Ricketts liked him, though Wells knew little about painting and hardly knew who Puvis de Chavannes was. When Wells blurted out that he had never heard of Degas, Ricketts was delighted and begged him not to find out: 'this proved him free at any rate from recent journalistic cant.' Ricketts found the young artist 'unduly attentive to the work of his contemporaries', and wondered whether he would continue to do good work. He was hard up, had a wife and child to support, and got little help or understanding from his family.[31] A few weeks later Ricketts and Shannon marched off to the Fine Art Society to make sure that Wells was included in a show of statuettes they were holding there. Shannon carried two small figures and Ricketts a bronze under his arm. These were substituted for the one thing by Wells that had already been included, while Ricketts and Shannon were 'jawing [the dealers'] heads off at the same time to prevent their giving an adverse opinion of Wells' work, which seemed to make their gorge rise'.[32] At the end of the year Wells sent them a bronze of a baby stooping, which arrived in a straw basket with a chicken and a rabbit. 'It is strange', Ricketts commented, 'that beautiful presents are being given to me at a moment when I feel

[29] CR Diary, 3 Oct. 1901.

[30] CR to Robbie Ross [*c.*1914].

[31] CR Diary, 19 Feb. 1902.

[32] Ibid., 5 Mar. 1902.

so lean and loveless.'[33] This may be the first hint of difficulties between Ricketts and Shannon that were to reach a crisis early in the New Year.

They were to see a good deal of Wells over the next few years. In October 1902 he started working on a bust of Ricketts. He measured Ricketts's face with a pair of wooden compasses, making comments about his personal appearance in a remote, abstracted voice. 'Some of these', Ricketts told the Fields, 'I like well enough, others plunge me into a vortex of astonishment and unpleasant afterthoughts.' Recalling Forbes's comment that he was such a modest fellow, he concluded: 'It is a hard thing for a modest man to be busted.'[34] The result, however, met with Ricketts's approval. It was a good likeness, and caught beautifully his noble brow and his fine sensitive features (Fig. 53). In 1904, when Wells was having difficulty with Van Wisselingh, to whom they had introduced his work, they commissioned a companion bust of Shannon. The two bronzes stood in their houses until the end of their lives, though Wells himself soon passed out of favour. Though able, he was not the man his work suggested. 'I think he wishes to do well', Ricketts wrote, 'but I am at sea over his indifference to the art of others, i.e. of the past or to do anything that does not corroborate his point of view.'[35] Harsher judgement followed. 'I fear he is impossible', Ricketts noted in 1904.

> Character counts enormously in an artist. No downright shit has ever been a great artist. The professional critic will try and disprove that, and say an artist depends on his art or his gift only.... Give an artist all the artistic virtues, his own personality counts in the background, and exceptions like Turner only prove the rule.[36]

How Wells had fallen from grace is not recorded, but his name rarely appears after 1904.

Roger Fry was another friend who became an 'enemy'. While their neighbour at Beaufort Street he had become a regular visitor at their informal Friday evenings when, Ricketts used to say, they were at home to their 'enemies'. At this time there was little to indicate Fry's later advocacy of Post-Impressionism. According to Rothenstein he

[33] Ibid., 12 Dec. 1902.
[34] CR to MF, 3 Oct. 1902; CR Diary, 25 Aug., 6 Sept., 3 Oct. 1902.
[35] CR Diary, 15 May 1902.
[36] Ibid., 9 Dec. 1904.

was 'shy, rather afraid of life, painting in the manner of the early English water-colour painters'. He benefited from Ricketts's and Shannon's connoisseurship: 'he, too', Rothenstein recorded, 'sat at Ricketts's feet; though he was never admitted to the inner circle of the faithful, to which Sturge Moore and the others belonged'.[37] Ricketts was astonished when a two-column article in the *Athenaeum* (23 Mar. 1901) on Shannon's 1901 show of drawings and lithographs, containing the 'most intelligent praise', turned out to be by Fry; 'certainly the last man I should have suspected'.[38] Clearly, he was already beginning to have his doubts about Fry and was annoyed by his 'silly' views and opportunism. At a Vasari Society meeting they had a discussion about whether some drawings should be attributed to Piombo or Michelangelo. 'We had the drawings out side by side', Ricketts noted in his diary, 'I found he could not grasp the resemblance in the underdrawing, or rough indication, the identical variation upon the contours (corrections), and the identity of angle in the diagonal shading. In fact, he has no eye, and sees retrospectively and by prejudice.'[39] A week later he dined with Fry, who showed him

> tosh Italian pictures of the type which should not be collected. Against all efforts to the contrary I found myself badly impressed. His effort to do well is very great, but the type of man is against it. He is minute, punctilious, and without general grasp. The result is a sort of intellectual dishonesty, which is not uncommon amongst the *cultured.*

Fry had been appointed art adviser to the Metropolitan Museum, New York: 'I can't think what sort of mixed and divergent rot he will advise them to buy, unless he follows the tips of Holmes and myself, which I shall avoid giving unless asked point blank. He reminds me of the rich young man who said "What shall I do to be saved?" '[40] Though he later decided Fry was 'gifted with a greater faculty for improvement and development than I originally thought', he still remained dubious.[41] In 1907 he met Fry at Carfax, 'I tried not to tear him to pieces several times; his inborn lack of intuition and sense of proportion have not diminished, but he is less conceited. He says America is awful.' Fry wanted him to ask the millionaire, Sir Edmund Davis, to back the Burlington, which was again in financial difficulty.[42] However, Ricketts

[37] *Men and Memories*, i, 176. [38] CR Diary, 23 Mar. 1901.
[39] Ibid., 25 Oct. [1904]. [40] Ibid., 31 Oct. 1905. [41] Ibid., 14 July 1906.
[42] CR to CHS [1907].

did not always find it possible to suppress his increased antagonism. One Friday a remark by Fry about the French writer Bréal met with such 'dismissive scorn' from Ricketts that Fry vowed he would never go there again. Nevertheless, they maintained occasional contact by letter or on committees.[43]

At the beginning of May 1900 Shannon and Sturge Moore had set off for Paris and Madrid to see the Titians in the Prado. 'These', Shannon wrote, 'surpassed anything I have ever seen in the way of pictures', and he encouraged Ricketts to undertake the journey to Madrid himself, as the travelling was quite easy, 'so long as you overeat ... & refuse to be hurried'.[44] Ricketts joined them on their return to Paris. In spite of his usual nerves and the absence of Shannon, he had had a fair passage across the Channel. Shannon had sent him careful instructions about the journey, showing once again his soothing and practical care of Ricketts:

> do take my advice & buy a book to read, a novel by Zola. It lessens the journey by $\frac{2}{3}$d. You won't believe it till you try—read the whole way & think of nothing else. Remember my instructions re Calais & mind you arm yourself with port. It makes one much younger towards the end of the journey.[45]

Ricketts dutifully re-read Flaubert's *L'Éducation Sentimentale*. He had come to see the Paris Exhibition that attracted some fifty million visitors that year. The day after his arrival he had a 'wild scamper' through the exhibition. The English section had won a good number of awards, but to him it looked 'singularly lifeless' and inferior to an ordinary show at the Royal Academy, because the selectors had been too 'democratic' and tried to include everyone. He liked works by Sargent and Orchardson, and found that Shannon's self-portrait, *The Man in a Black Shirt*, which had been hung well, looked like an old master. Ricketts, though invited, had not sent in. The honours of the show went to the French: Puvis was 'unexpectedly tender and adorable'; Fantin, excellent; Manet, 'at once common and good'. Five pictures by Gustave Moreau, who had died two years before, were exhibited. These pictures, Ricketts felt, were failures, the *Salome* the beginning of the end. Only Puvis had 'all that success of self realization denied to the other great frenchmen of the century, Delacroix, for

[43] *Affectionate Cousins*, p. 208.

[44] CHS to Mr Dowdall, 10 July 1900 (Huntington Library, San Marino, Calif.); CHS to CR [17, 19 May 1900].

[45] CHS to CR [17, 19 May 1900].

instance'. As for Moreau, he had 'all the nobler aspirations of the century in his heart & hardly realising anything at all'.[46]

In April 1901 Shannon cycled with Moore to Wells and Marlborough, and it was while they were on another such jaunt, to Salisbury, Glastonbury, and Winchester, in August, that Ricketts secured another prize for their collection. This was a large, highly-finished drawing by Puvis de Chavannes of *Work*, one of nine 'stupendous sanguines' for the Amiens Museum decorations. So great was his lust for possession that Ricketts 'burst into perspiration' at the sight of it, and bought it for the huge sum of £134. It had been refused by the British Museum, owing to size and cost, but Ricketts was sure they could not have realized how rare and important a drawing of that order was. Only one other of the nine was in private hands; the rest were in the Luxembourg. When Shannon returned, Ricketts said nothing. At dinner, he made Shannon drunk, having slipped into the next room to set up the drawing so that Shannon would see it when he went in. At first Shannon 'blundered' into the room and 'burbled' over his coffee without noticing it. When he turned and saw it standing between two candles, he 'suddenly became quite sober & serious in an instant, he almost white with astonishment and pleasure'.[47]

Not long after this Shannon began to teach Ricketts to ride a bicycle. The old machine that had served to teach Sturge Moore and Shannon was passed on to him. With his haste and impatience he had no more aptitude for the bicycle than the piano, but perhaps he minded not being included when his two friends went off on their frequent cycling trips. He certainly missed Shannon. When Ricketts was 'wicked', Shannon used to threaten to go off cycling with Sturge Moore. On his first lesson, in July 1901, Ricketts did 'unexpectedly well', and in August he was still 'improving wonderfully'. Shannon even bought him a new bicycle. But by October the next year the bicycle was for sale and Ricketts's cycling was over: 'I collapsed with nervous exhaustion at Cambridge', he told the Fields, '& I fear shed tears upon the Trumpington Road & for the first time the bike has passed into history.' From then on, when Shannon 'biked', Ricketts 'trained'.[48] Shannon cycled and played ping-pong or tennis but the only exercise

[46] Lucien Pissarro to Camille Pissarro [June–July 1900]; CR Diary, 17 Apr., 23 May 1900.

[47] CR Diary, 7, 10 Aug. 1901 = *SP*, pp. 65–6.

[48] CR to MF, 24 Oct. 1902; CHS Diary, 8, 22 July, 21 Aug., 2 Oct. 1902.

Ricketts took was walking. He often walked the three miles from Lansdowne House and back when going into town to auctions or the theatre.

In spite of the artistic isolation that was to grow over the years, Ricketts and Shannon always attracted to their Friday evenings at home a great variety of visitors. This very international group included the opera singer Blanche Marchesi, who became a close friend for a while, Toulouse-Lautrec, Count Harry Kessler, a key figure in the art world in Germany, Count Keyserling, and Durand-Ruel, the French art dealer, as well as numerous Dutch, German, and American artists and dealers. Whether drawn by the work, the collection, the conversation, or the chicken and whisky, they provided a steady stream of new interests for Ricketts and a number of friends. People fascinated him and, with the zest of a zoologist on the trail of a new species, he loved to study new specimens of humanity. One of the most intriguing was the Japanese expert Kohitsu, with an ancestry of art experts going back centuries: the Fields described Kohitsu's ancestors as having 'played Berenson to the Shogun'.[49] Binyon, who was looking after him, introduced him to Ricketts at the British Museum. Kohitsu knew nothing of European art, and Ricketts was intrigued to discover that the Japanese found all European painters indistinguishable. To test his response, Ricketts took him to the National Gallery to see the Italian masters. In his diary he carefully recorded Kohitsu's responses; how he was interested in some artists like Fra Angelico and Piero della Francesca, but found others like Botticelli incomprehensible. Ricketts was fascinated by the delightful and stimulating experience of trying to give an idea of the sequence and relative importance of the great European masters to the Japanese expert:

> Under his questions the strain put upon my habitual admirations was not unlike that which a man might experience were an angel to question him about his beliefs, or the reason for wearing our modern clothes. I felt bashful at times about the art of many masters, and some nations even, whilst the greatness of some famous Japanese seemed to him almost too difficult to convey

When Ricketts described the Japanese artist Korin as one of the greatest in the world, it was the turn of the Japanese to be astonished.[50]

[49] MF Journal, 1902, fo. 146^{v}.

[50] *Pages on Art*, pp. 179–80; CR Diary, 19 June 1903.

Kohitsu repaid their hospitality by inviting them to a Japanese meal at his club, where the ebullient Ricketts found it difficult to manage the chopsticks while talking. When the Japanese, who, he decided, was 'a curious self centred litle cuss', went off to the Continent, Ricketts gave him a copy of Berenson's *Florentine Painters* with those passages crossed out that might 'delude' him.[51]

Though he could analyse new specimens of humanity with great interest, there were other things that Ricketts found it very difficult to speak or write about. Whilst he could express anger or anxiety with an abandon that embarrassed him later, he rarely wrote about love or tenderness. Of these there are remarkably few expressions in his diaries or letters, at least until the end of his life. He was aware of this inhibition, and observed:

> It is curious the feeling of shyness one experiences at times with those one loves most. I believe men confess themselves more openly to those who are indiferent to them. This feeling of shyness affects the very character of ones speech even on unimportant matters. It is a curious fact in life that we should hide and guard feelings of which we are most fond, but one makes no disguise of feelings of which one might be ashamed. In common speech between men, they often tacitly admit vices or tendencies which they do not quite possess.[52]

Ricketts and Shannon always addressed each other by their surnames, in the formal manner of the time, when they were not indulging in the equally 'period' habit of nicknames. Shannon was known as 'Chubbers' and, in addition to the more ordinary greeting ('My dear old chap'), Ricketts's letters to him sometimes opened with 'Dear Old Chunkey' or 'Dear Old Rotter and Soaker'. Significantly, most of such teasing was on Ricketts's side. Today it seems stilted and embarrassing but, in that curiously indirect and rather gauche English manner, they were expressions of a deep care and commitment that could be communicated only if they were first trivialized. Similarly, Ricketts wrote and talked little about his relationship with Shannon, the most central and supportive one in his life. When circumstances forced him to deal with it, he found it easier to do so in Italian. Expressing his emotions in a foreign tongue somehow made them more impersonal, so that they could be faced up to and considered.

Early in 1903 Ricketts sensed a threat to their relationship. On 22 January he recorded—in Italian like all subsequent entries—that

[51] CR Diary, 19 June 1903; MF Journal, 1902, fo. 146v. [52] CR Diary, 23 Jan. 1900.

Shannon had said without any reason that he wanted to leave him. Two days later he wrote that his heart and head were empty, that it was curious that the time had passed for regrets and that life seemed empty when ... Here he broke off. The reason for these painful anxieties is finally given at the beginning of February: 'My friend wants to spend Sunday evening with a model called H. D. I hope it is not the beginning of the end of our life together.'[53] Over the months to come this fear must never have been far from his mind, but there are almost no surviving references to it.

However, the anxiety told. Hetty (Esther) Deacon first appeared in Shannon's diary, as 'Miss Deacon, a model', in early January, but she became 'Hetty' before the end of the month.[54] On 6 February Shannon went to Brighton for a few days but Ricketts did not go with him. Whether Hetty accompanied Shannon on this visit or not, she was certainly envisaged as a danger by Ricketts. Such nervous crises usually led to a physical collapse; a few days after Shannon's departure Ricketts felt unwell all day. On the eleventh he was drowsy with influenza and 'susceptible to creature comforts'. When he wrote of his illness to Shannon the latter returned immediately, to find him better but still confined to his room. 'Dear old chap returned', Ricketts wrote in his diary, 'quite concerned at my indisposition ... He brought me a ton of flowers and seemed hungry in his face with affection.'[55]

Hetty's name figures prominently in Shannon's diary of this period. Later in the year he spent a fortnight with her at Broadstairs. Ricketts, who was abroad, sent letters to him full of the usual long descriptions of the pictures and places he had seen, but he also added some boyish teasing: 'I hope you are both good & go to bed early' and 'pinch Hetty in the most pinchable place. Be always good and upright.'[56] When Shannon did not write, Ricketts hoped his silence was not due to 'babies or other accidents'.[57] Afterwards, Shannon often dined with her, took her to the theatre and exhibitions, sometimes with Ricketts, and made her small loans of money. Somehow, she acquired the name 'Smudge' so that references to her in his diary, such as 'Smudge sat

[53] Ibid., 1 Feb. 1903.

[54] CHS Diary, 31 Jan. 1903; the context and sequence of entries suggest that Hetty was later dubbed Smudge.

[55] CR Diary, 12 Feb. 1903.

[56] CR to CHS, Hôtel de France [Aug. 1903].

[57] CR to CHS, Sun., 23 [Aug. 1903].

... took her to the Oxford [restaurant] etc.' become frequent.[58] Though she usually arrived late, and sometimes failed to turn up at all, Shannon used her for most of his pictures of this period, both figure subjects and portraits; sometimes only for an arm or a leg. The portrait Shannon did of her is called *Lady with a Feather*, now in the Accademia, Venice.[59] It shows that she had charming, rather strong, features, large eyes, a wide sensuous mouth, dark-chestnut hair, and a somewhat *retroussé* nose. Although she was physically attractive she did not share Shannon's artistic and intellectual interests, and he was bored to death during a day spent with her when she was ill.

When the Fields were at tea with the two artists some time in 1906, they recorded their impressions of a model who was probably Hetty: 'A girl, obviously Shannon's model, comes in. Fay [Ricketts] has a slight truculence in his respectful greeting and a smile thrown—for the rest he seems a little chafed at the assured voice and too open manners of the girl whose face I think so often spoils Shannon's pictures.'[60] Ricketts's reaction was not the shock of a man being betrayed by his lover, but suggests a prior acceptance of the relationship, however much he may have felt threatened by it. Such liaisons had almost certainly happened before. As early as 1899 the Fields recorded that Ricketts seemed distracted and withdrawn into himself; an indication, they felt, that Shannon was involved with a woman.[61] What brought this relationship with Hetty to an end is unknown, for Shannon ceased keeping a diary after 1906, but Ricketts probably had something to do with it. Shannon later told a friend that he had wanted to get married but that Ricketts had prevented him; Ricketts's argument, according to a tradition in the Sturge Moore family, was that it would be impossible to divide their collection.[62] This probably in fact applies to Shannon's next relationship, but indicates that Ricketts was prepared to interfere and to fight to save their fellowship. If Shannon, however, had really wanted to marry, he would not have been put off by such a reason. It is quite possible that the gentle Shannon realized how much Ricketts needed him, how he was Ricketts's only

[58] CHS Diary, 19 Sept., 1 Nov. 1903, etc.

[59] I am indebted to Miss Mary Laura Gibbs for this description of the painting in the Accademia, Venice, which I know only from a black and white photograph.

[60] MF Journal, 1906, fo. 201.

[61] Ibid., 1899, fo. 82^v.

[62] Shannon's confidante was Mrs Laurence Irving, who told this to Prof. Carl Woodring; information from Miss Henriette Sturge Moore.

real ally in life, and did not have the heart to abandon him. Be that as it may, the relationship with Hetty must have been over by 1907, when another woman presented a much greater threat to Ricketts's and Shannon's life together.

At the end of March 1903 Ricketts and Shannon set out on their second trip to Italy. At the Hôtel Britannique in Paris, where they had stayed at least a dozen times, Ricketts was mistaken for a commercial traveller: 'such is fame & remembrance'. They set out, via Basle and Lucerne, for Milan, Bergamo, and Parma, 'where the night turned the beautiful cathedral into bleached coral, and the baptistery into basalt, the stars were so large they were positively indecent'.[63] Night had a special fascination for Ricketts,

> when the spirit comes out from the past of a place, when the shadows are so black—they are neither blue or green. The air too of an Italian night is so gushing and large in its freshness—the solitude, save round the social lights of several cafes, is complete.

While Shannon was often tired at night on their travels, Ricketts's 'Latin restlessness' came upon him, and he went out for long walks.[64] After a few days in Florence they set out for Rome, their final destination, where they arrived on Sunday, 29 March.

Three things fascinated Ricketts in Rome: 'the perpetual water, constant, generous, large, the vigour of the vegetation, the magnificence & copiousness of the roman ruins & relics, or rather one thing delighted me beyond measure, that is the mere roman brick itself; it looks like old honey-comb plus imagination.'[65] It was the archaeological sites that most interested him, that most stimulated his imagination. He claimed he hated the Rome of the Popes, but was still fascinated by the Sistine Chapel, for, with the halls and chapels of the Vatican, he saw it as 'the largest and most noble expression of continued effort that has come down to our time, revealing as it does the interchange between the artist doing his utmost and the magnificent patron for whom nothing was too fine or too ambitious'.[66] Its 'almost perfect preservation', however, spoiled for him the decorations by Raphael, for the great paintings had sustained many restorations over the cen-

63 CHS to MF, p.m. 20 Mar. 1903.

64 MF Journal, 1904, fo. 29^{v}.

65 CR to MF [9 Apr. 1903] = *Letters CR to MF*, p. 11. This account is based on Ricketts's travel diary (BL Add. MS 58095), part of which is *SP*, pp. 83–96.

66 *Prado*, p. 3.

turies: 'the repainting ... is often so cynical that one does not know if one feels inclined to laugh or cry.' During their week in Rome they kept up a frantic pace, visiting the many churches, the Doria, Borghese, and Vatican galleries, studying the pictures, making notes, comparing the originals with the reproductions they had known before. The necessity of tipping the custodians at every turn Ricketts found irritating and enervating: 'You cannot look at a damaged work of art with human eyes full of twopence watching you all the while.' Though Rome was crowded and uncomfortable, Ricketts liked the city itself, but not always the art: 'what is good is overloaded and smothered by what is bad.' Moreover, there seemed no way of discovering whether the galleries were open, and their 'smart German Hall Porter proved a fountain of fallacious information and we constantly took drives to places which were closed. I do not think the Pope could have known we were coming.'[67] Even the usually dependable Baedeker was unreliable. Then, when the galleries were open, it was often only in the mornings, and Ricketts was 'intelligent only in the afternoons'. On the whole, it was an ordeal, and they left two days early.

Florence, to which they returned, was entirely different—'like a jewel casket'—a place where you could not stay too long. Their favourite spot was the Convent of San Marco, which delighted him 'beyond expression'. However, Ricketts did not care for 'the unpleasantness of live Catholicity'. The church that delighted him least in Florence was Santa Maria Novella, 'at once in the hands of the tourists & the Catholics, a disgusting combination'.

In Florence they dined with Max Beerbohm's friend Reggie Turner (who regaled them with stories about Oscar Wilde), visited Miss Morris (who was also staying in Florence), and had tea with Herbert Horne in his charming rooms overlooking the Arno. Horne was one of those collectors who, Ricketts thought, 'sees entirely with his ears & even then in a crablike sort of way at an oblique angle'.[68] Apart from one or two fine pieces of old Italian embroidery, he had 'one or two wretched pictures by Tomasso Rotto or Bernardo di Fiesole', Ricketts told the Fields. These fictitious names were glances at Bernard Berenson, who lived in Fiesole and who of course was already deeply involved in discovering and attributing early Italian paintings. The real Berenson, whom Ricketts had never met, turned up unexpectedly

[67] CR to Robbie Ross, p.m. 21 Apr. 1903.

[68] CR to TSM, 27 Apr. [1903].

at Horne's, 'pink and panting after one hundred and ten steps'. They were exceedingly polite to each other, 'extremely haffable', recorded Ricketts. Berenson invited them to lunch, and Ricketts teased the Fields: 'I shall eat his lunch and if I don't like it I shall eat him.'[69] Ricketts never missed a chance to chaff them about his rival: when he saw Berenson's photograph on a table at The Paragon he commented that it was a head to be put on a charger.[70] They missed getting lunch at 'I Tatti', Berenson's lovely villa, however, because (owing to the sudden death of his mother) Horne failed to turn up at the rendezvous to take them to Fiesole. Later they called on the Berensons.

> The villa itself is nice. He has furnished and clothed it admirably, beautiful old furniture and mantel-pieces, too many pieces of old brocade perhaps and one or two pictures too many by Bernardino I Tatti, who is a late painter I have discovered [another Ricketts invention]. The drawing room where the Domenico Veneziano (he calls it Baldovinetti) hangs is an ideal room, beautiful and soothing to the eye, filled with flowers and perfume and the hum of insects.

It was the most charming house they had seen so far. Berenson struck Ricketts as 'rather mushy . . . a man who was not sure where he was—who lacked foundation and real seriousness'.[71] Berenson felt that the impression had not been entirely favourable on either side, but he recognized that in Ricketts he had met 'a man who has a universe of his own, always a rare thing'.[72]

On this trip they were amazed by the number of friends that they met, or just missed. In their hotel they found Fairfax Murray dining just a few tables from them. They met the Conders at the station, and learnt that their friends Edmund and Mary Davis had been there that morning. Dining with the Conders in Venice, 'some funny wines were given to us under swell French names'. They had arrived in Venice in torrents of rain, but after that the weather was perfect: 'Venice is a place of moods, sometimes it seems a paradise, sometimes a prison.'

'I should have left Venice in tears', Ricketts told the Fields, 'had not the fuss of the Railway station made me forget to do so.' They stopped overnight at Verona and set off very early for Munich, with a brief pause at Brescia. The journey was horrid. Ricketts confessed

[69] CR to MF [9 Apr. 1903]; CR to TSM, 27 Apr. [1903].
[70] MF Journal, 1900, fo. 58.
[71] Mrs Mary Berenson to MF, quoted in MF Journal, 1903, fo. 49.
[72] MF Journal, 1903, fo. 49.

that he had raged in mixed German and Italian at the guard in such a brutal way that he 'quite shocked & astonished' himself. 'God thereupon prepared a grit', he continued, 'which tortured my eyes from Ala to Innsbruck.' He had to sit still with his eyes shut and miss all the scenery. At lunch in Innsbruck Shannon had to put the food on his plate and cut it up while Ricketts remained 'conceiled behind a handkerchief'. Eating was a matter of hit-or-miss, as he picked at his plate with varying success. Then a young German came to his rescue, 'seized my lids, handled them rapidly once or twice and bade me get up and walk'. He was cured. Ricketts thanked him in three 'mutilated' languages, and later, when the German rose to leave, got up and thanked him again. The young man replied politely, 'Never matter, sar'.[73]

They were impressed by Munich and vastly so by the Pinakothek. There was a room full of first-rate Van Dycks: 'No nucleus of Van Dyck is comparable to this in its sustained level of excellence.' Portraits there by Dürer, Titian, and Rubens were among the finest pictures in the world. Ricketts was fond of making such vast judgements—which convey the excitement he felt before a great masterpiece—but they were not unconsidered. Some months later he was to describe the Dürer self-portrait as one 'before which one forgets all other portraits whatsoever, in the sense that this perfect realization of one of the world's greatest men is equal to the occasion'.[74]

Back in Paris, Ricketts saw Shannon off on 27 April and continued his journey alone by the Sud Express to Madrid. He had been commissioned to write a book on the Prado. Originally, he had been working on a *Dictionary of Italian Masters*, in which the lesser masters were to be allotted 'their appropriate award of space only'.[75] He disliked the scholarly practice of treating all artists as equals, of giving as much space to the lesser as to the great masters. There was also a plan to collaborate with Holmes on a book on the National Gallery, for which Holmes was to write the modern part. Eventually Holmes was asked by Constable, the publishers, to do a book on the Madrid gallery, and he generously handed the commission over to Ricketts, who had thought the Spanish show at the Guildhall in 1901 'one of the best

[73] CR to MF, p.m. 23 Apr. 1903.
[74] *Prado*, p. 10.
[75] MF Journal, 1902, fo. 42^{v}.

shows I have ever seen'.[76] An advance from the publishers paid for his trip to Spain.

In Madrid, he had difficulty finding a hotel, for, because of a medical convention, the city was full. He found a small, rather unsatisfactory room in a hotel 'always full of Germans & women, a combination which does not tend towards Silence & Contemplation'.[77] Eventually he asked for, and got, a better room. His waiter spoke Spanish at him 'very much and very loudly' and Ricketts, still surprisingly timid, was so abashed that when asked for his room number at each meal, he humbly wrote the number down, though he knew how to say it in Spanish.[78] He loved the food, the Spanish cigarettes, and the hot, sharp wine which reminded him of El Greco.

I am happy in Madrid [he confided to his diary], I like the air, the food. I move under ideal conditions (for me), being active, yet active in a sort of comfortable vacuum. Nothing intrudes on one; the streets are broad, the people well-bred ghosts. I like the cigarettes. I have rarely in my life been so happy as I have been here & in Florence.

The Prado at once delighted and disappointed him. There were tons of rubbish, but what was first-rate was really so. However, the Titians did not all please him, and he could not find some of the pictures Shannon had mentioned, as the galleries had been rehung since his visit. Van Dyck bowled him over, as did Velazquez, for he had been prepared by Shannon for the over-cleaning his paintings had suffered. On subsequent visits he found that his first impressions stood up. He made copious notes in the excellent catalogue, hunted up the Spanish and Flemish schools until his joints ached, and finally found all the missing masterpieces in a little gallery all together. On his last visit, he was disappointed to find the gallery was closed because it was a feast day. 'I stroked my moustache with a five-peseter piece, but the attendant became unexpectedly blind; perhaps the magnitude of the sum left him in doubt as to my intentions.' By 7 May he was back in London.

On 14 August 1903 Shannon saw Ricketts off for Berlin. For his book, Ricketts wanted to look up related works in the German and Austrian galleries. Shannon went off to Broadstairs with Hetty for a

[76] CR Diary, 30 Apr. 1901.
[77] CR to CHS, Sat., 2 [May 1903] = *SP*, p. 95.
[78] CR to CHS [29 Apr. 1903] = *SP*, p. 93 (misdated 30 Apr.).

fortnight. Ricketts found the 'vaunted luxury' of Berlin only a legend, 'like the progress of America'. The hotels and restaurants were disappointing, not like 'the awe inspiring, gilded, & glorious haunts of luxury and ease' of Paris and London. While the Rococo and Empire buildings were charming, 'like mantlepiece clocks', there were no book or photograph shops. 'I have not seen . . . a single beautiful or costly thing excepting a Panama hat which cost £8.'[79] The Egyptian collection was marvellous, and there were two or three late archaic Greek bronzes 'that appeal to one's palms and the tips of one's fingers. I think a great deal of my "exceeding lewd soul" (this is like Shakespeare) rests in my palms & fingers'. There was also a wonderful collection of Greek vases, including one 'so rude in subject that I shall have to register the letter which contains the sketch; it is quite obvious that the Emperor can not have seen it, or he would order the museum to be closed forever & Greek art to be forbidden in Germany'.[80]

From Berlin he went on to Dresden, where the galleries were more instructive than in Berlin and the food less good, but he loved swilling the blond beer, 'which tastes like melted topaz, while the sweat beeds off my noble brow and the walls wave about'.[81] Vienna, 'a beautiful phillistin town', seemed at the outer edge of Europe; a sign of which was that people understood his German perfectly 'whilst in Prussia people turned faint as if I had insulted them'. The shops seemed more civilized than in Berlin, and the first book he saw there was a German translation of Oscar Wilde's *Salome*. The Imperial Gallery was simply magnificent, and full of surprises: 'I do not think one knows all Titian, Velasquez, or Correggio without coming here ... Old Breugel simply flattens one out ... Holbein's young man ... (finger in book) brought tears ... the beauty of some of these pictures fills one with a sort of bashfulness'. However, the Academy was 'the worst collection of piffle I have seen outside Italy. Moore ought to be kicked for having advised me to go there.' He did not like Vienna as much as Berlin, and was bored by being told repeatedly by waiters and shop people that 'the town was one of pleasure & so dear'.[82] The people looked stupid:

[79] CR to MF [18 Aug. 1903]; CR to CHS, Hôtel de France [Aug. 1903].

[80] CR to CHS, Hôtel de France [Aug. 1903] and Hôtel de France, Berlin [Aug. 1903].

[81] CR to CHS, Dresden, 25 [Aug. 1903].

[82] CR to CHS, Vienna, Fri., 28 [Aug. 1903]; Vienna, 29 [Aug. 1903].

'I suppose they are realy half Turk. The men have slim waists and fat hips, which I think rude. The ladies wear their chests too low down and their backs too high up, but that is a matter only of local importance.'[83]

On this trip Ricketts enjoyed eating and drinking more keenly than ever before, perhaps as a compensation for the lack of companionship. He missed Shannon, especially in the mornings. He found the white German, Austrian, and Hungarian wines light and delicious, and on his return to Berlin announced that he was 'plunged in the fleshpots of Germany, caviar, peaches, etc.' He was dining in swell restaurants where he had learnt 'to bob automatically, touch my hat & say *Bitte*, in a strangled voice, just like a German'. His only complaint was that he tended to suffer sporadically from sleeplessness, 'owing to the inhuman German bedding or else to excitement over what I had seen'. Because of the good meals, he gained a lot of weight. 'The wise should be round', he wrote to the Fields, 'I myself am growing fat & brown from eating, drinking, & sleeping too much: all practices of wisdom, to which I add seeing too many pictures, which is, I fear, only American.'[84] On his return to England, Shannon could not get over his stoutness, and remarked that he looked like John Lane, 'which is an unexpected definition of the Beautiful', he observed.[85]

By the end of October he had handed in the last proofs of his book. As might be expected, it is no ordinary art book. Ricketts gives his qualifications for writing it as 'a wide knowledge of pictures and their technique and a good visual memory'. His aim was to record his impressions of the paintings as one who was 'unable to recall a time' when he did not enjoy looking at pictures.[86] The book makes no claim to be thorough or impartial and, as with his unfinished *Dictionary*, the artists are allotted space according to their merits. Some paintings—indeed, some painters—are dismissed in a sentence. And, as painting was a personal matter to Ricketts, value-judgements abound: Herrera, for instance, is a 'tedious painter of preposterous pictures' and his brush stroke is like 'macaroni in tomato sauce'.[87] Nor is chronology always carefully respected.

83 CR to MF [31 Aug. 1903].

84 CR to CHS [Aug. 1903]; CR to MF [31 Aug. 1903].

85 CR to MF, 7 Sept. 1903; CR to CHS, Hôtel de France, Berlin [Aug. 1903].

86 *Prado*, Preface, p. [viii].

87 Ibid. 71–2.

It was, no doubt, because Titian and Rubens were so well-represented in the Prado that Ricketts undertook this book. He also drew on his knowledge of their work in British and Continental collections to provide comparisons and fill in the background. Indeed, the book is really an excuse for Ricketts to declare his passionate love for the work of these great masters.

In addition to the painter's knowledge of his craft and a connoisseur's expertise, Ricketts brought an extraordinary verbal felicity to the task. His observations on the austere and melancholy Spanish landscape provide a striking way of describing the character of the Spanish School of painting. He likens

> the Spanish aptitude for what was sombre and solid in art and architecture to the character of the country itself: and ... the contrasting outbursts of florid effort—such as the composite Cathedral of Seville, some late painting of the seventeenth century, and the outbursts of art patronage that rarely lasted more than two sequent generations—to the sudden breaking out of the spring in that country.[88]

Titian's marvellous *Garden of Loves*, with its abundance of putti tumbling over one another, is likened to apples rolling down a slope; its workmanship has the 'edge and crisp tender quality of a petal or a shell'.[89] The French artist David, the founder of 'the heavy, emphatic, but not ignoble classical school of painting in France' with its strict art-regulation, had the 'mind of a policeman'.[90] To express the poignant mood of Titian's *Entombment*, where the seared body of Christ is being laid in the tomb by his disciples, Ricketts writes, 'those who are old are burying the young'.[91] In such telling phrases we can hear Ricketts's own voice, not only describing the pictures but making us share in his own rich and perceptive response to them. His wit is also occasionally apparent: 'I have heard Murillo praised for his *naïveté*: not often has a French word been so misapplied, even by Germans.'[92] At times we can sense the Ricketts who did not suffer fools gladly:

> The principal merits of the Flemish Primitives are a great sincerity, and the rich and instinctively harmonious colour we rarely miss in their work; both qualities attract the intelligent and foolish alike, and the latter are still further

[88] Prado, p. 22.
[89] Ibid., 133–5.
[90] Ibid., 60.
[91] Ibid., 147.
[92] Ibid., 49.

charmed by the elaboration of detail and the cleanliness and gloss displayed in the workmanship.[93]

As a general work on the Prado this book has its shortcomings, but few books ever written about a public collection express a profound appreciation of painting in such vital and evocative language.

What Ricketts admired above all else in an artist was creative power; a quality he thought neglected in his day. It existed in Titian and Rubens (and Rossetti) but not Velasquez; it is the ability to extend not so much the subject matter as the range of expression in painting, with its concomitant development of technique. It is not just originality, which is too often merely a search for novelty, a gimmick or an unheard-of subject; and it applied only to figurative art. For Ricketts, painting widened the viewer's experience and perception of life:

It is incredible what we owe in the matter of vision and experience to the greatest masters. Without Rembrandt, for instance, a whole range of emotion would have partly failed us, or would at least have been seen and known less readily, and we might not understand those beauties which are the fruits of experience and disillusion; we might have remained indifferent to the tide-marks of passion upon a human face ... Without Titian other experiences might have failed us, and that deep sense of crisis we each carry in us at times in our life might not have found its visible expression; we might have remained insensible also to the beautiful and unique significance of outward things, accepting them more casually.[94]

The second thing Ricketts valued in an artist was a nobility in the artist's thought or aim, in the values he expresses; what was coarse or ugly had no place in Ricketts's conception of great art. Sturge Moore recalled that Ricketts 'could not forgive Holbein for painting anyone so ugly as Archbishop Warham ... And the portrait of a man with a bulbous nose in the Prado counted against the same master as though his sitter's defect had been his invention; Raphael would not have accepted the commission.'[95] Thirdly, there was the skill to express fully the artist's meaning. This included the intellectual effort inherent in design and in draughtsmanship as well as technical proficiency with the brush; but both of these must be subservient to expression, as must be the use of line, tone, and colour. Technique was never an end in itself; a misconception characteristic of Impressionism, Ricketts felt.

93 Ibid., 154. 94 Ibid., 150. 95 *65 Ill.*, pp. [ix–x].

In both Titian and Velasquez, moreover, Ricketts praises the unstinting effort at perfection throughout their lives, and the ability to sacrifice parts of a painting that are irrelevant to the whole. It is no wonder that he found so many of his contemporaries wanting.

The Prado was published on 6 May 1904, just after news of a fire at the printer's had sent Ricketts into a panic—but there was no damage to his books. As might be expected, he was concerned that review copies be sent to the right places. Since Holmes was keen 'to puff it' in the *Burlington*, a copy should be sent to that 'pompous publication' instead of to the *Saturday Review*. There 'I have a real, red hot enemy & old friend in D. S. M[acColl]', Ricketts told his publisher, 'I do not think he would dare to bark or bite (I usually bite him) but he would damn the book on its format or character or illustrations, &, if he wants to be unpleasant, he will look it up at Kensington & you will be richer by a review gratis'.[96] When Binyon, who later said it contained 'some of the finest and acutest criticism of art ever published in this country', sent an appreciative letter about the book, Ricketts replied:

> I am delighted & flattered by your charming letter which made me glow & blush with pleasure, even down to my very navel. I agree with you that the Rubens chapter is the best & that the Raphael page or two count as the most significant and new portion of the book. The Greco portion was easy. I think the Velasquez chapter turgid and tiresome. His life and surroundings bored me. The Titian chapter is also not up to the mark though it seems to please people, Titian being quite the latest fashion just now.[97]

The critics praised Ricketts for his independence of judgement, his readiness of perception, his breadth of view and catholicity of taste, and criticized the incoherent form and some of his attributions. There was an almost universal feeling that this was an important contribution to contemporary understanding of the great masters treated. One of the best reviews, incidentally, was by Roger Fry in the *Athenaeum* (6 Aug. 1904). The book had a *succès d'estime*, but did not make the money Ricketts had hoped for: he said it 'killed the four-guinea art book'.[98]

Other planned books, like one on Watteau, did not materialize. A volume on Raphael was abandoned because it would provoke so much

[96] CR to Mr Killmann; a draft of this letter is in CR Diary, end of Apr. 1904.

[97] Binyon's review appeared in the *Saturday Review* (9 July 1904); CR to L. Binyon [May–June 1904].

[98] TSM to Kerrison Preston.

hostility; for a number of attributions then made to Giulio Romano were to be given back to Raphael, and most of what Morelli had said would be 'unsaid'.[99] Ricketts was learning not to be too provocative. In *The Prado* he kept himself in check, and curbed any extravagant declarations or side-swipes at 'enemies'. When the Fields wondered why Ricketts was taking time from his painting to write, they were told that there was money in it. Ricketts hoped to earn an income from writing which would permit him to carry on with painting and collecting, just as the Vale Press had done previously. This new worldliness in him was attributed by Sturge Moore to the influence of Berenson, whose books had made him not only an authority but, indirectly, a rich man.[100] Ironically, Ricketts, who could write beautifully, was never to achieve the success of Berenson, who wrote so badly that his wife had to rewrite everything. The massive catalogues that Berenson published were more useful as reference works to scholars than Ricketts's books. Those, however, who want to enhance their love, understanding, and appreciation of painting will turn to Ricketts.

Between 1900 and 1903 Ricketts also carried on with his publishing. After the fire, there had been more than three years of work ahead before the current commitments of the press could be completed. Perhaps because of the destruction of so many of their decorative devices, these later books tend to be spare in appearance, more reliant for their effect on the simple relationship of type to page. This spareness might also suggest that Ricketts, always rather restless, had lost interest in book design. However, though he gave some of his designs to be cut by a wood-engraver, he still cut a few new borders himself. Three were designed for the huge Shakespeare edition, one each for the comedies, tragedies, and histories, together with a large number of half-borders; but, for the first time, the latter were cut professionally in brass rather than in wood. He was also engraving five illustrations, mostly old designs, for the Latin edition of *Cupid and Psyche* and, inspired by Millais, he did ten woodcuts for *The Parables* (Fig. 23). Each of these took about three days to engrave. The results pleased him, but Shannon did not approve of all of them. His most ambitious project was the designing of an entirely new fount, the King's Fount, in which he tried to free the lettering from some of the innovations

[99] MF Journal, 1904, fo. 12^{v}.
[100] Ibid., 1903, fo. 163.

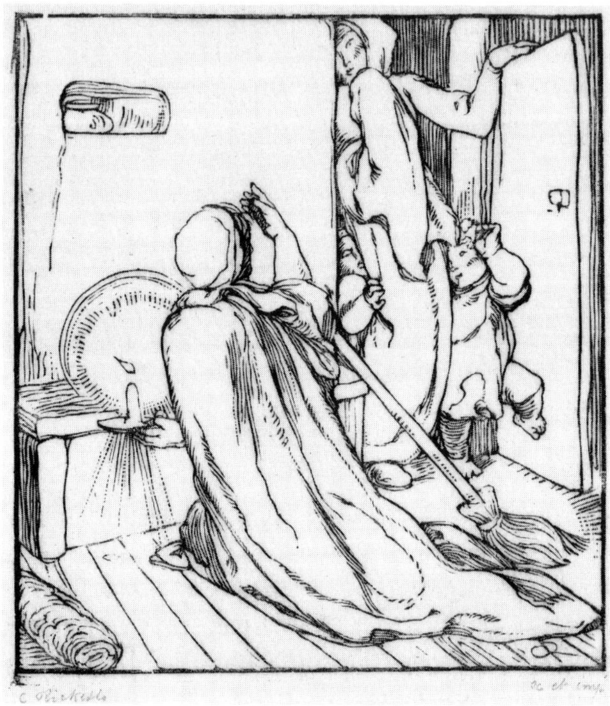

FIG. 23. *The Lost Piece of Silver*. Proof woodcut designed and cut by Charles Ricketts for *The Parables* (Vale Press, 1903; Colnaghi's)

made by scribes to facilitate their work. First used in the *Ecclesiastes* in 1902, it was, of the three founts he designed, the one he loved best, probably because it was received with almost universal execration. At Ballantyne's there was a virtual strike over their vellum books, which Ricketts had to smooth over. Then there were difficulties with the Shakespeare edition. Sturge Moore, the editor, resented the 'interference' of Holmes, and Ricketts had to mediate between them. On the receipt of an 'emphatic and pompous and irritating' letter from Sturge Moore about Holmes, Ricketts replied accusing him of 'high stumick', & reminding him that Holmes could be of assistance to him, and pointing out that the archaic abbreviations which Moore favoured and which Holmes apparently had emended were 'odious'.[101] In his atobiography Holmes only says, 'over the Shakespeare proofs I got into hot water once or twice with the patient editor Sturge Moore'.[102]

Ironically, the Vale Shakespeare, for which Ricketts designed a special fount, also led to the only row that Holmes had with Ricketts. It entailed the issue of thirty-nine volumes between 1900 and 1903, in addition to their other books, and increased Holmes's burden considerably. He asked for an extra office-boy. 'Ricketts, usually generous almost to a fault', Holmes recalled, 'on this occasion, as on one other connected with payments to an editor, proved inexplicably blind. On the previous occasion, I had succeeded in convincing him; now I failed.' Ricketts could see no need to pay ten shillings a week for another assistant when Morelli, the present office-boy, was perfectly capable.[103]

The inception of the row was noted in Ricketts's diary on 17 September 1901: 'An almost unaccountable quarel with Holmes in which he was unable to conceal his contempt for the Vale books, or to see them in any other light than as a speculation advantageous to the bookseller.' Ricketts was prone to consider any disagreement as a personal attack, but never did his touchiness produce a shot wider of the mark than this. Holmes had not only managed to rescue the press from insolvency but, by suggesting two alterations to their list, had brought them an additional £2,000. To ascribe to Holmes a 'contempt' for Vale books was absurd, and illustrates Ricketts's lack of judgement when he felt threatened. However, at this time he did not realize how

101 CR to TSM [*c.*1900].
102 Holmes, p. 175.
103 Ibid., 174.

great the financial success of the firm was, and could only see Holmes as another 'sad specimen of our modern middle classes corrupted by culture' and a dangerous friend for the future.[104]

The row came to a head on 10 October. Holmes so lost his temper that he gave notice that he was leaving. Ricketts was shattered. The same sensitivity and vulnerability that had produced so strong a reaction against Holmes now turned against himself. Filled with fear of change and of uncertainty about the future he fell ill, and he gave such raw expression to his feelings in his diary that he afterwards tore out the whole week following the row. Even a week later he was 'still a little feverish, partly I think from the vague gusts of anxiety that the matter is not quite arranged'.[105] When further discussions showed that their financial position was far better than Ricketts had realized, he not only granted Holmes's request but also increased his salary to £300 a year.[106] Two months later, looking back over a largely happy year, Ricketts wrote: 'A shadow has crossed the end of the year in the breaking up, to me the unaccountable breaking up, of an old friendship, & the discovery that after ten years, one was still and had always been an utter stranger.'[107] The isolation of his inner self was always ready to overwhelm the outer man.

During the months he was working on *The Prado* he was also overseeing the winding up of the Vale Press. Its last publication was Sturge Moore's *Danaë*, with three woodcut illustrations by Ricketts. He had hoped to complete his list of favourite books with an edition of Swinburne's *Poems and Ballads*, but the author refused permission.[108] Though on 22 May he wrote, 'Last Day at the Ballantyne', and underlined it, various other tasks occupied him until the end of the year. He was also preparing a bibliography of the press, for which he engraved a woodcut frontispiece based on the sign that Shannon had painted for their first shop in Warwick Street. In this book he included only the publications of Hacon and Ricketts, and he outlined the aims and achievements of the press. It did not appear until 1904, although the office of the press had closed on 23 June 1903. Some of the office

[104] CR Diary, 17 Sept. 1901.

[105] Ibid., 17 Oct. 1901.

[106] Holmes (p. 174) says £200 but Ricketts's contemporary record says £300 (20 Sept. 1902).

[107] CR Diary, end of Dec. 1901.

[108] Holmes, p. 175.

furniture was sold to the Pissarros; the founts were melted down and the lead sold to pay for the *Bibliography*; and they dropped most of the punches and matrices into the Thames, as Morris had done with his. Wood-blocks which had survived the fire or were cut subsequently were later bequeathed to the British Museum.[109] On the last day of 1903 Holmes called to give up the books, and Ricketts gave him a Centaur painting, as a token of gratitude.

The closure of the press upset Ricketts far more than he had anticipated. He felt sentimental, melancholy, full of regret, even though he realized that his life was starting up all over again. His attitude to Hacon was equally mixed: 'Hacon nervous with sense of indebtedness and general crookedness', and 'Hacon again apologised for being a drag on firm resources', he recorded.[110] This resentment towards Hacon, whom Ricketts had once dubbed 'a mere collectioneur', was caused not by a quarrel, as with Holmes, but by a sense that he was benefiting overmuch from Ricketts's skill and achievement. Here Ricketts minded sharing the profits, not the credit, from his work and was glad when their accountant, 'dumbfounded at the success of the firm', which had had a turnover of £36,000, found the arrangement hardly fair to Ricketts, and pointed out that once Hacon was paid out and his interest paid, he could not see what he left behind him in the firm 'to count against Mr. Ricketts's work & brains'. There is no doubt that Ricketts's ability in book production made the firm's success, but it is difficult to see how it could have been set up without Hacon's help, just as it is impossible to imagine how it could have done so well without Holmes's management. Hacon was never more than a sleeping partner, but his capital investment had had to be repaid with interest. The return on this was high, but Ricketts too had benefited. The press had given him a comfortable living for four years and, even though the final division was somewhat to his disadvantage, he still found himself in a position to say that 'with caution there is enough now for three years fodder for me'.[111]

Moreover, there were signs that the closure of the press had been well timed. Following the Boer War there was a drop in the prices of

[109] BM Acc. No. 1937:323–552. There are 229 woodcuts by Ricketts and Shannon in the British Museum.

[110] CR Diary, 24 June 1903; see also *SP*, p. 106, n. by TSM.

[111] CR Diary, 24, 25, 29 June 1903; Holmes, p. 211.

Kelmscott and Vale books.[112] As early as May 1902 Ricketts was distressed to learn of 'a fall off in value & sale of Vale books'. The next day he rushed to town 'to see Holmes and argue down panic' and that night he had a dream that Vale books were selling at auction under their published price and he was incapable of preventing it.[113] A year later Holmes told him that several Vale books had sold under price. At the next sale the slump was confirmed, and Ricketts himself bought in most of the vellum books.[114] In future, he made a practice of buying up any Vale books that were going too cheaply at auction.

'The Vale Press died this week', wrote Michael Field in June 1903, 'and the death will be hard. None can leave the death bed, neither Holmes nor Ricketts.'[115] There were plans afoot for a new firm, for which Holmes took the initiative. Ricketts was hurt and surprised to learn that he had approached Hacon about this first. A first project by Holmes to purchase the *Art Journal* 'at a tall price' was disliked by Ricketts, who felt the magazine was 'discredited', but he approved of a plan for a new limited company. Holmes had found a barrister called Squire who was willing to put up £5,000 but when Ricketts interviewed him he was disappointed, as 'a father has been sprung upon us who has to be consulted'. They approached Edmund Davis who, to their surprise, offered to advance £1,000 for the new firm; a 'red letter day', Ricketts noted in his diary.[116] He told the Fields that Davis was going to start a journal to be called the *Glass* or the *Frost* or the *Bun*, and jotted down two pages of ideas for articles in his diary.[117] Nothing came of all these negotiations.

Holmes, however, was soon drawn off into other ventures. With Roger Fry, he managed to rescue the *Burlington Magazine* from financial collapse and became joint editor as well as Managing Director. At first, Ricketts saw this as no more than a delay to their scheme, but it really proved to be its death. However, with Holmes at the helm, Ricketts went on the consultative committee and became a regular contributor, producing some eleven articles between 1904 and 1909. In that year Fry, who took over as editor from Holmes, published an article on

[112] *65 Ill.*, p. [v].

[113] CR Diary, 26, 27 May 1902.

[114] Ibid., 20, 27 May 1903.

[115] MF Journal, 1903, fo. 73.

[116] CR Diary, 30 June, 17 July, 29 Sept., 2 Nov. 1903.

[117] MF Journal, 1903, fo. 25^{v}.

Cézanne that so incensed Ricketts that he resigned from the committee, and refused to have anything to do with the *Burlington* as long as Fry was involved. He never contributed to it again.

Hacon and 'Ryllis soon disappeared from his life at this time. 'Ryllis had grown fat and Hacon was drinking too much. They took to quarrelling in front of guests. At the time of the press's closure Hacon made a confession of failure to Ricketts, 'the failure of affection, lust, companionship, daily habits'. 'I believe', Ricketts recorded, 'I said things I did not quite believe as consolation, though I have felt also at an age when one puts ones house in order.'[118] In about 1906 the Hacons, finding London too expensive, moved to Darnoch in Scotland, where they had built a house and he could enjoy his favourite sport, golf.

The Hacons' marital difficulties must have confirmed Ricketts's hostility to the idea of marriage for artists. Marriage was a form of death to him. The time for creation was short and should not be wasted on family responsibilities, which took three or four years at least away from work and achievement. Adding, perhaps, a touch of misogyny to male assumptions of the period, he felt that in marrying a man allied himself with an inferior mind which then had access to his every thought: 'as soon as a woman is concerned in a man's life, it becomes unintelligent and trivial from the senseless marriage festivities onward. Marriages ought to be forbidden by law!' No doubt he never thought of applying these strictures to his own mother, whom he loved and with whom he had more in common than with his father; and since he was addressing these remarks to the Fields, he may have exaggerated somewhat—but it was a view he often expressed.[119]

At this time there was a spate of marriages among his acquaintances. In September 1902 Holmes wrote to say, 'I hope you will not be horribly surprised and shocked if I tell you that I find myself most unexpectedly and suddenly engaged to be married'.[120] Ricketts was stunned. He recorded in his diary (and afterwards attempted to delete): 'A sudden letter I had not expected. A touch of pain in an old forgotten spot, the good reason now arises to quite bury away an old [illegible word]. A touch of age! A touch of regret! Have written . . . congratulations.' Next day he broke into Italian to describe how his

118 CR diary, 25 June 1903.

119 MF Journal, 1903, fo. 165.

120 C. J. Holmes to CR, 22 Sept. 1902.

feelings were both placid and painful, how he had passed a sad night full of memories.[121] Even more than a month later his emotions were still on the boil. He awoke at dawn, full of regret about an

> event that distresses me ... It stirred my blood and brought tears for the first time. I had apparently been suffering in my sleep over that which I am able to control in the day time, & the return of old, dead feelings, & an almost passionate regret, obliged me to get up and look at the dawn; it was as if the last 3 years had been without effect.[122]

Such outbursts indicate that Ricketts had been in love with the young Holmes. There is a hint of this later, when Holmes, who (incidentally) resembled Ricketts's father, was elected to a chair at Oxford, and Ricketts congratulated his wife: 'No one so young and beautiful so far has been Slade Professor.'[123]

However, it was not for these personal reasons only that Ricketts had reacted so strongly to Holmes's marriage. Other marriages, where there was less personal interest, produced the same shock, though without the same painful emotions. Laurence Binyon wrote 'a charming letter' in May 1903 to announce his engagement. Even the faithful Sturge Moore became engaged, 'against all expectation', to his cousin Marie Appia. He had been unable to tell Ricketts the news in person, though they had met since the engagement, as he feared that Ricketts was not 'in the right mood to hear such a shocking confession'.[124] Feeling 'quite dazed', Ricketts wrote congratulations, and assured Sturge Moore 'my old affection for you is so sound that your marriage can make no difference, at least to me, & I am looking forward to seeing your bride and falling in love with her myself'.[125] He approved of her, because at least she was French. Then even their new servant went off to get married. Ricketts complained to the Fields that he had no one to talk to now except Robbie Ross: 'I wish Chamberlain would tax our daily marriages.' The Fields commented that Shannon no longer needed to threaten to go off cycling with Moore when Ricketts was 'wicked'. Now, he would threaten marriage.[126]

All these weddings entailed presents. For the Sturge Moores

[121] CR Diary, 22, 23 Sept. 1902.

[122] Ibid., 30 Oct 1902.

[123] CR to Mrs C. J. Holmes.

[124] TSM to CR [13 July 1903].

[125] CR to TSM [July 1903].

[126] MF Journal, 1903, fos. 212^{v}, 165, 105.

Ricketts, after he had got over the shock, offered (with his usual ready generosity) 'to turn jeweller on my own account and make for your wedding the pendant of pendants which will tear all lace and scratch babies'. The design was based on one of Ricketts's illustrations for *Cupid and Psyche* of the Descent of Psyche into Hades, a theme—the love of a woman for a god—that fascinated Sturge Moore. It was completed in March 1904, and Ricketts liked it.[127] This was not the case with the jewel he produced for Mrs Binyon, a version of the blue-bird brooch, in white with a blue spray in its beak. It arrived back from the jeweller's, Giuliano's, where Ricketts had most of his designs carried out, 'such an unrecognizable lump of white enamel & hideous gold studded with gems (just as a bun is studded with currants) that it was handed the afternoon it arrived to a model whom I rather dislike, and the four pale sapphires & one chrysolite like a star will glimmer unseen in the studios of sundry bad landscape painters'. The model was Hetty Deacon.

'There has decidedly been a spell over these connubial jewels of mine', Ricketts told the Fields, 'which has cost me trouble and sleepless nights, instead of the respective bridegrooms.'[128] After 1904 Ricketts abandoned the art, as it was proving far too costly. The pendant for Mrs Sturge Moore, for instance, had cost forty pounds (though Ricketts assured Shannon it was only twenty-five). Nevertheless, Shannon calculated that £100 had been spent in a single year on gifts of jewellery.[129] Besides, the execution of the design always seemed to coarsen it, and a piece of jewellery was all too easily lost; indeed, this was the fate of a second piece of jewellery for Mrs Binyon.

With the closure of the press and Ricketts's first appearance as an oil-painter, 1903 had been a year of endings and beginnings. Another great change had already occurred in their lives; they had left Richmond. The departure was painful for Ricketts, who was 'broken-hearted', he told the Fields, while Shannon was 'panting to go'. Ricketts hated change, thought they would find that the four happiest years of their lives had been at Richmond, and feared that by moving to central London they would lose their privacy, by being more accessible to others and to the attractions of music and entertainment.[130]

[127] CR to TSM, Sept.–Oct. 1903; Scarisbrick, p. 168.

[128] CR Diary, 1 May 1904 = *SP*, pp. 106–7; CR to MF [24 May 1904].

[129] *SP*, p. 107, Scarisbrick, p. 169 n. 62.

[130] MF Journal, 1902, fo. 23.

On their last visit—before their departure—to the Fields, who were greatly distressed at losing them, they brought 'a sacrifice to the poets'. It was an exquisite silver sugar-bowl which had stood on their central satinwood table. 'We bought it in Richmond, we have always taken sugar from it at Richmond; we felt we must leave it in Richmond', they declared. The Fields were deeply touched.[131]

After weeks of clearing and packing, they were ready to move. The move was so well prepared that it went more quickly and smoothly than anticipated, and the furniture suffered merely insignificant scratches. Only their servant Marie 'behaved outrageously, falling ill, out of wantoness, ditto the cats, who wailed loud & long and still lift up manifold toned voices, if spoken to through the key hole; when let out, they behave like a tragedy by Euripides, all tears'. He told the Fields: 'I am still somewhat ruffled in my mind, as if I had crossed the Channel, but much of this should be put down against Marie on the judgement day.'[132] The first day in the new home was 8 May 1902, the Feast of the Ascension of Our Lord. As there was a lift to carry them up to their flat, Ricketts jested that he would be making his own ascension on that day.[133] The Fields were shocked.

[131] MF Journal, 1902, fo. 67.

[132] CR to MF, p.m. 9 May 1902.

[133] MF Journal, 1902, fo. 53^{v}.

7 *The 'House Beautiful'*

THE 'House Beautiful' like the 'Book Beautiful'—the inversion was essential—was one of the passions of the Nineties, and one to which Ricketts and Shannon subscribed. In spite of their association with the Arts and Crafts movement, however, it was not an 'Aesthetic' house that they made for themselves. Rather than furniture designed by themselves, by William Morris, or by the other Arts and Crafts designers, they preferred to furnish their houses with antiques. As soon as they could afford it they had replaced the scrubbed wooden chairs and tables of the Vale with furniture from the eighteenth century, the Regency, and occasionally from earlier periods, which better suited their catholic taste in pictures and *objets d'art* (Figs. 24 and 25). Aristocratic elegance was their hallmark rather than folksy craftsmanship or Art Nouveau extravagance.

To one of Ricketts's sensitivity, refinement, and visual preoccupations, beauty in one's surroundings was essential. It was a temperamental need, and never merely an attempt to impress, or to make up in possessions for any deficiency in his personality or his artistic gifts. In some ways, however, it can be seen as an escape from life. Beauty not only created a pleasing setting, but also a haven of order and delight away from the shocks, strains, and disappointments of existence. These often overwhelmed Ricketts, whose vulnerability was the obverse of his sensitivity to objects of beauty. Ricketts's and Shannon's houses became an artificial world where they could enjoy each other's companionship, contemplate their exquisite possessions, receive their friends and 'enemies', and carry on with their work. This need for a secure setting explains Ricketts's fear and hatred of moving house, or of any change whatsoever. Change meant trouble, uncertainty, unpredictable consequences. His unsettled youth and his deep pessimism made him see any change as for the worse. However, while his beautiful setting could often soothe, protect, or delight him, it was really no proof against the misfortunes every man is heir to. 'Lives devoted to beauty usually end badly' Kenneth Clark commented, and

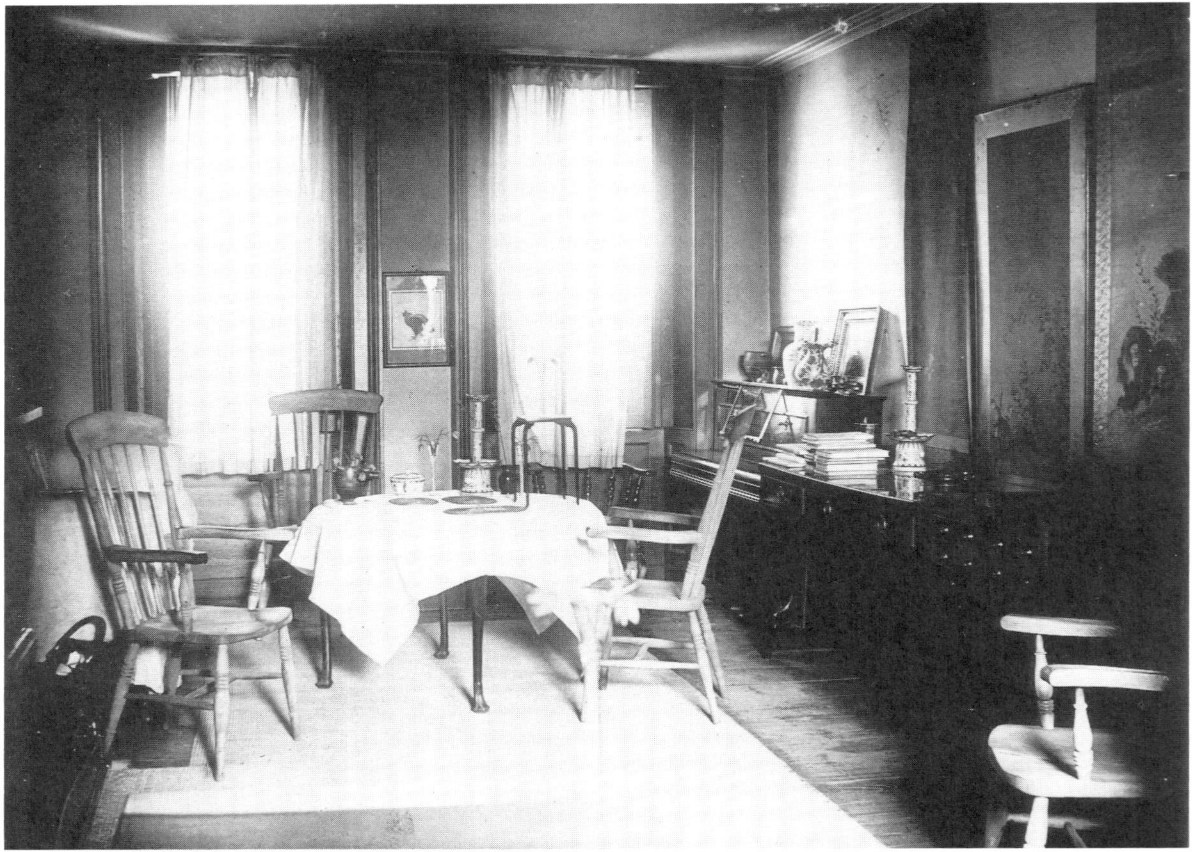

FIG. 24 The Vale, dining-room (*c.*1890)

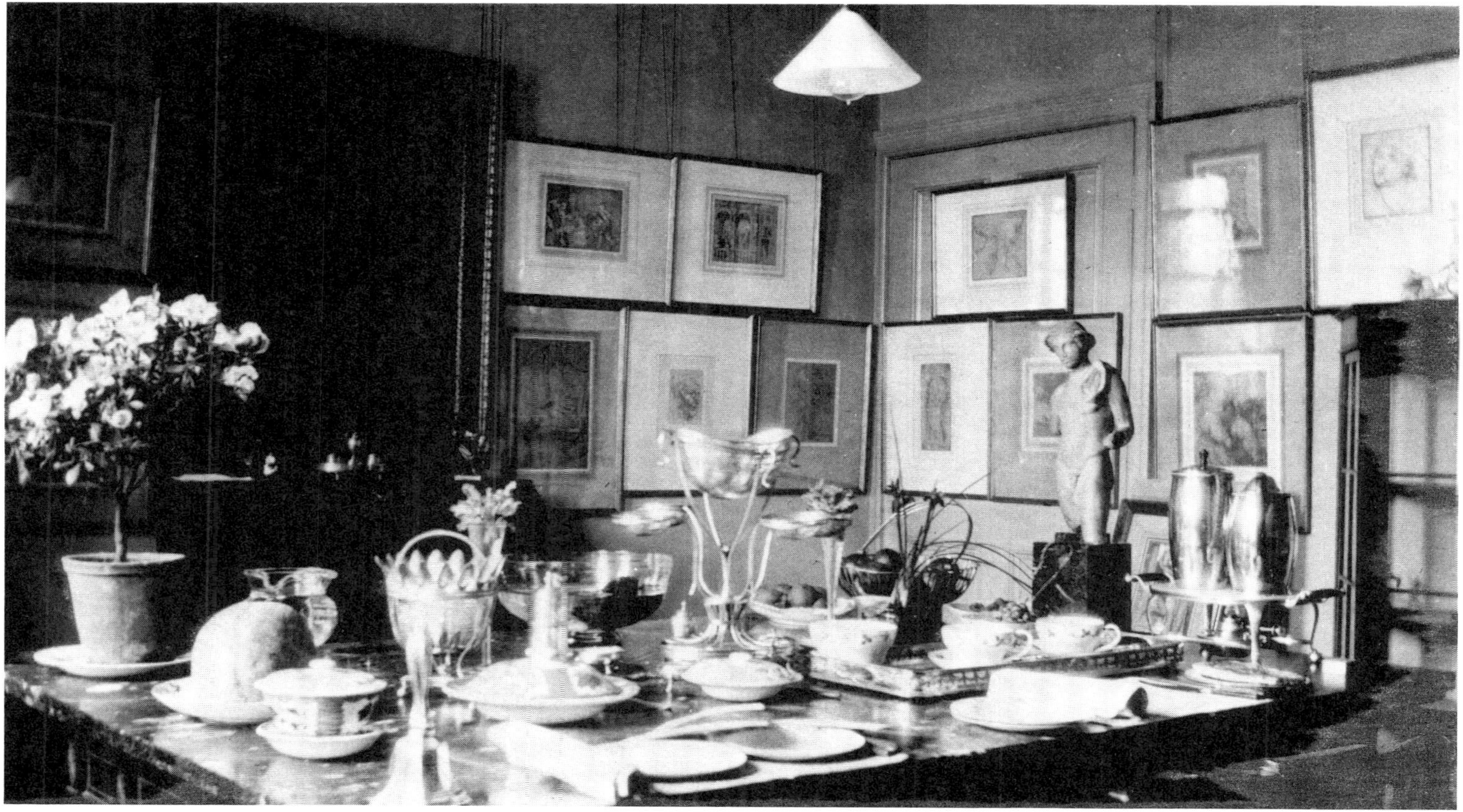

Fig. 25. Lansdowne House, dining-room (Fitzwilliam Museum)

whilst this may not always be true, he was right to apply it to Ricketts and Shannon.[1]

Their new home, Lansdowne House, was the brain-child of their friend and patron Edmund Davis, who had made money in South African gold mines. His wife Mary had been given art lessons by Shannon. A motherly person, she brought out the best in Ricketts, who described her as 'a delightful & tender hearted woman; I think on the whole the kindliest nature I have yet met,' and his letters to her are among his most personal and delightful. When she eventually insisted on being addressed by her first name he wrote, 'I hope your disclaimer of the Mrs. does not mean that you are not married to Edmund?'[2] Soon Ricketts and Shannon became unofficial and unpaid advisers to Davis on works of art, which he purchased, one suspects, as much for investment as for enjoyment. As a millionaire, he was often able to buy things that they had found but could not afford; like a marvellous French eighteenth-century bust of a woman that they discovered in a shop near Piccadilly. Its price of eighty-five pounds was beyond their means at the time, and Ricketts for a moment even contemplated selling the rare vellum copies of some Vale books in order to raise the money, but he knew Holmes would not permit such a rash act. So Davis was told, and the bust turned out to be by Houdon. Davis was later offered £20,000 for it.[3]

In return for such generous expertise, Davis bought a number of their own works, and offered them a flat in a new block he was building, not far from his house in Lansdowne Road, Holland Park. The block was conceived as a home for artists, and consisted of four large flats, each having a studio with a good north light. It was tall for 1902, and dominated the skyline. Ricketts and Shannon were given the top flat, built on two floors, on a twenty-one-year lease, for which they paid a low rent. From the beginning they had been consulted, and during construction they had clambered up ladders and suggested alterations. Thus adapted to their needs, it was to give them more scope for creating a palace of art and beauty than they had ever had. Holmes was so impressed that he hardly knew how to fit his behaviour to such space and magnificence.[4] Even the removers were so overawed that

[1] Clark, p. 179

[2] CR Diary, 31 Dec. 1903; CR to Mary Davis [1915–16]

[3] CR Diary, 30 Jan. 1901 = *SP*, p. 52.

[4] MF Journal, 1902, fo. 140^{v}.

they slightly damaged the satinwood furniture in minding the parquet floors. The Fields dubbed it 'the Palace'.

Long before the move they had been collecting satinwood furniture, Hepplewhite chairs, Adam mantelpieces. They also secured a superb ebony and tortoiseshell cabinet which they had advised Davis to buy at the Battle Abbey sale at Christie's in March 1902. It had gone to a dealer for £60, and when they saw it in a shop window for £130 they sent Van Wisselingh to get it for the trade price: £100. With such precious objects, and their collection of prints, drawings, and paintings, they arranged the flat according to their own unfailing taste.

As soon as the hall door was opened one felt an air of culture and refinement [a visitor recalled] in the surprising and eclectic choice of drawings by Turner, Hokusai, Gainsborough, J. F. Millet and Daumier.... In a white and gold drawing-room with a set of chairs of rather pretentious elegance there were shells, corals, and other sea treasures lying in crystal bowls in which floated cyclamen petals.

These arrangements, which later became fashionable, were originally devised by Ricketts and Shannon. The dining-room was a museum with drawings by Leonardo, Watteau, and Puvis de Chavannes:

There was a tall Roman table whose lilac brèche marble top reflected a silver candelabrum of the Dutch eighteenth-century style with Spanish influence. There were filigreed bread-baskets, some of which held bitter apples, small gourds, and moonwort. In others there were water-melons and tomatoes; in an Adam's water jug were piled various kinds of fruit in season.

Ricketts loved these arrangements of fruit and vegetables. He often used lettuces, his 'green roses', for decorative effect. On a table there would also be pieces of jewellery, sometimes of Ricketts's own design. 'Everywhere there were shells, mother-of-pearl sweet-dishes and Venetian finger-bowls, whose edges were festooned with baroque pearls standing on Queen Anne lace mats.' All of this can only be described as precious, but it was in no way a pose: beauty was their way of life.

Yet, after the magnificence of these rooms, the bedroom was simple and austere. 'Twin bedsteads of spotless white, two washstands, and, on the wall, drawings or prints of masters—such was ... the cubicle shared by Shannon and Ricketts'.[5] The flat, moreover, had, in spite of its luxury, little sense of comfort. 'I live like a grandee', Ricketts would

[5] Jacques-Émile Blanche, *Portraits of a Lifetime*, trans. and ed. Walter Clement (1937), 130–2.

say. 'Their reception-room would be rich and luxurious, filled with noble and beautiful things, fit places in which to dispense hospitality and receive their peers; but the rooms in which they habitually passed their days would be almost poverty-stricken puritan in their simplicity.'[6] His studio, therefore, was not a place where you could sit at ease. The chairs and sofa were hard and covered with cheap blue cloth with white stripes, such as butchers' aprons were made of. Ricketts could not relax; to sink back in an easy chair was anathema to him. He was ever on the *qui vive*. He perched on the edge of chairs, as he held forth, cigarette in hand, his hands as talkative as he was. Shannon's studio, much larger than Ricketts's, was a grand room with a gallery and contained the larger paintings and pieces of sculpture from the collection.

Hardly had they settled into their new home than their servant Marie became the only casualty of the move. She gave her notice. 'One shrinks with all the cowardice of ones nature before new faces and new habits', Ricketts typically noted. This shock, on top of the strain of the move, threw him into an 'execrable' temper for several days.[7]

Once Ricketts and Shannon had got the flat into shape their life settled into a pattern, combining old habits with new. Friday evening was still their time for receiving guests. In December 1903 they inaugurated 'tea-fights' to show Shannon's work. About sixty people turned up, including the Fields, Roger Fry, William Pye, Laurence Binyon, Robbie Ross, Selwyn Image, the Pickford Wallers, and R. C. Trevelyan. 'Models with eager eyes and movements offer us tea and sandwiches—impossible sandwiches devilishly combined of banana, cucumber and tomato by the arch-fiend [Ricketts]', observed the Fields. They found Ricketts's behaviour disturbing. He 'darts & blunders & circles & frets like a moth or butterfly—he has no dignity but his wings are felt. His frail presence is the really disquieting thing in the room.'[8]

A change at the International now enabled Ricketts to send in paintings there. Whistler's death in 1903 had led to the appointment of Rodin as President. Though Pennell and Ludovici, both still 'enemies', remained on the committee, Ricketts wanted to associate himself with Rodin, whom he regarded as the world's greatest living artist. He first met the great man at the Davises', who owned some of his work.

6 *SP*, p. viii.

7 CR Diary, 18 June 1902; MF Journal, 1902, fo. 103.

8 MF Journal, 1903, fo. 203^{r-v}.

Scrutinizing the sculptor's appearance and character, Ricketts found that he was less interesting to look at than he had anticipated, and said 'nothing fine or even intelligent'. Once or twice—as when Ricketts mentioned seeing his exhibition in Paris in 1890—the Frenchman's face lit up, but he looked as if he were prepared to be bored for the evening. Although Rodin was very impressed by the French paintings Ricketts and Shannon had advised Davis to buy, Ricketts was hurt by Rodin's polite apathy, and felt that the sculptor was 'hardly conscious . . . that I was a rather more than ordinary man, who probably knows & understand[s] his work better than any man in Europe'.[9] Nevertheless, with Rodin at the head of the International it had the stature to realize the international aims which had inspired its inception.

Its fourth exhibition opened in January 1904. Shannon, who had started exhibiting there the previous year, sent in two pictures, including the portrait of Hetty. Ricketts, whose bust by Wells was also exhibited, sent in one of his Depositions, entitled *A Burial*. He 'stormed' to the Fields that it had not been given a chance, as it was 'skied', but the Fields, after seeing the show, decided this was an exaggeration.[10] In spite of this bad beginning, and the fact that Ricketts came to consider the picture 'a cough drop and I should like to be rid of it forever', it was eventually sold to Dr Elizabeth Garrett Anderson, who had already bought one of his paintings, for £110. She 'devotes her money to woman's suffrage', he wrote, 'and to the purchase of my pictures; otherwise she does not deviate from a normal course'.[11]

While visiting the International Exhibition the Fields bumped into Ricketts and Shannon, and found that Ricketts was now storming about something else. There had been a fiasco with Rodin. One of the exhibits at the show had been a 'marvellous bronze study' of St John. When Shannon had enquired about the cost of it Rodin had offered it at the reduced price, £150, to which Shannon was entitled as another member of the Society. Then, when Shannon went to the dinner at the Café Royal in honour of Rodin, he was told by Derwent Wood that Rodin was willing to make an exchange for a picture by Shannon. Rodin had been shown, and been impressed by, his pictures. Shannon, thrilled by this 'great and undeserved honour', went to check out the story with Francis Howard, Secretary of the Society, and found it to

9 CR Diary, 3 June 1903 = *SP*, p. 97.

10 MF Journal, 1904, fos. 12^{v}–13, 31^{v}.

11 CR to R. Ross (n.d.); CR to M. Davis (n.d.).

be true. Next day, Shannon was astounded to receive a letter from Blanche, which had crossed with his own letter of acceptance, announcing that Rodin would accept a reduced price, £100, for the bronze; so they telegraphed acceptance and posted a cheque. Shannon was even more astonished the following day to receive another letter from Blanche saying there had been no misunderstanding at all and that an exchange had been proposed. To extricate themselves from this confusion, Ricketts and Shannon felt obliged to buy the bronze at the original price of £150, which they had never intended to do. Shannon, moreover, missed the honour of an exchange with Rodin, and may have seemed rude in buying rather than exchanging. The blame for the mix-up was placed on 'a set of misunderstandings by a third person' (i.e. Blanche). Ricketts decided that the moral was 'Never mix a third party up in any business, though I have heard of some marriages in which that arrangement is highly satisfactory.'[12]

Later in the year, all the same, Blanche painted a double portrait of Ricketts and Shannon. For two days they sat for him in the studio at Lansdowne House, and the portrait was finished. Ricketts found that the likenesses were 'striking & the heads well-constructed'.[13] This seems to have inspired Shannon to embark on a double portrait of his own. In contrast to Blanche's speed and sureness of touch, Shannon worked for weeks on the portrait, glazing and making alterations, and eventually destroying it. They also had themselves photographed by a professional photographer on 13 Sept. 1903 (Fig. 26). They were 'photoed 84 times between us. The man must think we are going to be famous', Shannon noted in his diary.

Shannon was still making great headway in his career. Early in February 1904 Hugh Lane called to ask him to lend pictures for an international exhibition at St Louis, Missouri. Shannon was offered a whole room to himself, but no sooner had his pictures started for St Louis than he heard that the Irish section was cancelled, and the exhibition was shifted to the Guildhall in London instead. For these shows the only condition was Irish blood, and Shannon consulted his father and aunts who confirmed his Irish origin. Ricketts, however, could make no such claim: 'I am furious with my parents for their

[12] CHS Diary, 12, 13, 14, 15 Jan. 1904; CR to TSM [Feb 1904] = *SP*, pp. 161–2; CR Diary, 12, 13, 14, 15 Jan. 1904.

[13] CR Diary, 11 July 1904 = *SP*, p. 109.

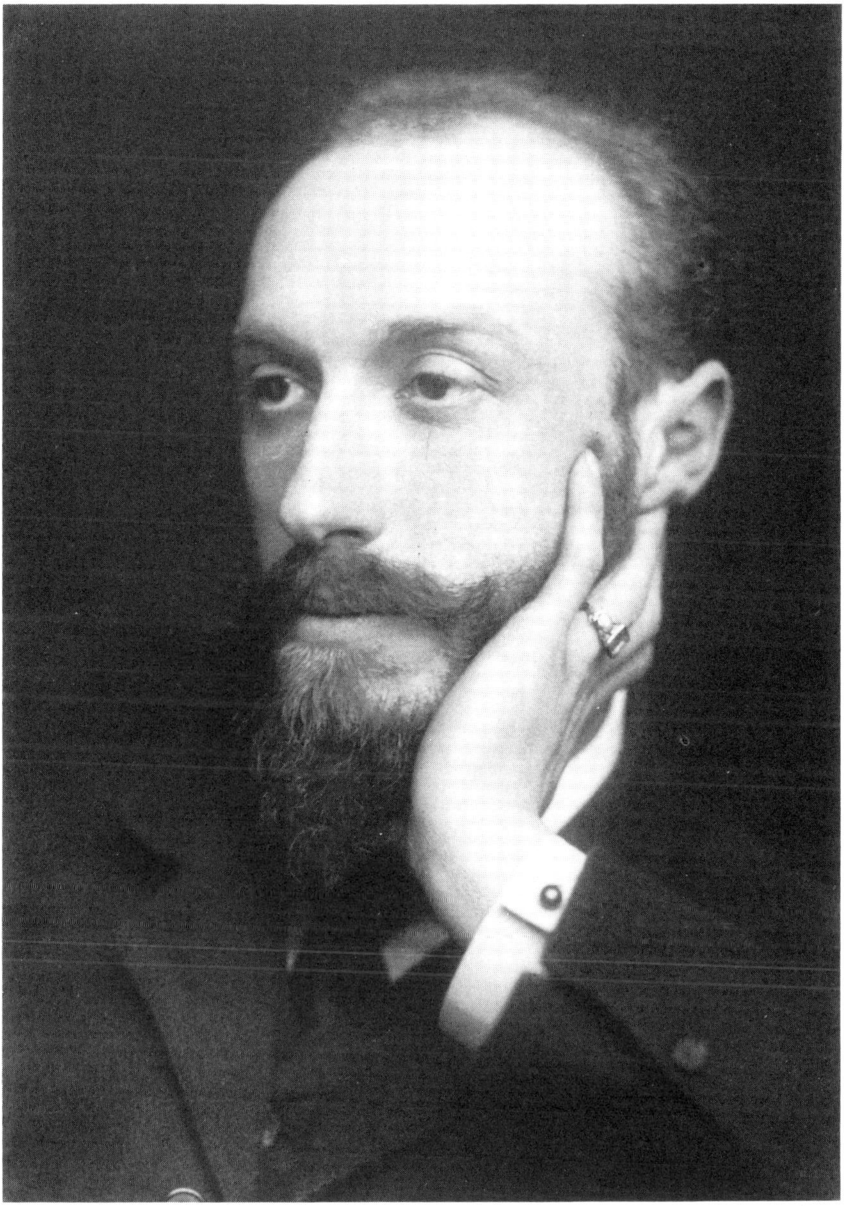

FIG. 26. Charles Ricketts (photo: 13 September 1903) by Beresford

oversight in not gifting me with at least one drop of Irish hot. I should have made a splendid Irishman with very long coat-tails.'[14]

Hugh Lane quickly became a friend. To the profession of a picture dealer Lane joined 'the magnanimity of a Medici', Ricketts recalled. His dream was to found a public gallery in Dublin, for which he not only bought pictures but persuaded his artist friends to make donations. Ricketts and Shannon thus made their first gift of their own paintings to a public institution. A hurt and depressed Ricketts later recorded (almost certainly with exaggeration) that Lane had admitted to hating Shannon's picture and also wanted to exchange the small oil Ricketts had donated, *Parable of the Vineyard*, for something else. Ricketts observed 'this is a lesson never to make presents to fools and to public institutions'.[15] Yet on the whole Lane's friendship was beneficial. To both Ricketts and Shannon he brought buyers for their work, and with his excellent taste he combined Irish charm. Once, insisting that Ricketts had seen a painting of which Ricketts had no recollection, Lane retorted, 'Yes ... I showed you its back.' Ricketts confessed he had a weakness for such blarney.[16]

'I have news', Ricketts wrote to the Fields in the summer of 1904, 'I have turned sculptor.' He had modelled in clay an actor with a mask, and a centaur with a sleeping baby fawn. '[I] ... now quite understand the rather youthful attitude of the Deity on the 7th day when he saw that it was good. Clay is a divine substance—responsive, attentive, definite, yet liable to improvement also. I think man must have been modelled in some less admirable and ductile stuff.' When, later in the year, the two clay models were cast in bronze, Ricketts had a sense of achievement he had never felt before, because they were 'in an inalterable medium'.[17] As early as 1900 'curious old wishes ... to do sculpture' had moved Ricketts on seeing Rodin's work. He thought that sculpture offered a larger original field for the artist, but he envisaged a different kind from Rodin's, one that 'would avoid all painter's tendencies to loose handling'. His ideal kind would show 'nothing since the traces of Myron, excepting Michelangelo & Barye', and would 'arrest the light, not play with it'.[18] Though Rodin was the dominant influence,

[14] CR to TSM [Mar. 1904] = *SP*, p. 103.

[15] CR Diary, 4 Mar. 1906; Joseph Hone, *W. B. Yeats: 1865–1939* (1942), 225–6.

[16] CR Diary, 23 Apr. 1904 = *SP*, p. 104.

[17] CR to MF [after 24 July 1904]; MF Journal, 1904, fo. 149v.

[18] CR Diary, 27 Nov. 1900 = *SP*, p. 49.

Ricketts's subjects are more dramatic and on a much smaller scale. He produced over twenty sculptures before the Great War.

His excitement over sculpture seems to have prompted him to try his hand at lithography, but he found the medium unsuited to him, and this brought on intense depression. 'The tiresome thing about a new medium', he noted, 'is that old failings return, old manerisms of ones boyhood'. As it was too 'disheartening, facing an almost certain failure', he decided to give it up, though later he returned to it.[19] Despite his quick mind, his virtual ambidexterity, and his great sense of materials and how things were made, Ricketts had to work hard to achieve his astonishing versatility.

It is curious [he acknowledged] that no medium has ever come easily to me, unless it is watercolour. I have to fight every inch of the way against an unexpected slowness, more especially in draughtsman mediums. Pen & ink, Pastel, Sanguine, each have been difficult in turn. Oil & watercolour painting came more easily, but with something offensive to one at the time or imediately afterwards.

The causes of this may have been the same haste and impatience that made it impossible for him to master the piano and the bicycle: 'I have ... a tendency to work too fast & feverishly.' However, he also recognized that he was perhaps too educated in art history and hence too self-conscious: 'I suppose the crytical faculty dwarfs the manipulative impulses belonging to simpler art temperaments'. He noted, in a clause more heavily deleted than the rest of this passage: 'I am always superior to my work'; a judgement with which many Ricketts admirers would agree.

At the end of 1904, while Shannon was away, there was a masked ball at the Davises'. Ricketts had always been interested in costume, and now had an opportunity to design something for himself. 'He is going as a pig with a trail-wreath of little pink roses, a white satin coat, gardenias plucked of their leaves', the Fields noted. On an exquisite Venetian chain he wore a little puff-ball, and he had one pocket full of pink carnations to give to the ladies. The servants laughed when he arrived, and Ricketts thought all would go well. He went in and bowed. No one took any notice. One or two people asked him if he was Comus, but on the whole people seemed to fear and avoid him. 'Their suburban minds seemed shocked, especially the men,

[19] Ibid., 10 Oct. 1904 = *SP*, p. 111; CR Diary, 24 Sept. 1900.

the women giggled as if against themselves.' He told the Fields, 'It was a frost. I was a dead failure.'[20]

In spite of his fondness for dressing up on special occasions, Ricketts was not much concerned with his clothes or appearance. He has often been described as a dandy, but this is wrong. Though his clothes identified him as a gentleman, it was a rather shabby one, his usual attire being a shiny old blue serge suit sprinkled with cigarette ash.[21] Money that might secure a splendid drawing or other treasure would not be wasted on clothes.

The start of 1905 brought Ricketts into political activity, even though he told the Fields that 'the minute an artist touches real affairs, he ceases to be an artist'.[22] News had reached England that the Russian writer Maxim Gorki was under arrest and in danger of execution. Artistic and literary circles in Germany were already organizing a demonstration. The image of the artist being persecuted by a brutal and philistine society was always one to stir Ricketts. Spurred on no doubt by memories of Wilde, he wrote to all his friends to enlist their support. Shannon tried to get the International Society to join in the protest, but he had first to ascertain Rodin's feelings. Rothenstein assured Ricketts that the New English Art Club would give its support. Ricketts thought of a petition to the King or the Queen. However, on 3 February a telegram arrived from Count Kessler in Paris ('an old personal dislike and enemy of ours') announcing that Gorki had been released. Ricketts then wrote to Binyon:

> The Gorki affair seems to have settled down, since he is now said to be out of danger ... The English painters' protest will be like the Carabiniers in an Offenbach comic opera who always arrive on the scene with the words 'Nous sommes les Carabiniers du 1er bataillon qui, par un heureux z'azard, arrivons toujours trop tard!'[23]

Increasingly Ricketts was a committee man. This was probably the only way he could defend and propagate the artistic ideals he cherished. As well as becoming involved with the National Art Collections Fund, he was on the committee of the Vasari Society, which issued fine collotypes of old master drawings. He was also acting as an examiner

[20] MF Journal, 1904, fo. 185v and 189v.

[21] R. A. Walker to Ursula Bridge, 4 Oct 1955 (Bodleian Library, Oxford).

[22] MF Journal, 1905, fo. 22.

[23] CR Diary, 30 Jan. 1905; CR to L. Binyon [Feb. 1905] = *SP*, pp. 116–17; CR to Bottomley, 29 Mar. 1915 = *SP*, p. 237.

for the London County Council. There were, too, endless meetings over a new grouping of artists. The 'Society of Twelve' had much the same aims as the 'Panel Society' which had been discussed a decade before. The exhibitions of the new society were to cover not only etching and metal engraving but also lithographs and woodcuts, 'thus for the first time giving representative effect to the revival of original work on stone and wood which has been one of the most notable symptoms in recent years'.[24] This not only stresses the particular achievements of Ricketts and Shannon, but also has the ring of Ricketts's voice. Drawings were also to be shown. At the first exhibition in 1904 the twelve members were George Clausen, Gordon Craig, William Nicholson, William Strang, Sturge Moore, William Rothenstein, D. Y. Cameron, Augustus John, Charles Conder, and Muirhead Bone, with Ricketts and Shannon. Legros was an honorary member. Since the Society's membership was wider than the usual Vale coterie, it had great appeal and importance. At the first exhibition Ricketts showed six of his Parable woodcuts and four drawings, while Shannon showed six lithographs and four drawings.

'Bottomley, a poet from the north and a great admirer, came to tea', Shannon noted on 3 June 1904.[25] Before calling at Lansdowne House, the thirty-year-old poet had sent them a copy of his book, *The Gate of Smaragdus*, which had just been published. It was inscribed to Shannon. Almost every aspect of the book showed its debt to Ricketts's and Shannon's early work. Gordon Bottomley had led a life of seclusion, owing to the dangerous weakness of his lungs. As a boy he had come under the influence of Wilde, Rossetti, and the Aesthetic Movement, and he was deeply interested in poetry and poetic and romantic drama, all of which commended him to Ricketts, who soon became his 'prince of men'. Not only did Bottomley fill his Lancashire house with works by Ricketts and Shannon, but for thirty years, he recalled after Ricketts's death, 'I never wrote anything without thinking what he would say about it'.[26] Ricketts was always prepared to accept disciples, but Bottomley's devotion both touched and puzzled him. After one of their early meetings Ricketts asked Shannon why Bottomley had such a high regard for him. 'He does not see you often', Shannon replied.[27]

[24] *Society of Twelve, Catalogue of the First Exhibition* (1904).

[25] CHS Diary, 3 June 1904.

[26] G. Bottomley to W. Rothenstein, 12 Jan. 1932.

[27] CR to Robbie Ross, p.m. 28 June, 1916.

On 23 February 1905 Ricketts's and Shannon's third trip to Italy began. On their way down to Pisa they passed Genoa, close to the graveyard where his mother was buried: 'I felt something like a pang on entering the tunel in the hill which is but a few yards from the rock which I shall never forget.' Pisa pleased him, even though the damp cold eventually drove him out of the Camp Santo. Owing to strikes they had to wait an hour in a cold, wet station 'full of complaining Britons'.[28]

Their first days in Rome were spent in 'mere lazings, looking at flowers and shops'. As before, they were plagued by galleries being shut, this time on account of the Carnival. As Shannon's throat was bothering him, Ricketts went off alone to the museums. He pottered about the ancient sites of the city, whose historical associations stimulated his imagination.

> The place where they murdered Caligula [he told the Fields] is not impressive, it is a sort of passage where any body might be killed . . . As often as not, the more majestic portions are without a history, or else are casual like the delightful little Romulus temple which commemorates some other one who has been forgotten owing to a blameless life.

He saw the ruins as evidence of the battle between 'the immortal earth & the immortal human energy'.

Shannon, meanwhile, had got worse, gone to a doctor, and been put to bed for two or three days. Ricketts was disappointed: 'We take every precaution but his throat wont travel—One wonders how we shall ever be able to travel to Italy, he is always ill.' After four days Shannon still had a fever and the doctor prescribed a tonic which contained strychnine. Ricketts believed he had given him this according to Shannon's instructions and the dosage written on the bottle, but mistook the amount and gave seven doses instead of seven drops. Then he went out for his afternoon walk. Shannon was 'seized with pains and a wish to vomit, got out of bed, & then in again owing to nausea and numbness, and vanished for about 10 minutes, after which he revived cold and numb'.[29] Two hours later, when Ricketts returned, he was better. The doctor was astonished that the poisoning seemed to leave no traces, and Ricketts was astonished that he had not felt

[28] The main source of this account of the third Italian trip and the quotations following, unless otherwise indicated, is Ricketts's Diary (BL Add. MS 58096).

[29] CR to MF, 9 Mar. 1905.

'anxious or fretful during the bother'. Indeed, he did not even mention it in his diary: 'so much for contemporary records', he noted later. Shannon wryly commented that, since he should have died of the strychnine on the spot, 'it's no use my thinking of suicide' by poisoning. The next day he was well enough to eat human food 'instead of nauseating slops', and to take a drive in the Pincian Gardens, which Ricketts did not like. The overdose appears to have helped him. They both liked Rome more this time, but found they suddenly got tired of the Eternal City, 'though that is a name I dislike, for in what other town is one less conscious of the eternity of any thing'.[30]

In Florence, where they spent a fortnight, they called on Reggie Turner and on Horne, who had been 'beastly ill' but was well again, and 'as usual doubts all pictures and drawings'. Both Horne and Berenson, Ricketts informed Binyon, 'confessed to hating art—I did not turn a hair but said I have always said so.'[31] Two days after their arrival, the Berensons left for Rome. 'I thought it looked like the passing of the moon before the advent of the sun', Ricketts teased the Fields, 'or like St. John retiring into the desert, but it seems B. B. has gone to hear Joachim play Beethoven.' On their return, Mary Berenson sent her carriage to collect Ricketts and Shannon. On their way to 'I Tatti', Ricketts noted that in 'the little village churches ... the holy pictures were covered with heavy wirework to prevent B. B. attributing'.[32] On this visit, they liked Mary more and Bernard less: 'There is an essential dryness or smallness of nature in him which warps a fairly good intelligence; he is not sympathetic & must give his wife bad quarter of hours.' However, when—over lunch—Berenson ran down Horne and Roger Fry, Ricketts was pleased to concur.

They visited and revisited the churches and galleries of Florence until they found the place was wearing out 'or, rather, just like a bottle of full liquid, it is difficult for us to hold more'.[33] After the cold, wet beginning to their trip, they had had a month of marvellous weather, but again the rain was setting in. Ricketts's fingers were beginning to itch for work. They decided to drop Venice from their itinerary and to start for home on Tuesday, 28 March, rather sooner than they had anticipated.[34]

[30] CR to MF [16 Mar. 1905].
[31] CR to L. Binyon [1905].
[32] CR to MF, 26 Mar. 1905.
[33] CR to TSM, Mar. 1905.
[34] CHS Diary, 1 Apr. 1905.

Music did not at this stage occupy Ricketts as much as it had done and was to do again. 'I hardly dare confess,' he wrote to Mrs Berenson, 'that I have let music almost slip out of my life, banished it as an exquisite but solitary need, when as it is I have too many other tyranous and sometimes angry admirations.'[35] However, the theatre still fascinated him. On 12 January 1905 he saw Laurence Housman's 'rotten play *Prunella*, it is at once philistin & "unwholesome". It dates circa *Yellow Book*; like many erotic sentimentalists, he points to the opening in his trousers as if it were a wound.'[36] In May he went to see the first presentation in England of Wilde's *Salome*, having been given tickets by Robbie Ross, whom he met by chance in the street. 'I was greatly impressed by *Salome*,' he told the Fields. 'I went there almost as a duty, fearing that the performance would be all that it should not be, but this was not the case. I was breathless with interest.' It was too long for one act, and the plot, which was 'strangely enough ... too horrid & too gruesome', combined with intense heat due to lack of ventilation to cause intense physical discomfort. 'I once or twice seriously considered the advisability of swooning back, but there was only the critic of the *Daily Telegraph* behind me, and one cant be too particular about such matters.'[37] Ricketts was the first to cheer at the end, just as he had done at a private performance two years before of Maeterlinck's *Monna Vanna*, which had been banned by the censor.[38]

At the end of June 1905 he attended concerts given by the Frenchwoman Yvette Guilbert, an exponent of French folk-songs and popular songs from the past. He found her wonderful, better even than all he had heard in her praise: 'Her quality is not mere charm but a profound intelligence.'[39] At one concert he was accompanied by the Fields, who did not enjoy his companionship, 'partly because he has in an acute form that masculine disease of fuss. Such struggles to get us a cup of tea', they recorded. They also found that his high critical attitude inhibited them:

As part of an audience he is very cold or very hot. Not a trace of applause if the piece is sentimental or if it is not all he admits as art ... Latin *bravos* for

[35] CR Diary, 6 Sept. 1904, draft of letter to Mrs Berenson. [36] Ibid., 12 Jan. 1905.

[37] CR to MF, 12 May 1905 = *Letters CR to MF*, pp. 17–18. The critic was Malcolm Watson (1853–1929), Scottish journalist and playwright.

[38] CR Diary, 21 June 1902 = *SP*, p. 78.

[39] CR Diary, 26 June 1905 = *SP*, p. 121 (wrongly dated 26 May).

the true and beautiful as *he* feels it ... ripples and freshets in the malicious laugh when the song is divinely naughty.[40]

Alive and responsive to every minute and nuance of a performance, he made his approval or disapproval felt.

Ricketts had always thought the theatre should occupy a high place in life, and that it was 'the least understood and native of our arts'.[41] On 27 January 1901 Sturge Moore discussed with him the possibility of the foundation of 'a Theatre Society for Romantic drama, etc. ... in which the scenery would be done on a new decorative, almost symbolic principle'.[42] This was a subject which had haunted Ricketts, who had thought of writing a pamphlet on it ever since his discussions with Wilde in the early 1890s. For *Salome*, 'I proposed a black floor', Ricketts recalled, 'upon which Salome's white feet would show; this statement was meant to capture Wilde.' He could not remember who suggested dividing the actors into separate masses of colour, but 'today the idea seems mine'. It was Wilde who wanted the Jews in yellow, the Romans in purple, the soldiers in bronze green, and John in white. 'Over the dress of Salome the discussions were endless: should she be black "like the night?" silver "like the moon?" or—here the suggestion is Wilde's—"green like a curious poisonous lizard"'.[43] It was along these lines that the symbolic stage decoration was to be developed.

Sturge Moore and Laurence Binyon had become enthusiastic about Yeats's ideas on poetic drama, and had begun to organize a new society to revive the 'lost' arts of musical speech and rhythmical acting as well as symbolic scenery. Ricketts was sympathetic to Yeats's attempts to reform the speaking of verse drama, because of the 'Church of England intonation' that marred many productions. Yeats, having seen Gordon Craig's staging of Laurence Housman's play *Bethlehem*, was eager to involve Craig in the venture and to stage his own play *The Countess Cathleen*. Yeats's play had—'to my astonishment and against all my suspicions'—impressed Ricketts when he read it (and he was also impressed by Yeats's plays given by the Irish players the following year).[44] In January 1903 Ricketts, with characteristic generosity, proposed to raise £600, with which Craig could stage the play, from the

[40] MF Journal, 1905, fo. 73v

[41] CR to G. Bottomley, p.m. 29 Jan. 1919.

[42] CR Diary, 27 Jan. 1901 = *SP*, p. 52. See Schuchard.

[43] *Pages on Art*, p. 243; See Holmes, p. 167.

[44] CR Diary, 22 July 1901.

sale of the Vale edition of Marlowe's *Dr Faustus*. Only £170 was actually obtained from this. Even after Yeats's English schemes had come to nothing he hoped to get the money Ricketts had 'for theatrical adventure' and to secure his services as a designer for the Irish National Theatre Company. He never got the money, but he did get Ricketts to work for him.[45]

Indeed Ricketts's first costume design was for Yeats who, in July 1904, thanked him for his 'admirable' designs for 'the Black Jester'.[46] Ricketts also did some designs for Binyon's *Paris and Oenone*, which opened on 8 March 1906.

> I think the people should look like figures by Gustave Moreau [he told Binyon] notably Paris who wears a traditional costume of great beauty. I hate tunics with key patterns round them and the common British Walter Crane notion of Greek persons, which is quite English & founded on Flaxman. Paris wore *boots* (a fact) and Helen probably stays and a crinoline with flounces.

Paris was to be in reds, Helen in golds, Oenone in dark green, the chorus in browns and purples. This is the first use of the colour principles that characterized his theatre work.[47] He designed no scenery, however; he later told Binyon that any decent scenery would do and 'the darkness of the stage would give people something to grumble about: "we could not hear Mr. Binyon's words because the stage was so dark" '.[48]

However, Ricketts's real work for the theatre began with the founding of the Literary Theatre Society. Once Yeats had decided to concentrate on his Irish theatrical schemes, which had proved more successful, the London group carried on without him. It included Sturge Moore, Binyon, William Pye, R. C. Trevelyan, and the actresses Gwendolyn Bishop and Florence Farr, as well as Ricketts and Shannon. Ricketts gave them the £170 from the Vale *Faustus*, and early in 1906 they set to work on their first production, which was to be Sturge Moore's *Aphrodite against Artemis*. Ricketts's set consisted of curtains between solid columns, and the dominant colour was yellow. Such curtain settings, with a prevailing colour, characterized these early productions. Ricketts felt that the principals should be dressed in vivid designs before a simple backdrop, and that there should be a subdued

[45] 'Heirs', p. 65; Schuchard, p. 416.

[46] *Theatre Notebook*, p. 7; WBY to CR, July 1904.

[47] CR to L. Binyon, 21 Aug. 1905 (2 p.c.s).

[48] CR to L. Binyon, 26 Sept. 1905.

entourage to give the illusion of space, and low lighting to give height to the stage.[49]

The play was performed at the King's Hall on 1 April. Ricketts was pleased with the performance, for, though it was bad, it was not as bad as the rehearsals had led him to expect. The good parts of the play told well, but the bad portions 'stood out in livid passages underlined by atrocious acting'. He was amused by Theseus, whose tunic was too short and whose 'hurried entrance and hurried intonation' set him off 'into a fit of convulsive laughter which nearly killed me in the effort to suppress it. Matters were made worse by two people in front of me on whose chair my foot happened to be resting, trying to stare me into order.' On the whole, the play was well received.[50] The next day, however, a highly critical review entitled 'A Rival to Euripides' by William Archer appeared in *The Tribune*. Ricketts agreed with many of his criticisms, but thought Archer's offensiveness showed 'the hostility of the modern mind towards all art effort'; he defended the play in a long scholarly letter, which he first showed to Binyon to have the spelling and punctuation corrected.[51]

A more ambitious project in the same year was a double bill of Wilde's plays *Salome* and *A Florentine Tragedy*. The production was plagued by bad luck. At the first meeting 'both goddesses from Moore's play [were] in ferment, that is Mrs Bishop anxious to prevent Mrs. Wheeler from having a chance of Salome'.[52] Ricketts did not like Gwendolyn Bishop: 'Candidly, I detest her! Once or twice ... I thought I should like to slap her fat bottom and then realised that I did not like her enough to do that. She was so bitchy that even T. S. M. [Sturge Moore] noticed it.'[53] Both Constance Collier and Mrs Patrick Campbell refused the part of Salome, which was taken by Miss Darragh, who acted in a number of Yeats's productions. Mary Davis was prevented from giving the play financial support by her husband, Ricketts noted acerbically, on the grounds that it was sure to be badly done.[54] As Ricketts disliked the King's Hall, where Moore's play had been given, he asked for the use of St James's Hall, but was turned down.

[49] MF Journal, 1906, fo. 208v.

[50] CR Diary, 1 Apr. 1906 = *SP*, p. 132.

[51] William Archer, 'A Rival to Euripides', *Tribune* (2 Apr. 1906); Ricketts's reply, *Tribune* (7 Apr. 1906); CR Diary, 3 Apr. 1905.

[52] CR Diary, 6 Apr. 1906.

[53] CR to CHS, 24–9 Apr. [1906].

[54] CR to L. Binyon [1907].

All the same, he threw himself into designing and having the dresses cut and stencilled, though he was suffering from gloom and mental fatigue. His scenery designs were entrusted to a scene-painter who then shut up shop and disappeared, so that he was landed with finding a new scene-painter at the last minute. Two days before the dress rehearsal there was still no scenery. He was visiting the dressmaker's four times a day on average and the new scene-painter's two to five times a day. Most nights he did not get to bed until 2.30 or later. Then the stage-manager quit without excuse or explanation, and Ricketts and Sturge Moore shouldered this extra burden. Ricketts had to take the rehearsals, and even to manage the lights, and the limelight man was 'the most absolute idiot I have met in my life'.[55] There can be little doubt that many of these troubles were inadvertently caused by Ricketts's own enthusiasm and perfectionism. His role was never merely that of designer, but branched out to cover managing, lighting, directing, and interpreting the play. Others were not as willing to devote time and energy to perfecting a production, and his determination to impose his own view caused friction and made him enemies. He would rail against people who got in his way, or failed to live up to his standards, and curse them in his conversation and letters, but later he would calm down and take a more balanced view. His friends, whether they were the hearers or even the victims of such diatribes, knew not to take them too seriously.

There was an angry scene with Robbie Ross, followed by a letter from him, which accused Ricketts of bad faith because a young friend had not got the part he had been promised. Ricketts, already under great strain, replied: 'I should be hurt and angry with you if I were not doing my utmost in manual work, time, and money to counter a malign influence I cannot understand but which has kept me on the *qui vive* for almost a fortnight.' It had been his 'express wish' that Ross's friend should get the part, but he wanted no more angry scenes, 'otherwise I shall take the part ... myself?'.[56] Then Salome's train, for which Ricketts had made forty-two blue foil and velvet butterflies, was lost in the building; so was the dress for Bianca in *A Florentine Tragedy*, and she had to go on in a borrowed costume. Finally, because of the ancient law forbidding the public performance of biblical drama, and the opprobrium that still clung to Wilde's name, the press boycotted

[55] CR Diary, 10 June, 1906 = *SP*, pp. 136–7.

[56] CR to R. Ross [*c.* June 1906].

the performance; *The Times* returned its ticket, and no illustrated paper would publish the photographs.

In spite of all this, the performance, on 10 June 1906, was a success. Dim, cypress-shaped bunches of curtain stood against a starlit sky. The cast was clothed in every shade of blue, deepening into dark violet with green, and the harmony of blues was relieved by the red lances of the soldiers. During the performance Ricketts was behind the scenes acting as stage-manager, a job he enjoyed. Owing to his nervousness, however, in directing the limelight man, 'the moon shone but very fitfully, generally after it had been mentioned in the text, and never once upon the floor!'[57] The experience made Ricketts suspicious of the possibilities of 'lighting effects'.

During *Salome* you could have heard a pin drop. Robert Farquharson's Herod was, according to Ricketts, the greatest piece of male acting he had seen in his lifetime—on a level with Irving's Shylock.[58] The play got four curtain calls, whilst *A Florentine Tragedy*, which was under par, got three. The house was enthusiastic and very distinguished, and included Shaw, Yeats, Hardy, George Moore, Max Beerbohm, Arthur Symons, Eleanora Duse, who had been approached to do *Salome* by Wilde, and Roger Fry, who wrote to a friend that *Salome* was 'superbly mounted and acted ... Charles Ricketts did the staging with ideas of colours that surpassed belief. I've never seen anything so beautiful on the stage.'[59]

During the preparations for the Society's first production Ricketts had, surprisingly, set off for Italy, though he returned in time to make last-minute adjustments and to oversee the draping of the stage. Before leaving, he had organized a show of his sculpture at the Carfax Gallery. An exhibition of his paintings at Van Wisselingh's had been delayed until after his return. He set off for Italy alone on 6 March.

Ricketts had agreed to write a book on *Titian* for Methuen and he wanted to look at the master's work in Florence, Rome, and Naples, the last of which he had never visited. During his brief stop over in Paris he found the Manet show at Durand-Ruel's better than he had expected. 'The Spanish Singer' was beautifully painted, and 'I only regret', he wrote to Durand-Ruel, 'that such charming works as *La Plage de Boulogne* and *Le Lapin* should probably be at prices for the rich

57 *Pages on Art*, p. 244.
58 MF Journal, 1906, fo. 136.
59 Roger Fry, *Letters*, ed. Denys Sutton (1972), i. 267.

only'.[60] In spite of a pledge of economy to Shannon, Ricketts purchased an exquisite fragment of a terracotta Hercules, a small, unusual kylyx depicting a figure full face, and 'an exceedingly beautiful small marble head of a severe pre Phoeidian type' with hollow sockets and a restored nose. 'They also gave me', he told Shannon, 'a tiny rubishy little toy pot which shows I did not haggle enough. I only got one pound off the head & 10 bob off the pot.' There was also a fine kylyx of a boy bathing, and, after getting the price down from fifty to forty pounds, Ricketts told them he was too poor to buy it. 'It is a thing worth having later on if we can afford it', he told Shannon, to whom he sent a sketch.[61] 'I expect we'll own it someday', Shannon replied, and he was right. These purchases were made on credit, and, asking Shannon to send cheques, Ricketts hoped he would 'not have a fit over my Greek purchases'.[62] His bronze show, he thought, would pay for such extravagances.

The journey to Italy was uncomfortable. 'I should have been quite happy if it had not been for the damned heating apparatus which set all my pet corns throbbing.' At Turin he made a dash for the gallery, which was interesting, but so cold that he got through it in half an hour. A batch of bad picture postcards, which 'cost agonies of calculation to the focils [fossils] who take care of the gallery', was despatched to Shannon. In Florence he did the museums more quickly than he had expected. 'As I had some time to spare', he told Shannon, '(in a weak moment) I called on Horne to ask him to lunch with me; before we left his house I was filled with remorse and convulsed with self compassion (nothing happened!). We had lunch; this blunted my sensibilities'. Horne then showed him a large drawing he attributed to Van Dyck (wrongly, as Ricketts thought) and a 'Michael Angelo' which was in fact by Rosso. 'I tried to be very gentle & finally got to the Boboli Gardens, where there were too many people & not enough wild flowers. I found two black irises, however.'[63]

He started for Sienna by an earlier train than planned, and did 'the whole humbugging place in two hours'.[64] He went to Sienna, he told the Fields, 'mainly to annoy my friends, who like to think that because

[60] CR to Durand-Ruel, 23 Mar. 1906.

[61] CR to CHS [7 Mar. 1906]; CR to MF p.m., 11 Mar. 1906.

[62] CHS to CR [Mar. 1906]; CR to CHS, Turin [8 Mar. 1906].

[63] CR to CHS [11 Mar. 1906].

[64] CR to CHS [12 Mar. 1906].

I have not been there, I know nothing about Siennese art, when it is a prolonged stay in Siena which really produces that effect'. The exteriors of buildings like the cathedral were fine, though the interiors were not worth seeing—except for the Donatellos—but 'the famous altar piece of Duccio di Siena ranks with any respectable Russian icon painting in invention and colour; it is a mere piece of furniture; to compare this trash to Giotto is a profanation so singular that one understands people admiring Cézanne or Gauguin'. The whole place, he felt, was fraudulent, 'a survival from the time of Ruskin when people said they liked what bored them profoundly'. As far as art was concerned, 'it belongs to the wrong side of Italy and to bogus Italian lovers'.[65] He left by an early train for Rome.

There, after sleeping 'like a log', he 'trundled to the Vatican where it was too dark to see', looked up the Titians in the galleries, and bumped into the Hodsons, and the doctor who had treated Shannon.[66] When Ricketts confessed in a letter that he was feeling homesick and anxious to get home, Shannon replied: 'I am glad to hear you express a feeling of homesickness; it shows you have a conscience after all, not to say a heart.'[67] Teasing remained an integral part of their relationship. At his hotel, the Victoria, he was overcharged for his *pension*, and on the last day they forgot to wake him, 'and I left, like Vesuvius, under a cloud'.[68]

When he got to Naples, Vesuvius, however, impressed him as 'a gentlemanly mountain'. Pompeii was at once interesting and disappointing—'a ruin and not a town of roofless houses'—but he liked the guide immensely: 'We got on splendidly together in a sort of horrid blend of French and Italian which produced rents and racks in the frescoes. He started by saying, "Monsiou veut voir les cochonneries? Si vuol vedere le porcherie voilà un falous monsiou".' This description to Robbie Ross was embellished with a drawing of a guide showing a phallus to a bashful Ricketts.[69] Significantly, this amusing anecdote was related only in a letter to Ross, who shared Ricketts's interest in naughty pictures, whilst to Shannon Ricketts wrote, 'I had no scarlet adventures'.[70] (This suggests not only that Shannon was more prim

[65] CR to MF [11 Mar. 1906]; CR to CHS [12 Mar. 1906].
[66] CR to CHS, 14 Mar [1906].
[67] CHS to CR, Sat., 17 [Mar. 1906].
[68] CR to CHS, p.m. 16 [Mar. 1906].
[69] CR to MF, Sun., 18 [Mar. 1906]; CR to Robbie Ross, 23 Mar. [1906].
[70] CR to CHS, Rome, 20 [Mar. 1906].

than Ricketts but also that their relationship was not as openly sexual as all that.) When the guide asked him if he was a student at the School of Rome, Ricketts replied that he was too old, but the Italian assured him that he had seen much older students: 'I tipped him so handsomely that Shannon will never be told and I believe he will remember me if I ever return to Pompeii again.'[71]

At his hotel in Naples he was bored at meals, as they usually sat him down 'with some other beast who is different each time'; he felt rather lost: 'I do not loom large in a hotel, whatever may be my position in London.' ('I have a tendency when *en voyage* to be come invisible,' the diminutive Ricketts confessed to Lady Davis, 'I have not the faculty for securing service & attention away from home.') During the day he visited the museums, but at night he paced by the sea and watched Vesuvius. 'I shall probably romp home as soon as I can', he told Shannon, 'as I am rather tired of sightseeing.' He was also tired of travelling, 'horrid departures & anxious arrivals' without Shannon's care.[72] Rome, to which he returned, was cold and damp, and he rushed off to Assisi, which was a disappointment. Back in Florence, time hung heavily on his hands, whereas before he had been too busy even to write his diary or about Titian. He found he could do a gallery in three quarters of an hour and get no new impressions.

By 27 March he was back in London, much sooner than he had expected. On his way through Paris he must have bought the Greek kylyx of the young man in the bath, for on 23 June he showed it to the Fields.[73] Before he had departed for Italy, one of the strangest turns in Ricketts's relationship with them had occurred. One of their dogs, Whym Chow, died at the early age of nine: the Fields were devastated. The extravagance of their distress can be seen in Miss Cooper's comment that it was 'the worst loss of my life—yes, worse than that of beloved mother or the tragic father'.[74] An intelligent and companionable dog, he had, in their isolated and rather stifling relationship, assumed the role of a beloved friend and inspiration. The Fields poured out their grief in poetry, which they had published privately at Pissarro's Eragny Press, as *Whym Chow, Flame of Love*; one of the

[71] CR to Robbie Ross, 23 Mar. [1906].

[72] CR to CHS, Naples, 18 [Mar. 1906], p.m. 20 [Mar. 1906]; CR to MF, Naples, Sun. 18 [Mar. 1906]; CR to Mary Davis, Sun., 20 [Nov. 1927].

[73] MF Journal, 1906, fo. 106^{v}.

[74] Ibid., 1906, fo. 14^{v}.

rarest and most bizarre publications of the revival of printing. During the dog's illness Ricketts had sent them a letter upbraiding them for 'morbid preoccupation' and suggesting they send the dog to the vet for treatment, but unfortunately the letter arrived after the dog's death. The Fields perversely put Ricketts's attitude down to 'jealousy of the dog', whose death also caused a row with Berenson.[75]

While in Italy Ricketts had received 'stilted and lacrymose epistles' from them, and more followed his return.[76] The Fields were exasperating and Ricketts was exasperated. He refused to see them again so long as they maintained this 'excessive and dolorous lamentation' about an event 'unpleasant & distressing in itself, but which in common decency should have been relegated to a volume of your diary'.[77] A week later he repeated his attack even more forcefully:

> It is now two months that you have bored & distressed me by refferences to the death of your dog. Not only do I dislike this degradation of the majesty of grief, but in the event of this so praying on your mind, it should be your duty to try and put order & a little silence in the place of this angry din of regret, this is your duty to yourself & to others.... Please consider that I am quite deliberate in saying that I hope you will soon regain the ordinary mansuetude of thought & speech common among the elect to whom I belong.[78]

In both these understandably angry, if rather cruel, letters, Ricketts went on to talk about such books and events as had always been an important element in their friendship, in an effort to bring it back to normal. When they finally came to the Palace, on 24 June, Ricketts was confiding and attentive, showed them all his new acquisitions, and gave them a cast of his bronze, *Mother and Child*. The loss of their beloved dog and the row with Ricketts shook the Fields deeply and weakened his hold over them. They were soon to find new mentors to replace him, the '*chers pères*' they turned to for spiritual guidance when they became Roman Catholics.

At the end of May—six months late—some twenty of Ricketts's paintings were shown at Van Wisselingh's, his first one-man show of paintings. In the window stood a painting of a figure Ricketts was to paint again and again: *Don Juan and the Commendatore*. The Don drops

75 Ibid., 1906, fo. 14.

76 CR to CHS, Florence, 23 Mar. [1906].

77 CR to MF, 9 Apr. 1906 = *Letters CR to MF*, pp. 19–21.

78 CR to MF, 16 Apr. 1906.

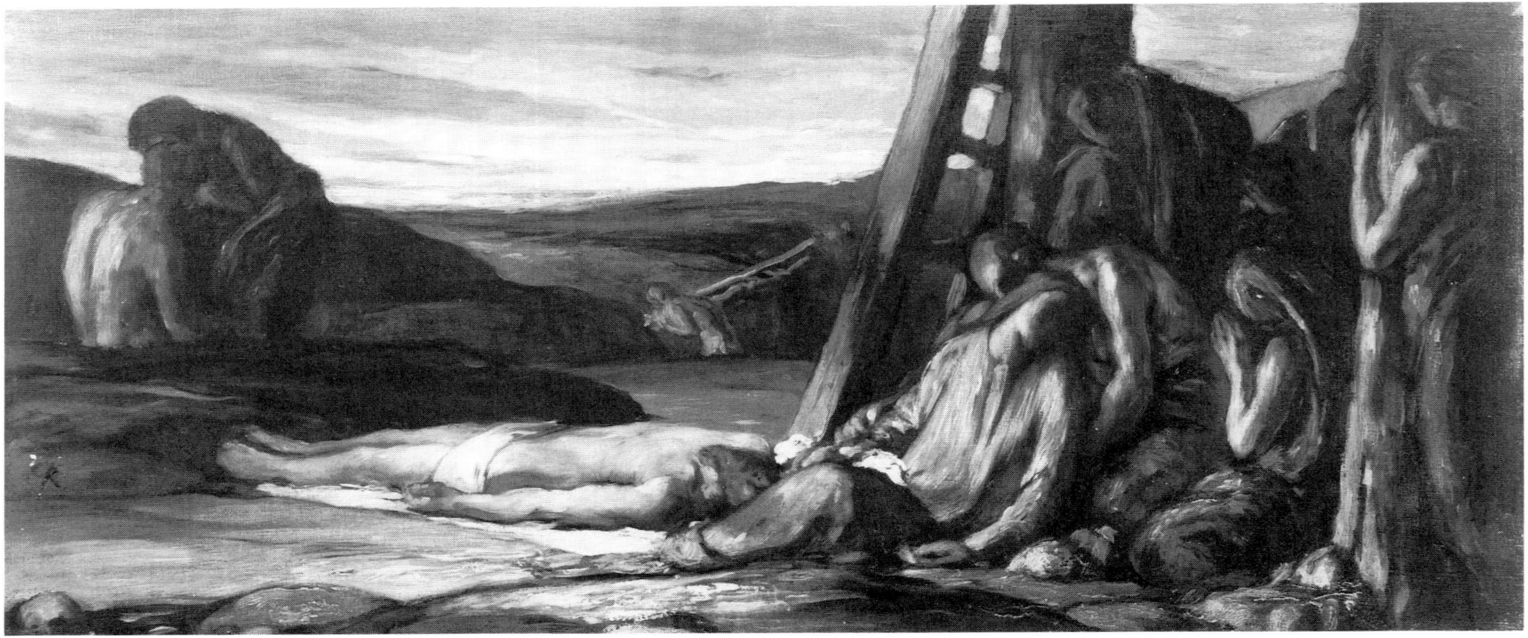

FIG. 27. *The Descent from the Cross*. Oil-painting (*c.*1905) by Charles Ricketts (William Morris Gallery)

his arm and sword before the apparition of the Commander in the cemetery. The inspiration for this picture was a poem by Baudelaire, as well as Mozart's opera, but the 'Don Juan in Hell' scene from Shaw's *Man and Superman* also influenced his later paintings of the Don. Ricketts admired Don Juan for being true to himself at the risk of all. He told the Fields that he wanted to write about him and that:

> Don Juan is a grand gentleman, the type of debauché who gives the world vast pleasure and takes on himself almost the pains of a saint. A creature enjoyed as a saint is worshipped for his idealism. He would make the world nearer to his heart's desires and Don Juan faced damnation for his dream. He never recoiled fearful.[79]

This unconventional interpretation of the Don once again sets man against his fate, man against the forces of death and destruction. This was part of that 'battle between the immortal earth & the immortal human energy' that Ricketts had reflected upon in the ruins of ancient Rome. Ricketts's vision of life was grand, heroic, and Romantic.

He was also exploring (like Delacroix) another favourite theme, Christ's passion, of which he had painted ten versions since 1903–4. An atheist and humanist, he did not regard these as religious pictures. Christ was seen as a man, not the second person of the Trinity, and was thus grouped incongruously in Ricketts's private mythology with such figures as Don Juan, Montezuma, Wilde, and even himself. When he did produce a painting which could not be understood in purely humanistic terms, like *The Resurrection*, he confessed that he could give no reason and pretended, in the vein of 'pretty fictions' that he enjoyed, that his brush was responsible. What interested him in Christ was the tragedy and pathos of the man who had stood up for his vision in the face of ridicule, torture, and death. Indeed, he is another example of the gifted individual persecuted and destroyed by the world's philistine authorities. Yeats, who admired Ricketts's paintings more than Shannon's, for 'their intensity of originality no less than Blake's', told him 'you paint the tragedy of Man. Most people only understand the tragedy of Woman, or the pittifulness of the tragedy'.[80] Several of his depictions of the Passion count among his best works, like the *Descent from the Cross*, in which the isolation of the dead body with its head in the lap of His Mother, who falls back in grief, is heightened by the

[79] MF Journal, 1908, fo. 198; 1904, fos. 23^{v}–24.

[80] CR Diary, 3 Nov. 1905 = *SP*, p. 126.

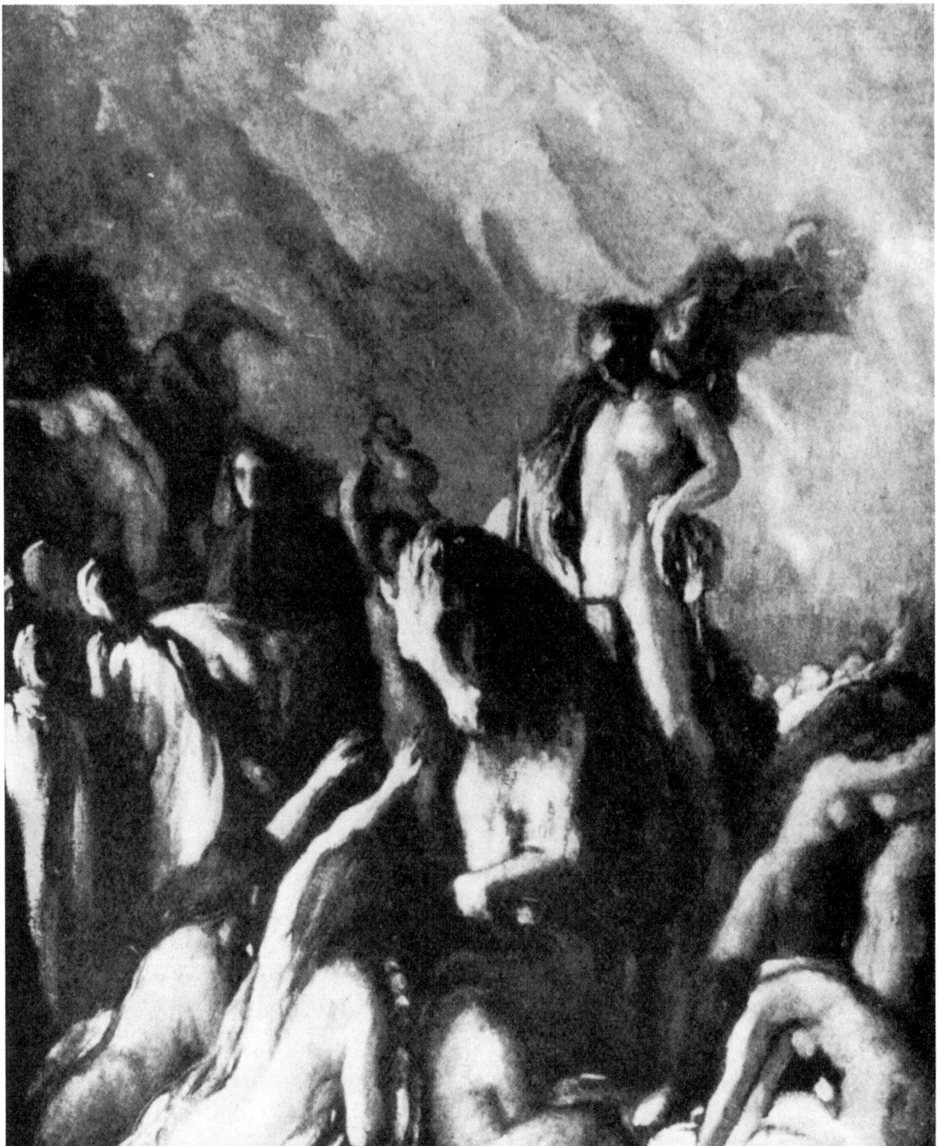

FIG. 28. *Don Juan in Hell.* Oil-painting (*c.*1908) by Charles Ricketts (Photo: Witt Library, Courtauld Institute, Univ. of London)

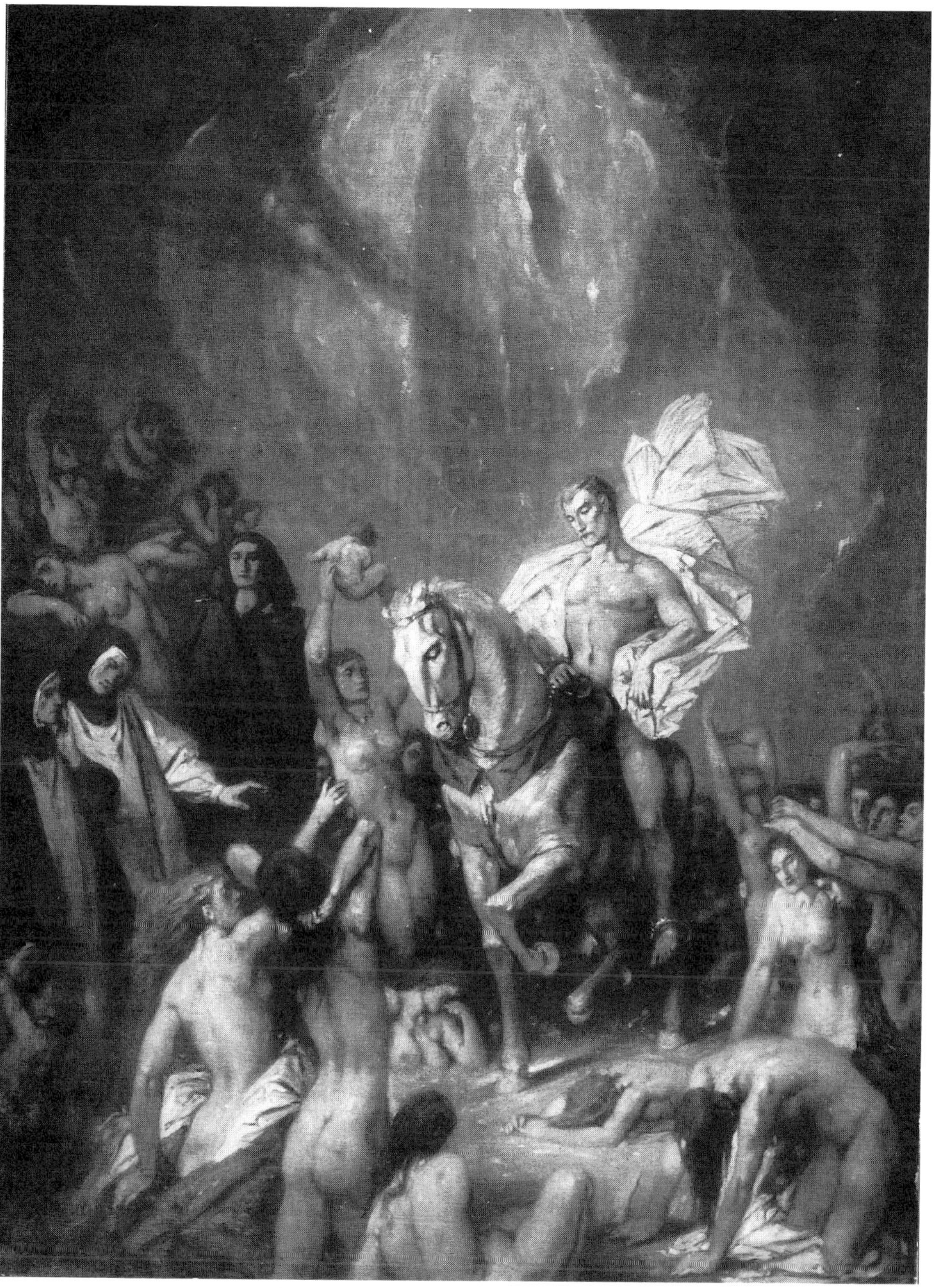

FIG. 29. *Don Juan in Hell.* Oil-painting (1931) by Charles Ricketts (Grundy Art Gallery, Blackpool)

long narrow format and the contrast between the pale body and the dark landscape and clothed figures (Fig. 27). Touches of drama in the wind-blown hair on the left and figures clutching the cross and looking at Christ on the right enhance the central stillness of the picture.

As an artist, Ricketts saw himself as having a split personality. On the one hand there was 'the born ornamentalist one finds in the details of my theatre work' (and book illustrations); on the other,

> the rather hectic improvisatore of my paintings and bronzes. Both currents run side by side and refuse to blend. My painting refuses to face detail and certain kinds of invention, which Moreau or Burne-Jones have at their finger-tips, and once the pencil or point is in my hands, I am incapable of breadth or suggestion and fall into filigree and can lapse into an almost Persian finick and finish![81]

The early influence of Daumier and Delacroix had encouraged breadth and looseness of handling. Moreover, he used a very limited colour range; Bottomley was probably right to suggest that the theatre gave Ricketts opportunities to satisfy his colour sense, so that it never became important in his paintings, most of which have a bluish or greenish-blue tone and an effective use of accents of red. He was (in his own words) a 'harmonist', who used subtle modulation and transition within a narrow spectrum, rather than a 'colourist', who makes use of a wider or bolder pattern of colours.[82] However, he was to evolve from the vague, passionate, flickering manner of these early paintings to a more precise style, with more detail, clearer outlines, and a firmer sense of design. If we compare two versions of the same subject, *Don Juan in Hell*, one (Fig. 28) painted in about 1908 (which inspired a poem by Yeats) and the other (Fig. 29) in 1931, this development can be clearly seen. The design is the same, but the vagueness in the earlier version, which probably arose from an inability to commit himself and, perhaps, from a lack of technical mastery of the medium, gives way to an assured and individual style. His vision too was personal and original, as Yeats pointed out. Nevertheless, since his work is not well known, he is often unjustly described as 'derivative' and a 'pasticheur'. Whilst he did borrow some themes and motives from artists he admired, in a way that was regarded as perfectly normal until modern times, he also painted others like Montezuma and Don Juan

81 CR to Cecil French (TS), 12 July 1927 = *SP*, p. 379.

82 *Titian*, p. 124.

that were entirely his own. Moreover, his work bears the stamp of his own style, is always immediately recognizable as his, and could never be vaguely classified as by an unidentified follower of Moreau or Delacroix.

Even in his most derivative pictures he transforms the borrowed material. *Heliodorus driven from the Temple* (Fig. 30) is closely based on the painting by Delacroix, just as the latter is in turn derived from Raphael. The incident depicted is the sudden arrival of an 'angel' on horseback with two attendants, in answer to the prayers of the high priest and people when Heliodorus, under orders from the king, attempts sacrilegiously to despoil the temple treasury. Ricketts has altered details, such as the head of the horse, the position of the attendants, and the placing and stance of the high priest, and has made some additions, like the sweeping clouds, the dramatic contrasts in lighting, and the vortex of smoke and flame. The event is seen from a much closer vantage-point and fills the canvas, so that the setting becomes vague and unimportant and the relationships in space are somewhat lost. The result is much more improvised, more visionary—like Moreau—and even more theatrical in comparison with the monumental quality of the Delacroix. It is not mere slavish copying.

The question of originality is a vexed one, full of ambiguities and questionable assumptions. It is only since the early nineteenth century that artists have been under such pressure to be original, to do or say what has never been done or said before. For Ricketts, this compulsion to owe nothing (or at least as little as possible) to the past was really a recipe for anarchy and regression, since the development of skill and culture depends on slow and careful accumulation, on one generation building on the work of its predecessors.[83] The aim of the great artists of the past was to emulate the master in whose studio they had worked; what was original in their work grew naturally out of their characters and particular gifts. Artists often chose a conventional subject, partly in homage, partly as a challenge, *because* it had been painted by a master they admired. Thus Ricketts gladly acknowledged his debts, and was delighted, for instance, when a Swedish visitor 'discovered the influence of Delacroix & Moreau in my work, as a compliment, whilst Holmes in his British way viewed it as a *score against me* and looked to see me blush'.[84] However, it is interesting to note that even Ricketts

[83] See *Pages on Art*, pp. 150–1.

[84] CR Diary, 13 Nov. 1903.

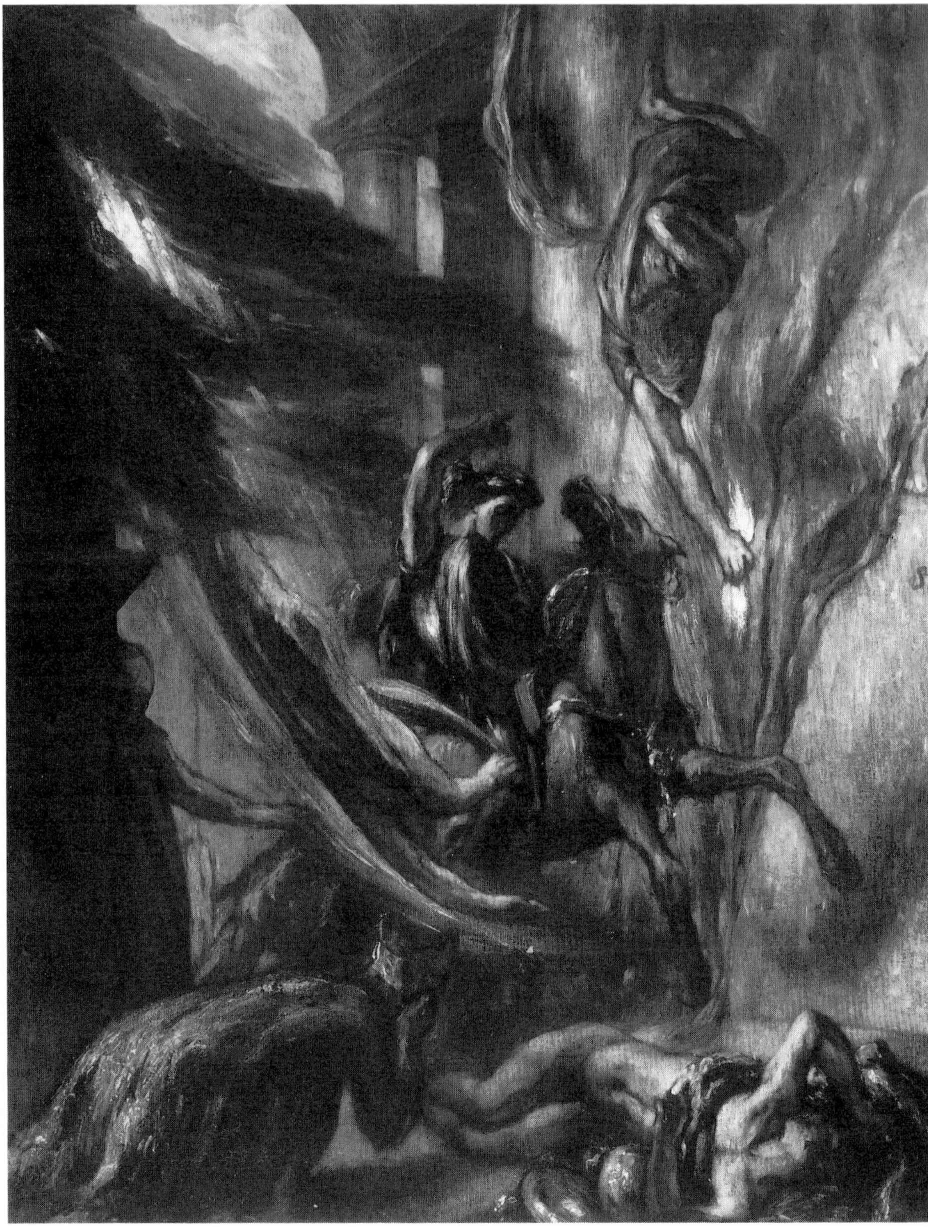

FIG. 30. *Heliodorus Driven from the Temple*. Oil-painting (1906) by Charles Ricketts (Photo: Fine Art Society, London)

felt inhibited from expressing 'the vein of Rossettian invention' that he possessed, for fear of revealing Rossetti's influence. 'In the past', he said, 'an artist would not have feared to own a debt to a great master of his own time.'[85] The youthful striving for originality that marred his early book-designs later gave way to a more genuine originality, both in his vision and in the qualities he sought to incorporate into his work.[86]

It is a modern illusion that the best work is always the most modern, the most advanced, the most superficially original. Some great artists, like Bach or Samuel Palmer, worked in styles considered old-fashioned in their time; and the course of artistic development usually contains several differing, and even conflicting, currents that not only reach full flood at widely different times but also come to be understood at different periods and are valued very differently by ensuing generations. One should not, therefore, damn an artist because he does not reflect the dominant style of his time. Ricketts's paintings have suffered from the fact that they have not conformed to the trends in art history that have mainly interested art historians. Now that Symbolist painters are becoming academically respectable, Ricketts's paintings will be taken more seriously.

Ricketts was not a great painter: he knew his limitations and acknowledged that the great artists compared to him were like 'Mount Everest compared to Primrose Hill'.[87] He was too uneven. His technique and draughtsmanship are too often faulty. His range of expression was too limited, and he lacked the 'creative power' he saw in Titian, Rossetti, and the world's greatest artists. Though some of his paintings are disappointing, and a few are failures, he did many fine ones. His achievement belongs within the classical tradition of European art, of which Symbolism was the swan-song, and must be judged by standards

[85] CR Diary, 9 Mar. 1900.

[86] It should be noted that the damning charge of derivativeness is usually only held against modern artists whose work derives from the European classical tradition, whereas to be derivative from contemporaries, as Camille Pissarro and others were from Seurat, or as many artists were from Braque and Picasso's development of cubism, or from primitive or foreign cultures, is thought to be highly advanced and original. Clearly, it is not being derivative itself that is regarded as wrong, but being derivative from the 'wrong' influences at any particular time. Yet, as Pietro Annigoni has pointed out, 'it is the destiny of every artist to say something new in an old language' (Annigoni, *An Artist's Life* (1977), 70).

[87] CR to Mary Davis (4 p.c.s), p.m. (envelope) 3 Oct. 1926.

that were falling out of use in his own day and have now largely disappeared from the aims of artists. Ricketts rightly described himself as 'an undiscovered Master' of the nineteenth century.

His was not a style that was liked, or indeed understood, by many other artists—the portraitists, for example, the landscape painters, and the other naturalistic artists (however poetic their evocations) like Steer, Tonks, Orpen, and the rest. Tonks, indeed, convinced that no modern artist untouched by Impressionism could be of any value, said of Ricketts's work: 'Of course, some people may think it very nice but it is just *not* painting.'[88] In their struggle for expression and recognition artists can become very bigoted. Now that the heated battles of that generation have passed into history, we can look at their work without the bias of the apologist or antagonist (from which Ricketts himself was rarely free: Laurence Irving once asked Ricketts who the best landscape-painter was, and got the reply, 'Wagner'[89]) and judge it on its own merits in terms of what the artist set out to do. Few other British painters of his period expressed (as did Moreau, Watts, and Burne-Jones) this kind of personal vision, which sprang from the deepest levels of his own nature and understanding of human life. Most of his contemporaries were intent on interpreting external reality in their personal idiom, rather than an inner vision. Nor do they consequently have the imaginative quality, the creative expression, that Ricketts (and to a lesser degree, Shannon) put into his work. An artist's choice of literary subject can be psychologically revealing: it can show the type of human situations that were most significant in his life. For the literary artist, the starting-point is the imagination rather than an external stimulus. 'Symbolism' was a vehicle of powerful personal expression, reflecting the artist's imaginative powers and, often, an interest in violent, dramatic, or even sensational action. All of these are found in Ricketts, whose major influences, even including Rossetti, can be classed as Symbolists. Moreau's rich colour and elaborate ornamentation, however, found its way into Ricketts's theatre work rather than his painting, which took from Moreau its fascination with violence as well as certain subjects and motives. Even here Ricketts assimilates his influences, so that his *Salomes*, for instance, are more objective, less visionary, than Moreau's, whose most famous *Salome* contemplates a

[88] Joseph Hone, *The Life of Henry Tonks* (1939), 245.

[89] Laurence Irving in conversation with the author.

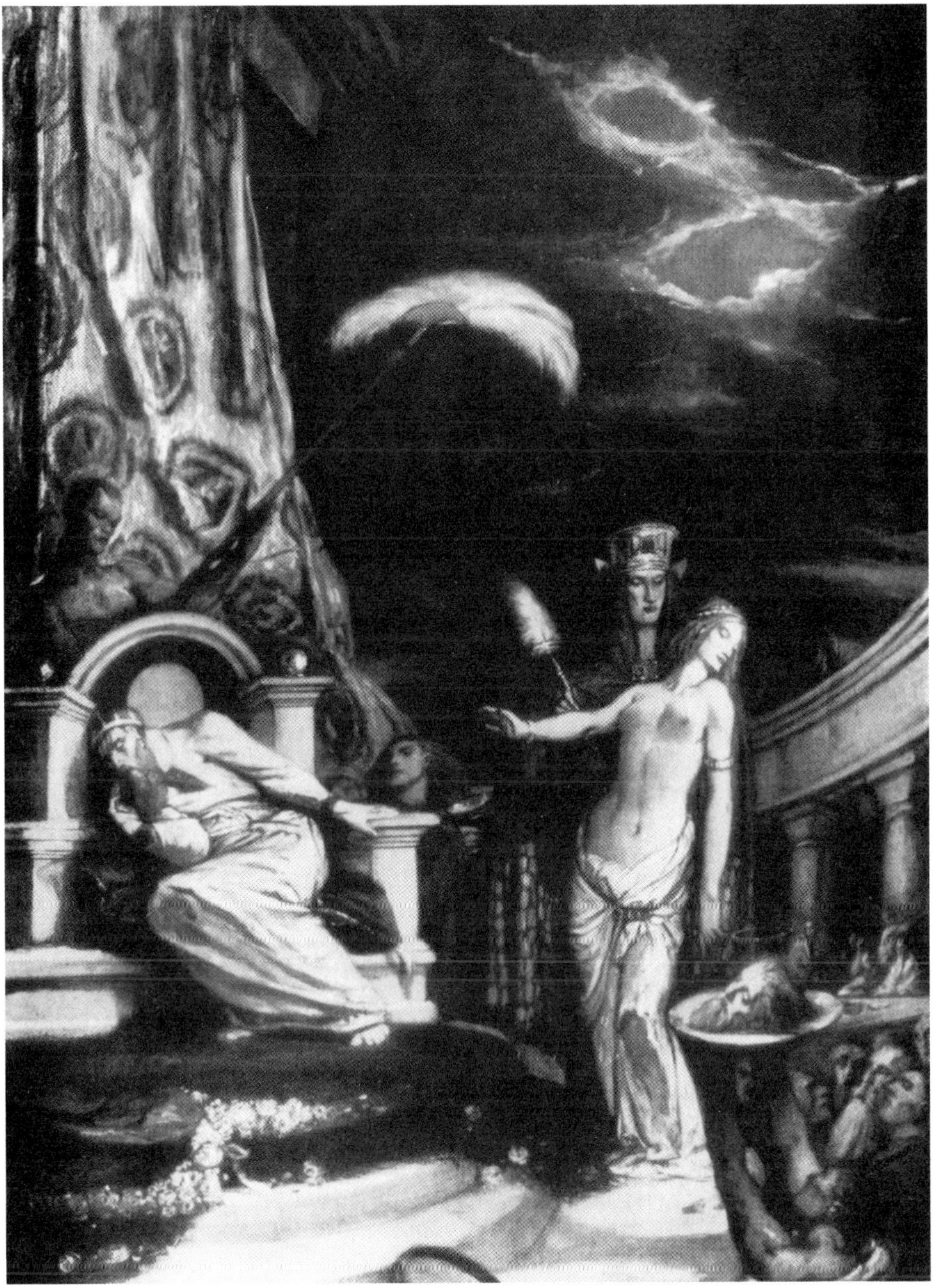

FIG. 31. *The Head of John the Baptist*. Oil-painting (1928) by Charles Ricketts (Photo: Witt Library)

head disembodied and suspended in mid-air; Ricketts puts the head firmly on a dish (Fig. 31).

Chimaeras—one of his most obviously Symbolist paintings (Fig. 32), though not one of his best—shows how his mind worked. It was inspired by a phrase he had read in Baudelaire: *chacun porte sa chimère*, and was sometimes given this title. It referred to classical art and involved one of the hybrid creatures (like the sphinx, centaur, and even minotaur) in his fabulous menagerie. Classical chimaeras usually had a dragon's tail, a she-goat's body, and a lion's head, together with the head of the goat and sometimes even of the dragon. His version has the goat's legs but also wings, and a human female torso with the curious cat-like face that he gave to so many of his figures. He described it as 'a sort of Dantesque scene of nude men with chimeras perched on their shoulders with their claws fixed in their head, and shoulders and back sides'.[90] It is a grim and disturbing picture. Though on a general level an image of the human psyche as a prey to guilt and terror, it evidently had a personal significance for Ricketts, haunted as he was by fears and anxiety. Nearly all his paintings had some personal significance for him, and deliberately sought to convey particular emotions. A 'sheepish American' collector brought by Robbie Ross so obviously disliked Ricketts's work that Ricketts, whilst affecting an easy indifference, experienced a new sensation of pride in it: 'I forgot its shortcomings, I became aware of its sense of design, & the emotions of awe, melancholy, and compassion which my work expresses.'[91]

A concern with a tragic fate characterizes most of his paintings. The best of these—among them *The Betrayal* (1904), *Ecce Homo* (1910), *The Death of Cleopatra* (1909), *The Flight of Cleopatra* (1912), *Jephthah's Daughter* (1924), *Orpheus and Eurydice* (1923), *Judith Returning to Bethulia with the Head of Holophernes* (1919), *The Holy Women and the Angel of the Resurrection* (1907), *Faust and Mephistopheles* (1909), and several depictions of Christ, Don Juan, and Montezuma—all evoke tragic situations, such as are rarely found so consistently within the work of a single British artist. Such a preoccupation was, as he said of his pessimism, more French than English. So, perhaps, are the qualities of expressive design and imaginative vision that he sought. Moreover, his painting reflected his inner self: its tragic and dramatic scenes

[90] CR to TEL [1917]. 'Chacun sa chimère' was a prose poem in Baudelaire's *Petits Poèmes en Prose* (*Le Spleen de Paris*).

[91] CR Diary, 15 Feb. 1905 = *SP*, p. 118.

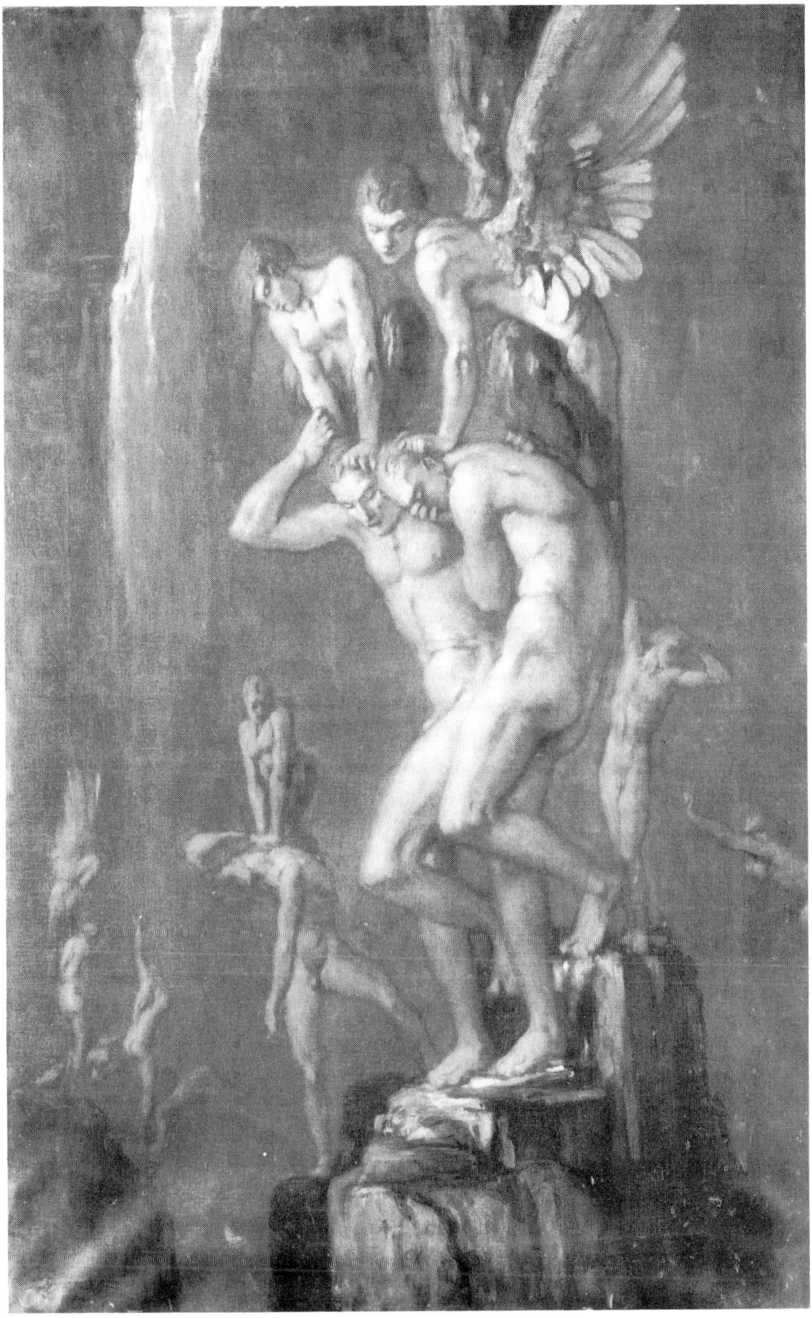

FIG. 32. *Chimaeras* (or *Chacun sa chimère*). Oil-painting (1917) by Charles Ricketts (Photo: Sotheby's)

express the deep pessimism in his own nature, his romantic view of man pitted against a hostile fate, his fascination with the darker elements in life: lust, cruelty, barbarity. Underneath these Symbolist themes lay a deep stoicism, a broad compassionate humanism, a fervent aestheticism that saw beauty as the only thing that mattered, and the only thing that lasted in human endeavour.

With his two one-man exhibitions, the first of his paintings and the first of his sculptures, Ricketts was making his bid for recognition as a 'fine' artist, as opposed to a decorative one, a book designer. Working in both new media at the same time, he noticed the different way the finished product in each reflected his original intentions. When painting he improvised, progressed intuitively, and usually felt as if another hand held the brush, so that he did not consciously know what he would express, what was in his thoughts. This frightened and often startled him. In sculpture, on the other hand, he expressed himself, could foresee his effects, provide for them, attain them. Working on canvas he was uncertain, whereas working in clay he was sure.[92] Yet painting remained for him the most desirable, if the most difficult, means of expression.

Throughout much of 1906 Ricketts was deeply depressed, partly over his work. 'There are moments', he recorded, 'when ones known limitations as an executant weigh upon ones shoulders and effort hardly seems worth while.' The world was very 'conscious of the short-comings of good work', but blind to bad or indifferent work which had a fashionable appeal. Moreover, he had reached 'the curious age when one seems to be parting company with most things, & only work remains; when this grows perplexing, one sinks into horrid depression. The evil spirit merely says, "Do you really think this worth doing? A quoi bon?"' He felt he was suffering from an attack of mental fatigue and hesitation, 'during which one sacrifices major things to small things. I am plucky at the attack, but slow or uncertain in following thing[s] up.' This, he felt, was racial, 'quickness of wit, lack of will due to a false environment or a dislike for my environment'. He did not think he would be different if he changed country, but might have been if he had never lived in England, where the climate depressed him when he was not actively employed and the people seemed 'dishonest in their indifference to fine things. They are vacant & grasping and unfriendly also'. In the whole of the year he was only

[92] MF Journal, 1908, fo. 10.

to complete two or three paintings, whereas in the previous summer he had done seven. 'The odds against the artist are too great', he decided.[93] When hanging pictures for his show he had sunk deeper into gloom. Then, not only had it proved a failure, with only one picture sold to a friend, Pye, but he had returned from Italy to find that he had sold only one bronze, again to a friend. It earned him twenty-five pounds, while the show had cost him fifty. Some years later Ricketts declared, 'I have often found in art, dislike is the sincerest form of flattery. This sounds like Oscar!'[94] Here, however, it was not so much dislike as indifference, which can be harder to bear, and much more undermining. 'I think something snapped after the failure of my two shows', he confessed afterwards.[95]

Whilst he felt he could earn some sort of living for himself as an art writer and journalist, this prospect was not pleasant. His responsibilities to Shannon had also to be considered (for it must be remembered that Ricketts had subsidized him for years). However, his writing on *Titian* was not going well. At first he found he could write, on average, no more than four pages of manuscript a day, no matter how energetically he set to work. It had been the same with *The Prado*. If he did more one day, then another day he did less.[96] By September his writing had 'got stuck'.

His inability to sell his work, moreover, meant worries over money, and not only for living, for there was 'no shirking the fact that one has acquired habits of extravagance or extravagant purchase of art treasure; these one would not like to have to sell'.[97] Suffering from nerves and despondency, he painted *Death in the Picture Gallery*. A skeleton removes a painting (rather like a Ricketts) that has failed to find a buyer at auction. Its meaning was clear.

Whilst doubts over his work and the failure of his shows certainly contributed to Ricketts's depression, a deeper cause is not hard to find. Shannon was in love again. Ricketts's anxiety breaks out in his diary in September: 'I have . . . been frightened by the question, "What should I do if Shannon got married?" '[98] This throws light on much

93 CR Diary, 2, 21 May, 28 June 1906 = *SP*, pp. 135, 138.
94 CR to Robbie Ross [June 1911].
95 CR Diary, 7 Sept 1906 = *SP*, p. 138.
96 CR Diary, 26 June 1906.
97 CR Diary, 28 June 1906 = *SP*, p. 138.
98 CR Diary, 7 Sept. 1906.

of his behaviour during this gloomy year. It may be no more than a coincidence, but it is tempting to speculate that Ricketts went to Italy on his own because he felt rather *de trop*. He had last travelled abroad when Shannon was involved with Hetty, though on both of these trips he was doing specific research for a book. Moreover, his cruel anger with the Fields over their dog (as they themselves realized later) may have been partly due to the strain of his worries about Shannon.[99]

Shannon had fallen in love with an attractive and talented sculptress, Kathleen Bruce, whose dark good looks and ardent romantic nature must have owed more to her half-Greek mother than to her father, who was a north-country parson. She had studied sculpture under Rodin, and had also nursed villagers in the Balkans savagely wounded by the Turks. When she met Shannon she was twenty-six; he was forty-three. She claimed to be the first woman Shannon had ever loved.[100] Without doubt he had never met anyone before whose artistic and intellectual interests, matched by beauty and a vivid personality, offered him anything like, or anything more than, his relationship with Ricketts. This is what frightened Ricketts so much. She sat to Shannon for several lovely portraits and, for her part, she produced statuettes of both Shannon and Ricketts. Throughout this crisis, and afterwards, Ricketts maintained friendly relations with her.

As before, Ricketts turned to Italian to express his fears in his diary. Miss Bruce was all the more dangerous because she was intelligent and '*simpatica*'.[101] 'They are foolish and go about together', he told the Fields, when they asked him brightly near the end of 1907: 'We hear Shannon is going to be married—is it so?' Ricketts thought not; but two years before he had been very frightened. 'And what would become of the collection?' he asked, just as Edith Cooper was about to put the same question.[102]

However, the danger was already past, though Ricketts may not have realized it. Kathleen Bruce had already met a man after her own adventurous heart, Capt. Robert Falcon Scott, who died tragically in the Antarctic only a few years later. When she told Shannon, at the beginning of January 1907, that she was to marry another, the 'gentle artist' blundered out into the street and into the path of a car that

99 MF Journal, 1907, vol. 2, fo. 15^{v}.

100 Lady Kennet, p. 84.

101 CR Diary, 17 Sept. 1906.

102 MF Journal, 1907, vol. 2, fo. 15^{v}.

knocked him down. He recovered consciousness only 'by contact with the ugly monument at Knights Bridge upon which he was propped', Ricketts told the Fields. He was taken to St George's Hospital, but his injuries were not serious. Bruised, scratched, and covered with mud, he returned home and was able to attend the International Society Banquet with Ricketts shortly afterwards, 'the only difference being that he had to sit down very carefully' at a table full of 'ennemies'.[103] Ricketts was shocked to think that Shannon might not have returned from the accident, and relieved, not only that he had done so, but also that their relationship had survived its most serious challenge. Shannon was to have other relationships with women (presumably, for instance, with the actress Hilda Moore, whom he painted some seven times), but none presented the same threat.[104]

Curiously, Ricketts was to look back to 1906 as the pinnacle of their careers.[105] In addition to his own two shows, he and Shannon had been represented at an important show at Agnew's, one of the first exhibitions of contemporary British art in Bond Street. Agnew's had bought Ricketts's *The Betrayal*, which John Masefield had described as 'the finest imaginative picture of our time'.[106] They had also been well represented at a large show in Bradford to mark the opening of the new gallery there. Never did Ricketts have so much of his work on show in one year again.

His depression and unproductivity had eased by early 1907. He was working again and 'like a cook with several sauces in the fire turn from *Death in the Picture Gallery* to *A Resurrection* and from *The Death of Cleopatra* to *Calvary*'. This last he felt was unsuitable for a church ('It is quite lay'), but it is now in York Minster.[107] Shannon thought it Ricketts's masterpiece. He went on to do *The Triumph of Bacchus* and *The Holy Women and the Angel of the Resurrection*. He was also deeply involved in theatre work for several productions.

Before any of these reached fruition, however, he had another chance to wear one of his own designs, at the Fancy Dress Ball sponsored by the International Society early in March. 'On that occasion', he announced to the Fields, 'I intend stepping out from the right hand

103 CR to MF [9 Jan. 1907]; Lady Kennet, p. 85.

104 I owe this suggestion to Prof. Carl Woodring.

105 CR to CHS [Oct. 1927].

106 CHS to CR ('My dear Ridgley'), [15 Mar. 1906].

107 MF Journal, 1907, vol. 1, fos. 25, 162^{v}.

of Cesare Borgia as he appears in the Vatican fresco'; a reference to a Greek or Armenian who 'stands with his hand on his hip and looks out into the future' in Pinturicchio's *The Dispute of St Catherine* in the Borgia Apartments. The underskirt was to be soft green covered with silver pomegranates, the overskirt of soft, thick cloth of gold, the sleeves of purple, the cloak of white watered silk and there was to be a curious cap of Samite on the crown of his head. In spite of a poorish supper and an idiotic human Punch-and-Judy show, the ball was a great success, as was the costume: 'My dress created quite a sensation, which surprised me.... Enemies & strangers were quite breathless with admiration, & I rather disliked having to appear in the ball room between the dances, as I was stared at too persistently.'[108]

White, purple, and gold, the dominant colours, were also the colours that Ricketts used in the costumes for *The Persians*, by Aeschylus, for the Literary Theatre Society, which was to be presented with *A Miracle* on 23 March. After *Salome*, which had lost thirty pounds, they had thought of winding up the Society but, finding that they still had fifty pounds, they decided to go on. For *The Persians* Ricketts also did a poster: 'It is my first, only, and probably *last* poster' (it was not).[109] Robert Farquharson directed *The Persians*, whilst *A Miracle* was done by Harley Granville-Barker; but the production was not a success. 'The press have yelled', Ricketts observed, 'with one voice of execration against Aeschylus, Farqharson, Barker, and my setting. At first I thought of rushing away to Spain or South America where I believe one can still be safe. But I found Farqharson & Barker were not going, so "Behold here am I".'[110]

Aeschylus's play had been given a translation 'filled with verve and muscle' (according to Ricketts) by Bernard Ryan, who, Ricketts told the Fields, was a friend of his.[111] They were quick to spot that this name combined the names of the two men dearest to them, Bernard Berenson and Dr John Ryan, husband of Edith's sister Amy. They strongly suspected the real translator was Ricketts and wrote to ask him. Binyon too wrote to Ricketts to say how good the translation was. 'Do you say so because Moore or Ross have blabbed?', Ricketts

[108] MF Journal, vol. 1, 1907, fo. 49

[109] CR to John Lane (TS) [Mar. 1907].

[110] CR to L. Binyon [Mar. 1907].

[111] MF Journal, 1907, vol. 1, fo. 36ᵛ.

replied.[112] His teasing of the Fields had combined with that love of pseudonyms he had shown in the Vale days. A gifted linguist, he had learnt Greek, along with Latin, Italian and German, largely through his own efforts, though in the modern languages at least he had had a few lessons, and for this Greek translation he had made use of a crib.

This was the end of the Literary Theatre Society. Their productions had been mounted with the greatest economy. *The Persians* had cost thirty-two pounds; *A Miracle*, only fifteen.[113] Ricketts, of course, had given his services free. However, funds were exhausted. For Ricketts, the Society had provided an outlet for schemes and ideas he had long cherished. Its scope, however, was limited and its appeal was to a small, if rather highbrow, audience. It did not really give full expression to his abilities, but it led to other things. Just as Wilde had fostered his early book designs, George Bernard Shaw was to offer Ricketts the opportunity to tackle more ambitious projects, to have greater resources, and to reach a wider audience.

112 CR to L. Binyon, [Mar. 1907]. The TS of Ricketts's translation is in BL Add. MS 61724.

113 CR to GBS [March 1907] = *SP*, pp. 140–1.

8 *The Theatre Designer*

'MY *Don Juan* is sold and sails for America', wrote Ricketts at the end of 1907. 'Mammon or Mephistopheles disguised as an American art dealer called this morning and tempted me with what was at first an insufficient sum of gold. I demured and pointed out that I could deprive England of this masterpiece for £50 more, and prevailed.' The American thought that his countrymen 'would just appreciate the modern character of my paintings', and asked for biographical details to introduce Ricketts to the art-loving public of the New World. 'I am preparing a lurid biography in which I shall be born in Italy. I shall knock ... 2 years or so, perhaps three, off my age. My grandfather shall be American. Any detail you can think of shall be written down', he told the Fields. 'Nothing too scandalous, however, as America is still somewhat Puritan, and Mephistopheles will [have] none of Shannon's pictures as they are too *nude*.' The dealer informed Ricketts, 'It is strange how particular the American public is on this point.'[1]

Another variation on the Don Juan theme had been played six months before. Ricketts had clothed the notorious lover for the London stage. For some time the Literary Theatre Society had thought of staging the 'Don Juan in Hell' scene from George Bernard Shaw's *Man and Superman*. Shaw and Florence Farr were invited to Lansdowne House to discuss the possibilities at the end of 1905, and, despite the fact that Ricketts and Shaw found they had different concepts of beauty and different interpretations of the characters, the meeting went well and 'the chatter was distinctly vital & vivacious'.[2] Whilst superficially the two men seemed worlds apart—Shaw a Fabian Socialist and Ricketts an aesthete and élitist—they had much in common at a deeper level. As well as being serious artists, and sharing a wide interest in art, music, and drama, they were both keenly aware of the dark side of life, the forces of evil and destruction. This made Shaw turn to social reform, and Ricketts to beauty. Through these different answers,

[1] CR to MF [Dec. 1907]. The dealer was C. W. Kraushaar, 200 Fifth Ave., NY, and the price was £250.

[2] CR Diary, 31 Dec. 1905 = *SP*, p. 128.

each sought to improve the quality of living. Wilde had first drawn Ricketts's attention to *Arms and the Man* in 1894, when Shaw was little known and, according to Ricketts, 'when the English stage had sunk to a curious level of meaningless dullness'. Soon he agreed that Shaw was one of the two contemporary 'Britishers' (along with Wilde) whose plays showed 'fundamental brain work or intellect'.[3]

Even before the collapse of the Literary Theatre Society the Court Theatre had won the right to stage the play, but Shaw asked Ricketts if he would still undertake the 'inscenation': 'Half the fun and interest of the exploit would be gone for me if it were done without you.'[4] Immediately Ricketts agreed to 'do sketches and provide tips', and presented such a reasonable estimate for his work that even the 'parsimonious management was enthusiastic'. Ricketts wished, however, to remain anonymous, since his work for the Society had raised 'a whole hearted howl in the press worthy of a masterwork by Wagner ... This I do not mind but it leads to difficulties with the actors'.[5] He did mind, of course, very much.

'Don Juan in Hell' was given on 4 June 1907, with another short play by Shaw, *A Man of Destiny*. Ricketts had the stage draped in black velvet, and the costumes were modelled on Velasquez. Lillah McCarthy as Dona Anna looked so gorgeous in her 'infanta dress'—rose silk covered with black lace and silver trimming and a flame feather in her hair—that Shannon painted her in it. Robert Loraine, however, as Don Juan, was 'simply wretched' in his exquisite suit of silver satin and purple and, said Shaw, thought 'Ricketts was making him ridiculous instead of immortal'.[6] Tactfully, Shaw left the problem to be dealt with by Ricketts. Nor did Shaw mind Ricketts's 'tips'. Ricketts objected, for instance, to the Devil's 'sheepish lunge' to get Dona Anna a seat, and suggested that he should deliberately fetch her a seat, and that when he ceremoniously offered it to her she should bow in an agitated way: 'I feel there is too great a lack of ceremony here and there and that Ana should listen to the Devil at times rather as if she were at a concert or listening to a cardinal.'[7] Every part of the play benefited from Ricketts's critical eye, attention to detail, and fertile imagination. The 'magical production' delighted its author-producer.

[3] 'Marginal Recollections' (MS); CR Diary, 23 May 1905 = *SP*, p. 121.

[4] GBS to CR, 29 Mar. 1907 = *SP*, p. 140.

[5] CR to GBS [March 1907] = *SP*, pp. 140–1.

[6] GBS to Mrs Patrick Campbell, 3 July 1912.

[7] CR to GBS, [1 June 1907] (Univ. of Texas).

It seems to me [he wrote to Ricketts the next day] that we (I say 'we' much as an organ blower uses the plural pronoun when speaking of the organist's performance of Bach's *Fugue*) hit on a most valuable and fascinating stage convention... If only we get a few plays with invisible backgrounds and lovely costumes like that in a suitable setting ... there would be no end to the delight of the thing.[8]

This production also brought Ricketts the friendship of Barker's wife, Lillah McCarthy, one of the most distinguished Shavian actresses of the day. She became devoted to Ricketts (who, she recalled, 'held my love and guided my tempestuous spirit over many rapids') and appreciated the scores of postcards he sent her 'each with a little word of advice or a phrase of congratulation'.[9] While Ricketts, for his part, admired her as an actress, he sometimes felt rather impatient with her mind for being so much slower than his: 'She is stupid but worthwhile', he told Bottomley. 'At rehearsals, one tells her a thing over and over, and it seems to make no difference. She continues in the mistake as though she has not heard—but if one keeps on long enough she suddenly forgets her cherished mistake and does something amazing.'[10] She often asked Ricketts to do her dresses for productions designed by other artists. Though he frequently had to work through the night, not only designing but stencilling the costumes (so that they arrived in the morning smelling of new paint), he usually refused any payment. For the 1907 revival of Shaw's *Arms and the Man* he produced two costumes for her that revived the bustle. She was rather apprehensive about this innovation, but, Ricketts noted, 'its success was immediate with the charladies who were dusting the seats during rehearsal. "Ain't she lovely?" they exclaimed, and I have ever since considered the employees of a theatre the soundest judges of scenery and dresses.'[11] The production was equally successful with the public; and the bustle became a familiar sight on the stage.

No sooner was this production off his hands than he was deeply involved in Laurence Binyon's play, *Attila*, the printed version of which was dedicated to Ricketts. His work for this production, given at His Majesty's Theatre, met with its share of misfortune. Having done twenty-eight designs for the properties in the play, he learnt that

[8] GBS to CR, 8 July 1907 = *SP*, pp. 142–3; GBS, *Letters*, ii (1898–1910), ed. Dan H. Laurence (1972), 698–9.

[9] McCarthy, p. 111.

[10] G. Bottomley to Laurence Binyon, 12 Apr. 1942.

[11] I cannot locate the source of this quotation.

the entire roll was lost and would have to be replaced rapidly: 'Six weeks have been wasted in mean and purposeless daudling.'[12] The setting was Hunnish, an architecture of 'huge, fiercely painted logs and crowned with bison's horns'.[13] In the costumes there was a savage grandeur, enhanced by the striking barbaric jewellery he designed—which was reproduced in a special article in the *Studio*.[14] As the first night, 4 September, drew nearer, Ricketts was rushed off his feet. 'We are in the last throes of *Attila*', he explained to the Fields, 'the period of genuine tragic effects and startling development of situation.' The play was to open on a Wednesday. In the previous week, 'On Thursday, every body looses their temper, on Friday, trying on day, Saturday Dress Rehearsal!! Perhaps you can advise me, should I commit suicide on Friday or run down to the seaside early on Thursday. There is something to be said for either course.' He was uncertain of his movements, however, 'since hairdresser, shoemaker . . . Costumier, Miss Stone, Scene-painter, Limelight man, property man, stage carpenter, two leading ladies and a popular boy actor, will all view me as "prepared" by the Lord for their own special good, comfort, and advancement'. Lily Brayton, the leading lady, went even further in her estimate of Ricketts's powers by sending him the following message: 'Tell Mr. Ricketts that he is not to forget that one of my hips is larger than the other, and that he must try and alter it.'[15]

Though Ricketts was 'plunged in grief' over *Attila*'s short run (only a fortnight), the 'masterful colours of the barbaric palaces and the savage grandeur of the robes' made a deep impression on those who saw it.[16] Here the 'arbitrary colouring' which he had spoken about in 1903 was given full expression. To suggest bloodthirstiness, Ricketts had a scene done entirely in blood-red, scenery, properties and dresses of the chorus. The only exceptions were the Queen, who was in gold, and the murderers in black. Another scene was in blue. The choice of such colours was, however, not really arbitrary: 'Certain colours for me have an almost symbolical significance', he explained years later. 'I have never put a good person in black or yellow, certain greens are gay, others sinister, blue holds both possibilities, but is mainly lovely

12 CR to MF [30 July 1907].

13 MF Journal, vol. 1, 1907, fo. 156^{v}; *Theatre Arts Monthly*, p. 382.

14 *Studio*, 42 (1908), pp. 137–8.

15 CR to MF, 20 Aug. 1907.

16 CR to W. A. Pye [*c*. Sept. 1907].

and persuasive.' This novel use of colour, according to Ricketts, 'caused only a momentary attention', but he had, after all, shown his marvellous designs in a major London theatre.[17]

As well as these two productions, Ricketts was involved in a number of other dramatic efforts in 1907. In the summer he was asked by Mrs Patrick Campbell to stage Hugo von Hofmannsthal's *Elektra* for America. Before the production crossed the Atlantic Mrs. Pat offered to pay him his passage and a fortnight's keep to look after his scenery. Though he realized he would only be 'valet to the scenery', he declared that had it been spring he would have gone.[18] He also had a hand in designing the scenery for Barker's production of *The Medea* which opened at the *Savoy* on 24 October. Though Ricketts went to the actual production 'prepared to hate play, players, and [scene-]painter', he found it 'quite fair on the whole'.[19] In July he had also done some thirty drawings for a pageant in Oxford. Such a spectacle was all too ephemeral, 'a thing forgotten as it passes along', the Fields commented and Ricketts felt his designs deserved better. One of them, *The Retribution*, in black with tongues of flame, scarlet veil, and copper wings, was so tremendous that 'it will reappear in a picture & justice be satisfied'.[20] No such picture, however, has come to light. At the beginning of July Ricketts and Shannon had run down to the Mitre Hotel, Oxford, probably to see the pageant, and for a little quiet. 'Shannon was tired and out of sorts and anxious for a change (this with him is a survival of religious observances)', Ricketts explained to the Fields, 'I have followed (out of weakness)'.[21] Later, he contributed to pageants in London and Cambridge.

In addition to the theatre work, other demands had kept Ricketts busy. On Friday 19 June, at one of their 'At Homes', he had embarked on a new venture: he had given a lecture with lantern-slides on Watteau (which he always pronounced in the French manner, Vatteau). This is all that came of his planned book on the French master, whom in his talk he described as 'the first artist of his time and the only poet of the eighteenth century in France'. When he came to consider the master's contribution he summed it up succinctly: 'he has found in reality

[17] Herbert Furst, 'Charles Ricketts, ARA, and his Stage Work', *Apollo*, 1 (1925), 332.

[18] MF Journal, 1907, vol. ii, fo. 21^{v}.

[19] CR to Harley Granville-Barker [Oct./Nov. 1907].

[20] MF Journal, vol. 1, 1907, fo. 156^{v}.

[21] CR to MF, 2 July 1907.

elements of beauty till then unnoted and with these he has created a world till then unseen': this was the 'creative power' that he admired most in great artists.[22] The Ricketts and Shannon collection was to boast a superb series of drawings by this master, including a Watteau copy of a drawing by Van Dyck, given to Ricketts by Robbie Ross.

Their friendship with Ross was flourishing. In July 1907 a show of bronzes and paintings by Ricketts and Shannon took place at Ross's gallery Carfax (at which they never exhibited again after Ross left in 1908). At the end of 1907 they attended a belated coming-of-age party which Ross had arranged for Wilde's son Vyvyan Holland. Holland had already asked to meet Ricketts, who had then invited him to the last night of *Attila*. A year later, on 1 December 1908, they attended a dinner in honour of Ross at the Ritz to mark the successful winding-up of the Wilde estate, the anniversary of Wilde's death in 1900, and the publication of the first collected edition of his works. At the dinner, an anonymous gift of £2,000 (from Adela Schuster) was offered for a monument to Wilde in Père Lachaise cemetery, on condition that it should be by Epstein (whose controversial work for the British Medical Association in the Strand had been defended earlier in the year by Ricketts, Binyon, and Holmes in a letter to *The Times*). At the earlier, 1907, party Ricketts met H. G. Wells, whom he liked immensely, William Archer, whom he found 'quite a nice chap' in spite of their earlier clash, and G. K. Chesterton, whom he liked but found 'much more common than his books'. When Robbie Ross revealed his true character 'by drawing a woman's face upon the table cloth with the charred end of a match', Ricketts proposed it 'at once be framed and presented to the nation'.[23]

Ricketts felt the greatest admiration for Ross's devotion to Wilde; it was largely through Ross that Wilde had regained his reputation. In jest he would say, 'like charity the name of Ross covers a multitude of sins', but he regarded him as 'the heroic, self constituted literary executor of Wilde, the man in fact who has helpt to negative the iniquitous consequences of the verdict', and until Ross's death in 1918 staunchly supported him against the 'persecution from ignoble quarters [i.e. Lord Alfred Douglas]' that was to make him 'a little uncertain in mental and bodily health, as a consequence an old man before his

[22] BL Add. MS 61724.

[23] CR to MF [Dec. 1907]; 'British Medical Association: New Building', *The Times* (24 June 1908).

time'.[24] Ricketts gave him the manuscript of *The Sphinx* with his early designs in 1904.

All the theatre work that Ricketts was doing in 1907 meant that he had neglected his painting. As winter approached, however, he began to get down to it. He was working on yet another *Good Samaritan*, a *Leda and the Swan*, which he later destroyed, and a new version of *Don Juan*. The *Leda* was an attempt at a lyrical picture, but Ricketts found he could only paint dramatic tragedy. As well as his *Don Juan in Hell*, he depicted the Don's female counterpart in *Messalina leaving the Pincian*, a subject no doubt suggested by his visit to Rome.

His new *Don Juan* was exhibited along with *The Resurrection* and four bronzes at the International Exhibition early in 1908, for which Ricketts designed a poster.[25] Shannon showed his only bronze, a lounging female nude called *Repose*, and two oil-paintings, one of them a portrait of Kathleen Bruce. These were such a success that Ricketts noted with satisfaction that there was generally 'a small crowd round them'.[26] The International was to be the main vehicle for exhibiting their work until the early 1920s. In 1907 Ricketts joined Shannon—who had been elected in 1905—on its executive committee, and they used their position there to further not so much their own work as the conservative artistic values they believed in so passionately. At the end of 1908, for instance, Ricketts explained to Cockerell that he could not miss the Society's AGM: 'This is the one meeting a year which I *must* attend. Each year there is a conspiracy to overset the present government which requires our personal influence to stem.'[27] Their plans and intrigues to stem the tide of modernism did not always go smoothly, however, and Ricketts's experiences at the International Society not only soured his relationships with his many 'enemies' but also disillusioned him with regard to artists he admired. 'All good artists are personally odious', he told Ross, 'I am founding my conviction on all the members of the International Society.'[28]

The end of the 1908 International Exhibition brought a new International Ball, an event to which Ricketts look forward with excitement. They were invited to dine beforehand by Mrs Barrie, the wife of the dramatist. The author of *Peter Pan* joined them for dinner (to their

[24] CR to MF, 17 Aug. 1908; CR to G. Bottomley, 27 July 1918 = *SP*, pp. 299–300.

[25] Copy in Staatliche Kunstsammlungen, Dresden.

[26] CR to W. A. Pye [*c*. Mar. 1908].

[27] CR to S. Cockerell, 29 Dec. 1909.

[28] CR to R. Ross [1910].

great surprise, as he usually avoided strange company). Ricketts told the Fields, 'Barrie is diminutive with a tragic face, the colour of rain water. I should fancy it constantly rains or drizles, in his insides.' When Barrie gave biscuits to his huge Newfoundland dog (the dog in *Peter Pan*), 'a faint smile glimmers across his face like a ray of drowned sunlight on a midsummer day in Scotland'.

Ricketts rather 'winced' at being introduced to the dramatist while 'robed like the rich in the Protestant Scriptures in purple and fine linnen and with a decided tendency to over décolletage, partly disguised ... by my headress'. He had attempted to surpass the previous year's dress. 'Imagine me a vision of perfection robed in Amethist & Sapphire silk ... You will have guessed that I am a new Byzantine Ambassador, a sort of illustration or comment on Verlaine's marvellous poem (Langueur): (*Jadis et Naguère*): "Je suis l'Empire à la fin de la décadence".' On the lined cloak he had painted two panels to look like woven gold tissue, the one representing 'Bacchus and Ambrosia' and the other 'Atys and the Lions'. 'I have promised the panels on my dress to three separate ladies', he told the Fields, 'there are only two of them (panels); this seems to be a problem for the mathematician.' Ricketts also wore painted gloves and 'many details too beautiful for the English language and which should be described in Italian'. At the ball, Ricketts predicted that Mrs Barker (Lillah McCarthy) would arrive late in the pink dress that he had designed (and Shannon painted) for Dona Anna, while Mrs Asche (Lily Brayton) would turn up early 'robed in her blue *Attila* draperies'. Mabel Beardsley, sister of Aubrey and a good friend of Ricketts, had ordered a dress that 'turned out to be so scanty' that she despaired of going to the ball: 'Shannon says that what he dared to see of it looked like a pair of stays under some transparent draperies & he thinks nothing else', Ricketts told the Fields, 'so I have lent her the Atossa dress from *The Persians*, which in a few minutes became a bundle of coloured strips by the mere fact of being tried on.'[29]

'Just now we are living at the theatre where Isadora Duncan performs almost nightly to us and a growing audience', Ricketts told Bottomley. 'It is like the reign of the Sicilian players [whom he had seen eighteen times], when an enthusiastic audience was as funny as the performance & the pit & the stalls were locked in each others arms

[29] CR to MF [11 and ? Mar. 1908]; CR to W. A. Pye [Mar. 1908].

during the entr'actes'. When she danced for the last time Shannon was 'reclining in the stalls', commented Ricketts, 'whilst I lead the applause in the pit'—'by far the most distinguished part of the house'. They actually met the great dancer, probably through Kathleen Bruce, and went back stage to her dressing-room after the performance. Miss Bruce also introduced Shannon to Ellen Terry. 'I hope the real Mr. Shannon', the great actress said, to show her awareness that there were two Shannons; Ricketts suspected her of having used that before. 'She is a sort of mother in law of Isadora Duncan who threw kisses to her from the front of the stage,' observed Ricketts. 'Some men in the stalls near Ellen blushed crimson with bashful pleasure.'[30] Ricketts himself met Miss Terry early in 1909, and 'with the deep discernment proper to the artistic temperament, she declares me "even more charming than Shannon". This is not the general impression!'[31]

In May 1908 Ricketts and Shannon went to Venice to stay at the Palazzo Desdemona, a villa purchased by Edmund Davis. 'I suppose it is the establishmen[t] she set up with Cassio after the republic had banished Othello, which is the true version of the story, it seems', he told Binyon.[32] During most of the first week Shannon again fell ill, but they liked the palace immensely and stayed on after the departure of their host. For nine days of perfect weather they had it to themselves. In contrast to their earlier hard-working visits to Italy, this time they took it easy. 'We are leading a bestial life', he told the Fields, not long after his arrival. 'I have not once been inside S. Marco & nearly fell asleep in the Accademia.' Ricketts had at first been charmed by their handsome gondolier, who looked like a 'well-dressed slave', slept all day outside the long, cool entrance, and took them on an enchanting tour by gondola at night. Soon, however, the roles were reversed: 'Now it is we who are slaves to his automatic hours when we are pulled about places we don't want particularly to see.'[33] The heat was such that they decided not to go on to Florence as they had planned, and came home early.

Painting had again suffered from the demands of the drama. In July Ricketts told the Fields that he had only one picture to his credit, the *Messalina*, 'not much on which to pay income tax'. He was also working

[30] CR to G. Bottomley, p.m. 20 July 1908; CR to MF, 30 July 1908.

[31] CR to MF [15 Jan. 1909] = *Letters CR to MF*, p. 26.

[32] CR to L. Binyon [June 1908].

[33] CR to MF, 21, 30 May [1908].

on an *Ariadne*, with Theseus kissing farewell as he descends into the labyrinth, 'an enclosure of circled roofs'.[34] However, as it was summer he was working at his sculpture. Earlier in the year he had given the Fields a bronze of *The Tragic Actor*, on condition that it was placed in the room where their tragedies were written. Of the pieces that probably date from this summer there was a version in bronze of *The Good Samaritan*, a *Paolo and Francesca*, depicting the two lovers clasping each other and floating in hell (a theme that was especially popular among Symbolist painters), and *The Deposition*; a rendering in bronze of another favourite theme.

It was impossible, however, for him to stay away long from the theatre. He was involved in *Lanval*, by 'T. E. Ellis', the pseudonym of Lord Howard de Walden. 'At the rehearsal', he told Binyon, 'de Walden came up breathless to tell me that George Moore thought my scene would be the ruination of the play & though that was my private opinion also at the moment, I rushed up to Moore & told him not to give assified opinions on what he did not understand.'[35] George Moore, as one of the apostles of Impressionism, was never a favourite with Ricketts, who had described his brain to the Fields as 'like Clapham Junction—all sorts of covered ways and stairs to everywhere'.[36] As for the play, Ricketts felt too demoralized to go and see it. He was soon again in the throes of two productions for Mrs Patrick Campbell (who was being painted by Shannon): Wilde's *A Florentine Tragedy*, which opened in Edinburgh on 19 September, and *Elektra*, which, having been given in New York and Edinburgh, opened in London on 27 September 1908. These productions were attended by the usual run of disasters, and Ricketts told the Fields, 'I am slowly recovering from too much Mrs. Campbell, over the preparation of *Elektra* & *A Florentine Tragedy*. If only the same atmosphere of disaster which surrounds the making of theatrical objects could ... pass across the footlights, we would witness a new birth of Tragedy.'[37]

Much of 1909 seems to have been given over to putting together or organizing exhibitions. In contrast to the previous year, only one play designed by Ricketts was to be seen in London, though in fact this was to make his name. At the International Society not only did he

[34] CR to MF [30 July 1908].

[35] CR to L. Binyon [June 1908].

[36] MF Journal, 22 Nov. 1900, fo. 157^v.

[37] CR to MF, 15 Sept. 1908.

throw himself into partisan politics, but he was also the moving spirit behind many of its annual exhibitions. Organizing these was something that he enjoyed doing and did extremely well, but usually with a lot of friction and frustration. He was responsible for the first and second 'Exhibition of Fair Women' (1908, 1909), at the first of which he exhibited some of his jewellery. Then, in June 1909, he and Shannon sent off forty drawings by Hokusai and Hokkei to be exhibited in Paris; visiting the exhibition themselves, they stayed at 'Oscars old hotel, famous in the past as the Hotel of De Musset & of Balzac'.[38] Throughout the summer there were two more exhibitions in preparation. The first was a major show in Manchester of oil-paintings, water-colour drawings, etchings, lithographs, woodcuts, and jewellery, by Ricketts, Shannon, William Strang, Charles Holmes, and H. Wilson. This was the largest show of their work during their lifetimes. Ricketts contributed six paintings and twelve bronzes, while Shannon sent in nearly thirty paintings, seven woodcuts, and nearly all his lithographs to date, almost seventy in number. Shannon was also responsible for hanging the show. Manchester bought one of his paintings, *The Mill Pond*; 'the first time England has done anything for me', he observed.[39] Then a show of old masters at the Grafton Gallery was largely Ricketts's work; 'hung by him and mostly procured through his influence', noted the Fields. Organizing this excited him greatly: it 'promises to be unrivalled in the history of loan exhibitions', he told them, in his most delightful vein of fantasy:

The Greatest Masters have accepted our invitation. Mantegna alone has refused so far, and Botticelli in his late Savonarolaesque mood, but we have invited Amico di Sandro to spite him, and Giorgione, Titian, Raphael, Rubens, Rembrandt, and Watteau are with us. The Grafton Galleries have been hung with gold and fair linnen to receive them. Fra Angelico curiously enough had disappointed us, probably to please the present pope he ran away to America with Pierpoint Morgan who has been useful recently to the papacy, Fra Filippo Lippi, however, was with us from the first. He was the first to accept, and Raphael, remembering my praises in *The Prado*, witheld only a peruginoesque work, which we had invited only out of politeness. Van Dyck, who probably remembered my chapter, was cold and distant in his answer, despite Shannon's politeness, and Velasquez has prefered the Academy. Some three or four pirate masters have thrust themselves upon us,

[38] CR to MF, 15 June 1909.

[39] CHS to R. Ross, 13 Oct. 1909.

because Cook was on the Committee and the Jewish influence has secured one or two places for low Dutchmen and time-serving Britishers.[40]

There were the inevitable battles with trustees and committees, whose members did not share or understand his values. In dealing with the official world, which he mostly despised, he was surprisingly successful, though he claimed he was frequenting bad company, such as the Trustees of the National Gallery and such collectors as 'have to be pushed, persuaded, or frightened into supporting the Old Master Exhibition, with which I am concerned, not wisely but too much'.[41]

He was also busy with *King Lear*, his only theatre work of the year, to be given at the Haymarket Theatre. Preparations for the play began smoothly.

I have been occupied, or rather preoccupied, by *Lear*, not indeed distracted night and day as on former occasions, since all was well in hand at an early date, but rather more anxious and distrustful over small things, aggravated by the telephone and something neutral and wooden among those with whom I come into contact.[42]

The scenery was conceived as a 'massive, mythical Stonehenge type of architecture' (Fig. 33). One or two scenes had backdrops from nature, such as the cliffs of Dover. To emphasize this mood, the play was staged in winter, so that Lear dies in a snow-blocked avenue. The tent scene was all blue, with the King in white on white fur, in order to give the scene a half dream-like quality. Through the use of great simplicity of line, and greenish-blue lighting, he was able to enhance the primitive and mysterious quality of the setting.[43]

An innovation was Ricketts's decision to vary the width of the stage opening. To give the impression of soaring height to certain scenes, the sides of the stage were blocked off, giving a tall, narrow opening, rather than the usual wider shape. 'With some difficulty, I have obtained an expanding and contracting proscenium', he told a friend, 'so the hovel won't be the size of Albany's Palace'. Such scenes were therefore of different dimensions, but to persuade the management to accept such an unusual practice he had to argue that it had already been adopted in Sweden.[44]

[40] MF Journal, 1909, fo. 172^v; CR to MF, 17 Sept. 1909.

[41] CR to MF, 10 Aug. 1909.

[42] CR to W. A. Pye (TS) [summer 1909].

[43] *Theatre Arts Monthly*, p. 382; Furst, *Apollo*, p. 331.

[44] CR to W. A. Pye [summer 1909].

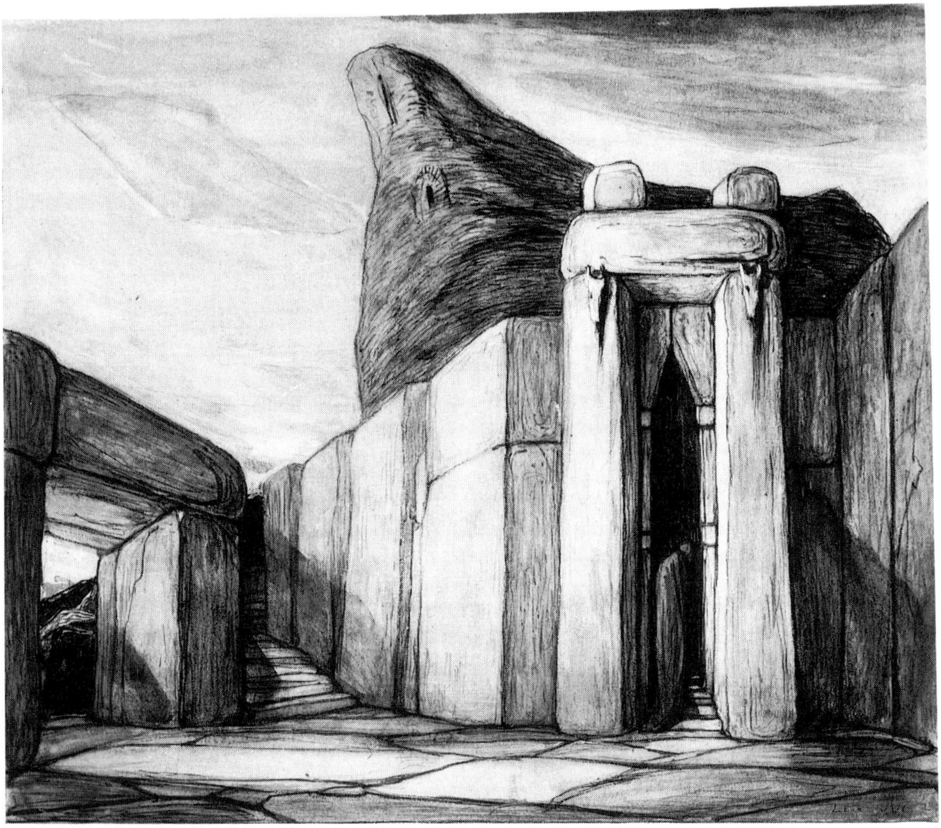

FIG. 33. Stage setting for *King Lear* (1909) by Charles Ricketts (Victoria and Albert Museum)

During the rehearsals he contemplated having his name removed from the programme. His share in the play, he felt, had fallen below standard, 'owing mainly to faulty detail or technique'. A large part of the blame for this Ricketts placed on the scene-painter, Joseph Harker. 'The scenery which is conceived finely has been abominably executed by a hack scene painter, who did not trouble to count the objects included in the sketches, or to keep the proportion of the separate parts.' Though the stage was too small, the use of tactful lighting and unobtrusive colours meant the result was not unpleasant, and the costumes were 'superb'; 'many have the tragic splendour which is becoming or proper to the major scenes'. As in *Don Juan*, however, one of the actors made difficulties about his costume. The actor who played Edmund had been given the 'most wild and lovely attire', but had 'cried out to have breeches, and would not cease till his legs were hidden'. Edmund (played by Dawson Milward) and the Fool, (played by H. R. Hignett, 'an unintelligent and agravating idot'), almost managed to ruin the play, Ricketts felt, and placed 'many scenes in the suburbs of London and carry their clothes like curates or Oxford professors'. However, Norman McKinnel's *Lear* was 'one of the most thoughtful and beautiful yet seen on the English stage', and 'far to good for London'.[45] It was his practice to write to congratulate actors for such performances, and no doubt he did so in this instance.[46]

Ricketts declared to the Fields that, in his anger at the 'meanness shown toward the scenery', he had burnt all his *Lear* sketches.[47] This was not wholly true, for he later presented some of the principal scene designs to the Victoria and Albert Museum. Despite his disappointments, Ricketts, according to Bottomley, 'to the end . . . looked upon this as his most complete achievement in the theatre'.[48] Many critics agreed, and Max Beerbohm commented:

> Mr. Charles Ricketts has risen to the level of his great opportunity. His scenery has a large and simple dignity of line and colour. It is a fit setting to tragic issues. It looms ominous in infinity. Darkling forests, sheer scarped cliffs, rude structures of stone—all are admirably right'.[49]

[45] CR to MF, 17 Sept. 1909.
[46] See McCarthy, pp. 161–2, 303–4.
[47] MF Journal, 1909, fo. 173^{v}.
[48] *Theatre Arts Monthly*, p. 382.
[49] *Saturday Review* (18 Sept. 1909), repr. in *Last Theatres* (1970), 483–7.

It also brought him international recognition, for in 1911 he was invited to design the first production of *Lear* in Japanese, in Tokyo: 'Shakespeare is delighted', he told Cockerell. For this he adapted the designs so as to hint at prints by Utamaro and Kiyonaga: 'I shall try and make it at once very strange, and in part familiar, by a hint of prehistoric Japan crossed with Celtic elements; they are easily amalgamated.'[50] Later he designed *Salome* for the same company, but his drawings were lost in transit in the New York customs and he had to reconstitute them afterwards.

The *Lear* production was not, however, without controversy. Several critics, with some justice, suggested that Ricketts had borrowed Gordon Craig's method of staging plays. Craig himself came to Ricketts's defence over this in a long, rambling, and ultimately rather ambiguous letter to the *Saturday Review*, which defends Ricketts's method as his own but concludes: 'Whenever an artist has stolen from another, the debt has been publicly acknowledged. Therefore, Mr. Ricketts had not, cannot, have stolen from me, for he has made no such acknowledgement, and that ends the matter.'[51] It appears that the thin-skinned Ricketts, an admirer of Craig's work, though with reservations, took offence at this, since no other occasion has come to light on which Ricketts might appropriately have told Cockerell: 'If Craig had not attacked me in the Press, I should like to write an article on his work.'[52]

Ricketts's diverse talents and interests meant that he was always to have many irons in the fire. 1909 was a particularly busy year. At long last the book on Titian, which had been commissioned in January, and promised by July 1906, was in its last stages, but this was unfortunately not the most convenient time. 'I have been in a whirl of work of an anxious and distracting kind in which one's personal responsibility does not count enough, and the individuality of others too much', he told Bottomley in August. '*Lear* is largely off my hands, but the Titian revises come at the wrong moments, and I am distracted from time to time by the claims of two exhibitions with which I am engaged.'[53] The writing had dragged on much too long. At the beginning of 1907 Ricketts, having received part of the proofs, had written to his publisher: 'I am glad you dont dislike the book; as for me, I feel that a

50 CR to S. Cockerell, 20 Dec. 1911.

51 *Saturday Review* (25 Sept. 1909).

52 CR to S. Cockerell, 17 Sept. 1911 = *SP*, p. 167.

53 CR to G. Bottomley, p.m. 9 Aug. 1909.

painter should not put down his brushes to write, and that the spelling of names should be a matter of "inspiration" just as it was in the time of Elizabeth.'[54] In January 1909 he was still making corrections, and explained that for the book he had relied on his extraordinary visual memory: 'I have not written the book before the pictures. A chapter I wrote in Naples was not retained, and the dullest portions of *The Prado* were written in Madrid. I take notes, sometimes small diagrams of damage, but these hardly count at all in the long run.'[55] In July, after these delays, he suddenly introduced a note of urgency: 'In October my name may be associated with a scheme which might give prominence to my book on Titian. I am anxious that it should appear about then.'[56] Ricketts hoped that his book might strengthen his chances of obtaining the Slade Professorship at Oxford, for which he had applied.

Ricketts states clearly in the first sentence of his *Titian*: 'My aim in this book is not to give an account of Titian's life but of his pictures, having no qualifications as a biographer, or indeed as a writer.' Where he differs from other authorities he promises to give technical reasons for his view.

> An attempt has been made [he explains] to record the different degree of 'authentic' preservation of each work, and to distinguish between small harmless mendings and partial or total repainting, since many excellent writers ... do not always use the term 'repainted' with sufficient care, whilst there are others to whom the total destruction of a work by drastic repainting presents no bar to enjoyment.[57]

Thus Ricketts gives clearly the gist of what he has been striving to do: to detail evidence of skinning and overcleaning, ironing or relining, patching or repainting, which presented difficulties for the art lover at the turn of the century, when many paintings were hung at a height or in a place where they could not be seen properly, and when dirt, dust, and darkening of the paint added to the difficulties.

Titian is seen as 'the father of modern interpretative painting'. He possessed that 'creative power' that Ricketts so much admired; he enlarged the syntax of painting by 'a new sense of pigment', a new 'use of chiaroscuro as a means of expression'; by developing its tech-

[54] CR to Mr Laing, 23 Jan. 1907.
[55] CR to Mr Laing, 15 Jan. 1909.
[56] CR to Mr Laing [July 1909].
[57] *Titian*, p. vii.

nical means, he broadened the subject matter of painting, making it possible for 'things evil and ugly, without losing their character, to become beautiful by mere interpretation'. With him, beauty became valued for itself.[58] 'The greatest artists', he tells us, '... never reflect their period, but remould it, regardless of its aims.'[59] On the individual works and on the general character of the different periods of Titian's *oeuvre* Ricketts is often at his best. Though he confessed that the artist's portraits of beautiful women often bored him, and were of value only to 'astonish the tourist and befriend the copyist', he was inspired to vivid and poetic accounts by the artist's greatest works, like the *Bacchus and Ariadne*:

> We owe to Titian the vision of the leaping Bacchus with his streaming hair, and the rout of fauns who stamp and chafe in a trance, active, yet held in check by some charm of the God, that rare and ardent Spirit of Delight, whose passion has flashed up for Ariadne, who turns fascinated in her flight, though Theseus' ship is still a wing-like speck against the sky ... A white wild rose of Sharon creeps among the grass, the iris and the colombine, the wild vine and the jasmine—each brings with its associated beauty an element of intimacy to a scene of splendour and passion at a crisis.

As evocative descriptions of painting such passages rank with the writings of Fromentin and Delacroix. What, perhaps, is even more typical of Ricketts is a little footnote that 'the white flower is not a Rose of Sharon, but the "Capparis Spinosa"', as if he was unwilling to spoil his beautiful description with a more ordinary, if more accurate, detail.[60]

Though Ricketts was inspired to a passionate admiration by many of Titian's paintings, his rather uneventful life offered little to inspire him; but Ricketts's writing comes alive when describing more colourful, more scandalous characters. The pages on Titian's friend Aretino, who is, surprisingly, given a whole chapter as well as numerous other references, vibrate with his interest and pleasure. It is easy to see why Ricketts's friend Cippico wanted him to write the biography of Aretino, and why his publisher suggested he paint 'The Death of Aretino', who had 'slipped out of consciousness in a peal of laughter'.[61] The Italian was the sort of ebullient scamp who delighted Ricketts and had that

[58] *Titian*, 7–8.
[59] Ibid., 68.
[60] Ibid., 78, 68–9.
[61] MF Journal, 1910, fo. 108; *Titian*, p. 136.

irony and bravado that he loved in the autobiographies of Berlioz and Cellini (books which he said ought to be read every two years).[62] Respectability was never a quality that he valued, and nowhere does his aesthetic creed sound out more eloquently than when he professes his profound admiration for the worldly popes like Julius II and Paul III, to whom the world owed a great debt. The latter was a patron of Michelangelo and Titian and 'humanity has benefited more by his foresight and magnificence than it has suffered by the actions of his family', the Farnese. To the saintly type—'the more approved type of little clockwork moral man', who was too preoccupied with himself—Ricketts preferred the worldlings who knew the right use of those 'two very admirable things . . . namely, Money and Power'. Concluding his paragraph, he observes, 'We, the wise sons of an enlightened epoch, have much to learn in the essential value of things from less enlightened men; we prattle about the proper use of things, but forget to make them.'[63]

Yet Ricketts evidently felt impatient and bored with the scholarly apparatus necessary in such a book. 'The size of the different canvasses', he told his publisher, 'I am unable and unwilling to give. It is an old fashioned form of information or compilation useful only (when it is accurate) in the matter of pictures still in private hands which are liable to loss.'[64] This perverse view did not prevail, for measurements were provided, either by Ricketts or by someone else. He suggested putting the illustrations together at the beginning, if they could not be placed throughout the text, but they were in fact put at the end.[65] Nor would Ricketts undertake an index: 'It is the dreary element of second hand compilation with which a book of this sort is usually associated'; the element, indeed, which had made him hesitate to accept the commission in the first place.[66] His publisher provided an index. Over questions of dating and chronology he agonized considerably, and often confessed to being perplexed and uncertain. The matter of attribution interested him more, especially when he disagreed with other current authorities, whose works he seems to have studied conscientiously. The uncongenial aspects of the work, the many revisions of the text, the length

[62] CR to MF, p.m. 19 Oct. 1895, 13 Aug. 1897.

[63] *Titian*, p. 109.

[64] CR to Mr Laing, Jan. 1906.

[65] CR to Mr Laing, 23 Jan. 1907.

[66] CR to Mr Laing, Jan. 1906.

of time between commencing and concluding the book, all suggest that Ricketts had more trouble over it than he had anticipated, and the result is in some ways less satisfactory than the marvellous passages on Titian in *The Prado*. It lacks some of the wit and liveliness of the earlier book, but was still a marvellously lyrical, perceptive, and stimulating commentary. Ricketts, however, never attempted a book of this nature again, for the fifty pounds he was paid for it left a profit of only some twenty pounds after the expenses of his journey to Naples.

Titian was published in May 1910, barely in time to help him in his bid for the Slade Professorship. He had been the first to learn, before it became official (some time before July 1909), that his friend and former disciple Charles Holmes, who had held the post for seven years, was giving up the professorship early the next year. Since closing the Vale Press Ricketts had felt the need for a permanent post, to give him a regular income, and this seemed very suitable. He knew the collection at Oxford by heart and had long loved the city. With Holmes, he spent several days walking about the countryside discussing what was expected of a professor, and is said even to have taken lessons in elocution. In December he consulted Cockerell: 'I dont know who are likely to be candidates also. I dont know what chance I run. I feel the salary would be very welcome indeed, and the event (one way or another) of interest in my career'. Ricketts hesitated to ask Holmes to back him, 'since a friend of his, Fry for instance, might try for the berth'.[67] In the end, his four referees were George Clausen, RA, Campbell Dodgson and Laurence Binyon, both of Prints and Drawings, and G. F. Hill, of Coins and Medals, at the British Museum.

Among the other candidates were Lionel Cust, W. R. Lethaby, Selwyn Image, and Roger Fry (recommended by Berenson), as well as a number of less distinguished contenders. Though most of these sent in a curriculum vitae, Ricketts said nothing of his past achievements. Perhaps he felt that these were well enough known. However, he submitted a detailed programme of lectures, under forty-eight headings, on the development of European Art since classical times. Though they reflect his own tastes and preferences, they attempt an unbiased view, ending up with Impressionism. He was keen, however, to keep things in perspective, to distinguish between the greater and lesser men.

[67] CR to S. Cockerell, 8 Dec. 1909.

I think the old *Plutarchian* idea of comparing and contrasting great men is first rate; it allows one, like a piano tuner, not to lose sight of the *tonic*; it keeps up the sense of proportion, which is the one virtue I have found in no living man. All extant criticism and collecting seems intended to destroy all traces of it.[68]

His place in the running interested him keenly. 'So far Oxford tends towards Cust, who is nothing. I am usually placed third, but this was before the sending in of my synopsis of lectures. Let us hope that *Titian* will come to the rescue. My book came out last Saturday, it seems to be liked.'[69] A month later he was writing to the same correspondent, Cockerell,

I have heard nothing from Oxford. I dont know if they let you know. From what I hear, I am always second in any proposal, which is ominous. Fry, whose chance was at first poor, seems to be top. I hear Image has a good chance; if I fail, I should like him to get it. Lethaby has no chance, he was too superior in his methods, probably he did not sign 'Yours obediently'.[70]

In the end the post went to Image, and Holmes recorded that both Fry and Ricketts had stood down in his favour.[71] Ricketts was relieved that Fry did not win, and joined in the protest when Image was removed from his post as an economy measure during the War. Soon he was able to view the competition for the post with his usual irony. To Robbie Ross he related a story told to Shannon: 'A house party had been playing "suck pencil" to find out who had been, in all history, the greatest intellect known to man. The fourth name in order on the list was "God". I believe that was the place in which my name figured for the Slade.'[72]

The only theatre production with which Ricketts was involved in 1910 was at the request of Shaw, who was putting on his *Dark Lady of the Sonnets*—a scene between Shakespeare and Queen Elizabeth interrupted by the Dark Lady—to raise money for the National Shakespeare Memorial Committee. Only two weeks before the matinée Shaw wrote to ask Ricketts to help. Ricketts rose to the occasion. The rush which led to difficulties with actors and rehearsals only brought

[68] CR to S. Cockerell, 15 May 1910.
[69] CR to S. Cockerell, May 1910.
[70] CR to S. Cockerell, 13 June 1910.
[71] Holmes, p. 274. [72] CR to R. Ross [1910].

out the best in him. For Elizabeth he had found, he told Shaw, a dress to rival that of Dona Anna:

> The stuff it is made of costs about one and six a yard, it should require about 13 yards; remain the velvet decorations and the peacock eyes, which I can make myself... I should like the entrance of Elizabeth to be made in a sudden oblique ray of soft light and to be accompanied by a rush from the wood instruments.

Ricketts went on to say that he wanted Barker, who was playing Shakespeare and 'who of course wants to show his legs, to appear *at first* wrapped in a large grey cloak, half statue and half Guy Fawkes; he can show his legs in white hose afterwards'.[73] The courtiers should wear black and red masks, like Elizabethan masquers, and yellow and black dominoes. Against 'an intense and abnormal sky of a fabulous blue', Ricketts dressed Shakespeare in greys and russets, Queen Elizabeth in silver and black, and 'the Dark Lady' in crimson and scarlet. 'The dresses (not including court) cost about £30; they would have cost less had Queen Elizabeth not insisted on crêpe de chine. I now realise the truth of history which states that Elizabeth left dresses to the value of two million sterling.'[74] In the end the acting was poor, and everything about the production showed signs of haste, except the costumes. Ricketts counted them among the best he had done.

Even with such demands, Ricketts was managing to get more painting done than in the previous year. The Fields, after one of his visits in the spring, had commented in their usual highly metaphorical style: 'Fay [Ricketts] is brightening from an eclipse of his very self... all last year he was not dwelling with himself & scarcely a bit of creation came into existence.' Now he was painting a picture of the *Danaïds*, 'weary pourers forth of the waters of doom'. Punished in the underworld for murdering their husbands, the daughters of Danaus were ordered for eternity to fetch water in jars that leaked like sieves. Was this another image, less conscious than *Death in a Picture Gallery*, of a sense of futility in striving to be an artist in a philistine society? It certainly reflects the deep pessimism that lay under all the wit and vitality Ricketts showed in his outward life. Equally gloomy, but perhaps even more tragic, was *The Plague*, which the Fields commented had become to him 'literally what he calls it'.[75] A large picture, it

[73] CR to GBS [Nov. 1910] = *SP*, pp. 152–3.

[74] CR to GBS [Nov. 1910] = *SP*, p. 153, where the £70 is an error for £30 in the MS.

[75] MF Journal, 1910, fos. 97, 112^{v}.

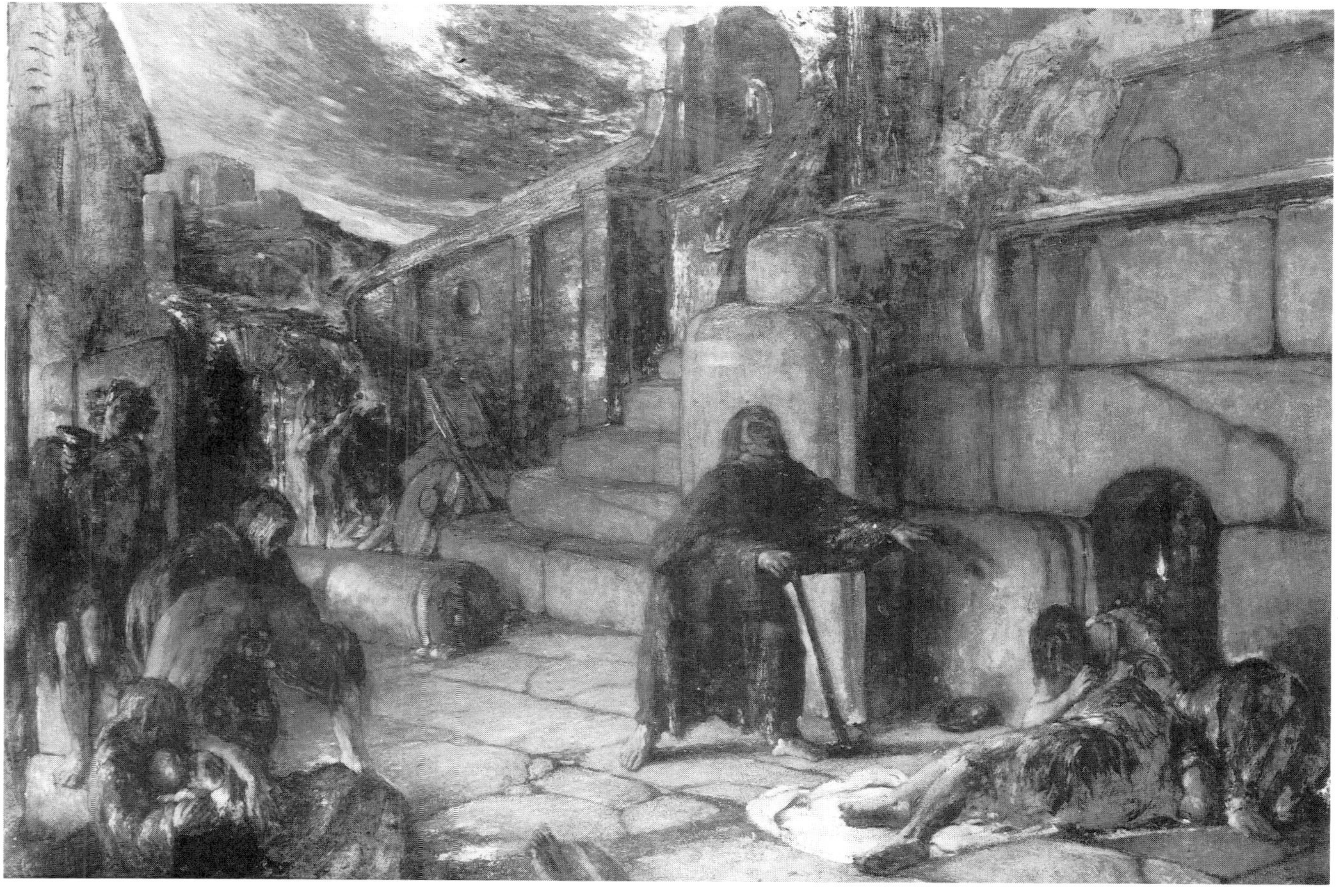

FIG. 34. *The Plague*. Oil-painting (1911) by Charles Ricketts (Musée d'Orsay, Paris)

depicts blind victims groping their way among the prostrate bodies of the dead and dying (Fig. 34). Such tragic pictures were not likely to attract buyers eager for something to brighten a wall, and the two pictures have ended up in public galleries. When the Fields tried to persuade him to choose a more inspiring subject, an Assumption, he replied: 'you know ... I cannot paint ecstasy—it is impossible to me. My note is always tragic. If I compose a theatre-dress, Mrs Pat Campbell takes it up, & says there is something of tragedy in it.'[76]

In February 1910 Ricketts published an article on Shannon in French in *L'Art et les Artistes*, reprinted in *Pages on Art*. This was not his first writing on Shannon: he had compiled a catalogue (with introduction) of Shannon's lithographs, published by the Vale Press in 1902. Their friend Count Cippico produced an article on Shannon in an Italian review, *Vita d'Arte*, in March. Shannon's international reputation was broadening, and he now had work in public collections in Bremen, Dresden, Paris, and Venice. Although Ricketts's painting achieved no such recognition, he never betrayed a hint of jealousy of Shannon's success. This had been foreseen from the beginning. His efforts to further Shannon's work came not only from personal loyalty, but also from an increasingly fervent (and eventually rather desperate) attempt to promote the right sort of art. This aim, however, received what was to prove a devastating blow with the Post-Impressionist Exhibition, organized by Roger Fry in September 1910. At the beginning of the year Ricketts had left the council of the Burlington Magazine

> owing to the Cézanne preface by Fry. This was the last straw after several equivocal articles tending towards the praise of the tragic followers of Neo Impressionism. So if Fry's friends speak of my brutality you will understand. There are frigid forms of mental prostitution which no lover of the old masters and fine moderns ought to abide.[77]

Now, with the opening of the exhibition that marked a new epoch in British painting and militated against so many of the qualities that Ricketts cherished, he was 'drunk with wrath', reported the Fields.[78] 'All London has been convulsed by an important Post-Impressionist exhibition', he told his Dutch friend Holst. 'People have gone in crowds to laugh. The success is one of scandal, but I have been distressed to think that people have crowded there, when they think

[76] MF Journal, 1910, fo.153.

[77] CR to S. Cockerell, 6 Jan. 1910.

[78] MF Journal, 1910, fo. 225^v.

it bad, who would not have done if they had been told ... the show was good.'[79] What also outraged him was the number of figures in the art and museum world who had lent their support to the exhibition.

Ricketts considered Post-Impressionism a disease that attacked all that was finest in the European tradition of art since the Renaissance. To the Fields he described it as 'a school, or rather an epidemic that belongs to 20 years ago in France, but has been imported into present England by Roger Fry'.[80] To Cockerell he wrote, 'the disease is wide spread. The Provost of Eaton has it; it might have spread to Oxford, had Fry been given the Slade. It is a blight which leads to the hatred of Art. It is passing in Germany, arriving in Russia, and touching Holland.'[81] To understand Ricketts's abhorrence of Post-Impressionism, a singular failure of sympathy in a man who appreciated every other form of artistic expression and described his own taste in art as 'omnivorous and insatiable', it is necessary to look at his character and his past.[82] What had fascinated him in painting ever since he was a boy was the classical tradition, which not only made use of the many technical developments since the Renaissance but also depicted 'literary' subjects from the Bible, the classics, or history. Impressionism, with its preoccupation with landscape and its obsession with a rather limited technique, did not interest him much, though he admired Degas (then classed as an Impressionist) and some of the work of Manet and Renoir. What he particularly disliked in Impressionism was its striving for novelty, which, he said, 'in itself is valueless', and 'the wish to owe nothing, save only to itself, or admit the debt which each artist pays willingly or unwillingly to his forerunners in the past'.[83] Earlier artists, however great, had readily expressed their debt to their masters. Ricketts noted, for instance, how Van Dyck had presented some of his finest early works to his master Rubens, as a token of esteem and gratitude.[84] With regard to Post-Impressonism, he wondered if 'this reactionary temper, this singular emphasis of personal limitations, [was] contained and foredoomed in the theory of Impressionism itself, in its advocacy of side issues, and contempt for its foes'. Ricketts noted how Roger Fry, 'who is now their champion, is without pity for

[79] CR to RNRH (TS) [late 1910].

[80] MF Journal, 1910, fo. 225^{v}.

[81] CR to S. Cockerell [Sept. 1910].

[82] Egyptian Diary, 1911, p. [6].

[83] *Pages on Art*, p. 150, 163.

[84] *Prado*, p. 180.

his contemporaries, who are not Post-Impressionists'.[85] Ricketts too showed little pity for his adversaries.

His opinion was not a hasty one. Sturge Moore recalled how 'Ricketts was interested in Gauguin, Van Gogh, and Cézanne before those who wrote them up in this country had ever heard of them.'[86] Whilst he hated the movement as a whole, his attitude to its adherents varied. Cézanne, its 'half-conscious founder or pioneer', Ricketts considered 'mentally deficient'; his pictures 'seem mere accumulations of thick pigment, applied to hack subjects designed in the style of our "proto-Byzantine" pavement artists; they lack only the written appeal, "Please remember the artist"'.[87] As for Gauguin, in spite of his technical shortcomings Ricketts thought he had 'flashes of real lyrical feeling and artistic originality in a welter of haste, bad taste, and occasional incompetence', and was attracted by his subject matter: 'I like the snake-like trees, the grape-coloured women; I do not hate their impassiveness and lack of "significant gesture" or something at once animal and vegetal in the life he paints.'[88] Ricketts found Gauguin's book on Tahiti charming, and Charles Morice's biography of Gauguin 'a most interesting character study'.[89] He was amused to record Strang's observation that Gauguin's *Christ in the Garden of Olives* represented 'Ricketts thinking about Post-Impressionism'.[90] Ricketts also found something to admire in 'a group of fantastic flowers' by Redon at a Post-Impressionist show.[91] Van Gogh's 'weird canvases, where the commonplaces of "painting from nature" have become tortured by some horror of vision' did not interest him as much as 'extracts from the interesting letters of this man who suffered under the stress of religious and artistic mania', which he had read in the early 1890s, when Émile Bernard first published letters from Van Gogh to himself in French.[92] Both the artist and his art were unbalanced. Only Maurice Denis, among the artists called Post-Impressionist, 'has brought a decorative or symbolic element to this "agony of Impressionism" ... his work is

[85] *Pages on Art*, pp. 152–3.
[86] Obituary of Ricketts, *Observer* (11 Oct. 1931).
[87] *Pages on Art*, pp. 154–5; CR to S. Cockerell, 3 Nov. 1921 = *SP*, p. 340.
[88] *Pages on Art*, p. 157; CR to S. Cockerell, 3 Nov. 1921 = *SP*, p. 340.
[89] CR to RNRH [*c*.Feb. 1923]; CR to Robbie Ross [*c*.1918] = *Friend of Friends*, p. 269.
[90] CR to R. Ross [Dec. 1910].
[91] CR to RNRH (TS) [?1916].
[92] *Pages on Art*, pp. 160–1.

allied to current tendencies in literature and music', but even he had 'come a cropper' in his decorations for Wagner's *Parsifal*, which were 'more unkind' to Wagner than anything written by Nietzsche.[93] For this reason Ricketts took a particular, if somewhat unenthusiastic, interest in his work. Summing up the spread of the 'epidemic', Ricketts observed, 'Lands without art have welcomed Cézanne; not all France, nor England, nor Holland, but Germany, Russia, and America.'[94]

This was to prove all too wrong. The 'French disease' swept France and England as elsewhere. It is ironic that this fatal malady had come from France, whence—after Italy—so many of his own artistic masters had originated. 'To me, at least', he told Holst, 'the French have suffered in all directions by the final overthrow of their academic foundation, which Puvis & Moreau had possessed, or Rodin, Degas, & Courbet in other fields of indeavour.'[95]

Ricketts's first response to the Post-Impressionist show was combative. He dashed off a protest to a newspaper, which made him feel guilty afterwards as it 'started a newspaper discussion which was foolish ... one longs for a little calm and no more French art'.[96] He also wrote a long attack on Post-Impressionism for the *Nineteenth Century* (an appropriate vehicle), but it was not published there (because he was absent in Egypt), and he later included it in his *Pages on Art* (1913). Ricketts was not going to miss the chance to have his say. His real answer, however, was an exhibition of nineteenth-century art that he had been organizing at much the same time as Roger Fry was doing his. This may not have been a deliberate response to the Post-Impressionist show, but it gave Ricketts the opportunity, as never before, to make a statement of his aesthetic credo, his artistic admirations, and his view of the development of painting over the previous century.

As Shannon was suffering badly from rheumatism, they had gone to Oxford in August, having left the house in the hands of house-painters, and their pet cat and jerboa to be 'prisoners' for a fortnight. The jerboa, called Hopkins, had been acquired a year previously, and soon became their 'new divinity, ... a wonderful creature that lives on

93 Ibid. 158; CR to RNRH, July 1914.

94 *Pages on Art*, p. 160.

95 CR to RNRH (TS) [?1916].

96 I cannot locate this letter.

oats & its hind legs. We live for nothing else.'[97] They had left London under the impression that one of Shannon's paintings had perished along with the crocodiles in a disastrous fire at the Brussels exhibition, but they learnt afterwards that the picture was safe. After only three days in Oxford Shannon was stricken again by rheumatism and they went on to Brighton: 'We would gladly return home, but cannot', Ricketts confessed. 'I hate British holidays. One exchanges home comforts for costly discomforts.'[98] The break, however, did not bring Shannon the needed relief, and the doctor ordered him off to Egypt and Greece for three months. Before leaving, they asked Cockerell to witness their wills, and Ricketts, 'with masterly boldness', asked Miss Bradley, 'Michael, would you like a scarab, a mummy, or a pyramid?'[99] Sensibly, she chose a scarab.

[97] CHS to G. Bottomley, 21 Jan. 1910.
[98] CR to MF, p.c. [Aug. 1910].
[99] MF Journal, 1910, fo. 228.

9 *Egypt and the* Ballets Russes

At the beginning of January 1911 Ricketts and Shannon, having put their smaller valuables in the bank and settled things with their executors, Ross and Cockerell, set off for Egypt. They went by train directly to Marseilles, where they boarded their ship, the *Macedonia*: 'she is vast, wet, only reasonably clean, and with the air of a huge second-class railway carriage afloat'. Sailing on the Mediterranean brought to Ricketts's mind images of the past, of the black waves that slapped the sides of the ship of Odysseus: 'During a gleam of sunlight seven dolphins leapt up close to the ship to warn us we were in Greek waters. They wore silver waistcoats cut to the pattern you see upon Greek vases.' On board there was a *cher père*, and an Anglican bishop who one evening 'preached to the females on the ship whilst all the males hid in the most distant part of the vessel or clustered in draughty passages'. Shannon managed to 'hide in the saloon behind a volume of Grote's *History of Greece*, which looks blameless and quite Anglican', whilst Ricketts was 'without resource, having a volume of Pierre Louÿs's *Aphrodite* with such a shocking and vulgar cover that I had to hide in dark corners where it and the illustrations cannot be seen'.[1]

The ship took them to Port Said, where they boarded a train for Cairo. They arrived 'tired and unnerved by the fuss of arrival' and, as the weather was cold, they lazed for the first few days. 'To Shannon's great satisfaction', Ricketts told the Fields, 'I developed a touch of rheumatism, strong enough to make rising from my chair a matter of careful consideration.' During these days they visited the museums and a 'dealer in dead divinities' whose shop was in two caves. The Cairo Museum owned the sarcophagus of Queen Nitōcros: 'she reclines in granite upon her granite bed; the expression on her face points to her knowledge that the enemies she drowned are unrepresented in the museum.' Its treasure, though abominably arranged, was 'too great

[1] CR to MF, *Macedonia* [*c.*8 Jan. 1910]. Ricketts's diary of this trip (BL. Add. MS 58097) and his letters to MF (known only in transcripts in Ursula Bridge's ed. of *Poets and Painters*—BL Add. MS 61721–3) are the sources of any quotations in this account, unless otherwise stated.

and at times overwhelming', and they visited it so assiduously that Ricketts discovered 'there is the possibility of dyspepsia in art'.

On their first visit to the Pyramids Ricketts was 'cowed with admiration', much to his surprise. Their approach, however, was marred by the clamour of guides, by the bank-holiday atmosphere, and by an idiot boy who wanted to tell Ricketts's fortune for a cigarette. Yet as they drew nearer the annoyance disappeared in the thoughts and emotions suggested by the massive monument, by the eternal struggle between man and the elements. 'The sand itself was once the granit of temples erected to the splendour of a life that thought it would never cease.' The monuments brought Dante repeatedly to his mind.

The first sight of the Sphinx was disappointing. 'It looked smaller and more trite in aspect than one had imagined.' It was, however, better preserved than Ricketts had expected, with traces of vivid colour and detail. The crowds of touts and tourists exasperated him; the latter seemed more 'bent on the rapture of donkey and camel rides' than on serious appreciation. Ricketts was glad to take refuge in the enchanting little temple of Khephren, where his admiration was 'complicated by a sense that it shows an affinity with my Theatre designs; it is in fact a sort of civilized Stone Henge, this and Fingal's Cave have constantly been present to my mind ever since my designs for *The Sphinx*'.

On 20 January they started up the Nile on a British steamer in a 'dense Scotch mist'. The journey, which took six tiresome days, was a sad disappointment, and hardly worth doing. The scenery consisted mostly of mud-banks. Ricketts only remembered

> some three superb quarries hammered out of beetling hills, a vista of desert under the soft drifting shadows of the clouds, some five or six mud villages in palm groves, at once ancient and tropical in aspect, a solitary pelican in a back-water, and a group of sulky-looking eagles on a sand bank, discussing cures for indigestion.

The people they passed were 'more silent than the beasts, less alive than the birds'. Ricketts never tired of looking at them

> with their beautiful draperies and movements, at once listless and dignified, or else sudden and quaint... Beauty of face, beauty of person, or charm of expression is as common here as it is rare elsewhere, even in Italy. In phisical presence, the men far surpass the women, though the latter present far greater variety of movement & pose. The children and youths are less interesting.

The quiet and impassivity of the people was rudely broken at the landing stages, in the din of donkey-boys and scarab-vendors. At Benihasan Ricketts was 'hoisted' onto a 'fierce beast' which suddenly broke into a gallop. In his horror and anguish he forgot the Arab word for 'slowly'; all he could remember was the word for 'quick'. Only cries of 'no baksheesh' succeeded in having the donkey slowed to a normal pace. His 'anger and a sense of humility at my annoyance' had obliterated all impressions during his ride, but he remembered a village of clamouring boys who followed the galloping donkey, holding on to the saddle for baksheesh and cigarettes. The tombs, moreover, were very disappointing, and Ricketts was depressed for the rest of the day.

> One has to make up ones mind that during the whole visit to a monument one will be shouted at, offered chips of stone, told names 'me Mahomet', 'he Alli'; this will be varied by yells of execration at new commers and, since one knows this, one ought not to mind; yet one minds greatly.

He had travelled thousands of miles to Egypt and had to see it in 'an atmosphere of fuss which can only be compared to the railway station at Florence'. This distressed him all the more because he rather liked the people. Shannon, however, did not mind, and would chat with any 'handsome ruffian' or 'venerable humbug' who attached himself to their party.

Five hours of indescribable boredom were spent at Assiut, the home of Plotinus 'whom I will never read'. At Dendera, the latest and most graceful of all Egyptian temples, he explored the temple of Hathor and the small charming temple adorned with figures of the goddess Bes, which was still half buried by debris. At its side, in the ruins of an early Christian church, 'lies the torso of an egyptian god with his sleak pagan balls basking in the sun'. Such a conjunction of the two religions not only amused Ricketts but also suggested to his imagination the ebb and flow of history, the rise and fall of civilizations. On this excursion they were accompanied by Kathleen Bruce, now Mrs Scott. Returning from Australia, where she had left her husband making preparations for his Antarctic trip, she was doing the Nile on a native sailing-boat.[2]

At Luxor, a fear of disappointment kept them away from the temple for three days, though it almost touched the garden of their hotel. Instead, they visited the wonders of Karnak they had glimpsed from the boat. The main block there, with its halls and temples, its colossal

[2] Kathleen Bruce 'Autobiography' 1910, MS (Cambridge Univ. Library).

figures, its colonnade after colonnade of pillars, was 'the hugest mass of buildings in the world', and one of its wonders, in spite of the destruction and pillage of millennia. The place conjured up the many historical figures who had visited it, even down to the 'footprints of Ricketts and Shannon', as Ricketts mused on man's brief span and the permanence of art. Luxor, too, when they finally saw it, was magnificent, overwhelming. Ricketts's response to such works of art was not only aesthetic but rich in historical knowledge and poetic feeling. His knowledge of Egyptian art and history, however, was always made to serve his philosophy of life, his deep feeling for beauty, his humanistic and highly romantic conception of man's place in time and the universe. After their last stop, at Sakkara, Ricketts decided that the strength of Egyptian art was 'in the round and the sense of mass, not in base relief'.[3] The colossi of Luxor, Karnak, and Memphis had astonished him by their 'real art merit'. These, the high points of Egyptian art, were also the high points of his trip.

They set out for Greece from Alexandria, where they had arrived in a downpour of rain and hail. While their ship was detained for thirty-eight hours in the harbour, a lighthouse was blown down, a buoy torn from its moorings, and five ships wrecked outside; there were even waves in the inside harbour. For much of this, however, Ricketts was below deck, 'with an extra dose of Mothersill's remedy inside, which felt rather like a small till full of large half-crowns'.[4] Smooth water was met only near the Cyclades. From Piraeus they travelled by carriage to Athens over a dusty road full of ruts and pits, which Ricketts surmised must be 'the footprints of St. Paul, or else of the last German archaeologist'. When, at the first glimpse of the Parthenon, they both exclaimed that it seemed diminutive after the colossal remains in Egypt, their largest portmanteau, containing their Egyptian purchases, 'goddesses with the heads of lions, hippopotami, hawks, and other engaging animals', was hurled into the road, but nothing, fortunately, was smashed. Ricketts blamed such mishaps on the Greek gods' anger at his and Shannon's 'whoring after the false gods of Egypt'.

During their fortnight in Athens they did the museums, which far exceeded expectation, in the mornings and the Acropolis in the

[3] CR to S. Cockerell, 21 Feb. 1911.

[4] CR, *Pages from a Diary in Greece*, ed. [J. G.] Paul Delaney (Edin. 1978), is the source for the Greek part of the journey.

afternoons. On their first visit to the latter, however, they found the iron gates locked, and so wandered about the Theatre of Dionysos, and it was probably then that Shannon took the photograph of Ricketts seated beside the Throne of the Priest of Dionysos Eleutherios; a photograph which Ricketts described in 1912 as 'first rate and ... in fact the "central" and final likeness of myself; it at once looks like myself today and twenty years ago' (Fig. 35). The Byzantine monuments and museums of Athens they left largely unvisited. Ricketts saw in them the eclipse by Christianity of the classical Greek heritage, and noted in particular how the tiny church of Haghios Eleutherios, according to Baedeker 'the earliest extant specimen of a Byzantine monument erected on Greek soil', marked a phase in the life of Athens 'when it entered for centuries into the night'.

The eight-hour voyage by ship from Piraeus to Itea, the landing-stage for Delphi, took the travellers through the Canal of Corinth. Their arrival was marked by Shannon's catching his foot and having to be supported as they waited to be 'trickled' from the ship into the waiting tender. Delphi was the high point of the Greek journey. Next morning the museum was, of course, closed, but a lazy guard sent a child to fetch the keeper. The bronze 'Charioteer' was the first thing they saw when the museum was opened, and prompted the usual vital and imaginative response:

> This is the famous portrait of the beautiful tyrant of some dead Sicilian town, whose hands still grasp the reins of the victorious horses who have galloped away for ever into the smelting-pot of some peasant, while a beckoning hand of bronze is all that is left of some satellite victory or slave who once stood at his side.

After this the ruins seemed disappointing, only 'foundations, "backing walls" and pedestals covered with inscriptions'; yet the valley remained 'one of the finest and noblest ... in the world'. At Olympia, the chief sanctuary of Zeus in Ancient Greece, while they were visiting the 'Bayreuth-Greek temple' which the Germans had built there to house the Hermes of Praxiteles and the other immortal fragments they had excavated, a thunderstorm broke, turning the roads and paths of the sanctuaries into torrents. It became impossible to visit the site, so that the travellers, after spending ten hours and two nights on their journey thither, returned to Patras, 'having seen the museum but not Olympia',

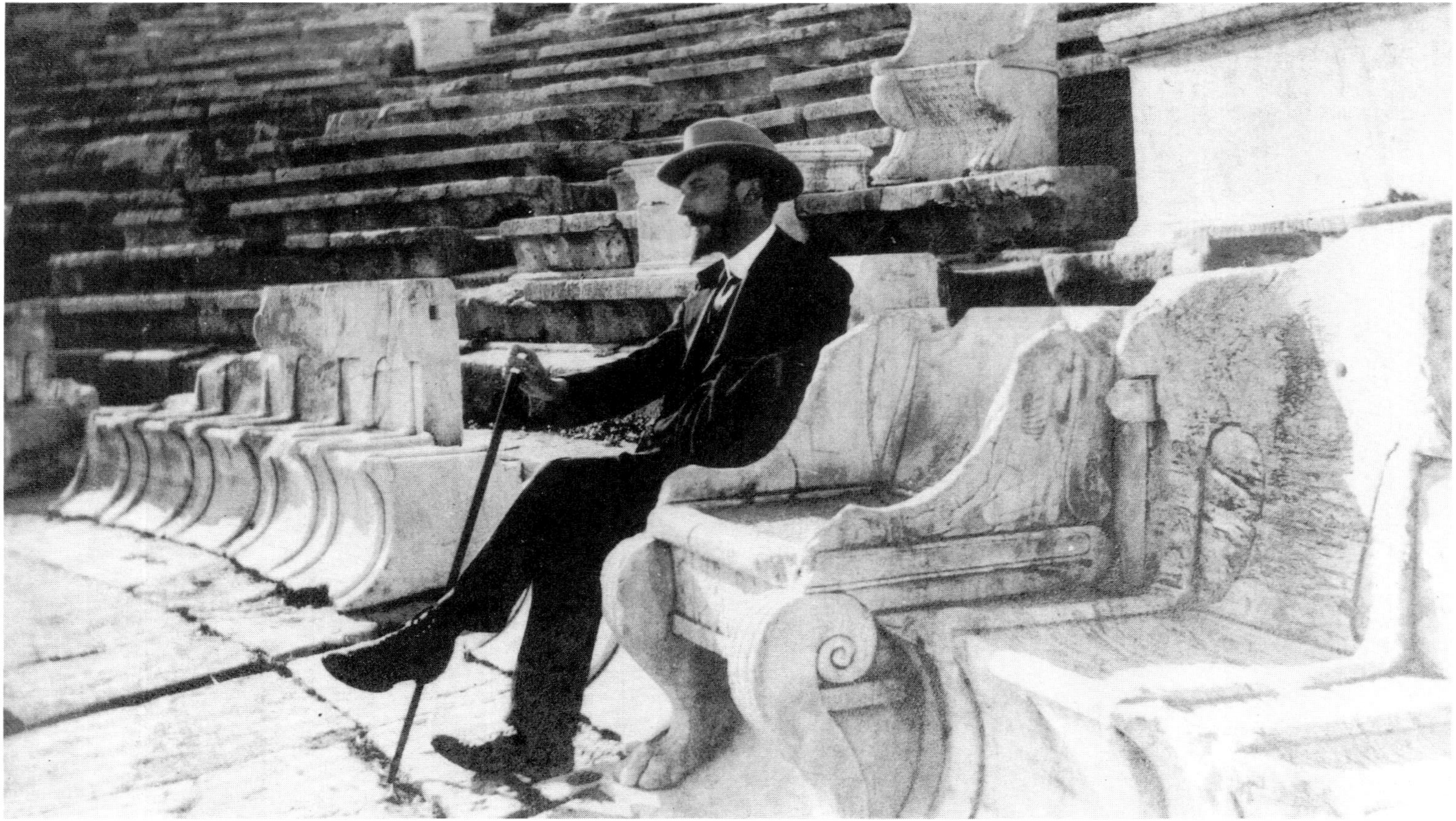

Fig. 35. Charles Ricketts in Greece (1911)

and left for Italy. Their journey to Greece ended, as it began, in a storm.

Ricketts, with a characteristic blend of fancy and irony, proclaimed the Greek gods 'false, treacherous, and fickle'; for the travellers had found the Acropolis closed, the museum at Delphi closed, and the site at Olympia unvisitable, so that each of the gods, Athena, Apollo, and Zeus, had betrayed them in turn, had 'slammed the door of his shrine' in their faces. Even Dionysos, exempted from the general censure, since 'the white wine of Hymettos' was excellent, had once 'parched' their throats 'with resinated wine when nothing else was to be got'. Though he conceded that the gods' anger might have been due not so much to the 'paltry portmanteau' filled with Egyptian deities as to a dull French book on Olympia he carried about with him, Ricketts declared that henceforth he would worship only the Pyramids. The following year they revisited Egypt, but not Greece. Even before returning to England Ricketts had intended to turn his diary and letters into an 'old-fashioned booklet upon Egypt and Greece'. The Greek part was finally written up as four articles, the first three of which were published in Italian in the magazine *La Ronda* in 1920–1.[5]

On their return they stopped at Naples and Florence, where it was so hot that they left early, travelling thirty-six hours without stopping, and abandoning the letters they expected to receive in Florence and Paris to return home on 20 March. 'Egypt can be summed up in a few words', Ricketts wrote to Holst; 'the people are fascinating; the art is *overwhelming*; the country on the whole disappointing. With Greece the verdict is different. The art surpasses one's expectation; the land is lovely; the people are kindly but they do not count.'[6]

Not long after their return Shaw turned to Ricketts once more for urgent help with one of his productions, *Fanny's First Play*, which had to be done in only seven or eight days. Shaw insisted on remaining anonymous, and so did Ricketts, who had refused another such request on 'the pretence that I have retired from art to paint pictures'.[7] The secret of the author's identity barely lasted the first twenty minutes of the play. 'It's Shaw. Of course it's Shaw,' was being whispered in the audience.[8] Its phenomenal success, however, did come as a surprise: it

[5] *La Ronda: literaria mensile*, 2 (1920), 284–8, 723–31; and 3 (1921), pp. unascertained.

[6] CR to RNRH (TS) [Feb. 1912].

[7] CR to GBS [12 Apr. 1911]

[8] McCarthy, p. 135.

ran for over 600 performances. Shaw remembered that Ricketts had asked no fee for his work, and, since the play had become 'the *Charley's Aunt* of the new drama', he generously offered him half the author's royalties for the first six months.[9] Shaw, who played up his public image as a tough bargainer, showed himself the opposite in his financial dealings with Ricketts, who replied with equal generosity:

> I have ... ascertained that the usual price for dress designing is so small that if I charged for it I should not be able to afford to live in Lansdowne House, so please allow me the cheap *beau geste*, and consider the dresses the natural neighbourly act of one art to an other.[10]

Shaw commented: 'If you persist in these *beaux gestes*, you will *mourir sur la paille*, and history, ascertaining what I have made out of *Fanny* and *Superman*, will execrate me as a callous exploiter of a genius infinitely more fascinating than my own.'[11]

With the arrival of the *Ballets Russes* in London in 1911, Ricketts's obsession with the new ballet was confirmed. He had been introduced to the Russian dancers in April of the previous year, when Michael Mordkin and Anna Pavlova made their début, and became overnight the rage of London, a year after the Diaghilev ballet had conquered Paris. Now the Diaghilev ballet had come to London. Beauty of music, of movement, of costumes and setting combined with the personal beauty and skill of the performers to enthrall him. He loved their variety of mood, 'their perfect *tragic* sense, the irony, gaiety, melancholy, brutailty and grace of their performances—I leave out many other qualities that have many names and many sexes'.[12] For the next four years they became 'a sort of malady' to him; their arrival was awaited with longing, their performances were attended with delirious pleasure, such as he had never experienced before, and their departures left him nervously exhausted. During their London seasons he was 'lost to all sense of duty, work, and honour'.[13]

Each year he noted their repertoire with his usual vivid and discriminating response. Of Schumann's *Carnaval* he wrote, 'I have seen nothing I like quite so much, and was quite idiotic with pleasure!' *Cleopatra*, however, was 'confused and overdone & not so improper

[9] GBS to CR, 7 Oct. 1911 = *SP*, p. 167.

[10] CR to GBS [Oct 1911] = *SP*, p. 168.

[11] GBS to CR, 22 Nov. 1918 = *SP*, p. 307.

[12] CR to RNRH (TS), [July 1913].

[13] CR to RNRH (TS), July 1914.

as its Paris reputation'. Nijinski, who had a minor role as a Negro slave, was 'consummate, and one understands the follies of the Romans over Pylades, Paris, and the lovely Mnester' (all dancers whose fame has survived from antiquity).[14] The ballets moved 'in a ravishing world of beautiful invention and sheer physical beauty', but they upset his work the next day, 'owing to the prolonged vibrations of a mental and muscular delight'.[15] To Albert Rutherston he strongly recommended the performances: 'I have been as often as I have found it possible and actually weep with pleasure and intoxication quite shamelessly and undisguisedly before one's best enemies of all sexes, Claude Philips, for instance, and Herbert Trench.'[16] So enchanted was he after Karsavina's first performance in the *The Firebird* that he walked all the way home afterwards in his pumps.

On their return in 1912 the Russian dancers thrilled him as before. On the last night they 'danced as they dance only in Paradise—Karsavina destroyed utterly all the diamonds which the Tsar had given to the rival lady! (I forget her name) and Nijinski never once touched the ground; but laughed at our sorrows and passions in mid-air'. In *Scheherazade* they had put 'such beauty in their deaths that we became amorous of death'. The audience applauded silently and

> that pimple of a French conductor struck up *God save the King*, which of course suggests to the Britisher coats, hats, and trains to the suburbs. I was white with rage and, the moment he ceased, I yelled 'Karsavina' twice at the top of my voice; there was a pause, then came the roar from the gallery, like the boom of a distant sea, and the house applauded for twenty minutes.

Robbie Ross wanted Ricketts to go to Russia with him to see the rehearsals for *The Fisherman and his Soul*, but Ricketts exclaimed, 'I dare not, I might never return'.[17]

At first Ricketts felt shy of meeting the Russian dancers and their group; afraid, no doubt, of disappointment—that meeting the artists would dispel the magical world of art they had created. 'These uncanny people have impressed me so deeply that I have avoided meeting them, though the millionaire who supports them has asked to be introduced, and I have been persecuted to meet Bakst.'[18] Nijinski, first met back

[14] CR to MF [8 July 1911].

[15] CR to G. Bottomley, p.m. 22 Oct. 1911.

[16] CR to Albert Rutherston (Tate Gallery Archives TAM 52 52/12).

[17] CR to Muriel Lee Mathews [1912] = *SP*, pp. 174–5.

[18] CR to G. Bottomley, p.m. 19 Sept. 1912.

stage after a performance, was initially indifferent to Ricketts's presence, but he was soon interested when Ricketts spoke animatedly of a ballet he had written. Under the spell of the Russians Ricketts wrote a libretto, never produced, for a ballet based on Keats's poem, *Lamia*.[19] He thought of sending it to Strauss 'who with all his vulgarity and impatience' was alive and had also fallen under the spell of the Russians.[20] Later, Diaghilev took Nijinski to Landsdowne House. When Bakst called on them, Ricketts found him 'as an art lover, superior to his work, sain on difficult points, where I would have thought him out of sympathy—the English school for instance—and quite passionate over the noblest art'.[21] His work enchanted Ricketts, 'whether I think it really good or not', and he and Shannon bought some of his designs and subscribed to a sumptuous book of his theatre work for the Russian Ballet. Ricketts considered him only an 'illustrateur, but one of genius'. With Maurice Denis (sometimes) and the sculptors Bourdelle and Meštrovíč, he was one of the few artists of the time who 'seem to me to count. All the Realists, Impressionists, Futurists, past, present & future, bore me to tears.'[22]

As designers for the theatre, Ricketts and Bakst have often been compared. When Bottomley suggested to Ricketts that the Russians were indebted to him Ricketts rightly denied this, and put down any similarity between himself and the Russian designer to a 'common indebtedness to Moreau, or I should say, to things discovered and initiated by Moreau'. Ricketts's stage work had anticipated much produced by the Russians, but included things that were unknown to them: 'the all-red *Attila* scene, the all-blue *Salome* and *Lear* tent scene, the all-green *Myracle* scheme, and countless details ... such as huge patterns & fantastic head-dresses.' All of these had been produced before the Russians had come to London, and most of them even before their arrival in Paris. Moreover, Ricketts had designed exotic oriental costumes as early as *Salome* in 1906. To Bottomley Ricketts underlined the lack of opportunity he had had compared to Bakst: while the Russian productions had 'benefited by a combination of excellence and success in drama, dancing, music, and the individual genius & beauty of exponents', his own work 'had no such advantages

19 BL Add. MS 61724.

20 CR to G. Bottomley, p.m. 19 Sept. 1912.

21 CR to RNRH [autumn 1913] = *SP*, p. 184.

22 CR to RNRH [autumn 1913], omitted in *SP*.

either on the stage or in my British audiences of vain, apathetic, and vaguely jealous London people'.[23] His productions had been greatly admired by a small discerning public, if not by the Press; had they been given in France, he always maintained, they would have received much greater acclaim. All this is true, but what also counted against Ricketts was his own versatility. If he had gone into theatre design in the 1890s, when he first discussed his ideas with Wilde, and not into book design, painting, sculpture, and writing, he would have had plenty of opportunity to make his name before the arrival of the Russians, who swept all, including Ricketts himself, before them.

For more than two years there had been a lull in Ricketts's own work for the theatre, which he attributed to the effect of the Russians upon him. The means at his disposal seemed 'like a quartet compared to an orchestra', and he did not feel there was any point in his own efforts, so circumscribed in scope and appeal compared with the work of the Russians.[24] Still, he suffered agonies in declining to do *Macbeth*, which he wanted to do; then he wavered and offered to do *The Winter's Tale* instead, having spent a sleepless night about it, but it was too late. The commission had been given to Albert Rutherston, but Ricketts may have had a hand in designing some of the costumes in this play and in *Twelfth Night*. For Reinhardt's production of *Oedipus Rex* at Covent Garden he designed a dress for Lillah McCarthy as Jocasta; but in writing to congratulate Miss McCarthy for her '*really notable* performance' in *Iphigenia*, he thought *her* dress—by Norman Wilkinson—in this production was 'the *only one* I have ever seen on the stage which really looked Greek'. He asked her to tell the designer that he liked all his dresses, but that this particular one 'filled me with bitterness and envy'.[25] He never baulked at giving praise where it was due, or at poking fun at himself in order to give a compliment.

With Maeterlinck's *The Death of Tintagiles*, first produced on 17 December 1912, Ricketts made his only appearance as the designer of a full London production between April 1911 and June 1914. For this production by Barker with Lillah McCarthy Ricketts reverted to his early method of using a backdrop of curtains only. He thought it one of the two plays he had seen in which the lighting was successful; and this must have given him all the greater pleasure when one recalls

[23] CR to G. Bottomley, p.m. 12 Jan. 1915 = *SP*, p. 227.

[24] Furst, *Apollo*, p. 332.

[25] CR to L. McCarthy, [Mar. 1912] = McCarthy, pp. 303–4.

his failures in lighting *Salome* and his doubts about lighting effects generally.[26]

An article on stage-decoration, 'written against over-dogmatism of a Craig kind' about four years previously, was among those articles and long reviews that Ricketts collected together for a book, *Pages on Art* (1913). When the book went to press Ricketts told Bottomley: 'On the whole, I am pleased with the temper, tone, and tendency of these pieces of Journalism, which cover 20 years, and were written without any idea *behind the head* save of teaching those who cannot learn.'[27] Included were articles from the *Dial*, the *Burlington Magazine*, and the *Morning Post*, and the unpublished review of the Post-Impressionist Exhibition. These are full of the erudition and the evocative style that characterized all his writing, but he made only six pounds from the book (which paid for the typing). The failure of this work was less astonishing than that of *Titian*, but after it he gave up books of art criticism along with his hope of making money from them.

On 3 June 1911 the 'Century of Art' Exhibition opened, the 'entire organizing' of which 'fell upon us', Ricketts told Bottomley. It was sponsored by the International Society, whose annual exhibition early in the year Ricketts had also hung. Hours of work and correspondence lay behind the nearly 300 works in the show: these 'represent 2, 3, 8 letters apiece, calls, lies, and a humbling state of anxiety. The works *not* there represent also wasted labour and anxiety.'[28] Watts's *Britomart*, for instance, which they thought they had safely secured, was suddenly to be sold at Christie's: 'the owner died ... out of sheer thoughtlessness'. Ricketts told Cockerell, who was to lend some works and use his influence over others, 'If you behave properly, I will do my utmost to get Davis to lend you his white Whistler later in the year.... But if you don't behave properly, I shall pip your chance.'[29] Behind the humorous bravado of this threat lay the steely determination of the perfectionist. For months Ricketts and Shannon had been involved in selecting representative exhibits, tracing pictures, contacting owners, arranging insurance and delivery—but it was a labour of love. It enabled Ricketts to show works by the Pre-Raphaelite, French, and

26 *Theatre Notebook*, p. 15.

27 CR to G. Bottomley, p.m. 11 Dec. 1912.

28 CR to G. Bottomley [May 1911].

29 CR to S. Cockerell [1 May 1911].

other European masters that they so valued, and to give his answer to Fry's Post-Impressionist Exhibition. It also included works by Hokusai, who was included because of his influence on contemporary art.

> The show is magnificent [he wrote] magnificent in the number of masterpieces, magnificent in the choice of each specimen even when it is not a masterpiece. Out of the 300 works, there are not more than 8 which I should like away, and these are not bad pictures, they include such names as Daubigny, Corot, and perhaps Watts & Rossetti, each with one minor work which hurts their overwhelming average.[30]

Some of their own collection was included, and Ricketts wrote a pamphlet on the exhibition which, as usual, shows the depth of his knowledge, the breadth of his culture, and the height of his enthusiasm.

The nineteenth century, Ricketts argues, was an epoch of great painting, in spite of the fact that the diminution of art patronage meant a harder and more insecure life for painters. It was characterized by 'the fruitful feud between realism under many guises and idealism under many names, or, to put it briefly, the struggle between observation put into immediate terms of painting and experience translated into terms of art'.[31] The latter is the 'rarer' and, to Ricketts's mind, the higher form of imaginative painting. (In his diary, he wrote, 'My personal belief is that great painting belongs to great artists, and that our best modern technicians are actually our imaginative painters, such as Watts, Puvis, Burne-Jones, Baudry, and Gustave Moreau; that their failures will even seem to possess a personal quality'.[32]) Though he dismisses Impressionism as a 'new academic tyranny', he finds things to praise in Manet and early Renoir, and, much less reluctantly, in the great precursors of the school, Turner and Constable. However, his heart is clearly with romantic painters, Delacroix and Puvis de Chavannes (surprisingly Moreau is hardly mentioned), with Blake, with the Pre-Raphaelites, especially Rossetti and Burne-Jones, with

> the ... original and varied art of Watts, who, we have been assured, lived only to say over again what has been said before: Alas that beauty of feeling, originality in design, an exigent standard of workmanship should often count for so little against chance efforts and experimental studies.

30 CR to G. Bottomley [May 1911].

31 CR's *A Century of Art, 1810–1910* (1911) is the source of this paragraph.

32 CR Diary, 13 Apr. 1914 = *SP*, pp. 192–3.

Watts had the creative power that Ricketts so admired, as well as that 'precious quality of personality' that counted most in an artist: 'it is the player himself who counts, not the rules of the game.' In addition to this tribute to Watts, Ricketts shows his independence of judgement by praising Wilkie, who was then unfashionable, and Chafflart, who is still unknown; and had not Paul Baudry been primarily a classical artist he would also have been included.

However, because of the coronation the exhibition hall was empty. The Society lost £400, mainly because the cost of transport and insurance was not recouped. Ricketts, moreover, was exhausted by it, the tenth show he had hung in four years. 'Neither Shannon nor I will undertake even a share in the work of organizing an exhibition again. It is our farewell to public work.'[33] Ricketts was to say this after almost every exhibition he hung in future.

Even before the end of the year they were off on further exhibition business, to the Low Countries for four days. Their visit was decided upon at only twenty-four hours' notice. Shannon was sent over by the Royal Academy to collect works for an exhibition of paintings by Josef Israels, and spent most of his time in Amsterdam, visiting dealers and private owners. Since Ricketts's visit to Brussels thirty years before, countless pictures had been added to its gallery, among them the Belgian Alfred Stevens's *Sphinx Parisien*, which Ricketts thought 'exquisite'. When he and Shannon had come to Antwerp twenty-eight years before, they were poor art students: 'Now we are doing it at the expense of the R.A.'[34] Shannon's election as an ARA a few months before had come as a surprise. They had only learnt of it when they read an out-of-date newspaper in Egypt.

Their collecting of Egyptian antiquities had been spurred to new enthusiasm by their trip. At an Egyptian sale in July 1911 Ricketts admitted he had spent too much, even though on the first day 'everything I wanted was bought by that swine Kalebdjian for Paris, at prices so indecent and unnatural that my Christian pen refuses to write them'. However, he did secure 'some nice odds and ends; two of them are more—in fact, of their kind, museum pieces'. One of them was a small lion in spotty stone, the other, 'a ring of Heru-en-Heb, the last King of the 18th Dynasty ... his effigy is one of the best things in the Cairo Museum'. Ricketts had photographs of this, as it used to fascinate him

33 CR to G. Bottomley [May 1911].

34 CR to R. Ross, Antwerp [late Nov. 1911].

'by a touch of melancholy and tragedy; he reigned a short time and his statue was found broken (intentionally) under the pavement of one of the Karnak temples'. He decided to have a gold hoop made for the missing portion and to wear it 'when I have to meet Impressionist & post-impressionist lovers. The ring is cut out of Emerald quartz. I am overjoyed!'[35] Though on a later day of the sale he was more successful in getting what he wanted, he still hankered after some of the things he had missed. 'After the sale, I shall call on the dealers—unless I am in the Work house.'[36]

One of their outstanding purchases of this year was a new pet. 'Shannon grew troublesome and rebelious the other day', Ricketts told the Fields, 'with a large desire for a tame squirrel.' Ricketts admitted that the fault was partly his own: his description of a cage of them he had seen was the cause of Shannon's whim. Carrots, as it was called, was 'so tame, affectionate, and so passionately attached to humanity that it has to be covered with a cloth to quiet its nerves'. It was very young and greedy and 'tests everything with its mouth, which is its intellect ...'[37] At first the squirrel found it difficult not to eat their fingers and ears.

Egypt, with its dependable sun and marvellous light, still beckoned to them, and they planned to return there at the end of December and come back to England in February, when 'the dear Russians return to show that Nietzsche's phrase, "the feet of flame" can be a reality, even in London'. Their last trip, 'in which every possible luxury was allowed', had cost far less (a third less, in fact) than they had planned, and they intended doing this one even more cheaply. Though tickets were dear, living there was less expensive than they had been told. A delay in their departure on the *Morea* from Marseilles enabled them to rush by cab to see the Puvis de Chavannes decorations in the museum. The passage was perfect; their train at Port Said punctual; the porter at their old hotel 'pretended' to remember them. They found the museums, of which Ricketts's memory was 'minute, almost uncanny', and the Pyramids, which they sketched and 'Kodaked', almost as if they had left them the day before, 'the difference being that one is no longer worried [by importunate natives] in street or by the monuments, as the Kitchener regime has put order where disorder had seemed

[35] CR to S. Cockerell, 13 July 1911 = *SP*, p. 165.
[36] CR to S. Cockerell, 19 July 1911.
[37] CR to MF, 14 Aug. 1911.

inevitable'. As soon as they arrived they rushed to the 'vendor of false gods ... and spent vast sums on Egyptian treasure'; prices had unfortunately gone up and the dealer at first refused to give them credit.[38] The books of Pierre Loti had persuaded them to visit the Mohammedan quarter this time, but they found the tombs of the Mameluke Caliphs sadly disappointing.

Almost at once they set off up the Nile for ten days, this time going much further up river than before. After pausing at Luxor, they saw the temple of Esna, the magnificent temple of Edfu, 'one of the finest and most complete in Egypt', and the great temple at Kom Ombo. At Assam they changed to another boat, and passed Philae, which, along with three other temples, was largely under water because of a new dam. Abu Simbel, the great rock temple and the goal of their journey, was disappointing: 'like most Egyptian art it looks smaller than it is. One does not focus the giants as 60 feet high; yet this is the case.' The famous bas-reliefs in the interior were also disappointing and difficult to see, as the sun in Nubia was 'so brilliant that one has moments of blindness; near a wall there are times when one dares not look up'.[39] In fact the whole trip was a disappointment. 'Our 10 days on the Nile count, with the exception of 2 superb temples and 2 interesting ones, as an elaborate and costly sell in which you cost yourself, or rather I costed Shannon, £3 a day without extras and tips.' They returned to Luxor, where they spent a week. In spite of all the attractions, Ricketts found he could hardly keep awake, owing to the air and light. 'Shannon says he thinks I may have the sleeping sickness and shall be competed for by the hospitals on my return as a rare and interesting case.'[40]

After three days in Cairo they returned home via Vienna, Dresden, and Berlin. It was probably on this occasion that Ricketts and Shannon visited the Albertina in Vienna to see the drawings:

> We were received very effusively by the Keeper ... Shannon has prints in the museum, but he seemed even more friendly towards me. 'We have one of your beautiful drawings.' I thought, 'How charming of these foreigners, they value me even more than Shannon ...' We sat down to the rarer drawings ... He returned wreathed in smiles, his hands fluttering like wings—'Here it is, your drawing.' The tissue paper is removed and I behold a thirty-pound drawing by Arthur Rackham.[41]

38 CR to G. Bottomley, 19 Sept., 30 Dec. 1912 = *SP*, pp. 179–81.

39 CR to G. Bottomley, Luxor, Mon. 20 [Jan. 1913].

40 CR to M. Davis, 17 Jan. 1913.

41 CR to R. Ross [*c*.1916–17] = *Friend of Friends*, p. 283.

Nowhere is Ricketts's ironic egotism more evident than in this story, which he told with relish to Robbie Ross.

After 'stuffy' Vienna and Dresden, where they were fined a mark each by a policeman for crossing the big bridge on the wrong side, they went on to Berlin, by which time they were 'getting fed up on pictures'. In German painting, moreover, the sense of beauty was too often absent. However, Ricketts liked the new architectural movement in Germany, 'the "*Echt*" style, which has little or nothing to do with that nightmare of Art Nouveau of 15 years ago'. Though modern paintings in Germany were disappointing—painting was 'a moribund art that languishes in France and England alone'—Ricketts found it in many ways the most interesting country in Europe.[42]

They were back in London by the end of February 1913, in time to catch the Russians and Strauss's *Rosenkavalier*, which moved and intoxicated Ricketts and often brought tears to his eyes. When the Russians returned in the summer Ricketts found that they were 'turning into Dead Sea apples now they are filled with the French poison of post-impressionism'. Ricketts was so annoyed about this that he wrote to a newspaper 'which had opened up a stupid discussion on the subject'.[43] Stravinsky's modernistic music for *Le Sacre du Printemps* made Ricketts want 'to howl like a dog. I thought of doing so, but realised I should not be heard.'[44] However, he was enchanted by the Russian operas, which were 'a revelation in frank, sincere music and in the fine tragic sense of their interpretation'. The great bass Feodor Chaliapin was 'an incomparable artist and *tragedian* of the first order'.[45] Ricketts saw him in *Boris Godunof* twice in the same week. When the Russians returned in 1914 he saw them twenty times in the one season.

On their return to London they had suffered an unexpected financial blow. Before their departure Yeats's friend John Quinn had agreed to buy an oil-painting by Shannon, probably *The Vintage*. On the strength of this arrangement Shannon, who had financed this trip, spent heavily in Egypt. Quinn, who in the mean time had received the picture in New York, did not like it and wanted to exchange it. To soften the

42 CR to G. Bottomley [Feb.–Mar. 1913].
43 MF Journal, 1913, fo. 64; CR to RNRH, p.m. 5 Sept. 1913.
44 CR to Muriel Lee Mathews [summer 1913] = *SP*, p. 183.
45 CR to RNRH (TS) [July 1913]; CR to W. A. Pye [June 1913] = *SP*, p. 183.

blow he promised to advance money for another picture, a considerate offer which Shannon accepted. By now Shannon was actually supporting Ricketts, who was reaping the benefits of the arrangement by which he had looked after the young Shannon. To his cousin Baroness de Sinner he wrote, characteristically, 'My success is of a very moderate type—as to money at least—but I live in luxury at Shannon's expense, and I think I am cheap at the price.'[46]

Ricketts had, like Shannon, suffered a disappointment when he spent money that was anticipated but did not materialize. In 1911 his painting, *The Plague*, had been offered to the Tate by Edmund Davis. 'At first I demurred,' Ricketts told Robbie Ross, 'then I remembered the *money*; in a few years time I shall complain that I am badly represented in the Tate.'[47] However, though Ricketts had thought the acceptance of the picture a foregone conclusion (and bought two Watteau drawings and ordered new clothes on the strength of the money), it was refused. The Trustees asked Davis to withdraw his offer, so that the refusal should not figure in the minutes of the gallery. Ricketts was deeply hurt and, with his usual bravado, responded, 'I must get into the Luxembourg instead. Alas!'[48] This refusal inspired Davis to make a collection of paintings to donate to strengthen its holding of modern British art. For the 1913 International at the New Grosvenor Gallery in Bond Street Ricketts had sent in his *Montezuma sacrificing to the Sun*. What fascinated him in the Aztec king-god was his barbaric splendour as well as his tragic fall. Because the label was lost, the picture was catalogued as *Hiawatha*. Ricketts, who had never read Longfellow's poem, paced the night away in a fit of nerves and fury, and at nine-thirty next morning was raging at the secretary of the gallery Francis Howard, who told him, 'You needn't make a fuss, you have sold the picture, & as it is to an American, probably the Longfellow title sold it for you.'[49] Howard, a year later, purchased Ricketts's bronze, *Orpheus and Eurydice*, and Shannon's *Mrs Patrick Campbell*, and later presented them to the Tate through the NACF.

'Both S[hannon] and I are ha[r]d at work on rather more ambitious

[46] CR to Florence de Sinner [Mar.–Apr. 1913].

[47] CR to R. Ross [1911] = *Friend of Friends*, p. 202.

[48] CR to R. Ross [1911].

[49] MF Journal, 1913, fo. 32.

pictures as to size and subject' Ricketts wrote in 1912.[50] He was returning to his old parables, four or five or which he wanted to turn into oils. One of these was *The Parable of the Vineyard*, which he gave to Hugh Lane for his Dublin Gallery. Shannon was working on a large picture of *Summer*, which was eventually bought by Quinn in place of the rejected picture. Since it was difficult to get a very large canvas into his studio, this was being painted as a triptych. Shannon was painting a series of canvases based on the seasons, which were to have been exhibited at the International Exhibition in the summer of 1914 at Karlsruhe, where a special room was being built for them. The outbreak of war put an end to this plan: 'the war', Ricketts told Ross, 'has therefore cut off from us a sort of outlet in Germany' where they 'even read my books'.[51]

However, 1914 brought instead the first American shows of the paintings, bronzes, and graphic work in New York and Buffalo. There were 'some very funny press notices', which Ricketts liked, 'in which we are described as the David & Jonathan & the Gilbert & Sullivan of modern British art'.[52] Some of the reviews classed them as belonging to the wane of British Victorianism and criticized them for not reflecting the life of their times. However, as such strictures were also applied to the work of Watts, Rossetti, and Burne-Jones, they did not pain him much. This was the occasion Ricketts saw himself and Shannon described as Neo-Pre-Raphaelites, when he thought they represented a reaction against that school. Moreover, their early work was so different in aim and character from their later that Ricketts felt it had been a mistake to include their woodcuts in the show. A visitor told them it was the wrong time to exhibit in America 'as the whole country has gone mad over Post-Impressionism, Cubism, Futurism'. The show, organized by Martin Birnbaum, had a *succès d'estime*, but was not a financial success. 'Yet, the good Birnbaum—whom we are treating heartlessly & abominably—seems quite affectionate & grateful!'[53] The net profit for both of them was forty-nine pounds.

Ricketts went back to theatre work when Yeats asked him to design *The King's Threshold.* Of Ricketts's dresses Yeats wrote, 'I think the

[50] CR to G. Bottomley, 12 May 1912.

[51] CR to R. Ross [Sept. 1914] = *Friend of Friends*, p. 266.

[52] CR to Mary Davis [Mar.–Apr. 1914].

[53] CR to R. Ross, 11 June 1914 = *Friend of Friends*, pp. 264–5.

costumes the best stage costumes I have ever seen.'[54] Yeats's *Deirdre* was also being given, and Lady Gregory sent Ricketts a copy of it, asking him for suggestions about 'shuffling' the costumes between the two plays. Ricketts designed them so that not only could whole costumes be shifted from one play to another but parts of different costumes could be recombined for various characters. This was in keeping with Yeats's own beliefs that the costumes should be for types rather than individual characters. So successful were Ricketts's costumes that they were used without difficulty for plays for which they were never intended. Ricketts was also asked by Lillah McCarthy for designs for dresses in Haddon Chambers's play *Tanta: or The Impossible Woman*. Ricketts promised original and charming dresses with a contemporary note and new effects in colour and form. 'I think it would be great fun', he told her, 'if the dresses are described in the programme as "Créations de C. R. Worch, 79 rue Royale"—who is of course the new rival of Poiré, Pagguin, etc. etc. and does not exist in that street or anywhere else.'[55] The 'idiot author' objected to his dresses, however, and they were not used. Ricketts also did some twenty-two 'modern dresses' for a revue planned by J. M. Barrie for the French music-hall star Gaby Deslys. 'I was trapped and intimidated ... and offered gold. I feel this is rather like Dannunzio writing the libretto for a musical comedy. It would not surprise me if ... I become known as the man who designs Barrie's dresses.' Soon he learnt that the revue was off—'and my future as the British Poiré or Pagguin dashed, ditto the cash, i.e. as far as I am concerned; this does not leave me cold'.[56] However, Ricketts was rather surprised that he did not feel particularly angry: 'In working for the theatre, I realise the technical difficulties surrounding every detail, owing to people who do not know their business, who wont learn or who cant afford to give the Time it requires to modify their routine.'[57]

On 3 July 1914 Ricketts and Shannon attended the 'Cinema Supper' organized by Barrie at the Savoy Restaurant and Theatre, along with Shaw, Yeats, Rupert Brooke, Chesterton, and the Prime Minister, Asquith, with his wife and daughter. After dinner, which was described as the first act, they took their seats in the theatre for the entertainment

[54] WBY to CR, 11 June 1914 = *SP*, p. 196.
[55] CR to Lillah McCarthy [?1914] = *SP*, p. 204.
[56] CR to R. Ross [July 1914].
[57] CR Diary, 14 July 1914.

by Barrie, which varied 'between quite charming sketches ... acted by very noted actors, and others that were far-fetched'. Altogether the evening was 'most entertaining and everybody fizzled a great deal'. On arrival they had been met by a fierce light, 'both Shannon & I shaded our eyes and made faces'. It was a cinema camera, which was supposed to take candid shots of the guests (and gave the evening its name). Later, Shannon was filmed again—leaving the German ambassadress to get a cigarette for Mrs Asquith. 'I have quite a childish wish to be visible in the cinema films', Ricketts confessed, 'I should like to see how I look and move.'[58] Barrie had hoped to use the film in his revue for Mlle Deslys, but it was never made public because Asquith was revealed in it to be tipsy.

News of the damage to the *Rokeby Venus* by a suffragette had left Ricketts 'astonished at my unfeeling reception of the news; a cold self-possessed stupor at the imbecility of the act, a dim stupid wish for retribution—which is idiotic'. He put such outrages down to 'the new element of danger, which is daily increasing, owing to the spread of education, which leads too often to an exagerated sense of the importance of the Ego'.[59] Two months later the suffragette attack on Bellini's *Agony in the Garden* upset him more. When Shannon told him, his knees began to shake, and he had to sit down.[60]

Ricketts was no democrat. He opposed universal education, and the policy of opening museums to all and sundry. The first, he felt, was only half education and brought no real appreciation of culture; the second filled museums with those who did not really love, study, or understand the pictures, but merely glanced at them in passing. In both cases the results were superficial and led to pretensions masking ignorance or indifference. However, Ricketts was an intellectual élitist rather than a social snob. He welcomed the gifted individual, whatever his social origins, into the company of the elect, so long as the latter showed real love and understanding of culture, or at least a real desire to learn and understand. He despised the fool, or the Philistine, even if puffed up with a noble title or a position of authority. The criterion of value was always culture, rather than class or status.

[58] Ibid., 3 July 1914.
[59] Ibid., 10 Mar. 1914 = *SP*, pp. 188–9.
[60] CR Diary, 22 May 1914 = *SP*, p. 195; 10 Mar. 1914 = *SP*, p. 188.

Even before the war, which brought so much death and destruction, Ricketts suffered the loss of many friends. Van Wisselingh had died in 1910 and, at his widow's request, Ricketts designed the lettering on his urn, but (as 'my hand & eye are not what they were') he advised her to have it traced by someone else.[61] In July of that year Hacon died, leaving Ricketts a bequest of £100, and Ricketts and Shannon bought his entire collection of Japanese prints. Mabel Beardsley was dying from cancer. Ricketts visited her regularly and made for her, with his own hands, four dolls based on characters from her brother's drawings, one of which was Mlle de Maupin; Yeats mentions these in his poems, 'Upon a Dying Lady'.[62] Against all expectations, she lingered on until 1917.

At the end of 1910 tragedy had also struck the Fields. Ricketts learnt that Edith Cooper, aged only forty-nine, was dying of cancer. Almost as if he had sensed the impending tragedy, he had been particularly generous to them since their quarrel of 1907. Perhaps their preoccupation with their *chers pères*, about whom he teased them endlessly, meant that they were less demanding and emotional, so that he felt less threatened. On one of his visits Miss Cooper absent-mindedly greeted him with, 'How are you today, Father?', which amused him vastly: he had been called many things, but not that before! The gifts he showered on them reflected their spiritual concerns; for, though he had no use for religion himself, he saw how it had benefited his two friends, made them more accepting and less bitter about their lack of recognition. He gave them his mother's coral rosary, fashioned for them a Madonna and Child, and presented them with a small pair of portable candlesticks, a Spanish crucifix, and a small Spanish reliquary, containing relics of saints and a tiny cross, 'in the past ... venerated as fragments of the true cross'. Its price was 'an interesting one', he told them, '40 silver shillings'.[63] In addition to these gifts, which show how deeply he could sympathize with their interests, he also gave them a plaster cast of *Silence* (1905) (Fig. 36), his memorial for Wilde, which, Sturge Moore told them, was modelled on Miss Cooper, whose face was also to be found in some of his woodcuts.[64]

On their account, though at first rather reluctantly, he had also

[61] CR to S. Cockerell, 16 Dec. 1912.
[62] See *Letters of Yeats*, p. 574.
[63] MF Journal, 1909, fo. 171; CR to MF, 27 Apr. 1910.
[64] MF Journal, 1910, fo. 228.

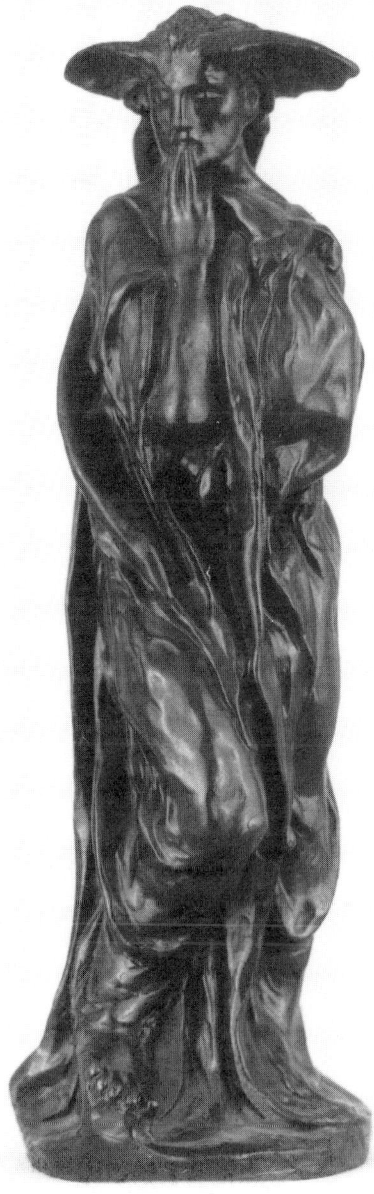

FIG. 36. *Silence*. Bronze statuette (1905) by Charles Ricketts (William Andrews Clark Library, University of Calif., LA)

returned to book design. He had designed emblems for a series of plays starting with *Borgia* (1905), which they had published anonymously in the hope that it might get more attention than their other work. Now they wanted more from him.

> Your request for a book cover [he told them in 1907] ... has filled me with grief and consternation. I hate book decoration ... Not only do I object to drawing all the little cells in the honey comb (the proper name is sells) but I find that I must go to the Natural History Museum to ascertain what a bee looks like; so far I can only draw what look like butterflys or wasps.[65]

The Fields had asked him to design the cover for *Wild Honey* (1908) and urged him to take as his model Rossetti's cover for his own *Poems* (1870). Ricketts had once placed a dead bumble-bee among a posy of flowers he had given them; now he sent them a drawing of 'a living bee' which he used, with the honey comb, to decorate the cover.[66] By the time that they asked him to do the later books for them the news of Miss Cooper's cancer had come out, and he was then as willing and generous as he had been reluctant before.

Ricketts told the Fields, 'there is in me something of the *garde malade*', and, much to their surprise, they had to agree that he was right.[67] Sickness in others brought out the great generosity, and a new tenderness, in his character, so that they felt he had become to them 'the Good Samaritan' he had so often painted. He had, no doubt, learnt this patience and sympathy at the bedside of his failing mother. 'I never thought the freakish aesthete of the Vale', noted Miss Cooper, 'could be so brotherly in sweetness of charity to those stricken by sorrow & disease.' His own suffering was also to bring out the best in Ricketts, and over the years this and the tribulations of his friends brought to the surface the deep but latent compassion that he had expressed in some of his earliest paintings, particularly those depicting the Good Samaritan, which Ricketts painted at least four times. (The subject was sympathetic to Ricketts because 'the Samaritan was civilized and intelligently human in leaving the two pence'.)[68]

He visited the invalid almost weekly, bearing gifts: of expensive chocolate, rose-leaf conserves, gentian roots. Having bought some flowering *iris pavona*, he realized it would be past its prime by the time

[65] CR to MF, 26 Sept. 1907.

[66] CR to MF, Sept.–Oct. 1907. The drawing is in BL Add. MS 58089, fo. 73.

[67] MF Journal, 1912, fo. 43. [68] Ibid., fo. 18^{v}; 1902, fo. 42.

she received it, so he made a drawing of it as well, which was hung beside her bed.[69] To entertain her, he vividly described the Russian ballets and operas he had been seeing, and lent her a translation of *Boris Godunof*, between each page of which he had put a fragrant leaf of lemon verbena. In 1912 he offered to design *Poems of Adoration*, largely the work of Miss Cooper, even though he detested religious poetry. Later he designed another religious work, *Mystic Trees* (1913), by Miss Bradley, and *Dedicated* (1914), a volume consecrated to Miss Cooper's memory. In December 1913 he brought her two clay angels and a camel laden with pearls, which he had modelled himself, for the Christmas crib. She died the next morning, 13 December, her last words being 'Not yet, not yet'. Ricketts was the last friend to see her alive.

However, Ricketts still found Miss Bradley difficult to cope with. Their relations over the past few years had often been strained. During the first Egyptian trip the Fields had been offended when Ricketts, tired of repeating the same impressions and experiences in letters to different friends, had asked them to forward to Gordon Bottomley the letters he had written to them. They refused indignantly. Then, before the second trip, Ricketts had quarrelled with Miss Bradley, and refused to give them his address. Though he sent them the usual letters and cards, they did not know where to reply, and finally chanced sending a letter to Cook's office in Luxor. On his return he had it out with 'Michael', shook his fist at her, told her she was deep in many sins—pride, untruth, jealousy, hardness, cruelty,—and advised her to put on her jewels and go to confession. Not surprisingly, Ricketts and Miss Bradley became estranged soon after Miss Cooper's death. She insisted, presumably for the good of his soul, that he should visit the 'Sun Room', the little sitting-room into which Miss Cooper had been moved during her last months and in which she had died. Because of its painful associations, Ricketts refused. After she had promised not to bring this up again, Ricketts visited her at The Paragon, but when she broke her promise he grew violently angry and left. Their correspondence was later renewed, and she wrote to Ricketts on the morning of her death, on 26 September 1914. (She died from heart failure on the way to mass.[70])

[69] Michael Richard Barclay, *Catalogue of the Works of Charles Ricketts from the Collection of Gordon Bottomley* (Stroud, 1985), 68.

[70] See *Letters CR to MF*, p. 7; MF Journal, 25 Feb., 22 Apr. 1914.

In spite of their quarrel, Ricketts was the chief beneficiary of the will she had made a fortnight after it; he was astonished to learn that his portion could amount to as much as £2,000; though it eventually became much reduced by the lowering of values caused by the war. In memory of them, he presented Rossetti's water-colour *Lucrezia Borgia* to the National Gallery in 1916.[71] Silly and exasperating as the two women could sometimes be, there was something truly noble in both of them; in Miss Cooper's refusal to take pain-killers during her illness so as to keep her mind clear for her writing, in Miss Bradley's silence, for fear of distressing her niece, when she learnt that she also had cancer. This nobility attracted Ricketts to them. He also admired their work, and after their deaths wrote, 'When we all come into our own, "Michael Field" will be remembered, when the Thompsons, Addington Symons, etc. are forgotten.'[72] It distressed him to see the jewels by him they had left to the Fitzwilliam Museum described as 'bequeathed by Miss Bradley' instead of 'Michael Field'. He had given the Rossetti in memory of 'Henry Michael Field'.

With the money he had inherited from the Fields Ricketts hoped to go to Ceylon, perhaps Angkor, perhaps Borobodour. As early as 1913 he and Shannon had wanted to go to Ceylon 'to visit the burried cities in the jungles and feed the tigers with buns', but now that they had the means to do this it was impossible, because of the war.[73] They had made their last visit to Paris, which they had rather avoided in recent years, in the autumn of 1913. They went to see Shannon's picture, which was hung in the Luxembourg, and Puvis de Chavanne's *Espérance* and his drawings in the Petit Palais; 'each year seems to make him greater'. They also went to see his murals at Amiens, for one of which they owned a study; a magnificent drawing. Ricketts was shocked, however, 'by the beastliness of the recent output in France' and believed that Impressionist pictures were darkening: 'their pictures looked like bluish slosh.'[74] This was to be their last trip abroad until 1923, as the war was soon to overwhelm them in its violence and the destruction of so much of that they loved and lived for. Afterwards, Ricketts often said, 'I died during the war.'[75]

[71] Acc. no. 3063, now in the Tate Gallery.
[72] CR to S. Cockerell, 6 June 1917.
[73] CR to R. Ross [late 1912].
[74] CR to RNRH, p.m. 29 Oct. 1913.
[75] *SP*, p. x.

10 *'I died during the war'*

'THE clock of civilization has been put back two centuries', Ricketts wrote early in the war.[1] Ironically, Germany, which had seemed to him the most civilized and promising country in Europe, was responsible for this. Now he worried at the way in which 'the organized forces of Germany, so admirable in all they have undertaken in late years (notably in art, about which I know something), contrasted with the passion for compromise & muddle for which we stand in this as in other matters'.[2] During the five years to come his attitude to the war itself did not change, but his interest in, and response to, events varied considerably. To a man as sensitive and imaginative as Ricketts, so aware of the powers of evil, so devoted to the actual physical inheritance from the past, the war represented a threat to civilization itself. Such a turn of events had seemed impossible, he said, to someone who dated from the last century like himself.[3] Nevertheless, he threw himself into his own war effort, and, as might be expected, it took the form of a battle for art.

Convinced at first that the Germans were bluffing, Ricketts thought that war would never occur. When he was assured by Lykiardopoulos, of the Moscow Art Theatre, that Germany would soon invade France he turned white at the threat to that country, 'the most civilized population in the world', and at the danger to 'the countless relics of the long civilized past. I felt as if I saw a mediaeval tapestry being used to clean out a pigsty.'[4] On 2 August, two days before war was officially declared, he realized he had been wrong. His friend and neighbour Edmund Dulac called round in a panic, but Ricketts felt curiously unmoved, even without resentment. Vague thoughts occurred to him of storing their treasure in the British Museum 'in case of tragic events'.[5] His mood soon changed. He thought of nothing but the war,

[1] CR to S. Cockerell, 19 Oct. 1914 = *SP*, p. 219; *Friends of a Lifetime*, pp. 342–3.
[2] CR Diary, 7 June 1915.
[3] CR to R. Ross [Sept. 1914] = *Friend of Friends*, p. 265.
[4] CR to François Monod, 17 Aug. 1916 (TS).
[5] CR Diary, 2 Aug. 1914 = *SP*, p. 210.

bought six or more newspapers a day, and developed 'an uncanny sense of hearing, for the least noise which might be a newspaper cry'.[6] With the onslaught on Belgian and French cities the full horror overcame him, and the realization that the newspapers were not telling the truth filled him with distrust and apprehension. Deliberate acts of destruction, such as the razing of the ancient library and university of Louvain, 'unhinged' him. When he learnt of the damage done to the cathedral at Rheims, which he had never visited, and thought safe, he was 'numbed with pain' and 'sniffed & cried outright for 2 days, sometimes even in the streets'.[7] With this, he began to feel an alarming tightness round his heart. In September he and Shannon escaped London on an excursion to Winchester, Stonehenge, and Romney.

After the siege of Antwerp he lapsed into a state of nervous apathy, with the physical stress showing itself in sudden gusts of temper. The next crisis occurred when Italy was pulled into the war, with the threat this posed to the art and architecture of Verona, Venice, and Padua, and, in particular, to the Arena Chapel with its frescoes by Giotto—after the Sistine, 'the second jewel in the great crown of Italian painting'. The thought made Ricketts 'brainless and effortless for days'.[8] The destruction of the Tiepolo frescoes in the Scalzi, Venice, distressed him deeply, and made him fret even more about future 'massacres'. Now he glanced at newspapers only to gather the trend of events, and any paragraph that might upset him was avoided. Similarly, when he met the Belgian refugees who thronged London he sought to divert or half hear any stories that might be too painful. 'In fact, I behave like a nice, selfish old gentleman which I believe I have become in spirit & in fact', he told his Dutch friend Holst.[9]

He saw the war, as he had seen the Wilde trial, as an attack on art itself. 'I know there is hatred and envy of the past and for all art in most human beings—latent somewhere.'[10] This was the only explanation that Ricketts could give for the forces of destruction the war had unleashed. What had taken centuries of effort to create could be destroyed in a matter of days, of hours, and could never be replaced. Even more than the surprising acts of cruelty and brutality to non-

[6] CR to RNRH (TS) [spring 1915].
[7] CR to R. Ross [Sept. 1914] = *Friend of Friends*, p. 265.
[8] CR Diary, 20 May 1914 = *SP*, p. 239.
[9] CR to RNRH (TS) [spring 1915].
[10] CR to R. Ross [Sept. 1914] = *Friend of Friends*, p. 265.

combatants, which made him reflect once more on the 'dark stain' in human blood, the idea of a 'War of Vandalism—of a set purpose' staggered him. Germany, he felt, should one day be made to pay out of the art treasures in her public museums for the 'abominable damage' she had done. Berlin, for instance, should be made 'to disgorge her Flemish pictures to make good the loss of Louvain'.[11] At the news of each outrage he would be too upset to work, and would go to Kew Gardens or the National Gallery. Beauty, both in art and the natural world, became ever more important to him.

Another irony of the war, moreover, was that Ricketts believed Germany was the focus of most art activity and had been a good market for their own work which was well represented in German galleries and was now also threatened with destruction. Berlin and Dresden had complete sets of Shannon lithographs, and Weimar a complete set of Vale Press books. When the foreign pavilions at Leipzig, which also had a large collection of Ricketts's books and Shannon's prints, were said to have been destroyed, their reaction was unexpected: 'this touch of comedy in the world tragedy makes us laugh and feel almost pleased.'[12]

Yet Ricketts did not let his emotions distort his appreciation of German art, or his support for individual German friends. At the news that the Wagner concerts at the Queen's Hall had been cancelled, in favour of French and Russian music, he was indignant. If the concerts had been of Strauss, who was a living German, he would have understood, for 'some fool or fools might demonstrate during the concert'. Wagner, however, was a classic; and the suppression of his music, 'one of those tactless things which will be quoted as evidence of shallow civilization & exagerated war fever'.[13] The newspapers, he felt, irresponsibly fanned such race hatred. Similarly, when the shop of a local Austrian butcher was attacked by a mob Ricketts rushed round to comfort the poor man. Another friend, a German painter marooned in England by the war, was Rudolph Hellwag, the delegate for the Karlsruhe Exhibition, who had been so co-operative over Shannon's decorations. Hellwag and his wife made repeated visits to Lansdowne House for advice and help. Their lawyers, as well as the painter Lavery and the journalist Wickham Steed, who lived downstairs at Lansdowne

[11] CR Diary, 30 Aug. 1914 = *SP*, pp. 215–16.

[12] CR Diary, 2 Sept. 1914.

[13] Ibid., 17 Aug. 1914.

House, all advised them to be naturalized as British citizens (as they feared imprisonment or public hostility), but Ricketts and Shannon advised against this course. When, however, the Hellwags decided to take it, it was the two English artists who signed the papers. Despite this precaution, and the fact that the Hellwags had lived in England on and off for twenty years, they had learnt by July 1915 that the husband was to be interned. By this time Ricketts had decided that he disliked them both, especially the wife, but he felt sorry for them and agreed to look after Hellwag's sketches and canvases. When the German wrote later asking for money they not only sent it to him but subscribed a pound or so a week to alleviate his condition in the camp.

Zeppelin and (later) aeroplane raids on London were a new and alarming aspect of this war—one which has become so familiar a part of warfare that we are apt to forget its novelty at that time. When walking at Kew in September 1915, Ricketts noticed with a sense of excitement planes circling overhead: 'We are still at the stage of the arab who watches the train', he noted.[14] Vivid accounts of the early Zeppelin raids are given in Ricketts's diaries and letters. Bombs fell in the vicinity of Lansdowne House, and all the occupants would shelter in the basement, where Ricketts passed around Bull's Eye mints. Their own treasure was also moved to the basement where, except in the event of a direct hit, they felt it was reasonably safe. Though the prospect of death by fire caused by bombing worried him, Ricketts's first concern was to safeguard as much as possible (not only their own collection but also Britain's artistic heritage) from the fate that had already befallen monuments and paintings across the Channel. Ricketts was one of the first to recognize this threat. As early as 11 October he wrote to *The Times* concerning the precautions that should be taken in the museums and at Westminster Abbey, where he wanted all inflammable stuff removed and the Plantagenet tombs to be protected with sacks of earth. His letter was not published. Shocked and incredulous at this apparent indifference, he carried on his campaign in letters both to the national newspapers and to museum officials. A private letter to the Director of the National Gallery Charles Holroyd, advocating the removal of paintings by Michelangelo and the great Italians, had some success, though Ricketts was furious to receive a reply saying that 'most' of his suggestions would be carried out. He was even more irked to discover that the first pictures to be removed were by Vermeer

[14] CR Diary, 16 Sept. 1915.

and Mabuse, and to learn that Holroyd had waited for a committee meeting to act at all.[15] The British Museum, too, had done nothing and Ricketts was responsible, together with Claude Phillips, for the removal from there of the Elgin marbles and some of the Greek vases. This limited action still left the major collections of the London museums unprotected, and he continued to bombard the press with letters. After an air raid in September 1915 he rushed to the British Museum to make sure that it was safe. What he saw as the dilatoriness and irresponsibility of museum officials made him furious. 'The scare has proved the utter worthlesness of our museum directors and trustees. The art world in England is headless, heartless, worthless in all matters of policy and responsibility; our press, venal and idiotic.'[16] In the midst of his efforts he was appalled and infuriated by the vulgarity and philistinism that the war brought out. When a newspaper reported a Scottish professor, Patrick Geddes, as describing the British Museum and the Victoria and Albert as 'loot-heaps' Ricketts had a bout of 'uncontrolled anger'. He felt that there was an 'essential lack of proportion & delicacy in average English thought' that terrified him.[17] The effect was to make him work all the harder for what he believed. His fervour, together with his lack of tact, caused him to offend and embarrass many of the powers that be, and he recognized that he was making enemies: 'of the reluctance, bad feeling, vanity, and bad faith which has been stired, the Devil only knows.'[18]

This experience deepened Ricketts's sense of alienation from his age and society. 'I believe I am conscious that in all I love most, Art, Music, Literature, Drama, there is a deeper cleavage between myself and the feeling of my countrymen than I had imagined before the war.'[19] His aesthetic creed, so profoundly and passionately held, had survived into a time preoccupied with other concerns. These were not only the immediate priorities of the war but also the current movements towards social and economic goals.

Moreover, the war was already resulting in the death of friends, which also deeply affected him. On the same day, 7 May 1915, that Ricketts read 'with some emotion' of the death of 'that beautiful yellow

[15] Ibid., 12 Oct. 1914 = *SP*, p. 218.

[16] CR Diary, 2 Mar. 1915.

[17] Ibid., 22 July 1915.

[18] CR to G. Bottomley, p.m. 9 Nov. 1915.

[19] CR Diary, 1 June 1915 = *SP*, p. 241.

youngster' Rupert Brooke, whom he had met, he learnt with distress of the sinking of the *Lusitania*, with the loss of his friend Sir Hugh Lane. Then, Oscar Wilde's elder son Cyril, who was charming and the more interesting of the two sons, was killed in action. Depression and pessimism about humanity and its future overwhelmed Ricketts, and the question whether he might ever regain his 'old cheerful pessimism and ironical outlook upon unessential or unimportant things' gave way to the feeling that he was slipping into 'the indifference of the elderly': 'I am growing grey and Shannon is quite white.'[20] The advance of age brought not only a lack of optimism but also a lack of memory and resilience. In the art world there was a sense of anarchy ahead, of 'vistas of conquest to the incompetent', of a younger 'arrived' generation 'with rough manners and rough paint' pushing to impose new values totally unsympathetic to him'.[21] The war in fact had brought to a head Ricketts's own middle-age crisis.

His depression, his deepened sense of alienation, his strengthened distrust of museum officials, all played their part in his decision to refuse the greatest opportunity that was ever offered to him. The first rumours that he was seriously being considered for a museum post—apparently the Directorship of the Tate for foreign work—came on 10 July 1914.[22] A few months later Yeats told him that Holroyd was proposing him as his successor at the National Gallery, and Robbie Ross confirmed this with the news that all those consulted by the Treasury had suggested Ricketts for the post. Holmes and Robert Witt, the other candidates, had even volunteered to withdraw to make Ricketts's election certain. From the beginning, however, he had no intention of accepting. Not only had he openly attacked some of the Trustees, though he was assured that they had no say in the election, but he also felt that he could not give up his work as an artist and sit in an office: 'The berth—as it now stands—cannot be taken by a man with any sort of personality.' Shannon was in favour of acceptance, as their financial outlook was not very secure, but Ricketts still refused: 'I should find it intollerable, I am unfitted to cope with the official mind. I should be unhappy, it might shorten my life.'[23]

[20] CR Diary, 7, 8 May, 1915 = *SP*, p. 237; CR to RNRH (TS) [June 1915] = *SP*, p. 240.

[21] CR Diary, 20 Nov. 1915 = *SP*, p. 250 (in part).

[22] CR Diary, 10 July, 1915.

[23] Ibid., 20 Nov. 1915.

When, early in 1916, however, two American museums approached him to act for them, Ricketts began to have second thoughts. These offers did not really interest him, as he felt he could not comfortably participate in exporting from Britain any key works of art, but it made him 'put a new light on the value of my "expertise"'. He wrote to Ross to inquire what the National Gallery Directorship would entail in terms of hours and circumscription.[24] It was too late, and on 22 July he learnt that Holmes had been appointed, which he felt was the best thing under the circumstances.[25] At least, some of his ideas would be gradually adopted.

Kenneth Clark suggests that Ricketts would have accepted the post, provided certain conditions had been met, such as the reframing of the pictures and the inclusion of flowers and Persian carpets.[26] However, there is no evidence in his writings at the time that he intended to accept the post (until the very end). Perhaps Lord Clark's account is based on Ricketts's descriptions of what he would have done, had he accepted it. Within a year Ricketts was suffering bitter regrets over his refusal, which he felt showed a lack of courage, insight, and public feeling: 'it would have been a dignified end to my life to have recast, rehung, and revised the dear old place.' A good example there 'might have given tone to the art feeling of our time', and in regrouping the pictures, giving them frames from their own time, changing the wall coverings according to school and period, and effacing the trade stencilling of the ceilings, he would have managed 'to slowly remould the place from a board of works aspect to that of a palace'.[27]

In spite of his refusal of the National Gallery post, Ricketts continued to be involved in art politics and exhibitions. When the sculpture of the Yugoslav Ivan Meštrovíč was sent to London to be exhibited, 'to create sympathy for Servia', Ricketts, who knew and liked his work, was asked to go on the committee. Meštrovíč was asked to dinner at Lansdowne House. Though 'quiet to a fault, unobtrusive, and child-like', he spread 'an aura of difficulty and uncertainty' about him, thought Ricketts, and was not only anxious and distressed over the war but also overworked, owing to difficulties connected with his

[24] CR to R. Ross, p.m. 6 Apr. 1916 = *Friend of Friends*, pp. 285–6.

[25] CR Diary, 22 July 1916.

[26] Clark, p. 178.

[27] CR Diary, 16 June 1916 = *SP*, pp. 261–2.

exhibition. However, he 'grew human in running down Maillol and his false archaism', and Ricketts warmed to him considerably. Meštrovíč's own work drew on the classical and Byzantine traditions.[28] When a letter from Selwyn Image appeared in *The Times*, attacking it as 'wilful, inchoate, amorphous, even monstrous', and, in at least one work, 'morally offensive', Ricketts, along with Sargent and Lavery, wrote defending the work, whose 'tragic intensity, austerity, and passion placed it apart from any merely morbid or pernicious tendencies, and in the tragic world to which these works belong, there is no room for what is trivial and morally offensive'.[29]

Ricketts was also determined to obtain for the nation any major works that came on the market. When the George Rae collection of Rossettis became available he threw all his energy into securing the best of them through the NACF. An offer was made to the Rae family which they rejected. At that moment a rarer and more important picture suddenly appeared on the market, Masaccio's *Madonna and Child with Angels*. Obtaining this was a formidable task, and involved a great deal of letter-writing and manœuvring behind the scenes, into which Ricketts entered with determination and zest. Once this was obtained for £9,000, half of which was paid by the NACF, he turned once more to the Rae Rossettis; but now they were short of cash. Over lunch with Lady Cunard and Lady Horner plots were hatched (to deal with the distressing news that the National Gallery Trustees openly disliked one of the paintings and were becoming 'tiresome & hostile'). Shannon brought Sargent in to testify, and 'all the arts of persuasion, intimidation, obsequiousness, and violence were used in turn'. Ricketts even entertained a wealthy benefactor, Sir Arthur du Cros, with 'stale scandals', to secure the disliked painting. It is not surprising that some of Ricketts's enemies accused him of being machiavellian. Finally, through 'a series of miracles . . . in which dear Lady Horner was most useful', some three important oils and eight water-colours (most of which had formerly belonged to William Morris) and drawings were obtained by the National Gallery with the help of a £3,000 gift from Du Cros to the NACF.[30] The Rossetti water-colour, *Lucrezia Borgia*, that Ricketts gave the National Gallery in memory of the Fields also came from the Rae collection.

[28] CR Diary, 24 Dec. 1914 = *SP*, p. 226; CR to RNRH (TS) [June 1915] = *SP*, p. 241.

[29] *The Times* (30 June 1915).

[30] CR to S. Cockerell, 23 Apr. 1916 = *Friends of a Lifetime*, p. 345.

No doubt because of the effects of the war, the market was suddenly flooded with desirable pictures. Some important Blakes, Alfred Stevens's portrait of Mrs Collmann, and, most important of all, the Wilton Diptych, became available at about the same time. To secure the Diptych Ricketts even considered trying to bribe the wealthy collector Herbert Cook to buy and give it, in return for a title: 'I should hate Cook being a Lord, but necessity knows no law.' He wished that an honorary order existed for first gifts to the nation 'with the chance of promotion later when more cash was needed'. Ricketts mistakenly believed that the Diptych had once been in Westminster Abbey, and he felt that an object of such 'historical and sentimental importance' should not be returned there because of the inadequate precautions against Zeppelins, and because 'our priests, despite their carefully evolved Family Butler mentality, retain that one characteristic of the priest—hatred of art'.[31] In the event, the Diptych did not become part of the national collection until 1929. Ricketts himself purchased the Alfred Stevens portrait for his and Shannon's collection.

'We are in the thick of a stupendous undertaking of national and European importance in which some hundreds of thousands have to be found, and imagination & decision instilled into our overworked government, besides countless Mandarins.'[32] In this breathless announcement, with its apparent relish for intrigue, Ricketts hinted to Cockerell of an even greater challenge. The Bridgewater Titians, which included key examples of the early and late work of the master, were reported to be in the offing. With an urgent sense of American millionaires waiting in the background to snap up these irreplaceable works, Ricketts, in conjunction with Holmes, his old 'enemy' MacColl—again necessity knew no law—and others, applied himself to the task. Drastic measures were necessary, and Ricketts felt strongly that the national collection should dispose of some of its Canalettos and Turners to raise the money. 'The spirit of Turner would exult if some of his works helped to secure Titian, but I believe there would be an uproar and questions in Parliament by men who had never looked at his pictures.'[33] When the proposal was discussed in the Press and Parliament Ricketts, undaunted, despatched a letter to the *Observer*, arguing eloquently for the disposal of redundant Dutch paintings and

[31] CR to R. Ross [Apr. 1916] = *Friend of Friends*, p. 284.
[32] CR to S. Cockerell, 23 Apr. 1916 = *Friends of a Lifetime*, p. 345.
[33] CR to D. S. MacColl [1916].

Turners to save a unique set of masterpieces by the incomparable father of modern painting.[34] Again, all the efforts came to nought, as the collection was not sold (and is still in private hands, though on loan to the National Gallery of Scotland).

The last of the great excitements came in March 1918: 'We are all of us in a state of agitation because the finest set of Blake watercolours in the world comes up at Christies next month, and I don't know if our public funds are good enough to secure them.' This was the set of illustrations to Dante. Ricketts had been ill, and during his 'eclipse' the NACF, he maintained, 'went to sleep' over the drawings: 'the lack of foresight and imagination amongst intelligent persons is unbelievable', he concluded. The NACF had raised £2,000 towards the 'purgatory set'—'not the best' one, according to Ricketts. He was fearfully upset to hear this, and agitated 'to raise cash left and right.'[35] Finally, a consortium of museums, along with the NACF and Ricketts and Shannon, the only private collectors, purchased the Dante series for £7,000. As with the money, the drawings were secured by the skin of their teeth: 'the American agent had wired to New York to know what was his money limit. The Blakes had just been knocked down to the N.A.C.F. when a wire was handed to the agent with the answer, "No limit".'[36] Ricketts throve on such dramas and triumphs. At the division of the drawings (on old packing-cases in the vaults of the Tate, while tube trains rumbled overhead) Binyon acted for the British Museum, Ross for the National Gallery of Victoria, Australia, Aitken for the Tate, and there were two other representatives, from Oxford and Birmingham. Ricketts and Shannon contributed £300, and were asked for another £50 later, which brought them three drawings, 'one first rate, a good one, and a bad one'. Ricketts felt they had done 'scandalously well—beyond hope. I should not have been allowed the *Lucia*,' but his conscience was assuaged by recalling their intention to leave their collection to the Fitzwilliam: 'we are Cambridge in the long run'.[37] He could not have foreseen the tragic circumstances that would one day force him to sell these drawings.

It was an immense satisfaction to Ricketts that, 'under penitential conditions', such important pictures had been secured—forever—at a

[34] *Observer* (3 Dec. 1916).

[35] CR to TEL [end Mar. 1918] and [Apr. 1918].

[36] CR to RNRH (TS) [summer, 1921].

[37] CR to R. Ross [Apr. 1918].

time when the 'idiotic and illiberal' Government had (from the start of the war) withdrawn grants, closed down museums, and prohibited importations, and when hundreds of people had cancelled their subscriptions to the NACF 'for patriotic reasons, beginning with Balfour'.[38] Art remained one of the few things of lasting value, and Ricketts felt that 'non-combatants should try and keep the house at home in order and not gnash false teeth or wave paper flags'.[39] Something had to be done for the future, for the day when the youngsters returned from the war. 'We owe an immense debt to the generosity & culture of the past and should carry on the torch to the next generation and not throw it away forever!'[40] In these young men in khaki Ricketts put a lot of hope, having been deeply touched by their cheerfulness after appalling experiences and their renewed interest in art. 'I grant that in a world conflagration ... like the present war it is not the time to think of the elevating influence of art; yet youngsters have been found in the trenches who thirsted for it and who liked to think that it was there for ever.'[41] He was delighted to learn, in a shop which sold reproductions of paintings, that there was a sudden demand for Watts and Rossetti, whereas there had been no interest before the war. He had noticed in the younger generation, especially the women, whose education had changed more dramatically, an amazing increase in good looks and good manners: 'the vulgarity of "society" is mainly the vulgarity of people who are over 40.'[42] As well as postcards of his and Shannon's own work, he sent news of the Rossetti purchases to young friends in the trenches, which was received with great enthusiasm. Victor Reynolds, a confirmed Rossettian who had visited Lansdowne House, died before the news of the clinching of the Rae collection could be given to him. A brother of Mona Wilson (who had bought one of Ricketts's Good Samaritan paintings) was killed at Ypres, with the postcards of Ricketts's and Shannon's paintings still in his pocket. This deeply touched Ricketts.

I did not realize that, to others, our pictures could be interesting in the war and its horrors, and bring charm to the monotony of tragic ... days ... The stress of the war and the gathering hostility to art has often made me brood

38 CR to Laurence Haward [1917].
39 CR to S. Cockerell, 29 Apr. 1916.
40 CR to S. Cockerell, 23 Apr. 1916.
41 CR Diary, 9 Oct. 1916 = *SP*, p. 268.
42 CR to RNRH (TS) [1915].

over the possibility of the future. After all, our work may mean something to someone else.[43]

With these heartening sentiments Ricketts concluded his diary for 1915.

Six wounded soldiers were invited to drink tea at Lansdowne House on 10 March 1916. What they made of its fastidious beauty, after their harrowing experiences, is not recorded. Ricketts corresponded with several young men at the front, commiserating with them, and encouraging their cultural interests with his quick sympathy and batches of postcards of the Greeks, the Italians, and his great hero, Puvis de Chavannes. To Robin Hollway he gave a copy of the Vale Keats, and wrote knowledgeably of French literature. Letters to Thomas Lowinsky (nicknamed Bengy—short for Benjamin—as he was the youngest of their friends) dealt with the young soldier's poetry and lithographs. At the end of the war Ricketts wrote to him, 'Don't call me Mr. R. any more, it makes me feel old, fat, and successful.'[44] Under Ricketts's influence he was to become an artist, as well as being a close friend, and one of Ricketts's executors. Cecil French bought Ricketts's and Shannon's work, and later wrote about it. All this was extremely heartening to Ricketts, who felt a great rapport with the young, and great promise for the future.

One young soldier-artist was for a time to receive the bulk of Ricketts's attention. Ernest Cole had first been discovered at the age of sixteen and a half, when Ricketts and Selwyn Image were acting as art examiners at the Royal College of Art. Hailed as a young genius, Cole joined the Society of Twelve, and exhibited drawings, sculpture, and copies after Leonardo and Michelangelo, which appeared to Ricketts to show a prodigious talent. Through the influence of his patrons, Cole was commissioned to sculpt two large groups for the new London County Council building. Here was Ricketts's answer to the Cubists, Futurists, and other modernists. Ricketts had no doubt that, if Cole lived, 'the Future of English art' would belong to him, and that he would soon acquire a national and European reputation.[45] At the prospect of Cole being sent to the front, Ricketts galvanized all his

[43] CR Diary, 31 Dec. 1915 = *SP*, p. 251.

[44] CR to TEL [1917] = *SP*, p. 309.

[45] CR to A. Cippico, (n.d.). See CR, 'The Sculptures and Studies of Mr Ernest Cole', *Country Life* (19 Oct. 1918), and Malcolm Salaman, 'The Sculpture of Ernest Cole', *Studio*, 78 (Jan. 1920), 140–8.

energy and influence. Edmund Gosse, Sir Edward Poynter, PRA, Lord Newton, even Lord Curzon, were approached in turn, as well as the more usual friends in the art world, like Ross, Holmes, Binyon, and Cippico. In his letters Ricketts was careful to mention the support of the Oxford and Cambridge Slade professors, and artists like Clausen, D. Y. Cameron, and Shannon. 'The Britisher always values the quoted opinions of others. I have gained immensely by Shannon saying "Ricketts thinks".'[46] Ricketts also wrote a highly laudatory article on Cole, published in *Country Life* in October 1918, and encouraged Edmund Davis to buy an unfinished statue of St John the Baptist which had been exhibited in the RA. Nevertheless, Cole was called to France as a Second Lieutenant, 'which is of all ranks the most deadly', and wrote back in his childish hand hoping that Ricketts's efforts would enable him to go back to his work.[47] Ultimately they did, and on 14 February 1917 Ricketts noted that Cole, at least, was safe. Cole was recalled and sent on a mission, thought to be less dangerous, to America.

The Cole episode is a sad and embarrassing one in Ricketts's life. The drawings which so impressed Ricketts have all the mannerisms of late Michelangelo without any understanding of the basic forms underlying them. Ernest Jackson, a brilliant teacher of drawing, maintained that Ricketts could be taken in by the presentation of a drawing, and this is probably what happened here.[48] Perhaps, too, he was so keen to find a representative for the kind of art he believed in that he was somewhat blinded. Nor does Cole's sculpture merit Ricketts's comparisons with Michelangelo, Alfred Stevens, and Rodin, and he never achieved a fraction of the distinction that Ricketts predicted for him. In America he met and married an American girl, and after his sudden return in May 1918 he contacted Ricketts again. In June, however, he was recalled to the front (in spite of the possibility of a commission to erect an important memorial to Kitchener) and called to say goodbye. With this, he passed out of Ricketts's life. After the war he was converted to 'modern art nonsense'. In Ricketts's eyes, he had 'gone over to the enemy', and Ricketts could never mention him again without a burst of rage.[49]

[46] CR to R. Ross [Dec. 1916].

[47] CR to F. Monod, 8 Aug. 1916.

[48] E. F. Jackson to TSM, 28 Nov. 1932 (Jackson Papers).

[49] *SP*, p. 275.

Ricketts also became involved in the publication of a new periodical—*Form*—devoted to art and literature. The guiding spirit was Austin Spare; 'a clever draughtsman', according to Ricketts.[50] Ricketts and Shannon both agreed to contribute, and Ricketts also approached a number of his friends, including Yeats, Binyon, Cockerell, Masefield, and Sturge Moore. Yeats offered eight unpublished poems, which Spare, without Yeats's authorization, printed in a separate edition of 200 copies. Yeats was furious, and tried to limit the edition to fifty, as he wanted the Cuala Press to have all his first editions and he disliked a 'vulgar' drawing of a nude woman, by Spare, which illustrated the pirated edition, and which Ricketts had not seen. Spare replied that this would ruin him; Ricketts, who had misunderstood the situation, offered to cover any losses which Spare incurred; then Yeats, who did not want his old friend to be out of pocket, relented.[51] When the magazine finally came out in April 1916, published by John Lane, it contained the promised prints by Ricketts and Shannon along with poems by Binyon and Sturge Moore as well as Yeats. Ricketts's lithograph, *Finis*, depicted a skeleton with a scythe stalking through a field of skulls; a comment on the war and its effect on his morale.

At the same time Ricketts was involved in the war effort, which he supported in every way he could. He and Shannon contributed generously to the various yearly war charities. The first huge Red Cross Sale, which took place early in 1915, however, Ricketts viewed 'with mixed feelings; that is, my feelings are not mixed at all', as 'some over-generous person' (F. Osmaston) had presented one of Ricketts's pictures, *The Good Samaritan*: 'the time is not a good one to sell modern works in the sale room and to avoid this I had presented a Palissy dish & Shannon a Gainsborough drawing, hence bitterness.'[52] The drawing fetched only £4 and Ricketts himself bought in the Palissy dish, which had earlier fetched £160, for £16, well below his reserve; Ricketts's painting was bought for £42 by Mary Davis. With old master drawings selling for almost nothing, they were at a loss what to send to the next year's sale. Ricketts thought of giving a copy of *The Sphinx* inscribed by Wilde, along with a letter from Wilde for which he made a vellum envelope; and, to tie it and the book together, he was presented with a ribbon by Mrs Davis painted by herself: 'this will, of course, sell in

[50] CR to WBY [1916].

[51] *Letters of W. B. Yeats*, p. 609; 'Heirs', p. 69.

[52] CR to RNRH (TS) p.m. 3 Apr. 1915.

America as *hand-painted* by me', he observed.[53] After one of the organizers, Edmund Gosse, had accepted the offer, and Ricketts had 'inflamed' a bookseller who had valued it at between twenty and thirty pounds, Ricketts changed his mind: 'Probably a touch of sentiment enters into the selfish reasons for my retention of the signed *Sphinx*—the other signed books are so much less *my* books.' He sent instead The Doves Press Bible and the Kelmscott *Love is Enough*: 'In theory, these should represent £30; in fact, they will probably fetch less.'[54] At the 1918 sale Ricketts was again distressed to learn that Lord Northcliffe, who 'thinks himself very generous', had sent in Shannon's *The Bath of Venus* (now in the Tate).[55] It was too large a picture for a time when the provincial museums were out of funds. Ricketts took the risk of sending in one of his own paintings, *A Deposition from the Cross*. At the sale, however, the Shannon fetched almost double its original cost, while the Ricketts, on which he had put a bid of £40 to prevent it going for nothing, went for £220. Ricketts was delighted at such high prices not only for their own work but also for that of other artists. It was a good sign for the future. Moreover Ricketts thought that 'we artists have behaved splendidly', during the war; Orpen, Lavery, Muirhead Bone, and D. Y. Cameron had contributed pictures worth thousands to war charities, and scores of artists had allowed £100 pictures to sell for £50 or less. Even then these damaging prices counted for more than 'the eight pound or twelve pound contributions by Lady This or Lord That duly noted in our flunky press'.[56] By March 1915 Ricketts calculated that, in sales and exhibitions, British artists had already contributed over £9,000; a low estimate based on incomplete knowledge.[57] This generosity could only confirm his belief in the value and the ennobling qualities of art.

Ricketts also made more direct contributions. Persuaded by Mary Davis, he gave his lecture on Watteau for a war charity, the French Wounded Emergency Fund, and promised, if this were a success, to follow it up with lectures on the French School: Ingres, Delacroix, Chassériau, and Puvis de Chavannes, and to 'have a whack' at the men of letters: Baudelaire, Flaubert, Verlaine, and either De Guérin or

53 CR to R. Ross [1916].

54 CR to R. Ross [Feb. 1916] = *Friend of Friends*, pp. 283–4.

55 CR to TEL [end Mar. 1918].

56 CR to R. Ross [June 1915].

57 CR to RNRH (TS), p.m. 23 Mar. 1915.

Fromentin.[58] These were the writers he knew 'better than most Frenchmen and I am sure better than any living Englishman', and he hoped to make the lectures up into a new book.[59] 'All depends on a start, possibly people wont come', he acknowledged to Mrs Davis and—as there is no record of any lecture after the Watteau—that, unfortunately, is probably what happened.[60] For The Cigarette Fund he did a large lithographic poster for Hardy's *The Dynasts* (Fig. 37); 'design good, drawing conventional, hell over the lettering', he noted in his diary.[61] The production itself, at the Kingsway Theatre, thrilled him greatly, and he considered it 'the most interesting thing of English growth I have seen upon the stage for years'. Fifty proofs of the poster, signed by Hardy (whom Ricketts had once described as a 'moulting old parakeet') and himself, were on sale at the theatre for half a guinea, together with twelve copies at two guineas, on which Ricketts had executed a small signed drawing or thumb-nail sketch of Belgian refugees, or else Napoleon in some sort of allegorical aspect, 'Napoleon and Death, Napoleon and the Sphinx, Napoleon as a Sphinx, Napoleon at St Helena, etc.'[62] Hardy was enchanted and made the first purchase; Ricketts bought three copies and gave them away.

Another charity project with which he was involved was a matinée production of Masefield's *Philip the King*, at Covent Garden in November 1914, directed by Granville-Barker. The setting was like that of a Renaissance theatre, with a stage within the stage. With his marvellous knowledge of costume and painting, Ricketts was able to produce austere and splendid period dresses and 'for Philip to pray to I painted two pictures by El Greco'.[63] One of these was a crucifixion, six feet by nine, a remarkable feat for a production mounted at only three weeks' notice. Overall the impression was rather severe, compared with the richly coloured effects of Ricketts's earlier work, with occasional 'touches of emphasis' in the costumes for the Spaniards, who were dressed in black, grey, silver, and gold according to character, and for the Indians, whose tawny yellows, reds, sharp greens, and blues

[58] CR to Mary Davis [Apr. 1915]; CR Diary, Apr. 1915.

[59] CR to G. Bottomley (p.c.), p.m. 17 Apr. 1915.

[60] CR to Mary Davis [Apr. 1915].

[61] CR Diary, 9 Nov. 1914.

[62] G. Bottomley to L. Binyon, 12 Apr. 1942 (Roger Lancelyn Green); CR to R. A. Walker, 25 Nov. 1914 (Princeton Univ. Library).

[63] CR to G. Bottomley, p.m. 1 Dec. 1914.

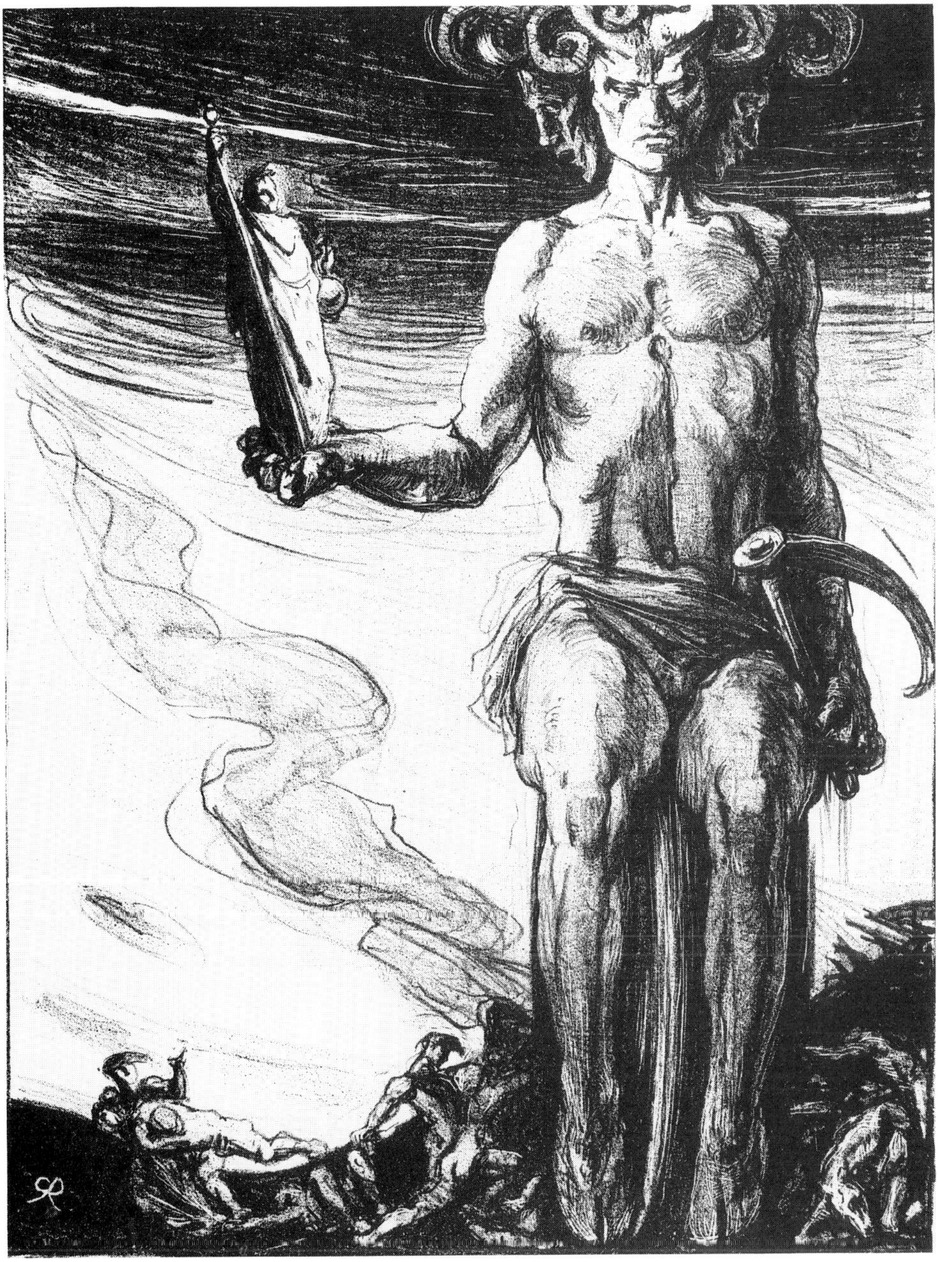

FIG. 37. Poster in lithography by Charles Ricketts for *The Dynasts* (1914), by Thomas Hardy

appeared only in a dim ghastly light.[64] Lillah McCarthy took the leading female role. Ricketts was touched when Henry Ainley thanked him for his costume and for his real assistance in the production. The 'supers' too were so pleased with their costumes that some even complained if anything in the sketch was missing. 'In any other country,' Ricketts commented, 'the thing would have caused a sensation. But tea time carried Queen Alexandra and the major part of society away from their boxes ... and the play was performed ... too rethorically, by famous actors to a partly empty house.'[65]

Not all Ricketts's theatre work was for charity. He continued to help his friends. When Yeats and Lady Gregory came to London for the Abbey Theatre's production of Yeats's *On Baile's Strand*, Ricketts did some new designs, and 'shuffled' some of his designs for *The King's Threshold.* As Synge's *Well of the Saints* (for which—at Yeats's request—Ricketts had already done some designs at very short notice in 1908) was also being given, Ricketts came up with some new designs for it, working at first from memory of the play. With his careful eye for economy, he had the dresses carried out by Mrs Dulac's seamstress under his supervision, which reduced the expense of the costumes 'to little more than cost of material'.[66] The two productions cost only eighty pounds.

For the Stage Society's production of Sturge Moore's *Judith* in January 1916 Ricketts went all out. It was a great rush, the actors only having twelve days to learn their lines and Ricketts hardly more 'to design, get materials blocked and delivered in wartime and carried through'.[67] To save time and money he himself did much of the sewing and stencilling, staying up until three o'clock every morning for a week. At the same time he was also doing a dress for Penelope Wheeler ('who is to recite Greek verses in America in vague Ricketts draperies') in a production of *The Bacchae.* He found this 'a pleasant break from painting and war news'.[68] To keep the cost down, he also gave most of the people bare feet and dressed them in shirts, gowns, loincloths, tabards, and so on. The costumes averaged thirty shillings to two pounds apiece; the scene was made up of thirty yards of canvas and a

[64] CR to Huntley Carter [*c.*1915] (Victoria and Albert Museum, 86 GG, Box i).

[65] CR to G. Bottomley, p.m. 1 Dec. 1914.

[66] CR to WBY [May 1915]. See 'Heirs', pp. 67–8.

[67] CR to L. Binyon [Jan. 1916].

[68] CR to R. Ross [early Jan. 1916] = *Friend of Friends*, p. 279.

column which cost thirty shillings: 'Remains an iron hoop, a couch and some cords, etc. It is always the etc. that costs!'[69]

A few days before the performance of *Judith* the chances of making anything of the production seemed to Ricketts to be five to one against. Though 'seedy and jumpy', however, he was not in despair, but thinking furiously how to deal with the problems.[70] With his temperament, he was bound to suffer, not only from over-reacting to the usual setbacks and disappointments in mounting a production but also from imagining other ways in which things might go wrong. At the same time, he had energy and determination enough not to sink under these pressures, but to fight and conquer them. He was battling not only with external circumstances but also with the element of pessimism in his own nature.

The production was largely a success. At the rehearsal Ricketts and Shannon dreaded a titter in the audience at more than one point; of this there was no sign 'though the Stage Society is of course a collection of all the disappointed cranks and pro-nothings in London'. Lillah McCarthy was intelligent and forcible as Judith, and would have been even better if she had been given more time. There were five curtain-calls, backed by genuine applause. Miss McCarthy asked Ricketts for her dress but, as it belonged to the Stage Society, he asked them to give it to her. Her acting in one scene was so splendid that Arnold Bennett later wrote a more extensive play about Judith for her. For Bennett's version, produced on 30 April 1919, Ricketts designed for Miss McCarthy a costume (Fig. 38), the top part of which consisted only of a strap across the breasts: 'the bare torso, covered with bands and straps of jewels, emerging from the fishtail skirt, the jewelled leg also bare, would have enchanted Gustave Moreau.' The dress was banned by the Lord Chamberlain, and Ricketts had to substitute another costume for it at the last minute.[71]

Ricketts made some more special dresses for Lillah McCarthy for productions in which he had no other part, one of the most magnificent of which was the *The Man who married a Dumb Wife* by Anatole France (January 1915). When Miss McCarthy offered to sit to Shannon again he painted her in this costume; Ricketts added the butterfly on her head-dress in the picture. Ricketts also designed dresses for her in

[69] CR to TSM [Jan. 1916]. [70] CR to TSM [Jan. 1916].

[71] CR to TSM [Jan. 1916]; CR to L. Binyon [May 1919].

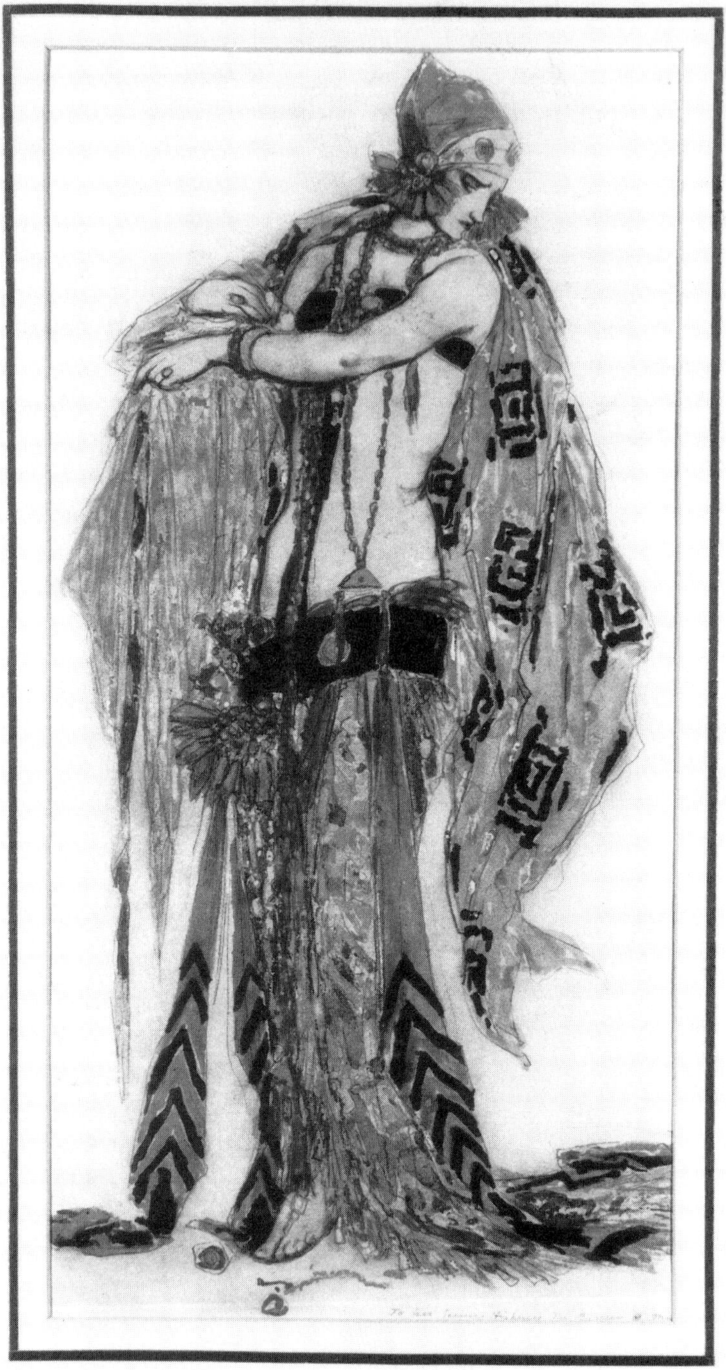

FIG. 38. Costume design for Lillah McCarthy as Judith (1919), in Arnold Bennett's play. Inscribed 'To her serene highness the Princess of Monaco' (private collection)

Rupert Brooke's *Lithuania* (May 1915) and Alfred de Musset's *A Door must be kept open or shut* (summer 1916), given in a charity matinée for the Italian Red Cross. For *The Admirable Crichton*, another war-time matinée in 1916, he imagined 'a cloak of grasses' and some 'jewellery of shells' for her, but this was ruled out by a swimming episode in the play.[72] The dress had to be improvised out of oddments of skin and canvas. An hour and three-quarters were spent in the actress's dressing-room 'sticking berries on her dress, then taking them off again, then sticking them on in the same place or another, then removing them again; her idea being the essentially feminine one that more things she had on, more beautiful she looked; mine the average male point of view which is the reverse'; Ricketts 'cut down and away enough stuff for a great deal of skin to slip in and out of view during violent movements. I wonder what Barrie will think.'[73]

During the breakup of Lillah McCarthy's marriage to Granville-Barker Ricketts tried to keep out of the way. 'I avoid all knowledge, but will remain friendly to both, hopeing for the best, which is not likely to occur.'[74] To cheer Lillah McCarthy up during these difficult days Shaw wrote for her *Annajanska*, a bravura piece about the Russian Revolution. For her grand entrance, disguised as a man, Ricketts designed 'a gorgeous white uniform, half covered by an enormous green overcoat trimmed with black fur'. Along with the costume he offered the usual advice—'the secret of male attire is perfect cut, worn with negligence'—and he sent her letters full of lively suggestions embellished with little drawings in the margin.[75] The costume was such a success that Margot Asquith, wife of the former Prime Minister, borrowed it to copy for her dressing-gown.

Again Ricketts had not asked for payment. Shaw wrote, 'I learn from Lillah that you having contributed the only element in the play that has made any sensation, the transaction has been left on the footing of a labour of love.'[76] She had botched the arrangements with the theatre management. Shaw had wanted the author's fee of fifty pounds a week to be paid by the theatre, but Miss McCarthy had let herself in for it instead. When she sent the first cheque for £150 to Shaw, he

72 CR to L. McCarthy [early 1916].
73 CR to Mary Davis [1916].
74 CR to G. Bottomley, p.m. 2 Aug. 1916.
75 CR to L. McCarthy [1918] = McCarthy, p. 112.
76 GBS to CR, 6 Feb. 1918.

proposed splitting it with Ricketts who, not to be outdone in generosity, accepted the money as a contribution towards a matinée concert by the Russian tenor Vladimir Rosing, whose career he was trying to further.

Ricketts first came across the Russian singer in the summer of 1917, and was absolutely thrilled by his art. (Out of 104 recitals by Rosing in London, Ricketts missed only 3.) He regarded Rosing as one of the three finest singers he had heard in his life. The voice itself he did not regard as first-rate—for it lacked the health and perfect balance and experience of Chaliapin—but it had, Ricketts felt, 'a nervous and magnetic force which upsets me at the time, and either prevents my sleeping or else plunges me into sleep afterwards, making me too tired to go to bed'.[77] In his singing Rosing moved 'in a world of art and ideas and emotions of which the Englishman seems incapable,' and gave Ricketts 'the keenest pleasure' he had experienced since the beginning of the war, such pleasure as had been occasioned by Wagner, Duse and the Russian ballet and opera. Though well known in Vienna and St Petersburg, Rosing was practically unknown in England when he arrived, and, when Ricketts met him, had 'remained in a sort of half recognition amongst smart women and in a cloud of silence from the press'.[78] Ricketts immediately set out to rectify this. Not only did he encourage everyone he knew to go to Rosing's concerts—'Send any man, woman, or child to the next Rosing Recital at the Aeolian Hall', he told Cockerell—but he set about trying to organize an opera for him.[79] Since Rosing excelled at rendering Russian music, Ricketts wanted to put on Tchaikowsky's *La Dame de Pique*. Any profits would be given to various war charities. To raise money he approached his friends and made use of contacts. Lady Scott and Sir Otto Beit, as well as Shaw, were among those who contributed, and Margot Asquith was interested. In the midst of this activity Ricketts was put to bed by his doctor, 'having had—I still have—a bad touch of Laryngitis which wanted to develop into something which I cant spell—newmonia'.[80] This was brought on by his neglecting a bout of flu in order to hear Rosing sing on a cold afternoon. Money poured in—he soon had £600 out of the £1200 he thought the opera might cost if properly managed,

[77] CR to G. Bottomley, p.m. 6 July 1917.

[78] CR to WBY, 6 Feb. 1918.

[79] CR to S. Cockerell, 6 Feb. 1918.

[80] CR to WBY [Feb. 1918].

and, despite the setback of his illness, he was well on with designing the sets and costumes, including a 'picture' he painted for it and 'two "overdoors" in cameo of bulbous cupids with faces like bottoms and legs like roly-poly puddings'.[81] However, difficulties soon arose: 'the times & conditions are heart-breaking'.[82] With the war, costs had soared, even scores having increased considerably in price. Drury Lane, where they wanted to stage the opera, could offer them bookings only in April (which was too soon), when they had asked for June. The opera had to be shelved. Rosing, moreover, sometimes got on Ricketts's nerves: 'his Russian temper and the times coincide in making him nervous and troublesome and he sometimes hands this on to me.'[83] In the meantime he did his best to promote Rosing, and wrote a signed article on him for the *Manchester Guardian* early in 1919.[84] When the opera was finally produced, in July 1921, as part of a Rosing week at the Aeolian Hall, it suffered the usual rows and disasters. When the sets were taken down for a recital in the middle of the week the manager and under-manager both went on strike. Ricketts and Dulac offered to be 'manager, call-boy, etc.' and, if it could not be set up again in time, promised to improvise new scenery with curtains.[85] This brought everyone to their senses, and the performances went smoothly.

Before Ricketts's plans for Rosing had materialized, other operatic schemes had intervened, though he felt guilty about leaving Rosing in the lurch. There was an offer to stage *Peer Gynt* for Sir Thomas Beecham, but Ricketts made conditions that he feared—rightly—would not be accepted. Granville-Barker asked him to stage Byron's poem, *Manfred*, for which Schumann had composed accompanying music—another scheme that never got off the ground. Lady Cunard was behind it, and amazed Ricketts with her power and vitality. With her, he went to see the Lord Chamberlain 'to try and help her in subverting the Law of the Realm by securing license to play *Manfred* on Sundays'. A one-eyed man—'the government probably likes people with one eye'—informed them that the Lord Chamberlain had just left. They had in fact seen him escaping in a taxi, and all the other officials were also in hiding. Finally, they were ushered into the office

[81] CR to TEL, 11 May [1918].
[82] CR to G. Bottomley, 9 Mar. 1918.
[83] CR to Kathleen Bruce (TS) [Feb. 1918].
[84] *Manchester Guardian*, 1 Apr. 1919.
[85] CR to G. Bottomley, 4 July 1921 = *SP*, p. 334.

of an under-secretary 'with the mind of a lead pencil', who assured them that neither the Lord Chamberlain nor the Prime Minister could give them permission, and that the law could only be altered by an Act of Parliament. 'Lady Cunard [wished] the law to be changed at once' and Ricketts suggested 'in my best revolutionary manner' that they should give the play, and have it forbidden by the police. This 'the lead pencil' took seriously, hinting that their licence to perform the play might be withdrawn at *any* time, and suggesting that they give it in matinées. 'Excellent idea, why not? Good-bye dear Mr. Lead Pencil!' And they were off to Drury Lane to see about another project that was in progress, a production (the first in England) with Beecham of Isidore de Lara's opera, *Nail*. Here, they tried to 'humbug' de Lara, who wanted 'cost at all cost' for his opera while they wanted economy; listened to a new American tenor accompanied by a young millionaire in anti-aircraft uniform; 'hoiked out' old scenery; were whisked off to Beecham's rooms; and then went to Lady Cunard's house. Ricketts was an hour late for dinner and 'had to drink a whole bottle of wine', he was so worn out. Lady Cunard, he decided, was 'a very capable and exceptional woman and should be made Minister of Fine Arts'.[86]

Beecham Ricketts found unreliable and uncertain, a difficult man to work with. So was de Lara, who was troublesome and wanted 'to alter and interfeer in a production which is not exciting in its conditions, Algiers, the desert—palms *obligato*! not in my line'. In fact Ricketts had evidently undertaken the opera to please Lady Cunard and, particularly, Princess Alice of Monaco, who was a patron of de Lara's. Ricketts and the *principessa* had met in 1916 and she soon became 'one of my dearest friends, a woman of the utmost kindness & generosity'. He wrote a tribute to her in *The Observer* after her death.[87] The prima donna Rosina Buckman, queen of the Arab nautch-girls, was like a barrel 'with a chest like a sideboard; the costume dictates masses of disk jewelry on that part of her person and I fancy she will look like a jeweller's *devanture de magazin* full of Corporation plate—this is not a pun'.[88] Ricketts had done 150 drawings against time, only to find that the arrival of the Christmas season delayed the opera for three months. There was a plan to start it in Manchester, but it finally took place in London in July 1919, having been produced and rehearsed in nine

[86] CR to Muriel Lee Mathews (TS) [Nov. 1917]: CR to Mary Davis [Oct. 1917].

[87] CR Diary, 9–10 Feb. 1916; *Observer*, 10 Jan. 1926.

[88] CR to Mary Davis [Oct. 1917].

days. During the last three nights, following the rehearsals, he did not sleep. He had to get scenery and 200 dresses finished in that time. On the first night they were still rehearsing at six o'clock before an eight o'clock performance. Non-singing 'supers' refused to make a costume-change for the last act. Forty dresses were 'carefully' lost. Ricketts was begged not to alter anything which might cause ten minutes' delay. During the performance sunsets occurred in the middle of the night followed by rapid dawns; the limelights played on unmarked spots. 'I don't think this was Bolshevism among the limelight men, but fatigue and temperament run riot.' Ricketts watched the first night 'grim, detached, amused, disembodied'. Some of his work looked surprisingly good, and the house was enthusiastic. He had developed a personal dislike of Beecham, liked all the singers, hated all the business-people, producers, and managers: 'these men deserve no mercy or forbearance, they could not sweep a crossing properly and should be put in the Foreign Office for life.'[89]

Music became his greatest consolation and relief from the war. It was the basis of some friendships, such as those with Sir Thomas Beecham, Isidore de Lara, and Mrs Muriel Lee Mathews, an accomplished pianist whom he had met through the Davises. At the end of 1914 Dulac had amazed him 'by the resources & technical perfection' of the pianola, but the war and other disappointments had put an end to his plans to acquire one.[90] However, with the inheritance from the Fields he ordered one to celebrate his fiftieth birthday (2 October 1916), and spent pounds on rolls of Chopin, Beethoven, Schumann, Wagner—'all the magicians'. Playing it for some five hours on the first day, he found it hard work—'comparable to riding a bicycle'—and his shirt got wet through. In the evening, with sore insteps and the backs of his thighs aching, he was prostrate.[91] The instrument at first gave him endless difficulties, because of a crooked spool, though there was probably an 'incompatibility of temper' between himself and the mechanical world—'I look upon all machines with awe and religious respect', he commented.[92] Many of the difficulties were, he felt, due to his having chosen an upright pianola, because of the narrow stairs. Soon he could play Schumann very well (though

89 CR to Muriel Lee Mathews, 20 July 1919 = *SP*, pp. 318–19.

90 CR to G. Bottomley, 1 Dec. 1914.

91 CR to G. Bottomley, p.m. 14 Oct. 1916 = *SP*, p. 269.

92 CR to G. Bottomley, 29 May 1917.

not *Carnaval*, which he regretted), astonished himself with his rendering of Brahms and the Russians, but was dissatisfied with his playing of Chopin. Because he believed that the latter's own playing was *piano pianissimo* he felt guilty 'in making the clatter and harsh *fortissimos* which please me who have been corrupted by Liszt & the Russians'; but others liked his playing.[93] Every morning he spent two hours at his machine, and even played to amuse Princess Patricia of Connaught, while Shannon painted her portrait. From the beginning the pianola was 'comparable to an exciting holiday': 'I walk with a spring, sleep like a hog, and am able entirely to forget the war & the general lowering of one's vitality it has brought with it.'[94] Music took the place of his 'pre-war normal life', and seemed more engrossing and dear to him than anything else; even more than the beautiful spring of 1917. He still read no newspapers, but only the headlines, and was curiously without emotional response over anything, music alone excepted. The pianola had created for him 'a second personality—the happy one'. When talking to others he was inwardly listening to Chopin: 'I wake up in the morning as I did during the Russian season with drifts of Schumann, Schubert, floating in my ears.'[95] The pianola, he felt, kept his brain in movement and replaced the book of poetry which he often kept open on the table in his studio for the same purpose.

When, in the last year of the war, Ricketts got wind of a plan for the Air Board to use the British Museum, and thus make it a prime target for air raids, he threw himself into more 'political' activity. Much of the Greek and the entire Egyptian collection were still *in situ*, and the addition of papers, desks, and so forth would add to the risk of fire. Letters were sent off to the *Daily Telegraph* and *The Observer* in which he reminded readers: 'We hold the masterpieces in the British Museum in trust for the world and for the future; their loss or damage would be more than a national disaster; it would affect the future of our civilization.'[96] In addition to such public appeals, he was busy manœuvring behind the scenes. Through his earlier protests and efforts over Cole he had acquired skill and contacts for such machinations. To Yeats he admitted, 'I have turned Politician and wire puller in real

[93] CR to G. Bottomley, p.m. 6 July 1917.

[94] CR to G. Bottomley, 14 Oct. 1916 = *SP*, p. 269.

[95] CR to G. Bottomley, 29 May 1917 and p.c. 9 Apr. 1917.

[96] *Observer*, 6 Jan. 1918; *Daily Telegraph*, 7 Jan. 1918.

earnest. You always said I was a Politician in disguise.'[97] He approached Robbie Ross and even MacColl to get them to use their influence. Lord Curzon was finally appealed to and the project was dropped. Hoping to benefit from the occasion, Ricketts wrote again to urge greater protection, especially for the Greek, Egyptian, and Assyrian collections.[98] In his earlier letters he had caused some ruffled feathers by accusing the British Museum's Trustees of apathy, since so many of the Museum's treasures were still there after two years of air raids. He was never one to let sleeping dogs lie.

At the same time he was immersed in more charity work for the theatre. For three Shakespeare plays to be given in YMCA hut performances in Le Havre and Paris he did a series of dresses which were interchangeable between the plays, making thirty dresses into fifty by transposing cloaks, doublets, and so on. By turning Malvolio's coat inside out, for instance, it suited Don Pedro or Jaques. Each play had five or six dresses, however, which belonged to it only, and gave it its period style. Those for *The Merchant of Venice* and *The Two Gentlemen of Verona* had touches of Renaissance paintings, like the striped dresses of *The Mass of Bolsena*; in *Twelfth Night*, the Illyrian dresses were in 'piquant contrast' to the Elizabethan costumes of Sir Toby and Sir Andrew Aguecheek, whose doublet was embroidered (by stencilling) with grapes, squirrels, and butterflies.[99] So clever were the changes, that Shannon could hardly detect the 'shuffling' and even Ricketts sometimes got confused about whom a coat or legs or cap had belonged to before. Both men agreed—as usual—that the designs were among the best that Ricketts had ever done, but there were the usual difficulties, as prices had soared and certain colours and materials had become unavailable: blocking materials which had formerly cost a shilling extra a yard now cost ten shillings, so he had to scrap two designs (i.e. four dresses). He stencilled day and night—the stencilling alone occupied him for four months—and covered 100 yards of material, made fifty jewels and 'countless other things'. He enjoyed the work immensely and joked while doing the stencils: 'I am the cheapest dress-maker in London.'[100] Lena Ashwell had asked him to do the design, but he had

[97] CR to WBY [1917].

[98] CR to *Observer*, 13 Jan. 1918.

[99] CR to TEL, Aug.–Sept. 1918.

[100] CR to G. Bottomley (p.c.), p.m. 31 Dec. 1918; See also Furst, *Apollo*, p. 332; *Theatre Arts Monthly*, p. 392.

really done all this work for Penelope Wheeler, who played Portia (in a gold dress covered with blue roses and mermaids); after the war, however, Miss Ashwell took over the costumes, and her players used them, though in rather topsy-turvy ways, as long as the group lasted.

When their friend Ernest Jackson asked them to contribute prints to a series he was editing, both Ricketts and Shannon complied. Jackson had first become a friend through his lithographic work and connection with the Senefelder Club, of which Shannon became a member. *The Great War: Britain's Efforts and Ideals* was another attempt by British artists to record activities associated with the war and the ideals which inspired them. Both artists chose allegorical subjects; Shannon did a female nude representing *The Rebirth of the Arts*, while Ricketts depicted *Italia Redenta*. The print expresses Ricketts's anxieties about Italy and shows six female nudes, representing Italian cities and provinces, rising to arms while one of their number meets the double-headed eagle of Germany in a mid-air battle. In an early version of the design the figures were identified with the names of the towns and provinces they represented. One of these was Spalato, an ancient Roman city which had recently been incorporated into Yugoslavia and renamed Split. Ricketts had been very annoyed to see it given its modern Yugoslav name in the Meštrovíč catalogue and committed an 'international outrage' when he included the Italian name here. He tended to be dogged and provocative about such things, but had to submit when it was suggested that 'the identity of the ladies who typify the Italia Red in your print should remain a mystery'.[101]

Even before the end of the war some aspects of pre-war life returned. In September 1918 the Russians returned to London, a rather scrappy company, with four good principals including Lopokova and Massine, but not 'on the supernatural level of the past ... not a "life-event"'. All the same Ricketts went six times, though Shannon did not want to go at all. The setting for *Cleopatra*, of such violently contrasted colours and cubists shapes that Cyril Beaumont thought they distracted from the dancing, was done, Ricketts felt, 'by a local genius [Robert Delaunay] who, before he turned Post-Impressionist, decorated ice-cream carts', and he disliked the dresses, mostly by Bakst for various other productions, which were *voyant* and lacked the unexpected. Though the men on the whole were better than the women, Massine was not on a par with Nijinski or Fokine, and lacked the gift of

[101] T. Derrick to CR, 31 July 1917.

imagination and artistic temperament. The whole venture proved once again to Ricketts that 'excellence in art is not a matter of intention or technical ability but resides in personality. It is a matter of character . . . and is of the spirit.'[102]

Shannon had been feeling rather seedy and went to Cambridge and to Lincolnshire to be with his sisters. Ricketts too was weary of the war and for months had felt like 'a stuffed beast'. He was without desire ('one of the springs of energy & success in life'), nerveless, and fatalistic about everything, though he was less anxious than before.[103] He hoped still that Paris and Venice might be spared destruction; the price that France had already paid was appalling. Yet small things continued to give them delight. Shannon had fallen for and purchased some new pets—Japanese mice. He soon feared, however, that they would present them with 'mouselets': 'though supposed to be of one sex, they have been seen in very intimate conversation, we hope it may have been unnatural vice, but I fear there is no chance, since one of them steals everything to make a nest, paper, string, stuff, even a cigarette'. Before long the mice presented them with a family of three 'mousekins': 'they are exquisite scraps of design and are always coming to grief in some way or another, and recovering from sudden deaths.'[104]

The end of the war came suddenly, as Ricketts had been told it would. With the first hooters announcing the Armistice he was 'crying like a pig'.[105] He had been working at a new version of *Descent from the Cross*. Madame Rose, a rather colourful neighbour downstairs, asked for a flag. They had nothing, only an old banner which Ricketts had painted for *Joan of Arc* (presumably in a pageant) months before. This, embellished with a tuft of arbutus, was hung from the balcony, and Shannon was as pleased with it as a schoolboy. This was 11 November, but it was not until the 16th that Ricketts realized that peace had returned. War no longer dominated their thoughts and conversation. Almost at once they started bringing up their treasures from the basement. At dinner gaiety returned, and for the first time in years they turned up the lights.

The effects of the war on Ricketts's character were never to be effaced. Like the Wilde trial, it marked a change in his attitude to life.

[102] CR to G. Bottomley, 25 Sept. 1918; CR to Mary Davis [Sept. 1918].

[103] CR Diary, 21 June 1918.

[104] CR to TEL, 29 May 1918 and [June 1918].

[105] CR Diary, 11 Nov. 1917 = *SP*, p. 305.

While a great part of Europe's artistic heritage was being wantonly destroyed, precautions in Britain against possible destruction of museums and monuments by enemy action had seemed to him shamefully insufficient, and brought home to him once again how what he cared for most was valued much less by his contemporaries. The forces of evil, ignorance, and indiscriminate destruction had been loosed; the 'dark stain' in human blood had shown its power over men and society; the ease and confidence of Edwardian England had been wrecked, and the way forward to the eclipse of the classical tradition in art, and the lowering of standards that went along with democratization, could only dismay him. Not only did this signal the end of the pre-war world, but also the end of his hopes that he and Shannon could make their mark on the modern world. The rise of the avant-garde art movements, his disappointment in Cole, the shock of his own mid-life crisis, had all combined with the effects of the war to hurt and disillusion him deeply. He recognized that what he stood for had little interest for the great majority of his contemporaries and for the most vital and dominant forces in the art world. The accelerated change and disruption he had witnessed made him long for earlier days, for more settled values, for a more congenial world. The modern movement in art was, as he saw it, every bit as futile and destructive as the war. Whilst he grew ever more conservative and backward-looking, he was too intelligent and self-aware not to realize the dangers of such an attitude: 'I am growing old', he told Holst,

> and so reactionary and so devoted to my habits & recollections, and I rather dread the future, which is the greatest possible sign of age. I dread the pace at which things go in the world of art, the lack of memory, intimacy, & sanity in those I meet: art should look backwards and forwards, but I look but one way.[106]

Even though the world war had been won (in his view) by the forces of sanity, he began to realize that he had lost in the art battle. Yet, with a deepening sense that he was regarded as an 'old fogey' or a 'back number', he carried on.

Shannon's *The Man with the Greek Vase* (Fig. 39), another portrait of Ricketts, was first exhibited in the RA in 1917. It depicts Ricketts holding a Greek vase while he looks out towards the viewer. His expression is at once intelligent, sensitive (even rather febrile) and sad,

[106] CR to RNRH (TS) [1916].

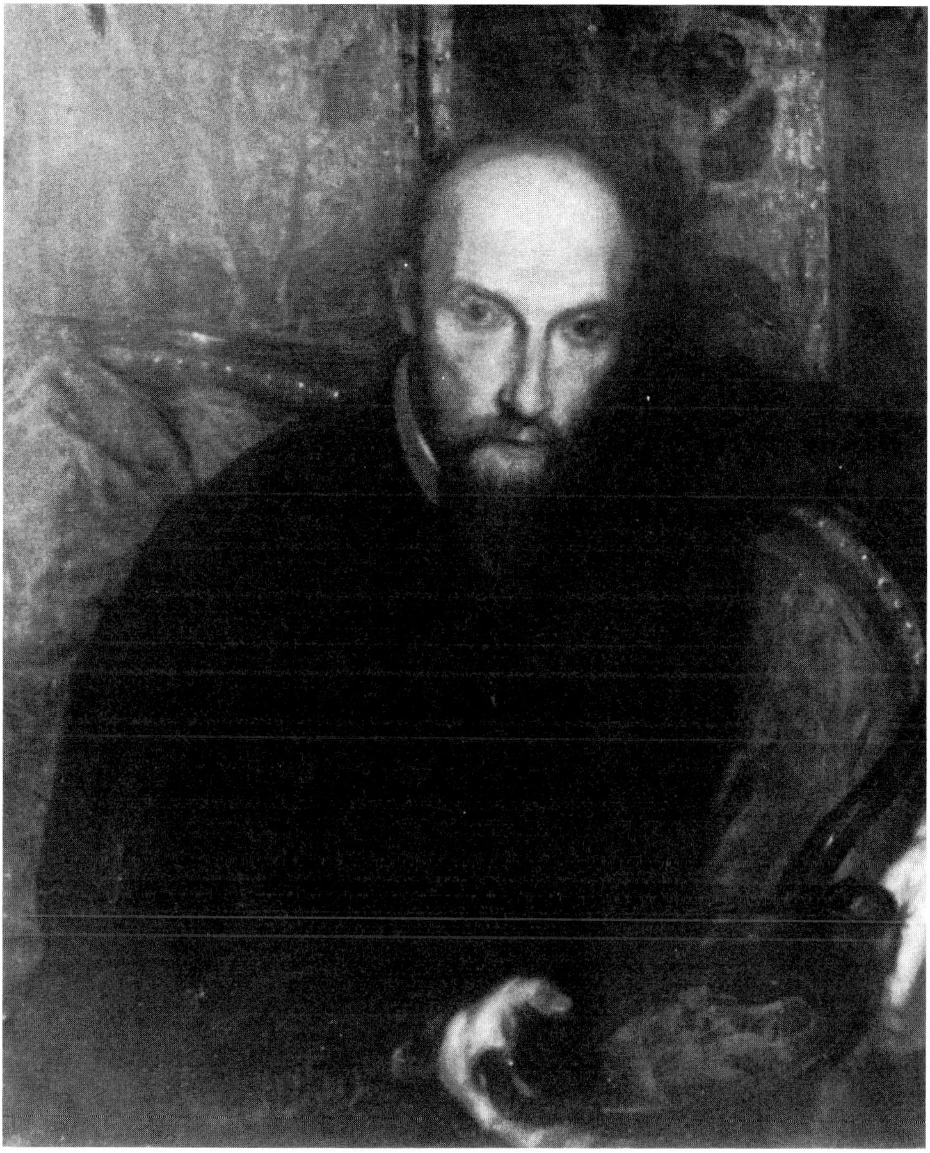

FIG. 39. *The Man with the Greek Vase*. Oil-painting (1917) by Charles Shannon (Art Gallery and Museum, Leamington Spa)

as if lost for an instant in dark, disturbing thoughts. The perfection of art and its fragility seem both to be embodied in the vase which has survived for so many centuries but could be so easily destroyed. Indeed, the picture suggests a plea for the value of art, of beauty, of the past, in an era dominated by death and destruction, not only caused by the war, but by the wilful onslaught on traditional artistic values. *Punch* recognized the element of sorrow and pleading, in parodying the picture with the subtitle 'The unhappy diner who had been refused a second helping.'[107] The portrait was bought by Mary Davis, who sold some of her jewellery to raise the money.

Whilst Ricketts had been distressed at the loss of so much life in the war, it is clear that the destruction of works of art affected him more. Unlike works of art, most ordinary human lives were to him expendable and, except for a few unique geniuses such as he had imagined Cole to be, replaceable within a generation. This view is consistent with Ricketts's élitism and materialistic philosophy of life, but it is not the whole story. He changed and developed over the years, and suffering, more than anything, opened his heart. When the time came for him to choose between the person he loved most and the collection they had made so many sacrifices to build up, he opted without hesitation or regret for the person he loved.

[107] 'Royal Academy Impressions', *Punch*, 152 (16 May 1917), p. 319.

11 *'Shannon's latest folly'*

BARELY a week after the Armistice Ricketts and Shannon acquired the country retreat they had long wanted. Since 1908 they had made sporadic attempts to find such a place; 'Shannon's latest folly', Ricketts called it.[1] Edmund Davis had purchased Chilham Castle in Kent, said to be the first house designed by Inigo Jones, with the money (£70,000) from the sale of three paintings—a Van Dyck, a Reynolds, and a Gainsborough—that Ricketts and Shannon had advised him to buy. As 'backshish' (as Ricketts put it), Davis offered them a cottage for life on the estate.[2] A lovely farmhouse with a big Gothic porch was first suggested for them, but they found it too big and too far from the station. Instead they opted for an ancient Keep, adjacent to the big house, which was to have been loaned to them as a studio. The Keep was Norman, with smooth stonework enclosed in an octagon of Roman brick, cased in flint and surrounded by crumbling walls and Roman foundations. It was full of the Romantic associations that so appealed to Ricketts. On this site the British army had camped when trying to repel the invasion of Julius Caesar, who had also lived on the spot which was, Ricketts said, 'the first page of English history'.[3] Fifteen skeletons had been found immured in the dungeon, where they had been left to die. There was a ghost: 'I wonder if we could persuade him or her to clean our brushes', mused Ricketts.[4] As the Keep was in a derelict state, there was much to be done and every opportunity to create a new house of beauty along different lines from the elegance of Lansdowne House. They had not expected ever to be able to furnish a new place again, and for two years they applied themselves with great zest and excitement to fashioning it after their heart's desire.

Having at first so looked forward to the task, Ricketts was then momentarily thrown off balance. When preparations became a reality, he found himself, with his nervous imagination, beginning to anticipate

[1] See CR to MF, 17 Aug. 1908: MF Journal 1902, fo. 146.

[2] CR Diary, Thur. [14 Nov. 1918] = *SP*, pp. 305–6.

[3] CR to TEL, 18 Nov. 1918 = *SP*, p. 306.

[4] CR to TEL [early Feb. 1919].

problems. Though Mary Davis, for instance, was 'a brick in every way', he feared she might 'wish us to be too clean and respectable'.[5] Much, moreover, had to be done before the Keep was even habitable. Windows had to be put in, floors laid, electric light installed, 'which will hurt the feelings of the ghosts, of which there are several, it seems—at least enough to make servants troublesome'.[6] They set about finding furniture or having it made. Ricketts found some late fifteenth-century French glass and, in a shop which sold geysers and gas stoves, a William Morris window based on Rossetti's unused frontispiece for *The Early Italian Poets*. He was thrilled when Cockerell (who had given him a locket containing hair of Rossetti and Swinburne) told him that Swinburne had 'sat or rather knelt' for the original drawing.[7] The top octagonal room—'the walls are the same in which King John has dwelt', Ricketts reported—had been panelled in the early nineteenth century in a Georgian style, and became their drawing-room. With the exception of a superb Japanese screen, a fine tortoiseshell cabinet, and four of the pieces of satinwood furniture inherited from the Fields, most of its furniture was eighteenth-century, and procured in their Richmond days; in the other rooms, the furniture was Tudor or Jacobean. Over the mantelpiece was a picture by Hoppner; over the sideboard two Wilkies; near the cabinet a Rossetti; in the alcove with the writing-desk two water-colours by Rowlandson. Evidently they had decided to house their English paintings and drawings here, adding later a 'nice tosh' picture of a 'Sir Philip Sidney type of Englishman in a top hat with a view of London in the background', though there was also an Italian classical landscape and Barbari's map of Venice.[8] The large octagonal room downstairs, twenty-four feet wide and twenty-four feet high, known as the Senate House, had once been a brewery and was now a dining-room. In it they placed four half-suits of armour, which soon looked as if they had always been there, as well as Burne-Jones's large unfinished painting, *Perseus slaying Medusa*, for which Ricketts painted a black and gold frame; there were also several drawings by Burne-Jones. In the bedroom, where 'the walls resist nails and the various nooks and niches dictate terms of their own', the two four-poster beds were draped with a cretonne, 'Dove and Fig', designed

[5] CR to Muriel Lee Mathews (TS) [*c.*Aug. 1919].
[6] CR to RNRH (TS), 13 June 1919 = *SP*, p. 317.
[7] CR to G. Bottomley, p.m. 16 Sept. 1922.
[8] CR to G. Bottomley, p.m. 15 Sept. 1919; CR to L. Binyon, p.m. 17 Mar. 1922.

by Shannon, which was also marketed commercially by Morris and Co.[9] The spare bedroom was the last to be completed. The day Ricketts looked forward to, when it became 'a really romantic and exquisite place', drew near.[10]

From the Keep's lichened battlements there was a lovely view over the castle grounds. The park surroundings were of exceptional beauty, 'real Tennyson landscapes with a touch of Millais added', Ricketts noted with delight. The Keep had its own garden, which had to be likewise enriched with 'floral treasure'.[11] It was their first opportunity of creating a garden, and their means and horticultural knowledge had progressed a great deal since Shannon had sown the marigold seeds at the Vale. The Keep's walls were soon covered with espaliered fruit trees and trim vines. A rose garden and a herb garden were made within the enclosure of the Keep. There was great excitement when a brick path was uncovered and pronounced to be Roman, but the narrow bricks turned out to be eighteenth century. This path ran through the (future) herb garden, which they wanted to contain only medicinal or aromatic herbs. Japanese irises, Chinese peonies, gentians, and many other 'queer dear flowers' were sent down from London.[12] Hundreds of bulbs were put in. Plants and trees were still difficult to obtain, because of the war, but they managed to make a promising start, though the garden looked bare and straggly for some time. Outside the enclosure another garden was formed with a sunk lawn and a pond full of lilies. Ricketts designed a fountain and a flight of steps for the Davises' garden. Though he and Shannon later had the services of a gardener, they did much of the initial planting, tending, and weeding themselves. 'Yesterday, I planted four dozen rare rock plants in the wall, watched by three peacocks', Ricketts boasted. It was a very Nineties picture.

The creation of the house and garden at Chilham brought life and delight back into their lives after the dreary years of war. Not only did it stir and occupy their creative imaginations, but it also shook them out of old habits, gave them a completely new interest, and took them out of London more frequently than ever before. 'It will be a constant

[9] CR to G. Bottomley, 15 Sept. 1919. The cretonne is reproduced in 'Tis', *Masters of Modern Art: Charles Shannon, ARA* (n.d.), p. xi.

[10] CR to Muriel Lee Mathews [Aug. 1919].

[11] CR to G. Bottomley, p.m. 20 Oct. 1919.

[12] CR to G. Bottomley, p.m. 23 Mar. 1919 = *SP*, p. 314.

pleasure to nurse the place', Ricketts told Bottomley, 'how it would have excited Oscar, the Michaels, & Robbie Ross!'[13] Within a few years their efforts had produced a remarkable new work of art, judged worthy to be featured in *Country Life*, where it was described as the oldest inhabited building in England.[14] It found its way into their pictures as well as their hearts. The huge wheel of the well house, uncovered when its ugly modern walls were removed, feature in the background of Shannon's diploma painting, *Vanity and Sanity*. Ricketts used the Keep in the designs for *Parsifal* (Fig. 40) (never produced) and in *St Joan*, his most celebrated theatrical production. To satisfy Ricketts's musical needs, a 'hideous' new pianola organ was installed 'on which I shall be able to grunt and squeal and get something back on the peacocks'.[15]

With the Keep, it was much easier for them to have friends to visit and even stay, and their social life became livelier. No longer were they the stand-offish young men of the Vale days. Many new young friends were drawn into their circle. Ricketts found that old friends and acquaintances seemed to avoid each other. Except for Binyon, whom he liked immensely, and the Davises, and Ernest Jackson with whom he dined once a week, he felt more at ease with, or interested in, 'strangers, that is, if I am interested at all: places wear well, better than human beings'.[16] Among the new friends was the musician Constant Lambert; and Laurence Irving—'a nice sensitive youngster', grandson of the great actor—came and sketched peacocks in the gardens.[17] Glyn Philpot and his friend Vivian Forbes came down, and were photographed with Ricketts and Shannon at the Keep. Particularly revealing about Ricketts's character is the fact that he preferred Forbes, whom he thought a most promising artist, to Philpot, who was actually of much greater artistic ability. Perhaps the reason was that Philpot, who soon became the youngest serving RA, had outgrown a strong Ricketts influence on his early painting and could paint the sort of pictures Ricketts and Shannon wanted to produce with greater ease and skill than either of them could command. Forbes,

[13] CR to G. Bottomley, p.m. 20 Oct. 1919.

[14] Christopher Hussey, 'The Keep, Chilham Castle, Kent', *Country Life*, 55 (21 June 1924), 1000–6.

[15] CR to Mrs C. J. Holmes, 15 Mar. 1921.

[16] CR to RNRH (TS), Christmas 1920.

[17] CR to WBY [1922].

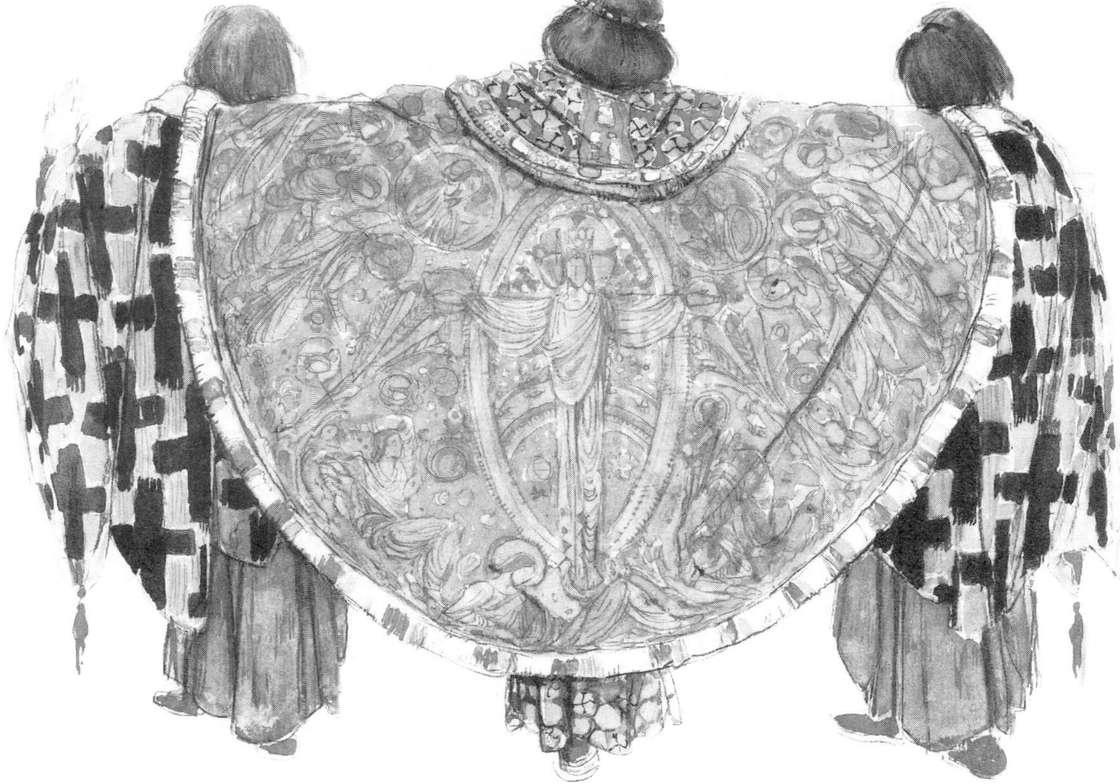

FIG. 40. Costume design for *Parsifal* (1921) by Charles Ricketts (British Museum)

on the other hand, was relatively unformed, and brought out that side of Ricketts which yearned to guide younger men through the thickets of art and culture. Writing of Forbes to Bottomley, Ricketts observed: 'The war caught him when hardly a man, and he is seeing Russian ballets, National Gallery pictures, and hearing Wagner or Chopin as novelties.'[18] This sort of *tabula rasa*, combined with an instinctive attraction to painting and culture, always caught Ricketts's interest. Through Ricketts's good offices with C. J. Holmes, an earlier protégé, a post in the National Gallery was offered to Forbes, but he refused it because he wanted to get on with his painting; a decision which even Ricketts could not but approve. He evidently also approved of the Philpot–Forbes ménage, and to several friends expressed his hope that they would become 'the Ricketts and Shannon of the near future'.[19] At the time this was a plausible hope, but it was not long before Philpot and Forbes ceased to share a house and their relationship came, by a very different way, to an end as tragic as Ricketts's and Shannon's. In spite of the many friends and interests they had in common, moreover, the two pairs of artists never became really close friends.

The young man who made the greatest impression on Ricketts at this time was Cecil Lewis. They met at a party at Vladimir Rosing's flat the year after the war. Lewis, aged only twenty-one, was already a war hero, having won the Military Cross and been one of the few pilots to survive from 1915 to the end of the war. Not long after meeting Ricketts he went off to China to help train the Chinese air force, and they began corresponding. So began what was to become, in spite of different and even conflicting interests, a key relationship in the lives of both men. Very tall and handsome, Lewis was charming and intelligent, and brought to the world of visual arts that Ricketts opened to him a freshness of response and the zest that had made him a hero. Lewis, moreover, wanted to write, and stirred in Ricketts that energy and sympathy which had exercised itself so often before in furthering the careers of good-looking young men. Ricketts corrected his writing, so that in three years he improved from 'flagrant journalese diction to clean speech', and they collaborated on several writing projects.[20] When, at Ricketts's suggestion, Lewis wrote a play about Montezuma, Ricketts designed a magnificent series of Aztec sets and

[18] CR to G. Bottomley, 29 May 1919.

[19] CR to WBY [late summer 1922]; see also CR to RNRH [autumn 1922].

[20] CR to Cecil Lewis, 14 June 1928 = *SP*, p. 400.

costumes, one of his greatest achievements as a theatre designer, in the hope that they would help put the play on the stage. Though they failed in this, the designs, many of which were given to Lewis, stand as a testimony to Ricketts's devotion. To their mutual friend George Bernard Shaw Ricketts described Lewis as 'that ridiculous & engaging boy', and 'a sweet natured & vivid youngster for whom I entertain a real affection'.[21] There was a strong element of somewhat flirtatious teasing: 'Let's go out and get drunk', Ricketts would say, 'You're only bearable when you're drunk', but they always returned sober. Or when Ricketts lent a book he would maintain, 'I'm not giving you that book, you beastly rotter!' And he would inscribe on the flyleaf, 'This belongs to C. Ricketts.'[22] (This was something he often did when lending books, sometimes even writing, 'This belongs to C. Ricketts and not to . . .')[23] They were, however, men of their own time and always addressed each other as 'Ricketts' and 'Lewis', except in letters where they resorted to French and Italian—'Mon cher Vieux' and 'Carissimo'—to express their genuine affection for each other.

Yet the relationship had its difficulties. Lewis himself, in writing the preface to the collection of Ricketts's diaries and letters which he edited in 1939, observed that Ricketts's affectionate nature craved for something which he was well aware that Lewis could not give. Sometimes Lewis even wished his 'nature had been otherwise'.[24] Clearly Lewis, who was quite a ladies' man, could not cope with the sexual element in Ricketts's love for him. For love—in the deeper sense too—it was, and Ricketts made no bones about it, even declaring one of his letters to Lewis a 'love letter'.[25] His feelings understandably spilt over into physical yearnings, which Lewis, equally understandably, found embarrassing and disgusting. This inevitably led to some painful situations. Nevertheless, for the next twelve years Ricketts never stinted in his generosity or in the spontaneous gaiety with which he gave. As well as helping and encouraging Lewis in his work, he showered gifts upon him: an antique intaglio ring, *Daphnis and Chloe* woodcuts, theatre drawings, a plaster cast of Orpheus and Eurydice, a Japanese screen, and an eighteenth-century French still life by Anne Velaya Costa, a

[21] CR to GBS [received 22 Feb. 1925] and [Aug. 1926].
[22] *SP*, p. x.
[23] One such, inscribed to R. Wells, is in Coll. Dr Robert Hillenbrand.
[24] *SP*, p. xi.
[25] CR to Cecil Lewis (TS, n.d.).

follower of Chardin (sold in 1983 for £21,000; Ricketts's generosity to Lewis thus continued long after his death). Lewis also often refused gifts, as he did not want Ricketts to think he valued him for these alone. A particularly striking example of Ricketts's devotion was his decision to leave Lewis the bronze bust of himself by Wells, thus separating it from its companion piece, the bust of Shannon. After Ricketts's death a strip of paper was found inside it saying, 'This belongs to Cecil Lewis'. The bust was later put to the best possible use by Lewis as a memorial to his old friend. Lewis was also incorporated into Ricketts's work. In addition to a drawing of Lewis in the nude (Fig. 41), intended as a study for a 'winged Hermes', Ricketts did a portrait bust (Fig. 50), which originally had wings, and later, satyr's leaves. This was Ricketts's only portrait in sculpture; it stands with the miniatures of Edith Cooper, 'Ryllis Hacon, and Mrs Green, among his few portraits of any kind. Three copies in plaster were made, with Ricketts teasing Lewis, 'I'm having you cast in plaster, you fraud. You're not worth casting in bronze.'[26] As with Miss Cooper, moreover, Lewis's face and figure found their way into some of Ricketts's paintings, including a late representation of Don Juan; the irony was probably intentional.

Among Ricketts's older friends, he was seeing Gordon Bottomley, who risked one of his rare visits to the capital in April 1920 and again in August 1922. These visits, however brief, always left him exhausted, but on the second occasion one of his plays, *Britain's Daughter*, was being given. Ricketts had also been engaged in designing book covers—free of charge—for Bottomley: *King Lear's Wife* (1920), *Gruach* and *Britain's Daughter* (1920), a *Vision of Giorgione* (1922), and *Poems of Thirty Years* (1925). The first had been designed in 1915, but because of the high cost of gold after the war only fifty copies signed by the author were published as Ricketts had originally envisaged, with the bulk of the edition using cheaper materials. Appropriately for collections of plays, the designs, especially for the first two, look very much like stage sets, thus showing, like the vivid colours, the influence of a new interest upon an old. *Gruach* was in fact dedicated to Ricketts and Shannon, though Ricketts asked Bottomley to use only initials, so that the dedication would be obvious only to the initiated. The last two books were published with a design blocked in blind on green cloth, rather like his earlier covers for *Hero and Leander* and the Vale

[26] *SP*, p. x.

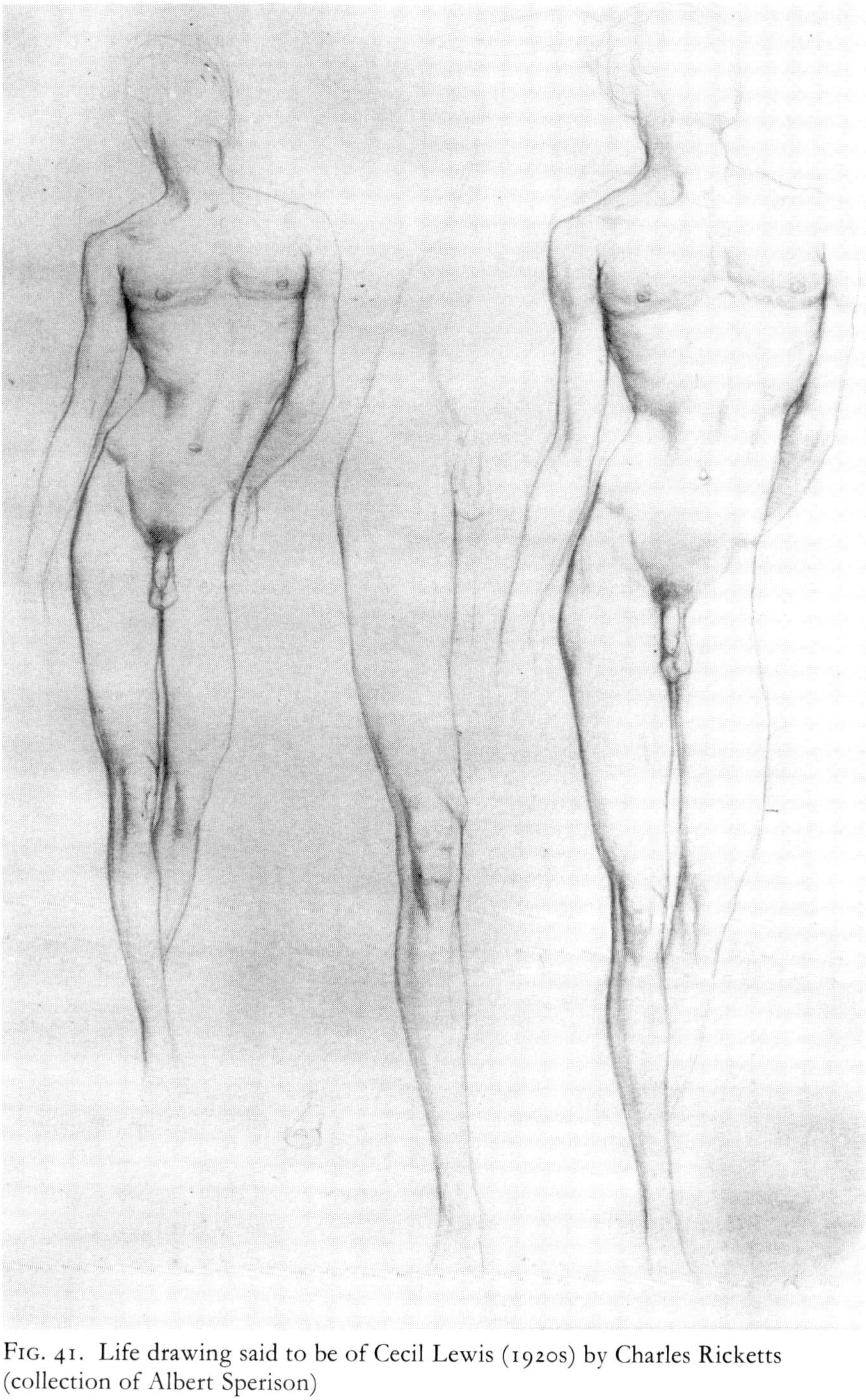

FIG. 41. Life drawing said to be of Cecil Lewis (1920s) by Charles Ricketts (collection of Albert Sperison)

Shakespeare. Ricketts liked blind blocking very much, though he believed it was not much favoured by the public, and even Shannon had his reservations about it. Most such designs by him are largely abstract and show his unfailing decorative sense; they are the only sort of abstract design that Ricketts ever practised or even approved of. This no doubt explains why he chose a blind-blocked design for the cover of Yeats's Collected Edition of 1922–6—which Yeats found 'serviceable and perfect'—and Binyon's anthology, *The Golden Treasury of Modern Lyrics* (1924).[27] Yeats too paid a visit to London at this time and on a later occasion was able to compare the Keep with his own Tower in Ireland.

Returning thus to his earlier interest in book design—for he had not designed any bindings since those for Michael Field before the war—seems to have restimulated Ricketts's love of book illustration. As Oscar Wilde too was much in his mind, he had begun to think of writing his recollections of his old friend, and in 1923 he took up a suggestion from his American dealer Martin Birnbaum to do a new series of illustrations for *The Sphinx* and to illustrate in his earlier style some of Wilde's *Poems in Prose*. The drawings have all the dramatic power and decorative quality of his earlier work, but the second Sphinx series is stranger and more elaborate. Both sets were sold by Birnbaum in America, but when one of the *Poems in Prose* was eventually returned Bottomley snapped it up, as well as some of the first sketches for *The Sphinx* designs.

In a curious way it was his old friends who were dead who seemed to preoccupy Ricketts at this time. Beardsley's mother was discovered to be very ill and poor, and Ricketts tried to arrange a sort of club to help her with a small annual income; his efforts here were more on account of Mabel than of her famous brother Aubrey. During 1923 he was intrigued by messages purporting to come from Oscar Wilde. Having followed what was published in the newspapers, Ricketts learnt, from a spectator of the seances, that some of the messages had not been made public, 'torrents of abuse of George Moore & Chesterton'.[28] All this intrigued him so greatly that he sent a few questions to be put to the mediums concerning the original scheme for *The Importance of Being Earnest*, the 'Crucifix Story', and the Centaur prose poem, but the answers were not very satisfactory. Another old association occupied him when he discovered that Michael Field's

[27] WBY to CR, 5 Nov. 1922.

[28] CR to TSM [Dec. 1923].

grave had no tombstone. One of the problems was that the two women were buried in different places, 'Michael' at Hawksmoor Priory where she had died and 'Field' at Isleworth, near Richmond. Immediately, he made a design and asked Fr. Gray in Edinburgh for an appropriate Latin inscription, which included the name 'Michael Field'. When Gray offered to help with the cost Ricketts refused, as he had been the chief beneficiary of their will. The tombstone caused no end of difficulty, reminiscent of one of Ricketts's theatrical ventures: 'a sort of fatality', he told Gray, 'has weighed on the "Michael Fields" monument'. Because Ricketts now dreaded artwork, 'with mechanical accuracy *obligato*'—eye and hand were no longer as skilful as when he had been a typographer—he handed his sketches to students at the Royal College of Art to turn into proper technical drawings. Nothing was done for months, and when the full-sized drawing was finally carried out it proved to be full of mistakes. Then there were difficulties with the mason, who kept changing his estimate. Yet it turned out well, 'austere & elegant, at once early Christian & early Renaissance', thought Ricketts.[29] Installed at Isleworth, the stone eventually cracked and had to be removed. Nothing now marks the grave. Bad luck often seemed to plague such ventures, though Ricketts's own versatility and tendency to branch out into fields in which he had no specialist knowledge may have counted against him.

His painting, however, was another matter, and here he was growing in skill, assurance, and reputation, at a time when the artistic values he stood for so passionately were being eroded by the modernists. During the agonizing years of the war he had kept doggedly at his painting, despite depressions, the demands of exhibitions and war charities, and the prolonged interruptions caused by his theatre work. The first hint of official recognition had come in 1915 when the Manchester City Art Gallery bought *Montezuma sacrificing to the Sun* (Fig. 42), a picture for which he had a great affection. He rightly thought it 'quite unlike any possible picture by anyone else and singular in temper and character'. Like most of Ricketts's pictures, it is immediately recognizable; it could not be by anyone else; it expresses the 'tragic splendour' that he valued most in art, though he recognized 'how small an appeal it has to most sane, average men in this country'.[30] Here, his particular aim was to convey 'an impression of brutality and

[29] CR to Fr. John Gray [1923] and [autumn 1923].

[30] CR to G. Bottomley [Jan. 1916].

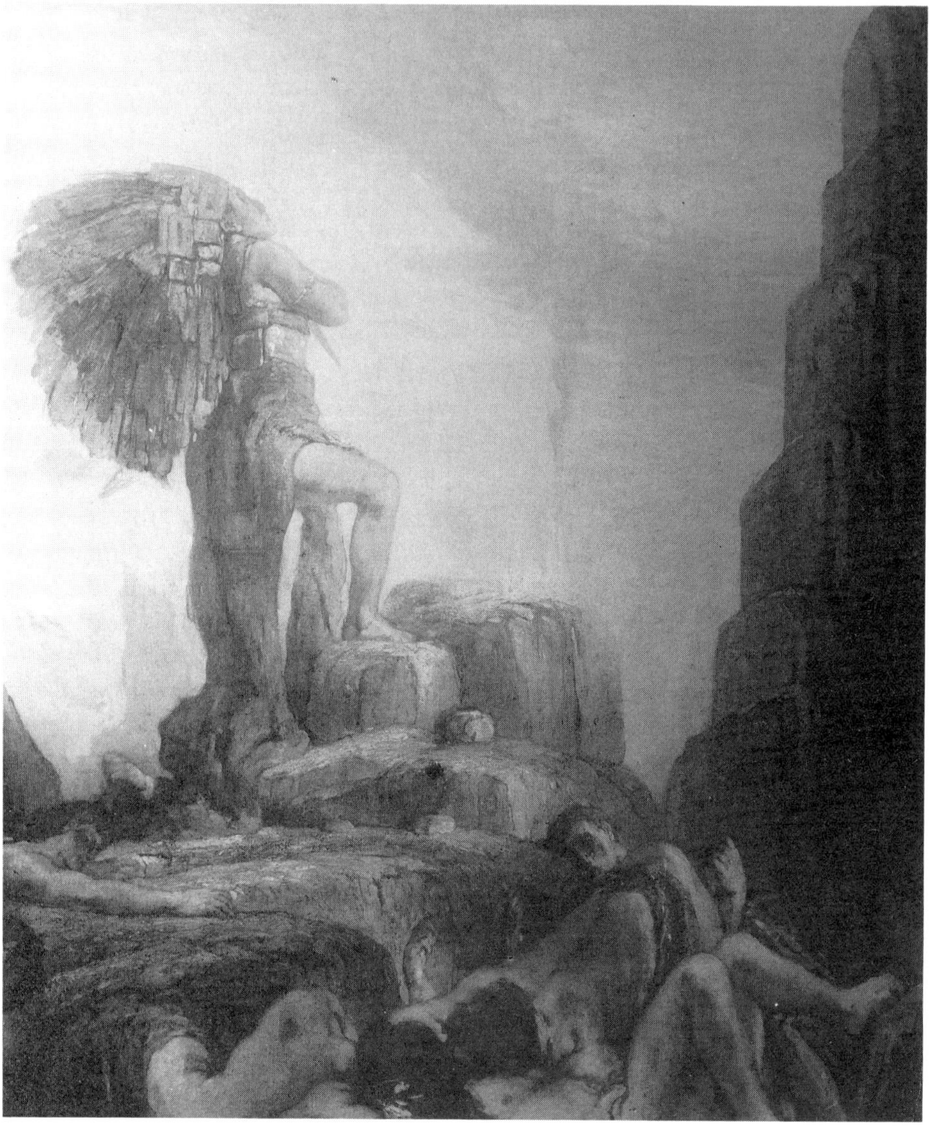

FIG. 42. *Montezuma Sacrificing to the Sun*. Oil-painting (1915) by Charles Ricketts (Manchester City Art Gallery)

magnificence, of savage splendour and savage fatalism', rather than to depict any particular instance of the annual human hecatombs offered by the Aztec monarchs to their bloodthirsty gods.[31] It reveals too his fascination with human cruelty and barbarity. Its date suggests a further point, for it was painted at the beginning of the war. Is there not perhaps also a connection with the war itself, with the thousands of lives that had by 1915 already been sacrificed in Belgium and France? If so, it was another, perhaps unconscious, form of war art, but one fitting for a Symbolist artist.

In September 1917 one of his *Don Juan* paintings was offered to the Tate, and accepted. It represents the Commander coming in through the window to claim the notorious lover, and was so closely modelled on Mozart's opera that the billowing curtain represented the rush of the wood-wind instruments in the overture. In addition to the honour of being represented in the National Collection, Ricketts received the substantial fee of £350, but he had no illusions about what this meant. He realized he was sending off his painting 'to the vast vaults underground where the ... pictures lie locked in each others frames regardless of age or sex'.[32] It was soon joined there by *The Deposition from the Cross*, which had been sold at the 1918 Red Cross Sale and donated by the purchaser. Shannon's Japanese patron Matsukata purchased Ricketts's beautiful and haunting *The Holy Women and the Angel of the Resurrection* in 1917, and a year later another lovely picture, a version of *The Wise and Foolish Virgins*, sold in Japan, where it had been sent on exhibition.

His painting was changing with his increasing mastery of the medium. About this he was surprisingly and uncharacteristically diffident. Regarding *The Danaïds*, he confessed he did not know whether he liked it or not: 'the subject probably lacks something, and may require better painting than I have at my disposal', but he started another version, as well as a *Siegfried and the Water Maidens* and a *Salome*. His return to this last theme seems to have been suggested by a current scandal surrounding a production of Wilde's play. Pemberton Billing, an MP, had alleged that Maud Allen's dancing was indecent. She sued for libel but, as Ricketts predicted, lost her case. Ricketts had learned to be cynical about the result of such moral disputes, especially where Wilde was concerned. His own painting soon got stuck; 'My *Salome*

31 CR to L. Haward [Feb. 1915]; CR to G. Bottomley [Jan. 1915].

32 CR to G. Bottomley, p.m. 19 May 1918.

deserves all that Mr. Billing would think of it and I feel too stupid to start something else.'[33]

With increasing mastery came the beginnings of success. *The Wise and Foolish Virgins* (1914), depicting the women dispersing beside a pond with a large rock in the background, streaked with bars of mist, was one of his most original and successful works (Fig. 43). He treated the subject many times, and always found himself concentrating on the Wise Virgins; he could never remember the number of the Foolish, sometimes having to add a foot or half a face to make the numbers tally. Yet his sympathies were with the Foolish Virgins. With his distrust of cool prudence and his love of spontaneous generosity, he rewrote the parable so that they found the Bridegroom by following the light of the stars, and presented him with garlands of flowers, whereupon he rewarded them above those who had brought their lamps but no gifts.[34] He was enchanted to learn that Lady Cowdray, discussing its acquisition when the visitors for the Chantrey Bequest arrived in the Grosvenor Gallery, bought it on the spot lest they should purchase it.[35] *Orpheus and Eurydice* (1923) was a new version in oil of a subject he had treated earlier in sculpture (Figs. 44 and 45). He played a new variation on this old theme: 'The Orpheus & Euridicy—how is her name spelt?—are not in the act of parting. She is a ghost following him in a spring-lit landscape, watched by Hermes.' Since the latter conducted the souls of the dead to Hades, he was a link between the living and dead, between the tragic Orpheus and his lover Eurydice—whom he lost, after rescuing her from Hades, when he broke the condition that had been imposed on him and looked back too soon as she followed him into the sunlight. Though Ricketts feared that the canvas was too large for the design, it is a highly successful work, which expresses again a sense of tragic loss. The telling poses of the figures may reflect his theatrical interests.[36]

With his 'moderate success as a painter', however, there was now the possibility of Ricketts also being elected to the Royal Academy of Arts. Immediately after the Great War this still had considerably more prestige than it commands today; the death of an RA could make the headlines of an evening newspaper even as late as the 1930s. Ricketts

[33] CR to TEL, 2 Sept. 1918; see also *SP*, p. 297.

[34] *65 Ill.*, p. viii; J. P. Raymond (i.e. CR), *Beyond the Threshold* (1929), p. 42.

[35] *SP*, p. 194.

[36] CR to G. Bottomley, p.m. 6 Feb. and 21 Feb. 1923.

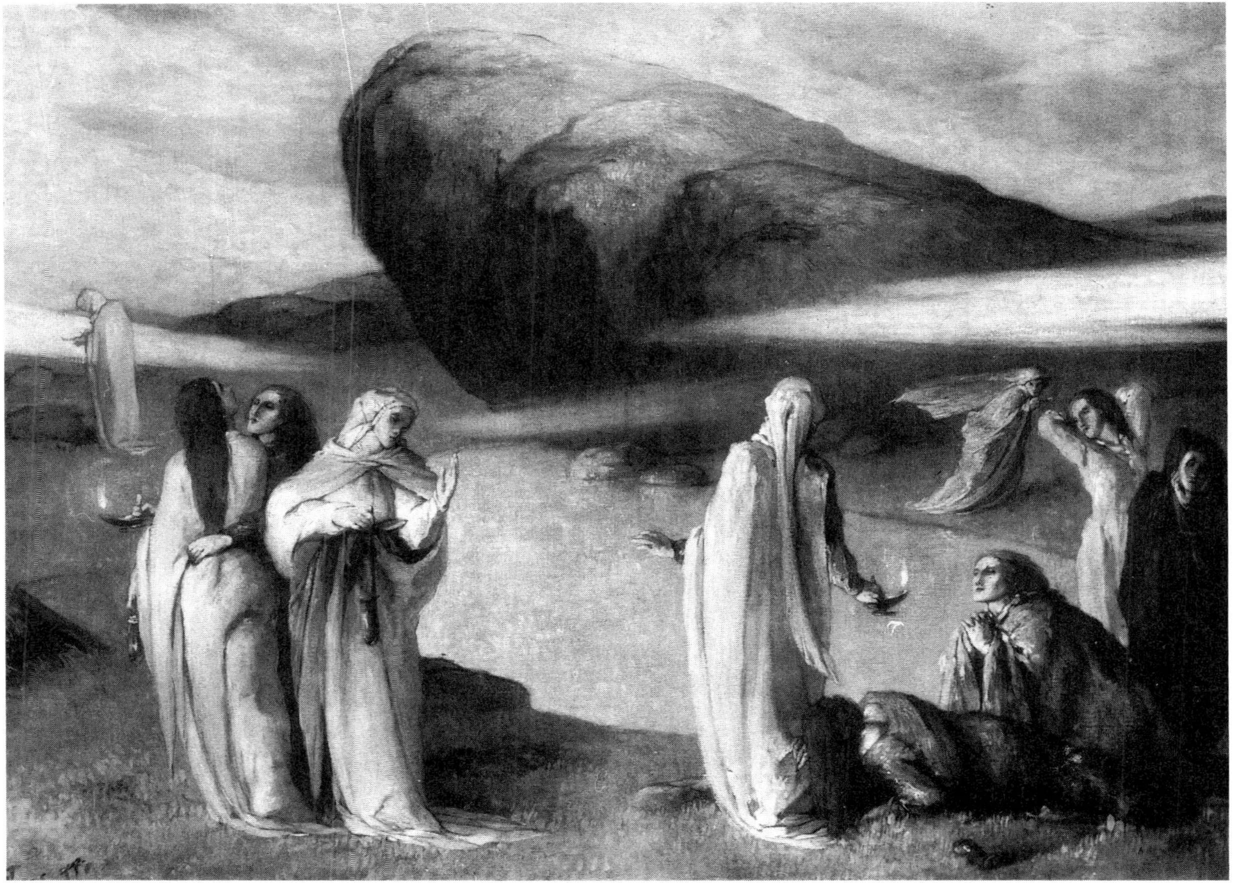

FIG. 43. *The Wise and Foolish Virgins*. Oil-painting (1914) by Charles Ricketts (private collection)

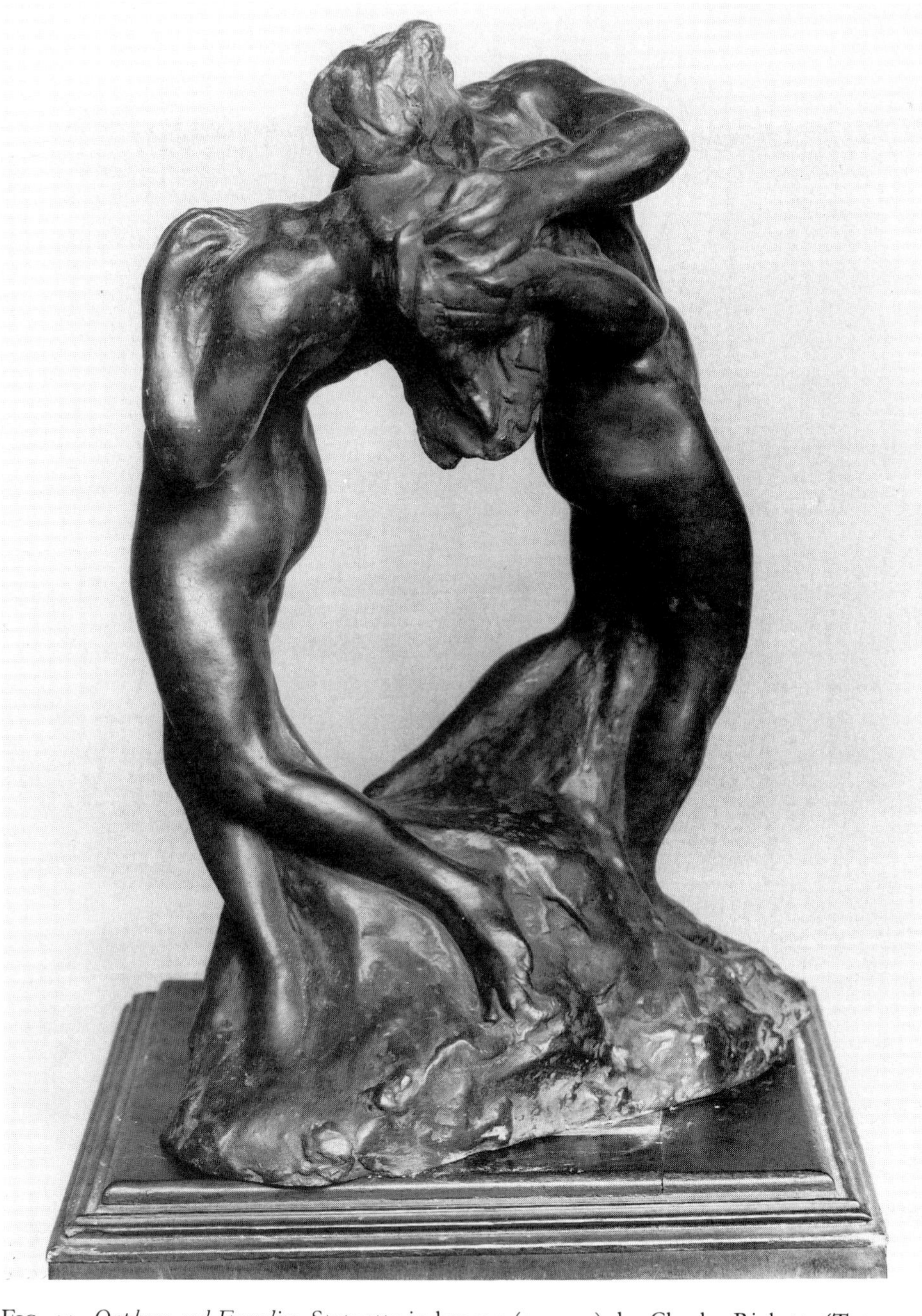

FIG. 44. *Orpheus and Eurydice*. Statuette in bronze (1905–7), by Charles Ricketts (Tate Gallery)

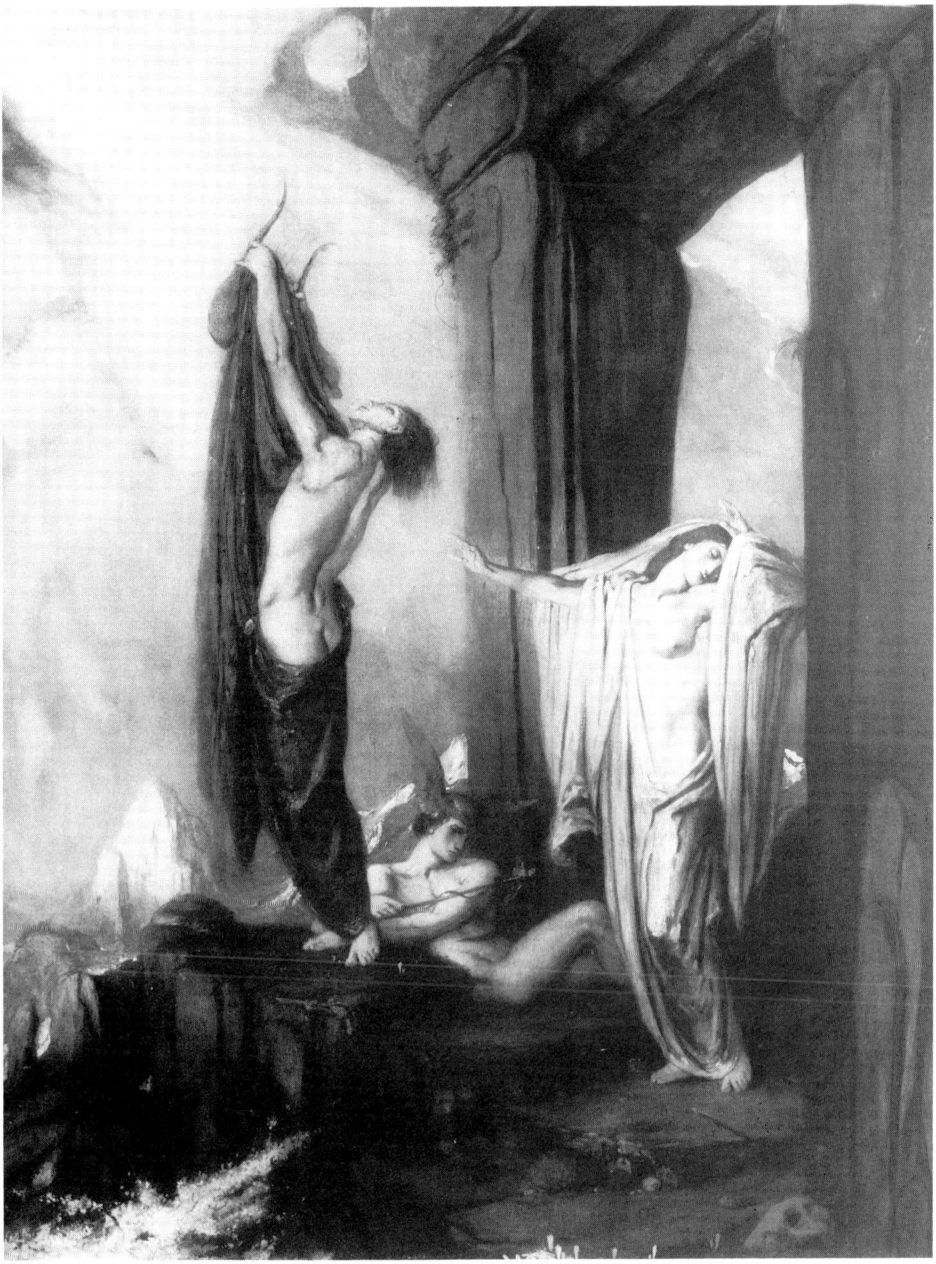

FIG. 45. *Orpheus and Eurydice*. Oil-painting (1923) by Charles Ricketts (Peter Nahum Gallery)

and Shannon, however, had always despised it; Rossetti had never been a member, nor had Alfred Stevens; Burne-Jones had resigned from it in disgust. What it stood for, as a body of older men, was inevitably somewhat behind the times, and in Ricketts's youth its stuffy Victorianism and its fondness for portraits, landscapes, and genre pictures had no attraction for a high-spirited young man devoted to Pre-Raphaelite and Symbolist art. When Clausen had first mentioned that he was recommending Shannon for the honour, Ricketts and Shannon had both agreed that 'the Society was obsolete, and would be ludicrous in a few years'.[37] However, there were still sound reasons for accepting election, as it brought not only increased recognition and power but also an annual opportunity to display your work at the year's most important exhibition. Shannon, in spite of an initial dread of ridicule, accepted his nomination as an ARA in 1911. His experience there was stormy, and he soon threatened to resign over elections of new members and the 'deplorable' Chantrey acquisitions; he was so disgusted, for instance, when James Pryde was defeated in three elections and the Chantrey refused to buy Philpot's masterpiece, *The Two Sisters*, that he determined to give up all official work. 'The bother is', Ricketts told Bottomley, 'that if the spirit of the R.A. is bad & retrograde, the press is worse & the public hopeless.'[38] With the demise of the International Society, the Academy offered them their only really suitable place for exhibiting their work or exercising their influence, so they had to turn to it *faute de mieux*. In spite of his mixed feelings, Ricketts therefore came to feel that to follow Shannon into the RA was the best course, so long as the election came as an unsolicited tribute from fellow artists. Ricketts had been amused to learn that the older RAs had been fascinated by his picture in the RA Red Cross Show and wondered why they had never heard of him. The younger ARAs thought 'differently of course, Strang roundly declaring the picture "affected footle", which shows he has a heart'.[39] When Ricketts's name was put forward that year, he was not chosen, Philip Connard being the one preferred, 'so my painting', Ricketts observed to Lowinsky, 'has not deteriorated at lightning pace, as is supposed to be the case once beckoned into the august house of Lords'.[40] A year

[37] CR Diary, 27 Dec. 1905 = *SP*, p. 127.

[38] CR to G. Bottomley, 7 May 1920; CR to Mary Davis [23 May 1923].

[39] CR to R. Ross [*c*.1918] = *Friend of Friends*, p. 268.

[40] CR to TEL, Mon. 29 [May 1918].

later D. Y. Cameron, who had supported Connard, began to canvass for Ricketts. In trying to persuade one old RA, who did not like Ricketts's painting, Cameron argued that the Academy would be stronger from the intellectual point of view if it included more men like Ricketts. 'But this is not an intellectual institution', came the rejoinder, which so delighted Ricketts that he always referred to the RA as 'that intellectual institootion'.[41]

Not only did Ricketts refuse to seek such honours for himself, but he was reluctant to accept them even when they were offered to him. He was never a respecter of titles or position, and no such props were necessary for his own self-esteem. When Cockerell, now attached to Cambridge University as Director of the Fitzwilliam Museum, mentioned the possibility to Ricketts of an honorary doctorate, Ricketts refused:

> It is most kind of you to wish me to become a Doctor, it is sweet of you but, please, dont think me ungracious, never think of it again; my imagination quails before any British honour; none were given to Rossetti; the little red French rosette is the only thing which might give me pleasure—or else, of course, the Garter with a tremengeous diamint star and a George the size of a saucepan—but this is improbable.[42]

Neither the Order of the Garter, with its wonderful opportunities for dressing up, nor the rosette of the Légion d'Honneur was ever offered to him, but in 1922 he was finally elected as an Associate of the Royal Academy. The level of culture he found there impressed him even less than the level of painting. 'They are so ignorant', he told Ernest Jackson.[43] This recalls how Sir William Llewellyn, PRA, commented on seeing Giorgione's seminal painting, *La Tempesta*, at the 1931 Italian Exhibition that it would not have been accepted by the RA hanging committee for the Summer Show.[44]

To the Baroness de Sinner's congratulations Ricketts replied that at the banquet, where he was placed (as a compliment) at the head of one of the tables, he had been introduced to some old retired RAs who were his father's contemporaries:

[41] CR to L. Binyon, 8 May 1918 = *SP*, p. 294.
[42] CR to S. Cockerell [summer 1924] = *SP*, p. 348.
[43] By oral account from F. E. Jackson to George Warner Allen to the author.
[44] Clark, p. 182.

How big the R.A. loomed in those days! with popular and rather splendid old humbugs, now no more; today, the humbug is neither popular nor splendid & the really famous men are actually rather common and shy, one looks like a pub loafer, another like a stable man [Munnings], another like a butcher, though this last (Sargent) has a fine voice, good pronunciation or diction, and a certain manner ... people are astonished by my manners & voice (which I consider quite ordinary, rather like your Father or mine).[45]

His diploma picture, presented when he became a full RA in 1928, was *Don Juan challenging the Commander* (Fig. 46), one of his best works, a telling design which once again shows man thrown face to face with his fate; indeed it depicts the Don's first realization of what lies in store for him. In it, Ricketts reverts to his practice of placing two figures in a dramatic relationship, though the head of the cowering servant Leporello is seen looking on from below. The design is built on the diagonal, on the electrifying current that runs between the outstretched hands of the antagonists as the defiant Don invites to dinner the statue which has just come to life. The colours are cool, as befits a moonlit night; the handling is crisp and sure; the obelisk that rises behind the Don (who resembles Cecil Lewis) makes a jagged silhouette that heightens the drama. It was a worthy offering to the Academy, where Ricketts had soon made his presence felt, hanging exhibitions, sitting on and manipulating committees, and playing art politics with great flair.

As well as the Decorative Arts Exhibition at the RA, where he was delighted to hang the Alfred Stevens room, he had been involved in organizing the International Theatre Exhibition at the Victoria and Albert Museum. Helping to organize such shows put him on the receiving end in dealing with spiky artistic egos. Just as Ricketts had refused to exhibit with Gordon Craig, who had 'attacked' him (or to let the Vale Press be illustrated in one history of book design published by his old 'enemy' the *Studio*, or in a second written by another 'enemy', the German Meier-Graefe), Bakst refused to exhibit in the same show as Craig, who, he claimed, had denounced him in a London art journal.[46]

Rather surprisingly, Ricketts found words of praise for some stage sets in this show by Paul Nash. Always on the look-out for artists

45 CR to Florence de Sinner [1922].

46 CR to Charles Holme and (reply) C. Holme to CR, 17 Apr. 1914; William Heinemann to CR, 1 Feb. 1906. The letters are in BL Add. MS 58090.

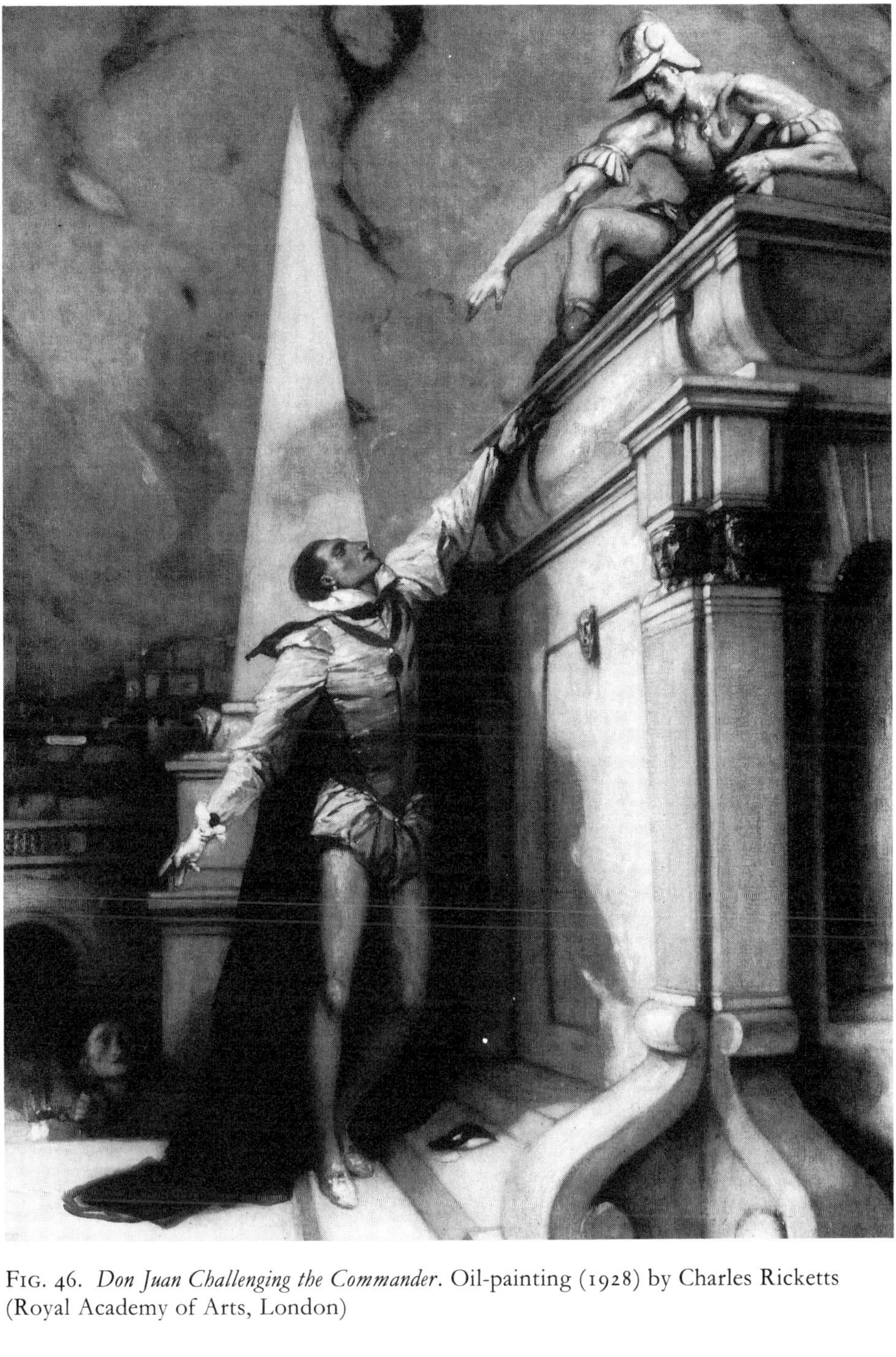

FIG. 46. *Don Juan Challenging the Commander*. Oil-painting (1928) by Charles Ricketts (Royal Academy of Arts, London)

whose work he could promote, he liked the 'stunning' war drawings and paintings of Eric Kennington and the 'neo-primitive' work of Henry Lamb and Stanley Spencer, and was so enthusiastic after seeing an exhibition of the designs of Nicholas Roerich, who had decorated *Prince Igor* and *Le Sacre du Printemps*, that he tried to get them shown at Manchester.[47] These he saw, quite wrongly, as hopeful signs for figurative painting in the near future:

> I believe modernistic tendencies and Futurism generally on the decline here as elsewhere, though I am told it is rampant in the 'Germanies'. There seems to be a wish to return to sanity and some sort of tradition, at least among the young. 'Les Jeunes' and 'les Féroces' all seem rather middle-aged than otherwise.[48]

He was also involved in organizing a major exhibition of British art in Paris. Despite his many protests and complaints, he clearly loved being at the centre of such things, as much because it gave him an opportunity of impressing his aesthetic values upon the public mind as because he enjoyed exercising the taste and expertise which he possessed in abundance. There was a sense too in which these exhibitions of the old and modern masters he admired were an insistent answer to the Post-Impressionist Exhibitions that had so deeply distressed him before the war.

Such exhibitions were often as frantic and exhausting as Ricketts's theatrical work, and his reckless expenditure of energy resulted all his life in frequent bouts of illness. In November 1920, for example, he was down with some mysterious illness which the doctor hoped, he said, might be something interesting 'like jaundice or some rare new plague'.[49] In March 1921 he was ill with flu, followed by a fortnight of laryngitis; in November he had two bouts of flu in succession. During these periods, Mary Davis ('who adores ill persons') tried to mother him and feed him up.[50] Shannon too was making Ricketts anxious, by a general deterioration in temper and vitality. His remaining teeth seemed to bother him and Ricketts hoped, in time, 'to

[47] CR to RNRH [*c.*1920?]; CR to Mary Davis [Mar.–Apr. 1914]; CR to TEL [Apr. 1918] = *SP*, p. 293.

[48] CR to RNRH, [early 1923]. See *Poet and Painter: Being the Correspondence between Gordon Bottomley and Paul Nash, 1910–1946*, ed. Claude Colleer Abbott and Anthony Bertram (Oxford, 1955); Nash to Bottomley, 27 Dec. 1913.

[49] CR to W. Rothenstein [Nov. 1920].

[50] CR to V. Forbes, p.m. 4 Apr. 1921.

persuade him to do without them', but he was assured that they were sound.[51] (Ricketts too had teeth trouble and made regular visits to a 'costly dentist who does not know that he assists in the restoration of an immortal', he commented with the usual ironic egotism.[52]) The trouble, he thought, might only be worry and the 'mysterious "unaccountable"' due to a transition in life.[53] Shannon was sixty in the spring of 1923, which Ricketts thought had distressed him unduly. Ricketts told Birnbaum of his worries over Shannon: 'Outside India no one has succeeded in realising a more perfect state of self-absorption or self-abstraction than he.'[54]

The fact is that Shannon's career was stagnating, in spite of an exhibition of his drawings at Colnaghi's in 1920, his first show for twelve years. While Ricketts was steadily gaining as a painter, and going from strength to strength as a theatre designer, Shannon, who became a full RA in 1920, had ceased to develop stylistically and had already painted most of his best pictures. There was, moreover, a slight but ominous decline in the lithographs that he began producing again between 1917 and 1920: they are rather coarse, and have less poetic quality than his earlier work. His paintings, which appeared annually in the Royal Academy shows, tended also to be repetitive in subject and sometimes clumsy in treatment. The decline was not sudden or striking; it was a question of increasingly frequent or obvious slips in draughtsmanship and more tired inspiration. As an established RA, however, he still had considerable standing and his pictures continued to sell as before. About the future, though, Ricketts was pessimistic, and thought that this would probably be Shannon's last exhibition of this kind, 'since another 12 years puts us possibly in another place'.[55]

Shannon, as a result of his feeling of stagnation, latched on to the idea of moving house as a change and 'a sort of renewal of life grown into a monotony of habits'. Ricketts, as usual, took the opposite view, dreading the effort and uncertainty and confessing 'to moments of acute depression and fear of the future which is not good for anyone'.[56] He was, he claimed, 'of a cat-like disposition, deeply attached to places,

[51] CR to Mary Davis [autumn 1924].
[52] CR to MF, 2 Oct. 1908.
[53] CR to TSM [Dec. 1923].
[54] CR to Martin Birnbaum [1924].
[55] CR to RNRH (TS) [*c*.1920?].
[56] CR to WBY [autumn 1922] = *SP*, p. 343.

which I generally like better than people'.[57] However, a move was necessary. Their twenty-one-year-old lease for the Lansdowne House flat was running out and Davis had decided to sell the building. Faced with the probability of a considerable rise in rent, they decided to buy a place of their own where they and their collections would be secure for the rest of their lives, and even Ricketts looked forward 'to the excitement of recasting and remoulding our "ambience" and the satisfaction of actual ownership to put an end to a deep indwelling sense of regret'. It was not easy to find a house to accommodate them, their two studios, and their large collection, at a price they could afford: 'Palaces are cheap, not so the smaller houses', they discovered, though the sort of smaller house they were seeking would seem more like a palace today. A month was spent house-hunting, and interviewing property agents and lawyers, before 'the possible and desirable place' was found. This required an outlay which intimidated Shannon, and they rushed to see their lawyer over mortgage and cash questions with the fear that it might be snapped up by someone else.

> Earlier in life [Ricketts told Yeats] we should have been more energetic & more easy to please, and more willing to take risks, but ... these many years of fear, anxiety, fatigue, & disillusion have had a wearing effect. I know I am a quite useless survival from another age and because the future has no use for me I look at it like a sheep in a railway truck.[58]

Ricketts admitted to a certain shame in writing this, as he recognized that his life had been sheltered and privileged compared to that of others. These were hardly desirable feelings with which to begin life in a new home. Moreover, whilst he was right to foresee that many of the values he stood for were on the wane, he was already launched on the period of his greatest fame and achievement.

The move entailed giving up a planned trip to Italy, emptying their bank accounts, and selling some of their treasures. They finally chose to part with their Daumier, and Puvis de Chavannes's *Le Pêcheur*, which they had bought from Durand-Ruel in 1907. Their collection was stronger in drawings than in paintings anyway, and these were things they could part with at the greatest profit, but they regretted the loss. The removal, redecoration, installation, all cost more and took three times as long as before the war; a change that was less

[57] CR to G. Bottomley, p.m. 17 Dec. 1922.
[58] CR to WBY [autumn 1922] = *SP*, p. 343.

familiar then than it has since become. The furniture, 'mainly Shannon's rubbish', Ricketts complained, took three days instead of one to move. The Greek things went in 'an old horse car accompanied by a wise old driver & Ricketts in the rear like a piano tuner'; nothing was broken. On 23 May 1923 they found themselves in their new home, Townshend House, a large Edwardian villa (since demolished) near Regents Park. Not a thing in the new house worked, not a window, tap, lock, stove, or geyser, Ricketts claimed, though they had been paid for weeks and months before; the painters had bolted before finishing the job, leaving their paint pots behind—Ricketts was clearly over-reacting as usual.[59] As soon as the stove in the dining-room was working, they made that their headquarters, as the weather was wet and wintry.

Before long life was back to normal. The new house was always full of flowers, which Ricketts loved as passionately as he did food, gaiety, and excitement. From the earliest days sprays had filled the house, though at the beginning they consisted of reeds, grasses, boughs of leaves, and the cheapest flowers, always beautifully arranged. His dearest friends—the Fields, Bottomley, the Sturge Moores—often sent him floral presents, and were thanked in his most delightful vein, as in this letter to Marie Sturge Moore:

> It is too kind of you to send me the superb silver leaves & the rare wild flower which is quite unknown to me. The latter looks like a thing which might grow in Italy, & the Acanthus—or is it a cardan—brings a feeling of design into the room. It looks carved by a great artist. The angel—for I believe in the plurality of gods of creation—who made it took great care in finishing it. This is not always the case with these creative spirits who sometimes get tired of their work. Witness for instance the legs of the peacock which were obviously intended for another creature.[60]

As they grew in prosperity Ricketts began to indulge his passion, and as early as 1912 confessed to the Fields that he spent on flowers in a year the rent they had paid on their house in Richmond. When Shannon later—probably in the mid-twenties—proved that Ricketts had spent £500 on them in one year, he blushed, but laughed impenitently. The dining-room at Townshend House was often like the Chelsea Flower Show, since Ricketts appreciated profusion as well as the individual perfection which he knew how to set off. He often played variations on the theme of 'the plurality of gods of creation' and would joke

[59] CR to Mary Davis [23 May 1923].

[60] CR to Marie Sturge Moore [11 June 1927].

about some fascinating oddity ('That is the work of a very minor—perhaps of a fallen angel'); or about some splendid plant: 'The Almighty would not trust even an Archangel with this, and designed it himself. His hand was even luckier than when he made Eve.'[61]

As before, the main casualty of the move was their servant. Ethel, the small, hunch-backed woman who had been found for them by Mrs Davis, left to emigrate to Australia, for which she was presented not only with £40 towards the cost of the passage, but with a leaving present of £100. She was too moved to express her thanks. A pair of new servants arrived, who at first seemed willing but soon turned out to be 'useless, ugly and unfriendly' and 'quite brainless'. The wife put an expensive *foie gras* into the oven 'and served it up bubbling hot in its *terrine* before two popular R.A.'s'. They left at the New Year (1924), and with their departure Ricketts felt that 'a vaguely hostile and boring element' went too.[62] After two weeks of discomfort, without any help, the two useless servants were replaced by 'two new treasures who, if appearances are a guide, seem nice and friendly'. They turned out to be treasures indeed. Percy Nicholls and his wife looked after Ricketts and Shannon devotedly till the end of their days. Soon Ricketts found that 'Our house is gradually getting accustomed to us', though for some time they missed the light and spaciousness of Lansdowne House and found that

> parts of our belongings and work trappings tend to disappear on another floor & to become lost forever in lumber rooms. Books still refuse to find comfortable homes & sulk in the wrong places; books, speaking generally, have bad characters, more especially if they are books of refference.[63]

As soon as they got their second wind they were on the look-out for new mantelpieces, secured a 'superb columned Georgian fire place ... in white marble with big spaces of grey-blue *lapis-lazuli*, which sounds like the Rothschilds or Pierpont Morgan', and were looking forward to creating a Hokusai room. 'We only think now', Ricketts wrote to Vivian Forbes, 'of what we can do to remould Townshend House as a new toy like the Keep.'[64]

Their Lansdowne House flat was taken over by Philpot and Forbes, which gave them great satisfaction and fed for the moment their hope

61 *65 Ill*, p. iv.
62 CR to Cecil Lewis [May 1923] = *SP*, pp. 345–6.
63 CR to G. Bottomley, Christmas 1923.
64 CR to V. Forbes [*c.* 1923].

that the two young artists would become a later version of themselves. Ricketts was particularly pleased that Forbes was taking his own bedroom and studio. It was disturbing, nevertheless, to be invited to dinner in the flat they had helped to make and lived in for twenty-one years. Philpot, who had very good taste, had got the place up very well, but they dined there 'with mixed feelings, or rather with a singular sense of unreality'.[65] Ricketts had these same mixed feelings about any house that was full of familiar associations to him, and he passed Spring Terrace or the Fields's The Paragon at Richmond

> with a curious mixed sense of distance and remoteness, even a slight angry sense that each place has seemed to forget too soon. Old houses and haunts of my childhood have even grown hostile and ugly: in thought one remembers only the romance of the past; in actual contact with its setting, age, decay, and a sort of tedium become present or remembered.[66]

To recuperate from the move, they made a rapid trip to Italy and Paris as the guests of the Davises, who had bought a villa at Baveno on Lago Maggiore. On 11 October they left London, travelling via Calais, where 'the smell of coal and damp and less describable smells typical of France' brought tears to Ricketts's eyes.[67] The Villa Fedora stood in terraced gardens in 'ten acres of Paradise', facing L'Isola Madre, the finest of the Borromean Islands. The gardener gave them pieces of whatever plants they liked or even asked the name of; Ricketts, full of his own new experience of gardening, decided to send him rare seeds and bulbs from London. On a day of superb weather they were able to visit both the Isola Madre and the Isola Bella, but cold rain accompanied their trip to Milan to see the Brera. Even the cold 'could not put out' the Bellini *Pietà*, the Mantegna *Dead Christ*, and the Crivelli *Madonna*.[68] In Paris, they stayed for four days at the Ritz (which Ricketts hated), again as the guests of Davis, and spent most of their time at the Louvre, where they were delighted by old friends and new acquisitions, though disappointed by the new arrangement of the pictures; Ricketts sensibly admitted, 'At my age, disapproval is often only resentment at any change whatsoever.' They also visited the Moreau Museum, which impressed them more than before, and were whisked off by a young keeper from the Luxembourg

65 CR to TSM [Dec. 1923].

66 CR to G. Bottomley, 17 Dec. 1922.

67 CR to Muriel Lee Mathews [Sept. 1923] = *SP*, p. 346.

68 CR to RNRH (TS) [Oct. 1923].

Gallery to see Moreau's *Jeune Homme et La Mort* in a private collection. With sentiment and pleasure Ricketts observed how the bookshops, furniture, bric-à-brac, and restaurants were the same as forty years before, though now there was 'diabolical traffic and the constant threat of death' from the motor car.[69] Nor was it as expensive as they had been told.

The journey, which Ricketts had feared on Shannon's account, seemed to have done him good, but he was still causing Ricketts worry. There were signs in both men of ageing. Ricketts's own anxiety over the future, his nostalgic response to Paris, and his increased harking back to the past, all point to the fact that he too was approaching sixty, though in some ways he retained the energy and enthusiasm of a boy. In Shannon, who was three years older, the changes were much more noticeable. For two years now his vitality had been low, and Ricketts had been anxious 'not about the slight touches of rheumatism & gum troubles which might have preoccupied him, but by something vague in his manner which might point to some trouble to come'.[70] These words were prophetic.

69 CR to TSM [*c.* Nov. 1923].

70 CR to G. Bottomley, p.m. Oct 1923.

12 *'My Canadian berth'*

As early as 1907 Ricketts had spoken to the Fields about his need to find a berth for himself. The Vale Press having closed, there was no sure income any longer to support him and Shannon through the uncertainties of an artist's life. At the idea of the mercurial Ricketts settled in a job the Fields had been greatly amused and sceptical, and their doubts were perhaps later borne out when he refused the Directorship of the National Gallery.[1] He had also failed to get the Slade professorship at Oxford. Now, in the early 1920s, as he entered his late fifties, he thought again about the desirability of a post with regular remuneration, which would enable him to enjoy life's pleasures and opportunities in the years that remained to him: 'I do not regret the asceticism & caution of my young days', he told Bottomley, 'but now I deny myself nothing in reason, since life is growing shorter, and tomorrow will bring what it will bring.'[2]

Eric Brown, the Director of the National Gallery of Canada, had been looking for someone on the spot in Europe to advise on purchases for the gallery. Before long, after his brother Arnesby Brown, RA (who first suggested Ricketts and Shannon) had sounded them out, he called on them to make a definite proposal. Ricketts alone accepted.

From the first the financial basis was ill-defined, and this led to recurrent difficulties. There seems to have been a gentleman's agreement that Ricketts was to take a 10 per cent commission on the price of paintings for whose purchase he was responsible. He soon recommended a 'ravishing' coloured stucco by Desiderio da Settignano (for which he was asked to design a case and pedestal), 'one of the finest Antonio Moros in existence', and a Moroni portrait.[3] As the Moro cost £5,000, a commission of 10 per cent seemed rather high, so Ricketts agreed to a reduction, and later suggested that 5 per cent should be paid on sums over £3,000 and 10 per cent on lesser sums. When the National Gallery then paid him 5 per cent on the Moroni,

[1] MF Journal, 1907, vol. 2, fo. 16.

[2] CR to G. Bottomley, p.m. 24 Jan. 1919.

[3] CR to L. Binyon [July 1924].

which had actually cost under £3,000, Ricketts was quick to protest. So began the refrain which was to recur in many letters over the next eight years, 'Canada is cheating me again'.[4]

One of his conditions for accepting the post was that he was not to be consulted on modern works. Such a stipulation shows how uncompromising he still was about 'modern rubbish'; indeed he told Brown, 'I fear all committees are wangled by dealer advice, & the Paris conspiracy to dump certain kinds of modern work is strong enough to intimidate the Louvre'.[5] What he advocated for a small gallery like the Canadian one was works 'highly characteristic or very typical of their kind'.[6] Having only £7,000–£10,000 a year at its disposal, Canada was restricted to works of the lesser old masters; for by the 1920s the greatest ones had, like the work of some early schools, already become so rare as to be prohibitively expensive. The Canadian gallery had arrived too late on the scene for these. Ricketts wanted the gallery nevertheless to reflect as far as possible the changes in, and development of, European art.

Not all his recommendations were accepted. Whereas works attributed to Sebastiano del Piombo (now given to Licinio), Castiglione, and Rubens, and portraits by Mabuse and Hans Eworth were bought, Stubbs's *Death of a Doe*, a Spanish Gothic Madonna, and portraits by Reynolds and Gainsborough, among others, were turned down. This was often because the gallery had spent all its allocation for the year, but Brown also became reluctant to buy portraits, as he feared the gallery would lose in general interest by having too many of them. About the Gainsborough—a portrait of Pitt—Ricketts replied: 'It is a picture concerning which I feel *very strongly* "Portrait or no Portrait"'; but it was not bought.[7] All the same, he did not give up, and later managed to get more portraits accepted, including, in 1926, a Tintoretto portrait of his servant, 'the most commanding picture that has been in the market since the Antonio Moro'.[8] Ricketts had evidently come round to this artist he had hated so much on his first Venetian trip. When the Trustees agreed to purchase a Botticelli, which Ricketts

[4] CR to M. Davis [*c*.31 Dec. 1928].

[5] CR to Eric Brown, received 11 Sept. 1924; CR to Mr. Yockney (Tate Gallery, Yockney Archive 72–4).

[6] CR to E. Brown, received 15 Aug. 1923.

[7] CR to E. Brown, 4 Jan. [1926].

[8] CR to E. Brown [Aug. 1926].

first attributed to Tura, he wrote to congratulate them on their 'courage and foresight'.[9] On the other hand, he rejected with characteristic wit some of the pictures he was asked about: 'The Flemish picture you sent a photo of is probably a late copy of a late composite picture of the school of Antwerp, the only fine thing about it is its price.'[10] Nor had Ricketts's earlier patriotism deserted him. Discussing G. F. Watts's *Daphnis*, from the Leverhulme Collection, he wrote: 'I imagine it is not fashionable enough for your Trustees; also I confess I should not like it to leave the country.'[11] Both Ricketts's praise of, and his doubts about, the Trustees show that his relationship with them was no easier than with recalcitrant stage-managers or Post-Impressionist critics.

Moreover, squabbles about money increased, owing mainly to the failure in their financial agreement to differentiate between works that had been found by Ricketts and those merely advised on by him, between what he recommended and rejected, and between paintings eventually purchased and those not. The vague terms meant that Ricketts felt he was entitled to full payment on any works over which he had been consulted, whether he had found them or not, and that he got nothing for his time and trouble when a work he had found and proposed, was not bought. By 1926 it became clear that a new arrangement was necessary. A retaining fee of £150 per year was set, to cover his expenses for his adverse judgements, and for his recommendations that were not accepted; 5 per cent was to be paid on all works discovered and recommended by Ricketts; and 2.5 per cent on works recommended, though not discovered, by him and finally purchased.[12] Ricketts agreed to this, but did not make matters easier by disingenuously asking the gallery to keep an exact list of purchases and advice given, as 'it would be hopeless for me to try to do so'; yet he was quick to jump on supposed errors, which, he noted sharply, were always in the gallery's favour.[13] Though the new system worked better, it did not eliminate either the problems or the rather petulant letters from Ricketts. These seem all the more surprising in view of his previous generosity and unworldliness. For years, after all, he and Shannon had advised Edmund Davis without any charge. However,

[9] CR to E. Brown, received 29 July 1927.

[10] CR to E. Brown [1926].

[11] CR to E. Brown, received 25 Sept. 1925.

[12] E. Brown to CR, 10 Aug. 1926.

[13] CR to E. Brown, received 6 Sept. 1926.

even though Ricketts's decision to act for a public institution rather than a private collector showed public-spiritedness, he had taken on the post because he wanted the money. His percentages, as he rightly pointed out, were nothing compared with what Berenson and others had been earning.[14] Nor was there ever any sign of trickery, though admittedly a tendency to overpraise is often noticeable; but that, as Yeats remarked, was characteristic of him generally. All in all, the gallery rightly realized how fortunate it was to have someone of Ricketts's acknowledged expertise and experience working on its behalf for relatively modest fees, and did not want to lose him. For his part, this new activity never became so important for him as his painting and writing, nor so time-consuming as his theatrical work.

'I dont wish to return to the stage (save to give a leg-up to a Friends performance) unless I have a good theatre, a friendly audience, and plenty of money', he announced after the war.[15] His reluctance to do theatre work had made him refuse to stage plays by Binyon and Sturge Moore, and Maeterlinck's *The Betrothal*, which he had promised Granville-Barker to do. For the moment he had had enough, and for a year at least he wanted to 'keep away from the Theatre & all its acts'.[16] It was in fact a little more than a year before he gave way, over *The Betrothal* (1921). Except for this, and the operas *Nail* (1920) and *La Dame de Pique* (1921), both of which had been dragging on for years, he kept out of the theatre from the end of the war until 1924. Then, as if refreshed from his rest, he produced in only two years a series of masterpieces which ensured his fame as a theatre designer.

He did not like Maeterlinck's play (a sequel to *The Blue Bird*) which he found pompous and sentimental. The play ran into the usual problems: 'Cost, time, every thing has so far been against me', complained Ricketts, '& none of the actors seems to understand anything.'[17] Because costumiers were busy preparing for Christmas pantomimes, the costumes had to be made in Paris, away from Ricketts's control. During the rehearsals he did not have a square meal for five days, and used to walk home from the Gaiety Theatre between 11 and 12 to plunge into Bach preludes on the pianola till 3 or 4 a.m.: 'I imagine that servants in the other flats woke up, decided to give notice, then

[14] See CR to E. Brown, 22 Nov. [1928].

[15] CR to TSM [1922].

[16] CR to G. Bottomley, p.m. 19 July 1919.

[17] CR to RNRH (TS), Christmas 1920.

fell asleep again and forgot. No one has complained.' In spite of this, and the fact that the stage-manager was sacked three days before the performance and the orchestra conductor only the night before, Ricketts felt that he had 'stood the racket and fuss of the production much better than the others concerned, though for three days I fed on eggs, slops, and aspirins'.[18] On 8 January 1921 the play opened and, while the Press attacked it, the public liked it, though Ricketts feared they actually liked what he hated in it. Because the theatre was wanted for a new burlesque, the play was withdrawn in full success, and put on again early in 1922. Ricketts was gratified enough to report to Cecil Lewis in Peking: 'There was a fairly solid body of opinion that nothing quite so beautiful had ever been seen on the English stage.'[19]

When Lewis Casson, who had produced *The Persians* with Ricketts's designs in 1907, approached him to do Shaw's *St Joan*, Ricketts again at first refused. Shaw, however, returned to the charge. Shaw wrote:

> I could get the dresses from Boutet de Monvel ... His picture book is ... scrupulously correct archaeologically, but it is lacking in force, contrast, and life: his Joan is not my Joan, and somehow even the soldiers and courtiers dont come out as I want them to come out.
>
> I have planned to do without scenery sufficiently to make the action continuous. A length of silk for the ... Loire, a Craig pillar for the cathedral, curtains for the palace and tent will enable us to whack it through within $3\frac{1}{2}$ hours. But this is just why I want an artist of your calibre in command. An Arboreal scenic production could be ordered in the Waterloo Rd without the least anxiety.

Ricketts capitulated, but at first undertook it only conditionally and on the understanding that his contribution to it should be anonymous. 'I have given my reasons', he told Bottomley, 'that if he can only write plays which take 6 days to act & with gigantic casts, I can only stage vast & expensive ventures like *Parsifal* & *Aida* on gigantic stages, regardless of expense.'[20] However, he later allowed his name to be used.

Ricketts plunged himself into the production as usual, though he had a bad cold and 'would doubtless mope, if I had not this work

[18] CR to Muriel Lee Mathews [Jan. 1921]; see also *SP*, p. 331.

[19] CR to C. Lewis, 15 Apr. 1921 = *SP*, pp. 330–2.

[20] GBS to CR, 13 Jan. 1924 (private collection); CR to G. Bottomley, p.m. 22 Jan. 1924.

which keeps me employed till 3 AM & on one occasion till $\frac{1}{2}$ to 5'.[21] He was working so hard that he scarcely found time to go to town for the materials. In the costumes he wanted 'an intelligent blend between Pol de Limbourg & the Van Eycks, avoiding the bright 14-cent colour of the first and the rather prosy phase in dress of the second'.[22] This scholarly approach extended to every detail of the production, and Ricketts consulted Cockerell, and no doubt also his friend and patron A. G. B. Russell, the Lancaster Herald, over the heraldry. Some of the props were of particular beauty, including a splendid mediaeval tapestry, which replaced his usual curtains as a backdrop for one act (Fig. 47). Ricketts took Shaw's suggestion of a 'Craig pillar' to suggest Rheims Cathedral, but he added curtains and a distant stained-glass rose-window, which enhanced the grandeur of the setting.[23] He always worked by first emphasizing certain elements which would suggest the scene, then excluding or simplifying non-essentials, and then finally introducing striking or unexpected detail. The bedroom scene, for instance, was reduced to a bed, a window, and a semicircle of curtains, needed for the action, but this was enlivened by a painted mediaeval chest and two pictures. The backdrop of the first scene, 'a sunny stone chamber' in a castle, was based on the kitchen of the Keep.[24]

Trouble arose at the rehearsals. Ricketts had always enjoyed this part of the work with Shaw, whom he had found 'a sympathetic producer, his little lectures in the wings & exaggerated renderings of what he wanted were delightful'. But with age or success Shaw had changed: 'Today', Ricketts wrote a few years later, 'he only thinks of his points &, I believe, dislikes good acting & production.'[25] The two men had collaborated closely over the costumes, with Shaw generally bowing to Ricketts on the visual aspect of the production, though stipulating what he envisaged for Joan herself. However, they soon found that their conceptions of the play differed. Whilst Shaw wanted it played realistically, Ricketts thought it should be played romantically and artificially. Ricketts, for instance, wanted the court ladies to make odd formal movements. At one point Shaw protested, 'Scenery and clothes have ruined my play.' There were numerous rows. At one

[21] CR to S. Cockerell, p.c. [Mar. 1924].
[22] CR to S. Cockerell [Mar. 1924].
[23] GBS to CR, 13 Jan. 1924.
[24] Furst, *Apollo*, p. 332; CR to L. McCarthy [*c*.1928].
[25] CR to L. McCarthy, 1 Sept. 1930.

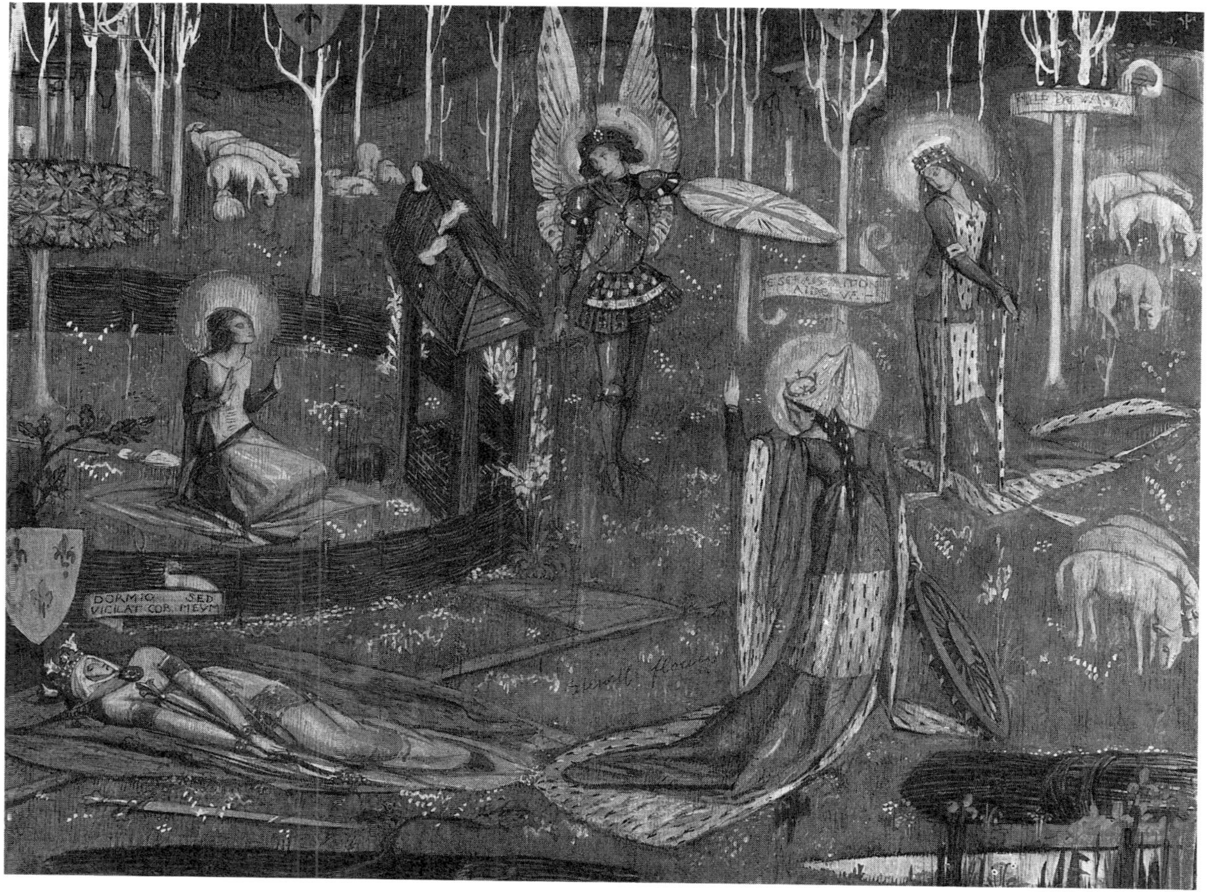

FIG. 47. *Joan and her Voices.* Design by Charles Ricketts for Shaw's *St Joan* (1924; Carlisle Museum and Art Gallery)

meeting at Shaw's flat Ricketts burst out, 'Shaw, you're the most ignorant man. Your wife is much cleverer than you'. Shaw roared with laughter. At the rehearsals Ricketts would lie down in the stalls almost kicking his heels in anger. As he got angrier, his voice got softer and higher till it reached a ferocious whisper. 'No, No, No, Shaw. Leave it to Sybil. She knows more than you do.' Casson's wife, Sybil Thorndike, who played Joan, which she considered one of the greatest roles of her career, also found herself at odds with Ricketts's interpretation. She wanted to play Joan as a rather dumpy, tough girl, but Ricketts wanted everyone to be tall and elegant, as in his drawings for costumes. 'It's a pity you're not six feet tall', he once told her. Dame Sybil felt he wanted everyone to look like Shannon.[26]

In spite of these conflicts *St Joan*, which opened on 26 March 1924, played to capacity audiences for six months before it set out on a successful provincial tour, and is probably the production for which Ricketts is best remembered. It was generally agreed again to be one of the most beautiful things that had ever been seen on the London stage.[27] Ricketts was making money out of the production, perhaps in royalties from the performances. Granville-Barker had suggested this form of payment for *The Betrothal* and Ricketts later regretted his refusal, as he would have made money out of it 'like the damned musician whose work does not count at all'. Now he was determined to make money out of the stage. 'The novelty of money has fascinated me', he told Binyon.[28]

Perhaps it was this that made Ricketts agree so readily to a sumptuous edition of *St Joan*, with his designs. The original idea seems to have been Shaw's. More than a decade earlier he had written to Ricketts: 'If millionaires had any sense, you would get £100,000 for a manuscript of *Man and Superman* with incidental pictures. Could we not do a first folio or something?'[29] The published edition of *St Joan* was put on a sound financial basis, with the profits being divided equally between publisher, author, and designer. Shaw generously saw to it that Ricketts's share should be paid before his own.[30] Ricketts felt

[26] Dame Sybil Thorndike in conversation with the author, 6 Feb. 1974.

[27] See George Sherringham and R. Boyd Morrison (eds.), *Robes of Thespis: Costume Designs by Modern Artists* (1928), p. 141.

[28] CR to L. Binyon [1921].

[29] GBS to CR, 12 Apr. 1911.

[30] A copy of the contract is in BL Add. MS 52746.

the publisher was inclined to overprice the book because of what he termed the 'showiness of my drawings', but 'I am informed', he told Shaw, 'that [the] dearer the book, [the] easier it is to sell. How I wish it were the same with modern pictures!'[31] Ricketts, who felt the book had turned out well, had evidently not lost his old publisher's instincts; because of its size and price (five pounds) it did not sell well. This put paid to another project, since, if the book had sold well, there was a chance of Ricketts's work, presumably his theatre work, being issued 'by a reckless publisher'.[32] Ricketts also worked at transforming the play into yet another medium. With Cecil Lewis he collaborated (under another pseudonym, Francis Strange) on a film scenario entitled 'The Maid of Orleans: A Nation's Martyr', which was never used.[33]

While he was working on the 'ill starred' edition of *St Joan*, Ricketts was approached (through a mutual friend) by Hugh Walpole, who wanted him to design a costume for his protégé Lauritz Melchior, the great Heldentenor, who was playing Siegmund in *Die Walküre* at Covent Garden.[34] Ricketts, who considered Walpole a 'charming humbug', offered to do it without charge, though he feared, as a Wagnerian, that it would 'be difficult to do something different from the door mats and feathered tin helmets' of most productions.[35] Ricketts found Melchior's figure unpromising: 'He had no forehead, no nose no chin . . . fat cheeks, no neck, fat shoulders, belly, and the heavy torso was supported on long thin legs: and I have to turn him into a tragic and romantic figure.' The costume consisted of wolfskins cut into strings, barbaric breeches and leggings, and a loose skin over his shoulders simulating a cloak, which tradition forbade him to wear. Ricketts also designed a superb wig with a false brow. Two hours were spent painting his face out and giving him a new one, so that he looked 'tragic and pictorial'; then his shoulders and arms were shaded down and black camouflage stripes were painted on his belly over the skins till, from the front, he looked more heroic in proportions.[36] Melchior's new appearance caused a sensation amongst the singers,

31 CR to GBS, p.c. [1924].

32 CR to G. Bottomley, p.m. 13 June 1924.

33 BL Add. MS 58093.

34 CR to G. Bottomley, p.m. 13 June 1924; *SP*, p. 348, gives another account of this episode.

35 CR to M. Davis, Sat. [June 1924]; CR to Sir Hugh Walpole (Univ. of Texas).

36 CR to G. Bottomley, p.m. 13 June 1924 = *SP*, pp. 348–9.

and Walpole recorded in his diary that he 'looked splendid' in the new costume.[37]

The summer of 1924 had been bad, and Ricketts and Shannon hoped to find some sun in Italy. On Monday, 22 September, they set out by the sleeper from Victoria for Lago Maggiore, where they spent three days with the Davises. From there they went on to Florence for a week, and then on to Rome. They broke their journey to make a day excursion to Orvieto, which was disappointing. So far, the weather too had fallen below expectations—thunderstorms in Baveno and deluges of rain in Florence, where Ricketts oscillated between happiness and fits of depression—but arriving in Rome on 6 October they found a fortnight of superb July weather and magnificent flowers. The galleries in both Rome and Florence had been on the whole improved, and reflected, Ricketts thought, 'the enthusiastic appreciation of the past by the Italians of today'. He felt that everything in Italy had more or less changed for the better. Ever since Mussolini had seized power in 1921 Ricketts had searched the morning papers for news of 'Mussolini & his incomparable *fascisti*'. They represented a sign of hope in a world that was growing ever more alien and hostile to him. 'Are they a sign of a world returning to order, duty, sense of real values, a return to construction & to veneration for firm things?'[38] Bolshevism, which Ricketts detected as much in art movements as in politics, meant for him the destruction of sound and well-tried values, the lack of respect for tradition and the achievements of the past, the dominance of petty egoism leading to rebellion for its own sake and disintegration. It must also be remembered that this was long before Mussolini made the fatal mistakes of invading Abyssinia and of allying himself with Hitler. In his earlier period he did much to strengthen and develop an unstable and faltering country. However, his most vaunted achievement, the symbol of the order and efficiency that he imposed—that he made the trains run on time—failed to benefit Ricketts and Shannon. Their train to Florence was an hour late and they had to stand in a crowded train from Orvieto to Rome. They returned to London after three further days of wet weather in Paris, and for a time Ricketts managed to keep clear of theatrical commitments, apparently returning to his painting.

[37] Hugh Walpole Diary, 28 May 1924. (Coll. Sir Rupert Hart-Davis, to whom I am indebted for this.)

[38] CR to WBY [autumn 1922] = *SP*, p. 343.

In January 1924 he attended a performance of Bottomley's play, *Gruach*, set in Scotland in the early Middle Ages. It went 'excellently well', with Sybil Thorndike 'generally quite admirable' in the leading role, Ricketts told Bottomley (who could not attend). The servants, however, were abominable; such roles required 'non-English' actors to bring them to life. English actors lacked the 'simplicity & roundness' of Irish, Russian, or Sicilian folk-actors, and were nervous, bewigged Cockneys dressed up and trying to look barbaric.[39]

Englishmen, Ricketts felt, also had 'a gift for looking foolish in costume. They stoop, tilt their hats or crowns wrong, & a first-rate costume can be made to look like nothing on earth'. This habit of blaming the English for not being like the French or Italians, with their sense of style in carrying clothes, was strengthened with his next play, where he wanted to make costumes to put 'Holbein on the stage' and to surpass all previous records. The play was *Henry VIII*, and it was being produced by Lewis Casson, who, 'good chap as he is', commented Ricketts, 'kills clothes & has a genius for casting elderly, stumpy men for elegant dresses & raw boys for venerable garments'.[40] Bruce Winston, a great friend who 'likes sweets & costs less than anyone else if treated nicely', and who had both worked on the costumes for and acted in *St Joan*, was carrying out the dresses, and Ricketts as usual was going to bed between 4.30 and 6 in the morning during the preparations for the play. Many of the elaborate properties were largely painted by his own hand, including a huge tapestry of *The Four Horsemen of the Apocalypse* (used as a drop curtain) and another of angels and a vast Gothic window for the trial scene. The scene-painter had disregarded Ricketts's sketch for the heraldry, over which he had taken a great deal of trouble, so that he was still repainting the mistakes two hours before the first night. The Westminster Abbey scene made him cry with vexation, for the scene-painter had 'introduced Anglican statuary of his own'.[41] On the last day Ricketts made a pomander, a rosary, two jewels, and a mask, as well as being involved in the rehearsals. The props, particularly, showed the fine, imaginative quality that marked his work. In the death scene of Katharine of Aragon the chamber was hung with a tapestry of angel-musicians, and it finally became apparent that the angels of the dying Queen's vision,

[39] CR to G. Bottomley, 22 Jan. 1924; CR to TSM [Feb. 1924].

[40] CR to G. Bottomley, p.m. 27 Oct. 1926 = *SP*, p. 367.

[41] CR to S. Cockerell, 23 Dec. 1925 = *SP*, p. 353.

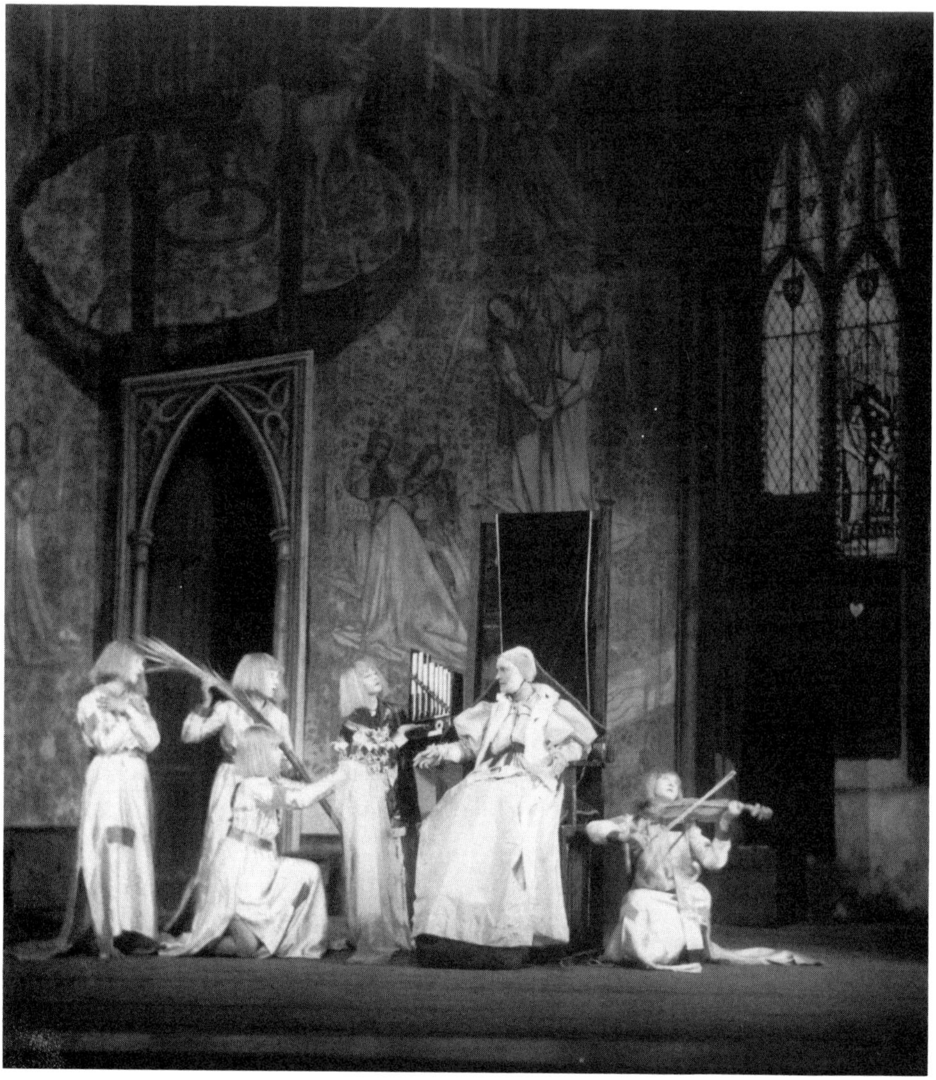

FIG. 48. *The Vision of Queen Katharine*. Production photograph with Sybil Thorndike as Katharine of Aragon in *Henry VIII* (1926)

bringing her the palm of 'the death annunciation' and holding a crown made of lilies and ears of wheat, symbolizing purity and resurrection, were the same as those in the tapestry (Fig. 48). In a prayer-book carried by Katharine this theme was also depicted in an illumination of the Virgin receiving the annunciation of her death, even though the book itself was never opened on the stage. The book was decorated with real gems, sapphires, amethysts, and grey pearls, which could not be seen by the audience. So deeply did Ricketts enter into the spirit and period of a production. An even more brilliant touch was shown in the play's final scene, in which the infant Queen Elizabeth is carried in. For this Ricketts had modelled a plaster baby, with a cap and shirt of Tudor embroidery, silver swaddling clothes, and jewels of pearl, coral, and crystal made by himself. No one had been allowed to see the doll, and its fame had spread through the company. At the dress rehearsal, when the curtain was raised on the final scene, all the court ladies crowded round the nurse to see the infant, as they might have done in real life.[42]

All the costumes for *The Mikado*, the next play designed by Ricketts, turned out perfectly in execution, and were regarded by Ricketts, yet again, as the most successful he had ever done. D'Oyly Carte had first asked him to design *Utopia Ltd.*, but Ricketts replied that 'there is, in fact, hardly anything to do, and I should be charging you for nothing if I undertook it'.[43] Ricketts then suggested *The Mikado*, and in a meeting with the impresario convinced him that authentic costume could be made theatrically effective. With his knowledge of Japanese art, he brought to the task a degree of scholarship that had been unknown in earlier productions, where the Japanese dress was made 'to look like a collection of dressing gowns'. Ricketts decided to set the costumes in a period—about 1720—when he felt the national costume was particularly beautiful (Fig. 49). The girls' dresses were very bright and gay. The Mikado appeared in the dress of a prince of high rank, the courtiers in proper Court costume, with fictitious banners and heraldry. They wore authentic-looking wigs and very wide breeches, more like divided skirts than trousers. Katisha's dress was particularly elaborate, with wider sleeves than was the current Japanese fashion. The women had to be made up very white and alter

[42] CR to G. Bottomley, 27 Oct. 1926 = *SP*, p. 367; CR to Cecil French, 12 July 1927 = *SP*, pp. 379–80.

[43] *Theatre Notebook*, p. 20.

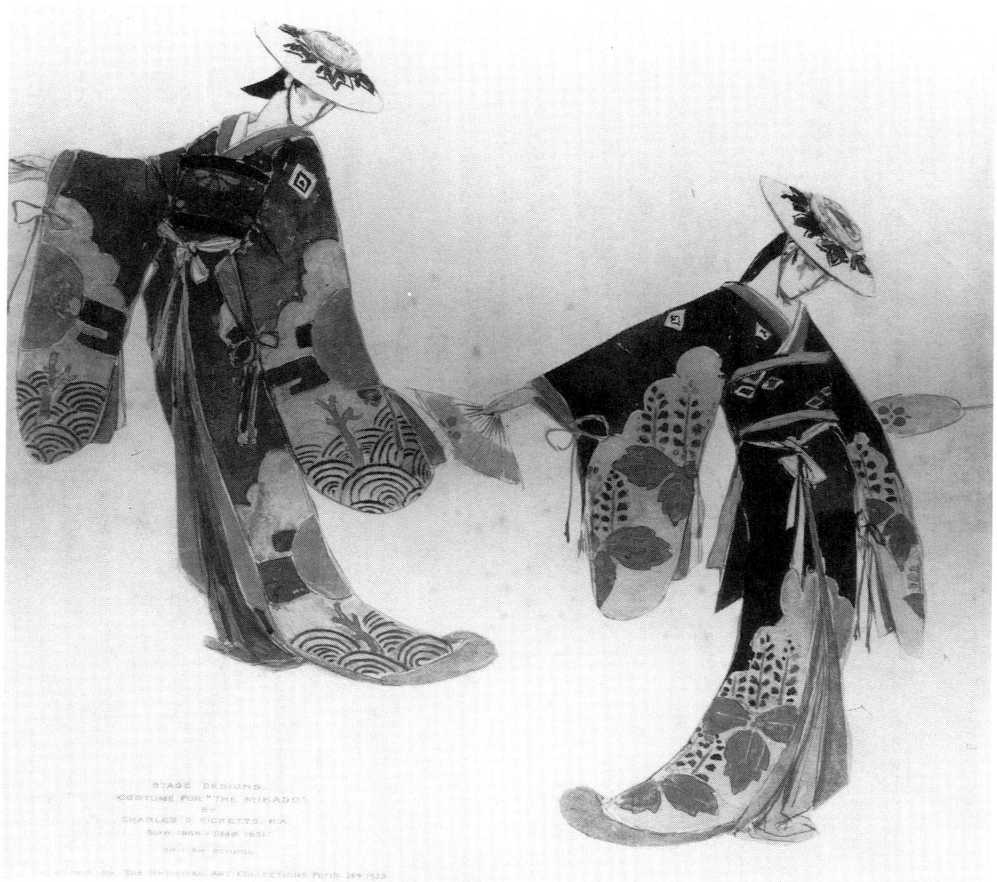

FIG. 49. Costume designs for female chorus figures for *The Mikado* (1926; Ulster Museum, Belfast)

their eyebrows.[44] All this was novel at the time and caused trouble among the actors. The men in the chorus viewed their unfamiliar costumes with fear and hostility, and all except Koko and Poo Bah looked paralysed by them on the stage. Katisha also hated her dress, and Ricketts had to remove a lengthy train decorated with a tiger design after the Japanese artist Korin, which he had painted himself. This dress was later replaced by another not by him. His hand-painted fans, with designs from Japanese artists, had to be scrapped, as one broke at rehearsal. The other female characters looked exquisite, and Binyon, who was the British Museum's expert on Japanese art, was overwhelmed, as were the public from the first night.[45]

The Press, Ricketts felt, was 'snappy and sniffy'. He attributed their hostility to 'some dozen interviews I gave to as many Pressmen at Townshend House in the dining room before D'Oyly Carte'.[46] There had been too much publicity of this type beforehand, stressing the fact that the costumes were being designed by an eminent ARA; for it was not common for members of the Academy to occupy themselves with theatre design. The critics had praised the superb drawings for *The Mikado*, which indeed are among his most beautiful for colour, design, and expressive gesture; they are works of art in themselves. Designing for the theatre Ricketts regarded as a 'holiday task', and in doing it he suffered none of the doubts and agonies he felt over his painting.[47] His finished drawings, which he liked to present to friends, show once again the marvellous talent for linear arabesque, for calligraphic rather than realistic drawing, that he had shown in his book designs, with the addition here of colour and even more dramatic poses. They are among his masterpieces, particularly those for *Montezuma*, *Salome*, the two versions of *Judith*, Wagner's operas (never produced), *St Joan*, and *The Mikado*. However, in the case of the last, the critics had reservations about the costumes when seen on the stage. There were, for instance, comments that the wide breeches of the male courtiers were like 'Oxford bags' or 'plus fours'.[48] Against such criticisms Ricketts felt

[44] CR, 'Why I Redressed the Mikado', *Daily Mail* (18 Sept. 1926).

[45] CR to G. Bottomley, 27 Oct. 1926 = *SP*, p. 369.

[46] CR to C. Lewis [1926] = *SP*, p. 369; CR to G. Bottomley, p.m. 27 Oct. 1926. See Percy V. Bradshaw, 'The New *Mikado*: Mr Charles Ricketts's Stage Designs', *Studio*, 92 (1926), 415–19; and Peter Goffin, 'Designing for Gilbert and Sullivan', *Studio*, 149 (1955), 161–7.

[47] *65 Ill.*, p. [xii].

[48] I cannot trace the source of the reference to 'plus fours' and 'Oxford bags'.

obliged to justify his work, in an article in the *Daily Mail* entitled 'Why I Redressed *The Mikado*'. His success was such that the costumes remained in use until the D'Oyly Carte Company closed in the early 1980s, and fetched the highest prices at the sale of the D'Oyly Carte wardrobe.

The Mikado had opened on 20 September 1926. On 24 December *Macbeth* opened on the same stage at the Princes Theatre. This production, too, had been designed by Ricketts, though he had been forced into doing it: 'The wicked Sybil Thorndike announced that I was to stage *Macbeth*, on the last night of *Henry VIII*. This I had told them I would *not* do, but man proposes & woman disposes.'[49] It was a large undertaking, with twenty-one changes of scenery and about seventy people to dress and make up. Referring to the separate activities of designing sets and costumes and then adapting them at rehearsals, Ricketts confessed to Bottomley: 'like all theatre work, it seems to begin again when one thinks it finished, & a vague dislike for circumstances of production & a general staleness has prevented my working with that sureness of touch I like to feel mine'. Still, he felt it contained 'many stunts and imaginative touches which would give world-fame to a German or a Russian', even though the shallowness of the stage made him scrap some first-rate ideas, and the concrete floor, which was without trapdoors, made it impossible for him to have the fighting in the battle scene take place below stage with hurrying lances and fighters appearing over the edge. He had intended Lady Macbeth to rise from the ground and sink back into it after the sleep-walking scene, 'as if she had approached the place up a flight of steps'.[50] His designs were published in the *Observer*[51]. When the time came for rehearsals he was suffering from one of his usual colds, but on one night, after finishing at 3 a.m., he never went to bed and was back again at 11 the next morning. 'Though I have been working hard & efficiently', he told Lady Davis, 'I have not once felt enthusiastic & am convinced that we are in for a failure, particularly as regards myself.'[52] His staleness and tiredness were understandable, as this was the fourth major production he had undertaken since 1924 and the second this year (1926). He took a violent dislike to everything about

[49] CR to M. Davis, 5 Apr. [1926].

[50] CR to G. Bottomley, p.m. 16 Nov. 1926 and p.m. 27 Oct. 1926 = *SP*, p. 368.

[51] *Observer*, 19 Dec. 1926.

[52] CR to M. Davis [Dec. 1926].

the production, 'saving Sybil Thorndike who is very good and wears the beautiful dresses superbly'; he swore it was his last theatrical venture, and even refused to go to the first night on Christmas Eve—Shannon assured him the reception was splendid. The play in fact turned out a huge success, 'so I shall make some money on it which I did not anticipate'.[53] He told Brown, 'I'm stupid with a cold, *Macbeth* rehearsals, & approaching Xmas', as a cure for which 'I intend spending that day in bed with an improper book'.[54] Earlier *Macbeth* designs were published in 1923 in the Player's Edition of Shakespeare, each volume of which was illustrated after the drawings of distinguished theatre designers, and in 1926 more of Ricketts's theatre work appeared in a slighter book, *Shakespeare's Heroines*, to commemorate a series of radio productions of the plays.

With all the theatrical work of recent years Ricketts had not found much time for painting. After *Henry VIII* he had returned to oil-paints, and started several new things before being swept back into the theatre for *The Mikado* and *Macbeth*. Now that these were out of the way he returned to his canvases, and finished another version of *Don Juan*. While modelling the armed head and the bloody child for *Macbeth* he had started—out of his imagination—a bust of Cecil Lewis (Fig. 50). 'I found my fingers quite happy to touch clay again after 14 years. Modelling does not tire or perplex me like paint. I shall probably do some more'.[55]

The punishing work which Ricketts had undertaken over the last four years had left him exhausted and dispirited, and he now faced another of the crises in his life. This was partly over his work: 'My trouble ...' he told Mary Davis, 'is not due to a lack of appreciation in others, nor do I envy the success of the top dogs of the moment ... What has knocked me sideways, for the moment, has been a feeling that the effort to paint was out of proportion to the result, in my estimation at least, & that I was *stale*.... My small upset coincided with one or two jars in my affections, & a sudden realization that time or life are quite definite quantities, like a banking account.'[56] Twice before he had had to face

this sudden sense, like Faust, that only ashes remained of years of effort, that one was suddenly attacked from within and from without by a spiritual

[53] CR to E. Brown [early 1927].

[54] CR to E. Brown [end 1926].

[55] CR to M. Davis [Jan. 1927].

[56] CR to M. Davis (4 p.c.s), p.m. (envelope) 3 Oct. 1926 = *SP*, p. 393.

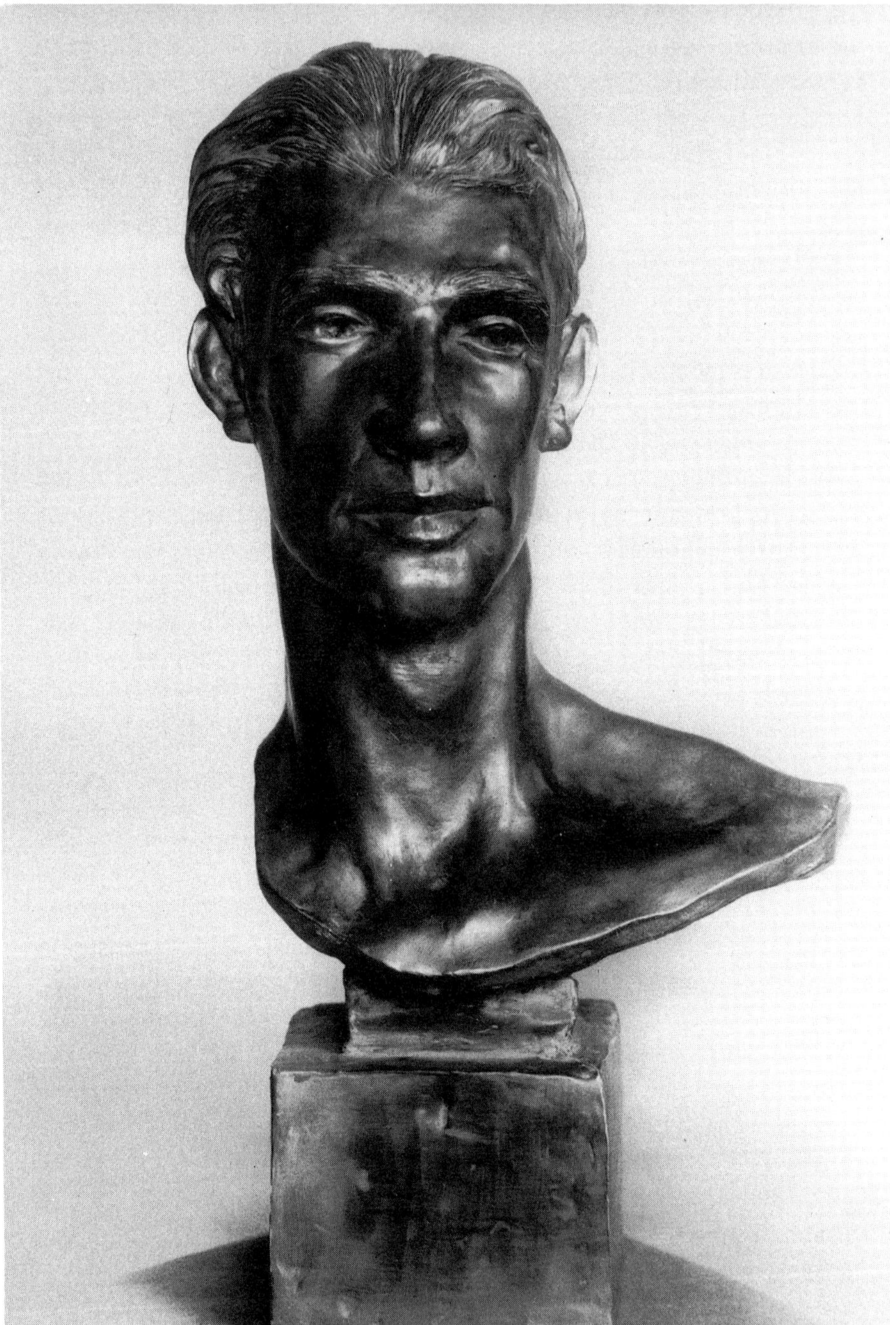

FIG. 50. Plaster bust of Cecil Lewis (1926) by Charles Ricketts (Witt Library)

'Termite' who seemed to have undermined everything. And the fact of one's age seemed to have made this crisis more difficult to cope with than others.[57]

He was now in his sixty-first year: his first major crisis had been at the beginning of his career, in the 1890s, the second, when he was approaching forty; now he was approaching another threshold, the onset of old age. These are the crisis-points in the lives of many men, especially those involved in creative work. For months he battled with this depression, and his trip to Tunisia in April was a vain attempt to escape from it.

There was at first some doubt as to whether the trip would take place at all; he feared 'that at the last moment, some dreary duty may intervene'. There was also the deeply-ingrained reluctance to spend so much money: 'I had intended this "escape" as a sort of new start in life—a sudden break with my ascetic past, but you know prisoners retain the habit of their chains and the money question often raises those imperatives, "Thou shalt" or "Thou shalt not".' However, Ricketts's new affluence, from his Canadian advisership and his theatrical productions, made it not only possible to go himself, but also to take Cecil Lewis with him as his guest. The tardy spring that year had 'put an urge in ones blood to move away from habits', and made him long 'for wild flowers and spaces full of light and clean air and possibly a scent of the sea which I have not been near for years, as it does not suit Shannon, whose tendency is all for solitude and silence to a point when I feel they become dangerous'.[58]

Almost from the beginning the trip went badly. The train journey to Marseilles was exhausting, with Ricketts suffocating from the heat in his sleeping berth, and this and the motion of the train afflicted Lewis 'with some sort of nervous affection' that later made him ill on the boat. Bizerta, the port at which they landed, was odious, and to escape another train journey they hired a car to Tunis. As there was a lawyers' convention in the city, they had to share a large room in the most pretentious and expensive hotel. Sharing rather 'unhinged our tempers', confessed Ricketts, and he snapped at Lewis the next morning for upsetting some coffee. Though afterwards they both tried to behave well, Ricketts felt this incident probably made a difference. Lewis was

[57] CR to A. Cippico [*c.*1927] = *SP*, p. 407.
[58] CR to C. French, p.m. 20 Mar. 1927.

still feeling seedy, so they hired another car—'Shannon would have a fit if he knew', Ricketts confessed to Mary Davis—to Hammamet, a seaside resort, where the hotel was too rough. They were charged a huge amount (sixty-nine francs) for an uneatable meal, and Lewis took a strong dislike to the place: 'We despaired & openly talked of a rapid retreat home, [but] this proved impossible, the boats being full.' While Lewis went off to investigate a suburban hotel which had been recommended, Ricketts explored the site of Carthage, where the museum later proved to be 'very dustbinney'. The hotel, too, turned out to be full, and they finally found rooms in a small, clean Arab village, La Marsa, from which an electric train left half-hourly for Tunis.

Tunis itself was 'a washout'. Ricketts could not understand why all its hotels were full. It was 'neither Monte Carlo nor an Arab city, & the souques filled with oriental & western trash soon wear out in interest. Nor are the people good-looking like the Egyptians, & Cairo compared with Tunis is Venice compared with Margate.' At other times it reminded him of Earl's Court. For much of the time Ricketts lazed, and he and Lewis met at meals and for occasional walks. The weather was perfect. Ricketts was often alone, and spent his mornings on the coast by the sea, reading Flaubert's *Salammbô*, which he accidentally dropped into the water while paddling near Carthage: 'I wonder if Flaubert is annoyed.' Lewis did not wish to bathe:

> He vetoed it absolutely to my intense satisfaction [Ricketts told Shannon] . . . I hope he will be happy here, but I am not sure that he has a happy nature. He is bright & gay when he feels inclined, but crytical when disatisfied which occurs unexpectedly. I fancy he is only happy with women & is rather slower in perception than one had imagined: this is not a complaint. We get on quite well.[59]

What can be read between the lines of Ricketts's letters is a great deal of strain between Ricketts and his young friend, which mainly originated in their differing sexual natures. Even though they had been friends for ten years and had worked out a *modus vivendi*, it had been tempting fate to throw themselves into such intimacy, especially at a time when Ricketts's morale was already low and his need for affection great. Lewis, for his part, obviously felt trapped by such close proximity

[59] CR to M. Davis [Apr. 1927]; CR to CHS, 11 Apr. [1927].

to a dear friend whose unspoken hopes he sensed but could not fulfil.[60] On the boat he had taken a fancy to a young, emancipated Frenchwoman, a journalist, whom he saw again in Tunis. Even before she left for China he had become involved with a cultivated German woman. Though Lewis enjoyed female company, these relationships were also a relief from the awkward intimacy with Ricketts who, as a result, was left on his own most of the time. Lewis may have behaved thoughtlessly, but the young man no doubt found the situation difficult to cope with. Never, then or later, did Ricketts mention to him the real cause of his misery.

'Fine day succeeds fine day and my longing to escape grows greater', Ricketts wrote sadly.[61] His attempts at sketching the sites failed utterly. He began a drawing, in local paints he had picked up at Marseilles, of a costume for Isidore de Lara, presumably for the 'new comic opera' *Prince Marcocana* at Aix-les-Bains, which Ricketts mentioned in a letter to Lady Davis.[62] In emulation of Lewis's attempts to reconstruct a prose poem by Wilde, Ricketts wrote some five prose poems, two reminiscent of Wilde, the others his own. 'To laugh down bitter and haunting thoughts' in his vacant hours, he also began to write his *Dialogues of the Dead*, later published as *Beyond the Threshold*. In it he put many of his own feelings about life, and the last speech by Epicurus (who was himself) was autobiographical and the 'final crystallization of my feelings in this thing, which was written in one of those crises in a man's life'.[63] It was in their hotel at La Marsa that Ricketts later set his moving book *Recollections of Oscar Wilde*—written under the pseudonym Jean Paul Raymond—where he feigns having met himself, Lewis, the German woman and her daughter, and a young Swede. Ricketts worked on articles on Botticelli and on Greek vases, but he could not tell if they were any good: 'My brain is not wholely inside its usual cup.' He read a great deal and found an early book by Proust he did not know, but liked it less than the others. Lewis too was writing, working hard at his play, about which he complained a good deal. 'I try and pour milk, honey, & roses about him, not always with success', Ricketts told Shannon. 'I do not know if he is profoundly bored or rather happy, possibly I shall know one day.' Ricketts himself,

60 C. Lewis to author, 27 May 1980.

61 CR to CHS, Mon. 25 [Apr. 1927].

62 CR to M. Davis [Apr. 1927]. See *SP*, p. 375.

63 CR to A. Cippico [Apr. 1927] = *SP*, p. 407.

who missed Shannon 'dreadfully', was bored to tears and sleeping badly, but he was not sure whether it was the mosquitoes, the lightness of the air, going to bed too early, or eating too much food (which was excellent). 'I hope the trip will seem better in after recollection', he said.[64] They returned on 3 May, with Lewis taking over Shannon's role of organizer. They stopped over in Paris, where he had one of the four nights of good sleep, and one of his few days of happiness, since his departure for Africa, enjoying 'my one morning in the Louvre as I did in my young youth'.[65]

The strains and disappointments of the Tunisian holiday left no permanent effect on the friendship between Ricketts and Lewis. Ricketts still wanted the best for his young friend, though his holiday experience probably strengthened his feeling that the young wasted too much time on amorous adventures. (Earlier, Ricketts had advocated freedom from society's sexual conventions, knowing as he did the sufferings of sexual frustration, but he eventually decided that this led to the young neglecting their work.) Although he complained in letters and conversation about how much he had hated Tunisia, within a few months he was writing in the old way—about how Lewis motored him down to the Keep, how he brightened the place up, and how he had 'struck for a tennis court and got it'.[66] However, on Ricketts himself the holiday had taken its toll. Although he was 'notoriously young & slim' for his age and was often told he had the secret of youth or must have been treated with monkey gland, his beard had suddenly started to turn grey on the trip.[67] His depression, moreover, carried on into the summer and, even though the flowers were first-rate and the Keep 'a mass of yellowish roses', he felt rather low and restless, with his left arm 'sulky with rheumatism'—though not enough to prevent work. He tried to keep himself occupied, but 'circumstances seem to reduce everything to the same level, and I am becoming a convert to the Einstein theory "that time does not exist", probably the most preposterous theory ever evolved by a man of intellect'.[68]

Shannon also was feeling rather depressed, and dissatisfied with his work. As an antidote, he agreed to do some teaching at the Byam

[64] CR to CHS, 14, 23, 25 [Apr. 1927].

[65] CR to Marie Sturge Moore [Apr. 1927].

[66] CR to Muriel Lee Mathews, 19 July 1927.

[67] CR to RNRH, 7 Dec. 1927 = *SP*, p. 390.

[68] CR to Muriel Lee Mathews, 19 July 1927.

Shaw School of Drawing and Painting, where he hoped that the contact with the young would stimulate him. Both he and Ricketts had taught as visitors to the Royal Academy Schools, where members of the Academy took month-long spells instructing students, but neither had much experience in this area. The Director of the School was their friend Ernest Jackson, who was keen to have a distinguished RA as a teacher of painting. Ricketts was delighted to report on Shannon's experiences to his Dutch friend Holst that 'the young at the moment are very promissing, & interested only in the Italians—mainly Piero della Francesca. They draw beautifully, find painting more difficult, & loathe & despise Impressionism & its sequel.'[69] This proved a false hope. Under Jackson's direction the school was a last bastion of sound training in the traditional skills of drawing, but it was not representative of the interests of the rising generation. When Shannon had tickets to private views he did not want to use he gave them to the students, and also invited them to Friday evenings at Townshend House. On one visit by Max Chapman, Richard Finny, and Bernard Pike, Shannon tossed each of them a little pen-and-ink drawing by Alfred Stevens from a portfolio full of them. Ricketts also took a great interest in the students and went as a visiting critic at the end of the year to judge their work and award prizes.

Shannon was invited to take part in the celebration of the Rubens Centenary in Antwerp, where Jackson was going to be decorated by the Belgian Government for his part in organizing the Flemish Exhibition, which Ricketts had found 'magnificent, stupendous, & overwhelming'.[70] So Shannon and Jackson went off to Antwerp together and then on to The Hague, Haarlem, and Amsterdam. Ricketts joined them in Paris, where he found the old Louvre more delightful than ever, but was distressed to notice how its stonework was becoming not only black but friable. 'I wonder if it is due to coal smoke & other ghastly chemical fumes of modern times.' Paris was 'grimey and sordid' and England seemed pleasant after its noise and glare, 'but England is a narcotic & we get cotton wool into our veins'.[71]

'I meant to die ignorant of America and Americans', Ricketts had told Cecil Lewis in 1920. 'I have never forgiven Columbus for his

[69] CR to RNRH, 7 Dec. 1927 — *SP*, p. 390.

[70] CR to M. Davis [Jan. 1927].

[71] CR to S. Cockerell, 5 Sept. 1927; CR to Marie Sturge Moore, p.m., 23 Sept. 1927.

discovery; an Italian should have known better!'[72] No one was more surprised than he to find himself engaged to go to the New World: 'I am leaving for *America* ... the day of myracles is not over.'[73] What he found there also astonished him. His departure was rather sudden, and connected with his role as adviser to the National Gallery of Canada. His purpose was to secure sufficient money from the Canadian Government to buy some of the pictures from the Benson collection, which Duveen had bought and taken to America. The possibilities excited Ricketts, who was inclined to try for the unique Dosso Dossi, or the Piero di Cosimo. All expenses for the trip were covered by the Canadians: the cost was shared between the Ottawa and Toronto Galleries, and Ricketts had £1,000 down to spend and, presumably, the possibility of more later.[74]

He sailed first-class on the Canadian Pacific liner SS *Empress of Scotland* on Saturday, 15 October, suffering from a cold. 'I received a blow to my self esteem at South Hampton where the nice servant from the ship asked me what class I was(!)', he told Shannon ironically.[75] After this things went smoothly though he found life on board tedious. With him were Colin Agnew, whom he liked greatly, and a cynical Frenchman (Count Sala, the head of Agnew's house in Paris), who 'knows everybody, hates America, & rather dislikes our third member, a noisy young American who loves art, has been over Italy with Agnew and who rags me'.[76] All three cared for art, and were well-read and fond of music, 'so our groupe represents civilization'. The young American liked to get a rise out of the others, and intrigued Ricketts by attacking Watts, Rossetti, and the Pre-Raphaelites, whom he found 'just comic', and admitting neither skill nor technical beauty in Millais's *Ophelia*. They all talked art so loudly and constantly that indignant Britons stared at them, rattled their newspapers, and even left the smoking-room with angry glances at their party. Agnew revealed stories about Berenson's crooked dealings with Mrs Isabella Gardner that the shocked Ricketts found 'impossible', but he had heard similar stories before and Agnew swore that 'enough is known in the trade to

[72] CR to C. Lewis, 24 Dec. 1920 = *SP*, p. 326.

[73] CR to S. Cockerell, 13 Oct. 1927 = *SP*, p. 383.

[74] TSM to Sybil Pye, 25 Oct. 1927 (Univ. of London).

[75] CR's letters to CHS (BL Add. MS 58085) are the main source of this account. Selections from the letters are in *SP*, pp. 383–8.

[76] CR to M. Davis, Thur. [20 Oct. 1927].

ruin B.B.' (During Ricketts's absence, incidentally, Berenson visited Shannon, won him over at once by showing a great interest in their work, and reattributed all their early Italian pictures, Shannon taking down the names out of politeness and commenting later to Ricketts, 'I take it these attributions disagree with his previous ones'.[77]) Aboard ship they had plenty of time to discuss the prices of the Benson pictures Ricketts had earmarked. As the journey progressed Ricketts found Agnew kindly and enthusiastic, the Frenchman rather 'a wet blanket' and tyrannical over food, and the American, who had at first got the party going, the least likeable. The food was varied and delicious; 'I have oysters every evening, lobster or crab, hors d'œuvres and light fishey food of the type I like.' At first he avoided wine 'to train for God's own country', which was still under Prohibition, but, pressed by the others, he eventually gave way on this and on smoking. So generous were his friends that the cigarettes Shannon had given him were hardly touched.

After a brief stop at Quebec City, they sailed on to Montreal. Ricketts was taken to see the Montreal Art Gallery and the collection of Sir William Van Horne (who had died shortly before) and found it showed 'considerable judgement along fashionable lines'. Lobbying people over the Benson purchases was one of the purposes of his visit, so he started as he meant to go on by winning over the Chairman of the Montreal Gallery, Dr Shepheard, who, as a Trustee of the Ottawa Gallery had rather got in his way with suggestions that it should purchase poor Courbets and Mauves. Ricketts soon realized he might be kept in Canada longer than he wished 'to meet the Prime Minister whom I mean to nobble'.

At the Ottawa Gallery Ricketts found himself 'immensely impressed by my purchases and several works dating before my time'. The effect of the damaged early *Pietà* by Veronese on a wall between Tintoretto's *Adam and Eve* and his portrait of his servant was overwhelming, and difficult to beat anywhere. However, there was a row with Brown because the latter had toned down the fine old frames with horrible brown paint. To his discomfort Ricketts found the Mabuse (now described as Flemish school) *Man wearing a Collar of the Golden Fleece* that he had recommended was actually very damaged, and the Monets, Pissarros, and Boudins, which he would never have recommended, were all good. On seeing their own works, his *Danaïds* (which had

[77] CHS to CR, 26 Oct. 1927.

been purchased in 1922) and Shannon's portrait of Princess Patricia (purchased in 1919) he was also disappointed, and told Shannon: 'both look laboured & purpleish, yours suffers from a nasty flat bright gold frame—which serves you right'. Though he did not approve of many of the other modern English pictures, 'still we escape from the dinge of English Provincial Galleries'. He was rather homesick and missed Shannon greatly, and when the band in his hotel, the Château Laurier, played one of Schumann's melodies, which he sometimes played at home on the pianola, he had tears in his eyes. He told Shannon, 'had I had an audience, [I] would probably have wept'.

One of the people whom Ricketts had to win over was the Governor-General of Canada, Lord Willingdon, with whom he went to dine one evening. Though not quite sure at first whether Willingdon was Governor of Canada or of Ottawa, he set off in a limousine, 'all ebony & rock crystal'. Surrounded by flunkeys 'like Roman Emperors and beautiful young men covered in orders', he felt, wearing Shannon's dinner-jacket, 'like Robinson Crusoe fresh from his island'. Feeling both trapped and lost, he 'ate nothing and drank less' until he was able 'to talk shop to the King of Canada', whom he found charming and well-bred. A fortuitous point of contact emerged when it turned out that both Willingdon and his wife were descended from Lady Dacre, the subject of the Eworth portrait that Ricketts had purchased for the gallery. They discussed the construction of a new gallery, the character of future trustees, and the possibility of a grant from the Prime Minister for the purchases from Duveen, and Ricketts was gratified to be told constantly 'you are the man to talk round the Prime Minister!' He found himself quite popular in the drawing-room afterwards; 'my torrent of jaw & lecturing faculty was made an open subject of chaff, even the beautiful young men seemed affectionate'. Shannon, of course, was confused with J. J. Shannon, and Ricketts gave up explaining the difference. He was made to promise to design a Chinese decoration for a 'hideous' Victorian banqueting tent for Lord Willingdon. That he should have been expected to do this for nothing again proved a favourite maxim: 'the poor always tip the rich', and it took him three weeks to carry out on his return.[78] He was later reassured that his meeting with the Governor-General had gone well: 'I understand I have done wonders. Curiously enough, I feel more empty headed than bored, but not unhappy as I was in Africa.'

[78] CR to M. Davis, 30 [Oct. 1927]; see also *SP*, p. 386.

His first glimpse of the Prime Minister, The Rt. Hon William Mackenzie King, had not been promising: 'he has a beastly face like a convict & I feel we shall dislike each other at sight since he is shy & looks vulgar & colonial'. His anxieties were somewhat assuaged when he learnt that King had made the elaborate carillon bells on Parliament House play good music—things which Ricketts himself played on the pianola. While waiting to visit him, Ricketts made an excursion by car into the beautiful autumnal forest, and, with Brown, whom he liked less and less each day, he went off to Toronto to see the gallery and meet its Director Mr Wood, 'a bullet headed man, exceedingly frank and simple'. The gallery also considered retaining Ricketts to act as its adviser. Wood showed him a Titian which was 'a wreck, flayed to extinction, not repainted but reglazed & gormed'. It had been purchased from Duveen, whom Ricketts felt had 'robbed' the gallery, and not only did he encourage them to return it but he also offered to press Duveen to exchange it when he saw him over the Benson collection. He found 'these half Americans easy to get on with & [I] do not feel the element of hesitation & suspicion I find in Englishmen with their various glazes, scumblings & patinas'. Having met the four Toronto Trustees, he felt his advisership there was safe, but he was rather put out when Wood hated a Titian because a man in the painting had a bushy beard, or the Eworth because the subject looked like a mother-in-law. 'I fear people hate Art, but my personal success is quite astonishing—so far', he told Shannon. Wherever he went people had heard of him and he was asked to sign books, and even some de Tabley illustrations. 'Unless I am mistaking American welcome for real interest, I seem to be viewed as a prodigy to be got hold of, & Brown openly speaks, so do the others, of my returning here.'

Finally, on Friday, 11 November, after returning to Ottawa and hanging about for several days meeting Trustees and advising on more pictures from private and public collections, he met the Prime Minister. He had had to spend three weeks in Canada instead of the nine days he had planned, and had grown 'sick of Hotels, sick of B[rown], sick of Trustees, sick of pictures', and even refused to see the prints and drawings at the gallery. He went to dine alone with Mackenzie King with the knowledge that he had only an hour before the Premier spoke at the Cenotaph. He also felt that the Trustees and Lord Willingdon—who 'had promised everything and, being English, did nothing'—had let him down. They were frightened of the Premier and had rushed through a grant of thousands for a new site for the gallery the day before Ricketts met the Prime Minister. This made Ricketts's appeal for money less likely to succeed. He discovered that the Premier had not even heard of the Benson Collection. Ricketts found himself possessed by 'the calm of despair or of renunciation. I was prepared to let things slide ... and I think I never have shown greater self possession, greater felicity of phrase, or a sweeter touch. I was able to hear my own voice'. He set out to fascinate. After a great deal of tactful fencing, inspired jawing, sympathetic listening, skilful leading, and effective 'stunts', he saw he had done the trick. An emergency grant of £10,000 was offered toward the Benson pictures.

Ricketts followed up this victory by a 'most charming' letter to King—'with luck [I] shall try & make the Benson purchase, if it happens, a personal affair of his'—and another to Lord Willingdon, urging him to be sympathetic, but he was not sure if the latter was a good move. The next day, under the first fall of snow, he set off for New York, which lived up to its reputation for luxury and vulgarity, but which he liked, much to his astonishment. He liked Fifth Avenue immensely, impressed by its wealth and tasteful luxury, its well-dressed women, and its well-managed traffic. He liked the bright lights and polyglot crowd of Broadway. He liked some of the skyscrapers, which at night looked like 'Babylonian buildings under an Eastern sky full of stars', and the large houses in clean, white stone, though he felt there was a lack of trees and monuments. He even liked 'better than I care to admit' the Manets in the Metropolitan, where the classical collection was magnificent and the paintings full of interest and old favourites. He was greatly cheered to hear that the museum was contemplating a show of his books. The Frick Collection, then still in private hands, overwhelmed him; the Winthrop Collection, which contained twelve Ricketts drawings, including the second Sphinx set, and several Shannons along with 'amazing' masterpieces by Ingres, Rossetti, Burne-Jones, Blake, Daumier, and many others, impressed him hugely: 'I could never have believed America could boast of an assemblage of art works like this one', he told Birnbaum.[79] In Boston too, where he went for a week, he found the public and private collections showed a level of taste in purchasing and arranging objects that was not only much higher than expected, but also showed that Americans were 'going ahead of us'. At the house of Mrs Gardner, 'I

[79] Birnbaum, p. 179; CR to Mary Davis, Sun. 20 [Nov. 1927].

trembled before the Giorgione Christ *unmistakably by him*', Ricketts told Shannon: 'the collection, though dingey as to lighting & condition, contains things which cannot be found again and of superb quality'. As he was running out of time, he had to cancel a proposed trip to Philadelphia to see the Widener Collection.

All this was fascinating, but Ricketts was feeling he had seen too many pictures, and the interview with Duveen, 'an amazing creature, a sort of Balzac personage', had gone badly. When Ricketts approached him over the Toronto Titian he refused to exchange it; he considered 'his opinion as good or better than mine', wrote a rather deflated Ricketts. Moreover, they were too late for the Benson pictures, and the Piero di Cosimo he had decided upon had been sold. There was nothing else he wanted in the collection that was available at possible prices, so he tried to divert the Benson Fund towards the purchase of a *Magdalen* by Veronese, 'with what success I cannot tell', as the fund had been earmarked for a specific purpose. With the apparent failure of his mission in New York Ricketts felt the whole trip had been in vain and that he had come to America for nothing: 'I cant say how angry & depressed I feel at times', he admitted to Shannon. However, the Veronese was purchased, and became one of the Canadian gallery's great pictures, and Duveen did exchange the Titian. Ricketts's visit was a story Duveen told against himself every time the English artist's name was mentioned.

After some difficulty with his clearance papers—'bother over passport business is all I have to complain about America'—he set sail from Boston on the *Laconia* for England on 27 November. In spite of the Benson disappointment the trip had not really been a disaster, like the Tunisian holiday. Not only had he enjoyed, and been stimulated by, the collections he had seen, but he also found the respect with which he had been treated almost everywhere and the success of his lobbying of Canadian officials a great tonic. After the vitality and wish to do well of the Canadians and Americans, the English seemed full of indifference and apathy and England still 'plunged in the squalor & stupor left by the war'.[80] Among the European countries he had visited, only Italy under Mussolini showed at the time the same wish to advance.

Early in 1928 came Ricketts's election as a full RA, which he greeted with rather jaded feelings: 'this follows automatically like a coffin after

[80] CR to RNRH, 7 Dec. 1927 = *SP*, p. 390: CR to M. Davis, Sun 20 [Nov. 1927].

death', he told French. 'Shannon's and Philpot's sales fell after their election, so have mine—one picture to an American in the last four years—*Glissons*.'[81] Life in England was too much the same as ever. With all his travelling now taking up his time, rather than the theatre, he had painted only one picture (which he did not like) in the last year. Now the theatre was again to demand his attention.

There was a plan to stage Masefield's *Coming of Christ* on the steps of Canterbury Cathedral. Gustav Holst was to supply the music, and Ricketts to design the costumes and scenery.

> Years ago, the idea would have thrilled one to the very toe-nails [he told Cippico], today—such is the effect of years—I do not feel very enthusiastic. It probably means a great deal of work, for nothing—by that I do not mean money, but the curious 'lack-lustre' of the English mind and its lack of appreciation... In any other country, Masefield, Holst, and Ricketts in combination in a cathedral would create a sensation and be reviewed—were it in Germany—by *The Times*'s correspondent.[82]

As well as designing the whole production, he made with his own hands many of the properties, including a doll of the Holy Child which he created 'without the assistance of a Virgin—[Bruce] Winston calls it Ricketts's Immaculate Conception'.[83] Having made the doll of Elizabeth in *Henry VIII*, the bloody child in *Macbeth*, and now the Holy Child in this play, he boasted: 'this is my third distinguished infant in the last two years, which is a good record for a bachelor.'[84] The performances were a huge success, realizing £500 in chance contributions, and initiating the series of Canterbury Festival plays which included Eliot's *Murder in the Cathedral* (1935). Photographs were featured in the *Illustrated London News* and the *Graphic*. When Ricketts was asked to sign a musical manuscript of the production, he wrote this aphorism: 'Art sees, music knows, literature remembers.'[85]

This he regarded as one of the best of the aphorisms he had been writing recently. In the course of a conversation with Cecil Lewis, while retouching a picture upside-down, he had quoted '*Les amoureux se recherchent à faire une chose très ordinaire avec beaucoup de mystère*' as a maxim by La Rochefoucauld. When searching for this in the relevant

[81] CR to C. French, 23 Mar. 1928 = *SP*, p. 395.
[82] CR to A. Cippico [spring 1928] = *SP*, p. 394.
[83] CR to C. Lewis, 28 May 1928 = *SP*, p. 397.
[84] CR to C. French, 23 Mar. 1928 = *SP*, pp. 395–6.
[85] CR to C. Lewis, 14 June 1926 = *SP*, p. 400.

volume, he could not find it and concluded it must be original, which amused him vastly and prompted him to write 300 more 'almost automatically in what I am told is classical French', within four days.[86] They summed up his philosophy of life. Cecil Lewis gives an amusing picture: 'While ... painting with his left hand (for he was ambidextrous), Ricketts found himself almost automatically, as he put it, scribbling aphorisms in French ... with his right.'[87] Among these there are some which sound autobiographical: *Dans l'amour qu'une mère porte à son fils, il entre tout ce qu'elle n'a pas pu donner à son mari.* These aphorisms show Ricketts's wisdom and acute perception of life: *Illusion a son prix, elle nous empêche de reconnaître toute la verité: la verité est sans pitié* and *Tous les philosophes voudraient mettre l'univers dans une coquille.* It concludes with a teasing *envoi*: *Mon meilleur ami n'a pas ri d'une de mes maximes; je l'aime déjà moins.* Ricketts hoped the collection, *Maximes en Marge*, said to have been translated into French from English by 'Jean Paul Raymond', would be published, and even left instructions to that effect among his papers, but they remain in manuscript.[88]

Ricketts also carried on writing his dialogues, which, he announced, were intended for men only.

> In the Roman banquet it contains, Nero, Catullus, Lucretius, speak, and Petronius reads a missing portion of the *Satyricon*, which is quite as improper as anything in that admirable classic. Oscar Wilde crosses the stage & recites 3 prose poems, one genuine, one part genuine, the third a fiction of my corrupt imagination. I enjoyed writing it immensely.

Designing a frontispiece to Humbert Wolfe's *Troy* in the Ariel Poems series prompted Ricketts to take up illustration again and 'return to the manner of my young youth'.[89] *Beyond the Threshold*, with five illustrations, and a superb abstract binding also designed by him, was published through A. J. A Symons by the First Editions Club in a limited edition of 150 copies in 1929 under the pseudonym 'Jean Paul Raymond'.

Reading, as much as writing, helped him through his depression in 1927. This was as varied as ever, with the emphasis on French and

[86] CR to G. Bottomley, p.m. 20 Mar. 1929.
[87] *SP*, p. 380.
[88] BL Add. MSS 58092 and 61724.
[89] CR to G. Bottomley, p.m. 19 Oct. 1928.

British writers, though he branched out into Latin, German, and American literature. His taste was his own, never merely fashionable, and he regularly returned to the classics of European literature as well as reading the latest books. Over the past two years, for instance, he had read Aldous Huxley, Gide, Proust, ('curiously interesting and almost intolerable'), Colette, and Paul Morand; Thomas Mann's *The Magic Mountain* and Emil Ludwig's biographies, *Goethe*, *Napoleon*, *Wilhelm II*, and *Bismarck*; E. M. Forster's *A Passage to India* ('a very notable book indeed'), Vita Sackville-West's *The Land* ('full of charming things and a great knowledge of the subject'), *Pipers and a Dancer* by Stella Benson—a favourite writer—Compton Mackenzie's *Vestal Fires*, Virginia Woolf's *Jacob's Room*, which disappointed him, Rosamond Lehmann's *Dusty Answer*, and Gerhardi's *Jazz and Jasper*, which made him 'shriek with laughter'. The novels of Sinclair Lewis he thought 'worth reading'. On the trip to America he had read Suetonius ('very dull and common') and Montaigne ('slower and more archaic than I remembered') and Norman Douglas's *South Wind*, which he found very amusing but 'possibly overdone'.[90]

Two other books he read at this time show his open-mindedness. Both he and Shannon were impressed by *Lady Chatterley's Lover*, then published privately, which he considered one of the best things Lawrence had done, though he disliked the tiresome reiterations, the hypnotic obsessions with certain words and ideas, and the crude sexual terms, which, 'too often associated with lewd stories, hurt & disconcert one even in scenes of sexual eroticism'.[91] He also later promised to testify for Lawrence, thirteen of whose pictures had been seized by the police at an exhibition in July 1929, for indecency. No record remains of his opinion of Lawrence's paintings, but the novelist's victimization by the philistines was enough to attach Ricketts to his cause. He feared that he would have to testify before Judges 'Darling, Avory, or Hewart, all of whom belong, in my opinion, to the sub-criminal classes'.[92] In the end he was not among those, including Orpen, Augustus John, Philpot, and Rothenstein, whose testimony for the

[90] Proust: CR to TSM [*c*.1924]; Forster: CR to M. Davis [Dec. 1927]; Sackville-West: CR to M. Davis [July 1917]; Gerhardi: CR to Cecil Lewis, 10 June 1928; Sinclair Lewis: CR to M. Davis [Apr.–May 1927]; Suetonius and Montaigne: CR to CHS [16 Oct. 1927]; Douglas: CR to CHS, Mon., [31 Oct. 1927].

[91] CR to G. Bottomley, p.m. 19 Oct. 1928 = *SP*, p. 405.

[92] CR to M. Davis [June 1930].

defence the magistrate refused to hear. The other book was the lesbian novel *The Well of Loneliness*, which Ricketts thought 'too sincere ... and so cautiously done that you could give it to a girl of sixteen'.[93] When the book was banned on the basis of 'one hysterical denunciation', and 'without the usual formalities or defence allowed by law', Ricketts took up his pen to write to the *Observer* (4 October 1928), where some months earlier (29 January 1928) he had published a letter prompted by the Lord Chamberlain's refusal to license a performance of *L'Amour Masqué*, by Guitry and Messager.[94] This censorship, he pointed out, had to its past 'credit' the suppression of such fine and serious works as *Oedipus Rex*, *Mrs Warren's Profession*, and Pirandello's *Seven Characters*, and, whilst in England 'we are accustomed to this sort of muddling, with its insincerity, clumsiness, and lack of right reason', such decisions were greeted abroad with 'derision and irritation'.[95] To Ricketts censorship of this kind only showed once again British 'hypocrisy and contempt for art'. He told Mary Davis that he had joined 'the Sunday Society for the production of improper plays'.[96]

In June 1928 Ricketts was off yet again on another trip. He went with Brown and his wife to Florence as the guest of the Canadian National Gallery, to look for possible purchases for the gallery. He enjoyed himself hugely in spite of frequent irritation with Brown, but the trip appears to have resulted in no purchases. Ricketts's relationship with the Canadian Gallery was still fraught with squabbles about money. Payment of his fees was far in arrears, though partly because he had asked for the 1925 payment to be paid (for tax reasons) in two parts; a decision he much regretted. A recurrent difficulty was over whether works purchased had been discovered solely by Ricketts or jointly. Brown usually came to England in May, when, Ricketts argued, most new pictures became available, and so they went to see pictures together. This was the case with Titian's *Daniel Barbaro*, which the gallery bought for £10,000 in 1928. Ricketts had seen it first and went with Brown to see it again the same afternoon and on a subsequent occasion, when Brown did the bargaining. Brown, however, claimed

[93] CR to G. Bottomley, 19 Oct. 1928 = *SP*, p. 405 (where it is attributed mistakenly to L. Binyon).

[94] CR to Editor, *Observer* (29 Jan. 1928) = *SP*, pp. 405–6.

[95] CR to Editor, *Observer* (4 Oct. 1928) = *SP*, p. 396.

[96] CR to M. Davis [Feb. 1928].

it could more fairly be regarded as a joint purchase, especially as he had gone round to see the picture before Ricketts had, but had not been able to examine it because the owners were at lunch. In such a situation, he maintained, even if Ricketts had actually been the first to see it, they were working together. So Brown insisted on 2.5 per cent: otherwise it was certain that the whole arrangement would be 'seriously, if not inextricably imperilled'.[97] One of his most frequent claims was that some of the Trustees would be only too delighted to make difficulties or even end the arrangement, which as far as possible he played down with them. Evidently rather unsure of his own judgement, Brown had to steer a perilous course between a rather petulant, grasping Ricketts and a committee of worthy men who were largely ignorant in art matters and did not see the need to waste money on an adviser, especially one as difficult and demanding as Ricketts. Though Ricketts did not get his way over the Titian, he did manage to keep them buying portraits, by Gainsborough, Cariani, and Moro again, as well as the Titian. He also recommended *The Wreckers* by Morland and Rubens's *Christ Carrying the Cross*, which were bought.

At the end of 1927 the Toronto Gallery completed its agreement with Ricketts. A painting by Paris Bordone was purchased on his advice for £12,000, as well as two sculptures by Meštrovíč. When he recommended paintings by Aert de Gelder, Ruysdael, and the School of Titian, however, the gallery found that it had no further funds for such purchases, and after little more than a year brought the arrangement to an end. Though he feared his Ottawa advisership might also be discontinued, he was still prepared to stand his ground on the question of his payments.

Towards the end of 1928 there was a lot of disruption in Townshend House. There was a major exhibition of Shannon's work at Barbizon House, and paintings and drawings from all periods had been borrowed and assembled for the show. This was his first retrospective, and his first exhibition for more than a decade. Shannon's native county of Lincolnshire also wanted to honour him, so it was arranged for the exhibition to move on to Lincoln. On the whole it was a success, both in the Press and with the public. Shannon's health was still a source of anxiety to Ricketts who, in August 1924, had told Bottomley that Shannon 'has grown more than ever dilatory & vague, & I believe

[97] E. Brown to CR, 25 Oct. 1928.

lives in a mist through which he dimly perceive[s] time & persons'.[98] Ironically, these worries were to be confirmed by the tragic accident that was the indirect result of these belated honours.

[98] CR to G. Bottomley, p.m., 31 Aug. 1924.

13 *'I'm tired of Art'*

EARLY in 1929, on about 12 January, Shannon was hanging some of the pictures which had returned from his Lincoln exhibition. While standing on a ladder on the principal staircase of Townshend House to hang *Myriam*, he lost his balance, struck his temple against the wall, and fell, hitting his head again on the marble flags. Nicholls, their servant, also fell, but, his fall being broken by a hired man, he escaped serious injury. Shannon, however, did not.

For nine days he was unconscious, hovering between life and death. Twice he seemed on the point of dying. At two o'clock one morning his doctor Sir Farquhar Buzzard, a distinguished neurologist and later physician-in-ordinary to the king, told Ricketts to be prepared for the worst; but Shannon survived. The doctors diagnosed severe concussion, but their examination showed no sign of his having had a stroke. There was no paralysis; nor were there any broken bones. A lumbar puncture to relieve pressure on the brain seemed to exert a favourable influence. Though the doctors could find no sign of haemorrhage or clot in the brain, they felt that some lesion or bruising might have escaped detection. What seems to have happened, in fact, is that the front part of the brain was extensively bruised, resulting in 'softening of the brain'.[1] In addition to Buzzard, there were three male nurses for twenty-four-hour care; and, of course, the servant, Nicholls. Ricketts did not himself try to nurse Shannon, but he was so upset and preoccupied after the accident that he did not take off his clothes for four days, snatching hours of sleep on chairs and couches in his studio, where Shannon was taken after the fall because it was on the ground floor. This was converted into his bedroom, while Ricketts, after nine days, moved to another room. The crisis threw him into a state of nervous anxiety and frenzy, and he got hardly any sleep until the doctor gave him a sleeping draught, after which he slept for twelve hours.

The considered opinion of the doctors was that there was no reason

[1] I am indebted to Dr Myles Gleeson White for medical guidance here. CR to G. Bottomley, Sat. [28 Jan. 1929]; CR to TSM [Jan. 1929].

why Shannon should not get better. After he recovered semi-consciousness, his pulse, temperature, and all physical functions (apart from bed-wetting) were normal, but he seemed to live in a remote world of his own. Most of his rambling comments were about art, sending pictures away, and so on, and incomprehensible even to Ricketts; but he ate and slept well, sometimes responded, and recognized Ricketts. He seemed like a child of two or three. It was a great consolation that he seemed to be neither unhappy nor in pain. Ricketts later commented to Sturge Moore that Shannon's behaviour indicated what a good-natured child he must have been.[2] The doctors advised a policy of calm and silence, so as not to excite or stimulate him, so the Rossetti and Burne-Jones pictures in Townshend House were taken down or covered up. By the end of the month he knew his own name at times and understood questions.

Over the weeks Shannon's condition seemed to show signs of progress, so that a new policy of gentle stimulus was recommended. Ricketts himself thought he saw signs of change but, curiously enough, as things improved (he told Bottomley) 'I find the situation more difficult. I am in fact tired out & make a point of sleeping enormously.'[3] As it was, he often 'turned night into day' and got up at eleven or twelve in the morning. There was 'an epidemic of small bothers, which, I suppose, have helpt to keep me alive', he told Holst.[4] During a cold period the pipes froze, and at the first hint of spring early in February they burst, pouring water through the ceilings. Nicholls, who was not well, was also a source of worry, and one of the nurses went down with flu. Ricketts also dreaded the letters of sympathy and enquiry, which added to the pain and sadness and, as they had to be answered, to the work. Daily reports were instituted, on the telephone at about noon. He felt very keenly the loss of Shannon's companionship, 'the lack of advice and the kind of protection one meets with when one is not entirely alone, as I am at present', he told Cockerell; 'it is strange how one can hesitate over quite common things & I have had my fill of every kind of anxiety & worry & even trivial annoyances'.[5] There were times he dreaded going into Shannon's room. The strain told. Nerves and fatigue caused outbursts of temper or irritation. At one

[2] TSM to M. Davis, 27 Oct. 1933 (Univ. of London Box 19/26 (ii)).
[3] CR to G. Bottomley, p.m. 3 Apr. 1929.
[4] CR to RNRH, 12 Mar. 1929 = *SP*, p. 412.
[5] CR to S. Cockerell, 6 Apr. 1929.

dark moment he gave way to his fears and frustrations and accused the doctor, who continued to be optimistic, of lying.[6] This may be why another expert, Dr Gordon Holmes, replaced Buzzard after a few months. Ricketts grew very cynical about doctors, especially the knighted and the 'big, i.e. expensive', ones.[7]

He was advised to send Shannon to a nursing home for a fortnight or longer. Though he had doubts about this plan, he recognized it as necessary, to give Nicholls and his wife, who had behaved admirably through this difficult period, a paid holiday, and to allow for repairs, redecoration, and readjustments to the house. Shannon duly went off to a Catholic nursing home nearby, and the Nichollses to Margate. Ricketts made twice-daily visits to the home, which he found dingy, gloomy, expensive, and not up to its reputation. Though Shannon's mutterings remained inscrutable, he gradually showed a tendency to respond and remember better. He could shave himself, and could now look at reproductions of Rossetti and Burne-Jones with interest and even criticism. Even when alone he sometimes looked at them. However, he did not always recognize his own work, and is said to have observed, when shown one of his paintings, 'I used to paint things like that once'; but he also told the doctor, 'I painted that a year ago. Ricketts retouched the curtain', which was true.[8] His refusal to eat as much as before, and a vague dislike of his surroundings, showed signs of a returning personality, but these merely added to Ricketts's strain.

Ricketts continued to fluctuate between hope and despair. Early in June he told Bottomley that he was just recovering from a low period:

> Nothing in Shannon's condition justified the fact, but the months of monotonous anxiety & isolation have had their effect. For about a week, I have insisted on isolation. I have seen Shannon less. I have even tried to work, but I am lacking in resilliance and response. Briefly, I am below par.

While Shannon seemed to be making slow progress, or so the doctors assured him, Ricketts found the situation even more difficult: 'I find his physical fitness and perfectly normal manner and appearance, contrasted with his mental state, more distressing than when it was associated with weakness and obvious illness.'[9] The isolation he suffered

[6] *Yeats and Moore Correspondence*, p. 143.
[7] CR to TSM [Apr.–May 1929].
[8] CR to S. Cockerell, 15 Feb. 1929.
[9] CR to G. Bottomley, p.m. 11 June 1929.

during these months was largely of his own choosing. He refused to see people and discouraged visitors. No doubt many concerned friends wanted to drop in, as did their neighbour Henry Winslow, an American who bought Ricketts's work and for whose house in St John's Wood he had done the interior design. It was probably at this period that the son of Ricketts's cousin Florence de Sinner called on him for the first time. Ricketts spoke briefly through a crack in the door without inviting him in.[10] Cecil Lewis stayed with him during part of the time Shannon was at the nursing home, and Ricketts commented: 'I shall miss his presence in the house. He was a change from the constant contact with persons who are paid, with whom normal or real contact seems impossible, & from the sympathetic callers who have to be told details of the case.'[11]

Lewis was on his way to Italy to stay on Lago Maggiore, to do some writing. On his return in June he talked of this then unspoilt corner of Italy, of how beautiful it was, and of how he had doubled his output and vitality there. Next day Ricketts 'trotted' into the room with a cheque for £350 in his hand. 'I am an old man', he said. 'I shall only buy more Art. I'm tired of Art, no—really—from now on I intend to live only for food.' The money was for Lewis to buy a patch of ground on the eastern, and then unfashionable, side of the Lake, and fifty pounds was included to cover lawyer's fees, deeds, and other expenses. When Lewis protested Ricketts replied, 'My dear boy, don't be foolish. Take it, take it! After all, don't let's be sentimental; but it will be something to remember me by.'[12] At first there was a plan to build a villa jointly. Ricketts and Lewis began to make designs for this, and 'knocked up a first rate Italian villa', harmonizing with the surroundings, both buildings and scenery. This ground-plan was sent to Edmund Davis, who thought it would cost about £1,000. Ricketts told Mary Davis, 'We could fight the ownership out later, or he could buy me out over a film [*St Joan*] we may possibly do together. Anyway the idea is pleasing us hugely for the moment & sufficient for the day is the pleasure thereof.'[13] This villa too was never built, but Lewis was able to buy a half-mile-long strip of wild land between cliff and lake near the tiny village of Arolo, where over the next few years he built

[10] Dr Vincent von Sinner to author.

[11] CR to M. Davis [May 1929].

[12] *SP*, pp. xi–xii.

[13] CR to M. Davis, 28 Aug. [1929].

a modest house. 'I shall be answerable for trees and bushes later—in fact, for anything belonging to the ground', Ricketts wrote to Lewis.[14]

The money for the gift to Lewis had come from the sale of four Persian paintings, including two of those bought in 1900, which Binyon valued at £1,000 apiece. To cover the enormous costs of Shannon's illness, with bills from doctors, nurses, and nursing home, Ricketts had decided to sell some part of his collection. For this he turned to their friend, the New York dealer Martin Birnbaum, who in happier days had organized their pre-war American show and later sold some Ricketts costume designs for *Macbeth*, the second series of *Sphinx* drawings, and those for *Poems in Prose* to American collectors. 'I want a fancy price', Ricketts told him, 'as I am spending money at ... the rate of 4 times more per week than in normal times.' This made it necessary to choose items that would fetch an 'exceptional price', and to seek American buyers, who could pay more.[15] As well as the Persian drawings sold to Philip Hofer of the Fogg Museum, Boston, he sold their three Blake drawings (£1,500), and two Alfred Stevens bronzes (£150) as well as Stevens's superb portrait of Mr Collmann (£2,100), to Grenville Winthrop, whose collection he had so admired on his visit to America, and a Persian manuscript.[16] This money was to build up a fund against future contingencies, but if necessary he was also prepared to sell the bust of Antinous, which had been carved for the Emperor Hadrian and found at Tivoli; Ricketts had been lucky enough to secure it at the Lansdowne sale early in 1930 when he left a bid of less than £600 on the 'odd chance' it might go cheaply. This, their last major acquisition, bought at a time when he was thinking more of what to sell than what to buy, fortunately remained in their collection. He knew Binyon of the British Museum would regret the loss of the drawings from their oriental collection, 'but there is such a thing as *force majeure*', he told Cockerell who, as Director of the Fitzwilliam, also had reason to regret losses to the collection; Ricketts went on: 'Please dont mourn loudly or lecture me, our Lansdownes, Holfords, Bensons, Yeates-Thompsons, sell without demur, when they are rich & I am poor & would not sell unless obliged.'[17]

During these sad and trying months Ricketts kept himself as busy

[14] CR to C. Lewis [1929] = *SP*, p. 416.
[15] CR to Martin Birnbaum [late 1927].
[16] Birnbaum, pp. 51, 199.
[17] CR to S. Cockerell, 3 Feb. 1929 = *SP*, p. 410 (in part).

as possible, to take his mind off the tragic situation. While Shannon had been prostrate and unconscious he had collected up scraps and fragments of designs from his early days and arranged them in a series of scrap-books; two of book designs for the Vale Press and earlier, one of costume designs, and one of jewellery designs.

Even before Shannon's accident he had commenced work on an opera by Donald Tovey to be produced semi-professionally in Edinburgh. *The Bride of Dionysus* was being financed by R. C. Trevelyan, the librettist, a classical scholar and poet and close friend of Sturge Moore. Ricketts had known him since the days of the Literary Society and Trevelyan almost certainly approached Ricketts over the designs for the production. The designs were actually planned 'before the crash', and Ricketts thought them 'good in themselves', but he found working at night 'a great solace' during the first weeks of Shannon's illness.[18] With this added strain, the preparations for the production led to inevitable difficulties. The stage-management was in the hands of C. E. Hedmondt, 'a genial old lunatic, an ex-Wagnerian singer', Ricketts told Bottomley.[19] Like many of his predecessors he eventually resigned, but his row was with Tovey, not Ricketts, though Ricketts believed he hated his designs as well as Tovey's music. Ricketts also clashed with the composer: 'Tovey has made me lose all interest in it. He is slow witted & selfish', he told Sturge Moore. He expanded on this to Bottomley: 'Tovey and I are temperamentally antagonistic. He writes five-page letters which do not answer my questions. He is slow and I imagine pedantic and confesses he does not visualize things.'[20] Ricketts adopted Tovey's suggestions, but though the stage sets were magnificent, and authentic in detail, one of them was completely inappropriate. On the panoramic backcloth Ricketts had painted a huge rock, which did not move as the action took the characters towards Crete, where they landed at the end of the act. An infuriated Tovey maintained that Ricketts had never read the libretto properly, a distinct possibility under the circumstances. In spite of the fact that Ricketts had been instructed to dress an elderly—rather than the usual young—Dionysus (so that his white beard, flowing locks and voluminous clothes made him look like Father Christmas), the cos-

[18] CR to G. Bottomley, 10 Feb. 1929.

[19] CR to G. Bottomley, 10 Feb. 1929. See Mary Grierson, *Donald Francis Tovey* (1952), 240–6.

[20] CR to TSM [Apr.–May 1929]; CR to G. Bottomley, 20 Mar. 1929.

tumes were on the whole 'beautiful in design and gorgeous in colour and material'. Ricketts never saw the production in Edinburgh, but he told Bottomley that

> some newspaper pictures shocked me so deeply that I kept no trace in sketch book of drawing of the production. Johnstone, my scene-painter, says the ship, cave, & labyrinth showed traces of me; over the actors & dresses he remains silent and embarrassed. I feel sorry for Trevelyan who, throughout, behaved splendidly.[21]

Not long after this was off his hands he was approached by the D'Oyly Carte Company to stage *The Gondoliers*; 'not greatly in my line, but this kind of work will distract & occupy me for a while'. The Savoy Theatre was entirely redecorated 'in modern style—all silver, looking-glass, & gold' for the production, which Ricketts felt would be liked.[22] He designed a 'Tiepolo ceiling' for it, thirty-two feet by twenty, for the Palace of the Gondoliers. 'So far it looks more like Delacroix cum Ricketts than Tiepolo. It contains Apollo in his chariot, Comedy, the muse of Painting, Music, Fame, Time & about 20 Cupids all showing their bottoms at once.' As well as this elaborate metaphorical backdrop, he produced seventy costumes from eighteenth-century Venice. While he was hard at work on this production, he confessed to Binyon:

> I am a little driven & my memory & other faculties have not [been] improved by the solitude and total isolation in which I have spent these tireing & anxious months ... With age & worry, all leisure seems to have gone out of life. I think in wonder of one's young youth, when one reclined for hours upon grass or went for long walks. I fancy the days must have been longer by several hours.[23]

Despite such dark thoughts and poignant memories, the designs for *The Gondoliers*, some of which were illustrated in the *Graphic* (20 September 1929), were gay and lively and showed no trace of them. However, he was working too hard, and wrote ominously to Lewis: 'The rate I am working at would be a strain on a young man.'[24]

This was not all he was doing. As well as the theatre, he was deeply involved in exhibitions at the Royal Academy. Early in the year he was rehanging the Diploma Gallery, where the diploma works of generation of RAs were still exhibited. 'Unless I am mistaken, it is a

[21] CR to G. Bottomley, p.m. 11 June 1929.

[22] CR to M. Davis [summer 1929] and [*c.*18 Oct. 1929].

[23] CR to L. Binyon [summer 1929].

[24] CR to C. Lewis [1929] = *SP*, p. 416.

transformation, & for the first time in the history of the Gallery, one is able to notice some good things.'[25] He was also elected to serve on the RA Finance Committee 'on the English theory that if a man is good on several things he is good on all'.[26] An even greater job was the hanging of the 1929 Summer Exhibition. He told Mary Davis that despite the opposition he met with, his 'pressure and influence had included 900 more works in the new black & white room & some 50 extra pieces of sculpture for which there was no room in the sculpture gallery'. He also secured important places for outsiders (i.e. non-RAs), which, he thought, was without precedent in the RA's history. Yet an RA student is said to have spread a rumour that Ricketts had been responsible for rejecting many works. Recognition for his hard work came from the RA's President, Llewellyn, who proposed Ricketts's health at the official dinner when the selecting and hanging committees met the Council, another action 'which, I believe, is without precedent', Ricketts noted with satisfaction.[27]

In addition to all this hard work, he threw himself, after the period of silence and isolation, into more feverish activity. He found he got used to this new life: 'I try & interest myself in the occupations of the moment. I am saturated and almost druged in music, concerts, pianola, gramophone, & I read a great deal', he told Holst in 1930.[28] It was while Lewis was staying with him, in May and June 1929, that he started going out to theatres again, and Lewis twice motored him down to Chilham, which he had thought of giving up because of the expense. One of his greatest pleasures was hearing Béla Bartŏk on the piano: 'I have never heard the piano sing before as he made it sing; in fact, I felt I had never heard the piano played before, just as I felt, with Rosing, that I had never before liked a tenor voice.' Chopin particularly absorbed him at this time: 'No painter and no poet has had the same hold upon me at any time of my life.'[29]

Shannon still seemed to be showing marked signs of improvement. He went for walks, insisted on exploring the house, and even cleared out his studio, but his conversation was still largely unintelligible except on art matters. Physically, he looked better than he had for ten

[25] CR to E. Brown, 29 Nov. [1929].
[26] CR to M. Davis [c. Mar. 1929].
[27] CR to M. Davis [May 1929]; CR to C. French [1929].
[28] CR to RNRH [Sept 1930] = *SP*, p. 421.
[29] CR to Mrs Muriel Lee Mathews [1929] = *SP*, p. 414: and [June 1929] = *SP*, p. 415.

years, an impression corroborated by one of his sisters. In the summer of 1929 Ricketts was able to dispense with the night nurse, but there remained two new female nurses. During that summer and autumn Shannon spent two long spells at Chilham with his nurses, and was able even to write letters to the nurse's dictation, read with greater success, and speak of Ricketts being hard at work in London. Soon he even tried to do some work in pastel. These improvements, however, brought little comfort to Ricketts, as both a nurse and Nicholls hinted that he exercised a disquieting influence on Shannon; so he stayed in London, coping with work, letters, and business matters, glad of the 'solitude & calm in a house empty of nurses & Doctors'.[30] His weekly visits to Chilham, with a two-hour journey each way, were tiring as well as very trying, and left him below the mark for days afterwards. On Shannon's return to London Ricketts took him to see the Wallace Collection, which he enjoyed, but, as he began to be able to join more in conversations, he began fussing about his work, so that Ricketts and Nicholls removed some unfinished pictures from his studio. A psychologist, Dr Bernard Hart, was brought in, since Ricketts thought the return to consciousness would probably raise psychological problems. The changes, slow and almost imperceptible, seemed to promise a complete recovery, so that Ricketts could write at the beginning of September, 'for the first time during these nine months, I have felt a gleam of hope'.[31]

Soon after the hanging of the RA Summer Show Ricketts found himself in the throes of another, much more important, exhibition. This was the great Exhibition of Italian Art 1200–1900 at the RA from 1 January to 8 March 1930. In spite of all his work anxieties, Ricketts, who was on both the selection and hanging committees, found himself caught up in the excitement: 'We are promised unbelievable things from Italy.'[32] The Pope and Mussolini had each promised important works, as well as private collectors and public collections all over Europe and the USA. Ricketts also borrowed three of his purchases for the National Gallery of Canada—the Botticelli *Christ and St John*, Tintoretto's *Servant*, and Cariani's portrait of a man (also exhibited at the Venetian Exhibition, London, 1983–4). Ricketts did not let pass an opportunity to justify his purchases and to remind Brown, 'Believe

30 CR to Fr. J. Gray [late 1929].

31 CR to F. E. Jackson, 3 Sept. 1929.

32 CR to M. Davis [summer 1929].

me, the Italian show will put all Italian works at a premium.'[33] Along with the great galleries, the rich dealers, the noblemen, and the millionaires, Ricketts and Shannon lent their prized Piero di Cosimo and two drawings. Ricketts thought of producing a handbook on Italian art, in which he wanted to reproduce two or three of Canada's pictures, and he entered into negotiations with Gollancz. By the time the idea was eventually dropped (because the printers were too busy to produce it on time), he had written 'a good half of it' and regretted the loss of a £100 advance as well as a percentage on the sales.[34] He had earlier contemplated a *History of the Art of the World*, similar to H. G. Wells's *History of the World*, but this too was never realized. An article by him on his Canadian purchases was, though, published in the *Observer* (19 Nov. 1930).

It was not long, however, before Ricketts was in conflict with the 'unfriendly and divergent committee', on which, ironically, he felt that 'Roger Fry & I alone seem to have any common sense'.[35] By the end of October he was feeling the strain of the exhibition on top of his other troubles, and he told Brown, 'I am going to drop the Italian show, that is, go easy. I think the Italians should do the work, & I am worn out'.[36] This uncharacteristic withdrawal was probably necessary for the moment, but it did not last long. Ricketts could not resist the opportunity to be involved in something that was not only so dear to him but so important in educating the public in a proper appreciation of art. However, he obviously needed a holiday, which the doctors had advised him to take six months before. Though he was worried that Shannon might sense his absence, which would not have been so likely before, he decided to go to Italy with a new friend.

'I may have a secretary in the house, or companion to keep me alive', he told Brown in September 1929.[37] This was Henning Nyberg (Fig. 51), a young Swede of twenty-six, who had lived in Canada and studied art at the Byam Shaw School and the Royal Academy Schools. After less than a year at the latter he had taken up a job as an apprentice at Alec Johnston's Scenic Paint Room, where, it seems, Ricketts met

[33] CR to E. Brown [Sept. 1929].

[34] CR to E. Brown, 24 Sept. 1929.

[35] CR to M. Davis [summer 1929].

[36] CR to E. Brown, 27 Oct. [1929].

[37] CR to E. Brown, received Sept. 1929. See J. D. Hodin, 'A Swedish Painter in England', *The Anglo-Swedish Review* (June 1954), 7, 8–10.

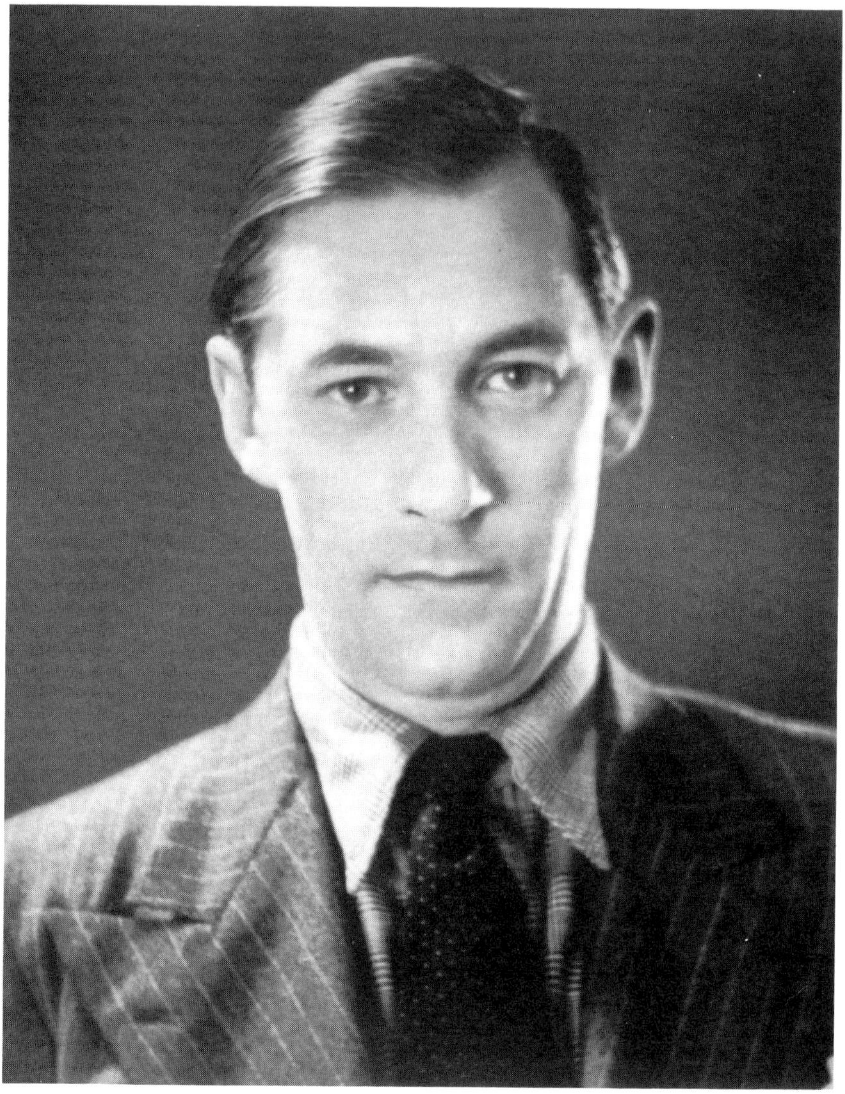

FIG. 51. Henning Nyberg (1903–64)

him. Nyberg was tall and masculine, with rugged good looks, but gentle and quiet in character. 'He is a very nice boy', Ricketts told Mary Davis, 'but too silent; he will probably always retain the "too well brought up" character stamped upon him by his Swedish parentage.'[38] Nyberg moved into Townshend House, where his presence was a great boon: 'If it is not unlucky to say so, he seems quite the companion I need', Ricketts told Jackson. As Nyberg was 'a whole-hogger over cinemas and theatres', he and Ricketts went out on average three times a week. It was Nyberg who imported the gramophone into Townshend House, which Ricketts also enjoyed.[39] There is some likelihood too that he was able to be more accommodating in sexual matters than Cecil Lewis, and may have given Ricketts some sort of physical relationship.[40] After Ricketts's death, Nyberg eventually married.

On 12 November Ricketts set off for Rome with Nyberg. Before leaving, he carefully made arrangements with Jackson and Lowinsky, in case anything should happen to him while he was abroad. In Italy Nyberg was 'almost beside himself with pleasure' as they visited the churches and galleries and explored the new excavations in Rome, which excited Ricketts's romantic imagination in the old way:

> they have uncovered the Imperial law court on the Palatine and the subterranean staircase for prisoners, down which St. Peter & probably St. Paul were led to execution ... They now have located the very site of Caesar's murder, it is under a modern theatre which they propose demolishing. Last year, they gave Shakespeare's *Julius Caesar* on the very site to a thrilled audience.

Cippico, who was in Rome, 'buzzed' them around the city in his car, driven by a handsome chauffeur, and Ricketts dined with him while Nyberg entertained some English art students who were at the British School. One of them was Richard Finny, who at nineteen was the youngest student ever to win the Prix de Rome and had been a student of Jackson's and Shannon's at the Byam Shaw School. Ricketts had been on the committee which selected him, and was instrumental in his getting the award. Ricketts invited Finny, who was unhappy in Rome, and another Byam Shaw student, Bernard Pike, to dine with him at the Hotel Hassler, where he was staying, and tried to draw them out about art. One night Cippico took Ricketts to see *Lucia di*

[38] CR to M. Davis, Tue. [May–June 1929].
[39] CR to E. Jackson [11 Nov. 1930]; CR to M. Davis [Feb.–Mar. 1930].
[40] This was hinted at by Nyberg's widow and two contemporaries.

Lammermoor, which enchanted him; and Ricketts and Nyberg enjoyed a Richard Strauss concert conducted by the composer himself. Nyberg was thrilled by being saluted in the Fascist manner, and Ricketts's observations on this visit only confirmed his good opinion of what Mussolini was achieving. Ricketts told Fr. Gray that, despite torrential showers, he had enjoyed himself hugely, but shortly after his return he claimed that the visit to Italy had left him unmoved.[41] Gloom had enveloped him again and obliterated the benefit and happy memories.

He had found Shannon in a poor state, as a bout of flu had set him back considerably, so that he was less conscious than a month before. Ricketts was overcome with despondency.

> I have courage to tell you [he wrote to Cippico] despite the doctors, how meagre is my hope that he will be anything more than the kindly and listless ghost he is at present. The conventions of ordinary life oblige me to exaggerate any small fluctuating progress. You are the only person to whom I have confessed what my feelings really are. It is one of the greatest elements of strain that I have to keep up this pretence of hope, and have never spoken of the tragic phases of these eleven months of monotonous anxiety. Please do not mention this to anyone, or, for that matter, to me.

Though he had given way to despair many times before, this collapse, after eleven months, marked the end of any real hope. Yet, however demoralized he became, he never suffered from self-pity. He struggled on:

> I find to my regret [he also told Cippico] that I have to be active in hanging the Italian show; this at some other time would have been one of the events of my life, but the other members of the hanging committee are lacking in experience, vitality, and conviction ... Possibly I ... may find the contact with these immortal masterpieces a tonic and stimulus; it should count as something after all, to help to lift and hang *The Birth of Venus* in its place, and to see that Fra Angelico and Mantegna are comfortable.[42]

Of course, the exhibition was the usual rush, and he threw himself into the work with the usual abandon, however unwise it was considering his age and his personal circumstances. Though the hanging committee also included the Italian delegate, Lady Chamberlain, Sir William Llewellyn, PRA, and Sir Robert Witt, the hanging of the exhibition fell mainly on Ricketts and its chairman W. G. Constable, of the

[41] CR to M. Davis, Sat. [Dec. 1929]; CR to Fr. J. Gray [Dec. 1929]; Mr Richard Finny and Mr Bernard Pike to author.

[42] CR to A. Cippico, 10 Dec. 1929 = *SP*, p. 418.

National Gallery (who was still weak from an operation), 'after two tiresome days when the Italian delegate helped & then withdrew, after many distressing scenes & complications'. Ricketts found himself in the role of 'pacifier as well as guardian against interferance from important persons', and characteristically rushed out and bought an enormous bouquet of Parma violets for one of the secretaries, to whom Witt was being 'horrible', but as he could not find her he gave them to Kenneth Clark's wife.[43] Altogether, the exhibition was 'the most hopeless muddle I have been in, not excepting *Macbeth*, where the impossible had to be done in a few days.... The organization was hopeless, the personel unwilling & incompetent.'[44] The Italian and German pictures arrived at the last moment, some after the catalogue had been set up, so had to be hung from memory. Many of the pictures already in place had to be unhung during the preparations, in order to be glazed, or photographed. Without the help of '8 angels disguised as workmen from the National Gallery staff' they could never have completed the job, though they worked thirteen hours a day, Sunday, Christmas, and Boxing Day included. Six weeks' work was accomplished in nine days. During four of the nights Ricketts was sleepless, due to 'anxiety, anger, & actual fear for the pictures due to radiators & rain coming through the roof'. On the last day he lost his temper, and afterwards looked back on the 'amazing experience of handling these unbelievable treasures, not with pride & excitement but with sullen resentment'. Yet he himself had hung one wall given over to Titian, Raphael, and Giorgione only, and another to Botticelli, Mantegna, and Piero della Francesca: 'these 2 walls are without a parallel in the world', he told Bottomley.[45] He was also responsible for hanging Fra Angelico, Masaccio, Fra Filippo Lippi, and many other masters. Even without the domestic strain he was under it would have been a very taxing effort but with it it was a remarkable achievement. Moreover, though it took him out of himself and the sad situation at home, it actually increased his stress.

As soon as his work was completed he took a rest-cure. His part in the organization was widely commended; the show itself was the highlight of the art world in Britain for 1930, and one of the finest exhibitions ever held. Attendance broke all records: 538,000 visitors

[43] CR to G. Bottomley, 1 Jan. 1929 [actually 1930]; Clark, p. 182.

[44] CR to M. Davis [Jan. 1930].

[45] CR to G. Bottomley, 1 Jan. 1929 [1930].

in just over two months; twice the number at the highly successful Dutch Show the year before, with over 150,000 catalogues sold. In early March it all ended 'in medals, dull banquets, and tedious speeches'.[46]

Not surprisingly, the Italian exhibition had left Ricketts rather unwell. There was a possibility of his going to Sweden to supervise there the production of his *Henry VIII*: 'the bother is it is a long way off & I should have to pay for Nyberg.'[47] He thought it would be fun to hear Shakespeare in Swedish, having read some of his plays in French, German, and Italian, but the trip was called off when the first night was postponed to an inconvenient date. He had hoped to take in Berlin on his way, and almost immediately he had an opportunity to go to Germany on Canadian business. As Canada was footing his expenses, he was able to take Nyberg along. They spent a fortnight in Berlin and Munich, going out to the theatre or opera almost nightly. These performances seem to have interested him more than the art in Berlin, though he found some marvellous things—Rubens, Van Dyck, Tintoretto, and a fresco ceiling by Tiepolo—at prices considerably cheaper than London. Though people assured him that the war had changed everything, he found the changes 'for the better as far as museums are concerned and, if life there has lost much of its energy, the theatres, as far as Berlin is concerned, are up to post-war standards and singularly free in their settings from exaggerated and unpleasant modernism'. Even the galleries had not 'swung round solely to Cézanne and his following but, to my surprise, had acquired pleasant and generally early specimens of nineteenth century painters who I thought had reached their nadir, such as Lenbach, Hakart, and Keller'.[48] Altogether the trip was a great success, in spite of a bad return crossing and Nyberg's bag being lost on the ship.

Relations with Canada were still strained. Another new arrangement had been worked out in June 1929, assuring Ricketts a commission of 5 per cent on all pictures purchased by the gallery, whether discovered or only recommended by him. Ricketts had come to the conclusion that 'the cheating by Canada is deliberate, not bad management', and no sooner had the new arrangement been made than there was a disagreement about when it came into effect.[49] Repeated irritation at

[46] CR to Max Chapman [end of Mar. 1930].

[47] CR to M. Davis, Wed., 23 [?Feb. 1930].

[48] CR to C. French, p.m. 18 May 1930.

[49] CR to M. Davis, Tue. [before Apr. 1929].

delays and errors in his payments led him to declare, in June 1930, that he was prepared to bypass Brown and put his grievances directly before the Canadian Trustees. In a press notice the gallery had sent him he had noticed the acquisition of a Cranach. 'Is that the picture we saw at Christies?' he asked Brown suspiciously; and if it was he insisted on his earlier fee of 2.5 per cent for expert advice, though he did not know what price had been paid for it.[50] He evidently felt entitled to his percentage even though Brown was also present, his own judgement had been unfavourable, and he himself played no part in the acquisition. His attitude seems particularly grasping here, but once riled he was never one to forgive and forget, and Canada had riled him almost from the beginning. However, whilst Ricketts was acting unreasonably, Canada was not without some guilt. In May 1929 Ricketts described to Brown a 'highly decorative, showey, & furnishing' altar-piece which he attributed to Benozzo Gozzoli. Brown telegraphed to say that there were no funds available for it, but that he had asked Durlacher, a dealer, to purchase it at a price up to £5,000, half of which was to be paid in 1930 and the rest in 1931. On 26 June, two days before the sale, Ricketts cabled that he had decided against the picture ('neither condition nor authorship satisfactory') and confirmed this view in a letter dated 8 July. On seeing it again later, however, he felt something of his old interest, and admitted to being 'a little ashamed at my wobbling in the matter', which he said was due to fear of public taste: 'the picture is *not* popular, I am told Duveen would be glad to get rid of it. I fear a special grant could never be obtained for it; yet some day it will be a famous picture.' In spite of his doubts over a special grant, he nevertheless wrote to the Canadian Prime Minister, Mackenzie King, to recommend its purchase.[51] In the end it was acquired out of the annual grant and, despite his uncharacteristic 'wobbling', which may reflect the strain he was under, Ricketts was assured on 19 March and on 3 October 1931 that he would get his commission. However, payment was later refused to his executors. Nor was the commission ever paid on several other late purchases where no 'wobbling' on Ricketts's part offered any excuse, in spite of persistent applications by Ricketts's lawyers. Here, the stock-market crash of 1929, the beginning of the Depression, and the drastic

[50] CR to E. Brown, 23 Jan., 3 June 1930.

[51] CR to E. Brown, 31 May 1929; E. Brown to CR, 10 June 1929; CR to E. Brown, 26 June, 8 July 1929, and [Mar.–Apr. 1930]; CR to Mackenzie King [30 May 1930].

reduction of the annual grant intervened to add to the gallery's financial difficulties. Whilst the museum was put to every sort of shift to obtain money for purchases, Ricketts showed pretty clearly his new fascination with making money, though, as he frequently pointed out, his commission was much lower than that which advisers to American museums received. On the financial side he and the gallery were to some extent at cross purposes.

On his return from Germany Ricketts found Shannon had improved during his absence, but almost immediately Shannon suffered another set-back, requiring a new specialist and two nurses. It is not clear whether he was suffering from gallstones or jaundice or both, but it shook and weakened him and kept him in bed for a month. It also led to 'fits of crying—this is a form of consciousness I dread most', Ricketts confessed.[52] An operation was contemplated but proved to be unnecessary. Ricketts insisted that the doctors give morphine to Shannon, to relieve his pain, and also to Nicholls, who came down with acute sciatica and then tuberculosis, which kept him in bed for three months. Ricketts decided that Nicholls needed some sun, and he was sent off first to a sanatorium and then, with his wife, to Bognor for convalescence, all of which added to the bills. While they were away Nicholls's cousin filled in as housemaid. Mrs Coombs, their admirable housekeeper from Chilham, also came up to London to help out. Nyberg too was ill, with 'a bad chill on the liver' and a bad foot, and required daily visits from the doctor. The cost of all this at the 'Townshend House Hospital', as Ricketts put it, was enormous, and so was that of the additional staff, with an extra cook, maid, and two nurses, as well as Mrs Coombs, in the house. Ricketts was answerable for the salaries and upkeep of seven people, on top of doctors' and the usual tradesmen's bills. 'Every day I seem to have to write a cheque; it gets on my nerves sometimes', he confessed to Mary Davis, who seems to have offered some financial help, but, though expenditure was greatly in excess of income, he assured her 'there is enough money for two or three years *with care*'.[53]

To top it all, in the summer Ricketts himself caught his left hand in a collapsible chair and severely injured one of his fingers. The doctor was called: 'I should like so much more to spend his fees on plants and grub.' While his swollen finger was still 'like a baked potato' and

[52] CR to S. Cockerell, 22 May 1930.

[53] CR to M. Davis [Sept. 1930, June–July 1930, and Aug.–Sept. 1930].

in a smart black sling, a boiler burst and flooded the scullery and, when this was seen to, a lavatory went wrong: 'this ... did not add to my good temper'. It was a decided advantage to be ambidextrous, and he was able to make his bed, plant six dozen chrysanthemums, and even write more 'improper stories'—for another collection—and some memories for a book on Oscar Wilde. These two works, *Unrecorded Histories* and *Recollections of Oscar Wilde*, were not published until after his death. Eventually 'the bloody hand', as he called it, got better; so did Nyberg and Nicholls. Even Shannon was gaining weight and making copies in pastel of old designs with a tremulous and, at times, uncertain touch, but he was also becoming difficult again; 'he is developing a will of his own, a subconscious resistence to doctoring; he refuses point blank to be punctured anymore & argues over everything. Nicholls seems to have the best influence over him & I the worst', Ricketts confessed.[54]

It was now openly admitted that Ricketts's presence exercised a disturbing influence on Shannon ('seems to throw him back'). How distressing this was to Ricketts can easily be imagined. Earlier, to John Gray, Ricketts had written, 'I realise that, viewed as a whole, this trial is small compared to the forty years of perfect companionship I have enjoyed.'[55] Those forty-six years had in fact, though, been bought at a price, and Shannon's sacrifice in this respect was far greater than Ricketts's. There is little doubt that if he had not met Ricketts he would have married, had a family, and led a far more ordinary life. It can be argued that without Ricketts's influence and support Shannon would have been a less interesting artist and had a less exciting life, but the price he paid was high. The fact that Ricketts, in particular, upset him after his accident suggests that Shannon may have been expressing a long buried, and perhaps never acknowledged, resentment about Ricketts's hold over him. (That such subconscious antagonisms should rise to the surface after brain injury is a recognized phenomenon in psychology known as disinhibition; it would also explain Shannon's fits of crying and his refusal to submit to treatment.[56]) It is a highly plausible and tragic explanation of Shannon's attitude to Ricketts. As often as possible Shannon was sent down to Chilham, where Ricketts's presence would not upset him. Anxious only for Shannon's good,

[54] CR to M. Davis [July 1930 and May 1930].

[55] CR to M. Davis, Sun., 30 [?Sept. 1930]; CR to Fr. J. Gray [late 1929].

[56] Dr Desmond Henry discussed Shannon's symptoms and disinhibition with me.

Ricketts accepted his hostility (which actually increased over the months) stoically, without ever, at least in his letters, analysing it. Yet it is hard to imagine that, with his quick intelligence, he did not consider its possible causes.

No sooner had Shannon returned from one of his visits to Chilham than there was another set-back:

> Gordon Holmes was here on Friday [Ricketts told Jackson] and considered there had been an improvement, with the result that on that afternoon there was a bad relapse, due probably to the strain of medical examination and to my deep conviction that Friday, being the day of the grave relapse months ago, has remained a bad day; just as Tuesday and Wednesday are good. I believe there is a clockwork sequence in hours of the day and in days of the week.

Ricketts was seeing another doctor, Dr Finney, whose advice he trusted more. The other doctors 'merely exaggerate what I think and say. Both drove me mad on Friday with impossible, premature, and costly suggestions. I fear I showed bad temper and intend editing their suggestions, but I feel the need of advice.'[57] Ricketts had wanted an X-ray to be taken, which the doctors had not thought necessary, and he felt vindicated when it revealed Shannon to have a dent in his skull near the temple. Nevertheless, it is hard not to reach the conclusion that Ricketts, with his nerves and temper, was making everything more complicated, however understandable his reactions were.

Through these sad and stressful days Ricketts struggled to cope. Tragedy, as before with the dying Edith Cooper, brought out the best in him. His noble and generous spirit sometimes rose above the nerves, the touchiness, the anxiety, the furious tempers that so often assailed him. Even with his enormous expenses, he did not curb his generosity, either in big or even in little things, as many would have done. When Nyberg, for instance, noticed that Ricketts never refused a beggar or gave less than a shilling, he remarked that this was perhaps excessive, but Ricketts replied, 'I can't help it.'[58] To his close friends he became ever more tender and gentle. Every time Bottomley was with him after Shannon's accident

> he opened up more and more; and I used to come away with awe mingling with my ancient admiration. There was one night, when he kept Emily and me sitting late and then walked with us to the bus station, that we shall never

[57] CR to F. E. Jackson [May 1929].   [58] *65 Ill.* p. [iv].

forget: his loneliness, his solicitude for Shannon, the dignity and beauty of all he said, made us often feel like tears, if he had not been on a plane above them.

Recalling the same meeting ten years after the event, Bottomley wished he had written down that conversation as soon as it was over: 'for he spoke as if we were sitting with him in eternity, and as if we were, all three, immortals'.[59] Earlier, his human qualities had been partly submerged under his aestheticism, and evident only to those who knew him well. Even the Fields had been surprised at them. Together with his deep vulnerability, they may help to explain why Shannon had stayed with him all these years.

As in his gift to Lewis, Ricketts's generosity actually grew during these dark days. A second instance occurred at about this time. One of the many committees he sat on (along with the Council and Schools of the Royal Academy, and the Royal Commission controlling the rebuilding of London) was the School of Rome committee, both Council and Executive. In this capacity he had helped Finny to win his Prix de Rome in 1929, and he tried his best to give it to another promising student of Jackson's the following year. Max Chapman had submitted some accomplished designs, so much in the line that Ricketts approved that he was determined Chapman should get the award. He thought Chapman's show 'the best sent up since that of [A. K.] Lawrence some years ago'. However, the other members of the committee felt that his work was too much like Finny's the year before, and he came third. Ricketts was furious: 'Jackson's pupil should have got it easily', he told Lady Davis, '. . . I shall resign from the thing later. Art committees are impossible.'[60] To Jackson he wrote, 'The whole affair has added to my sense of vicious and exasperated indignation against all R.A.'s, Philpot included.'[61] He invited Chapman to visit Shannon with him and Jackson at Townshend House before having dinner at a restaurant, Canuto's. Ever firm in his conviction, Ricketts decided to make up himself for the blindness of the committee and gave Chapman a private scholarship of £200, so that he could visit Italy.[62]

[59] G. Bottomley to TSM (Univ. of London, Box 17, 212) and G. Bottomley to M. Davis, 22 Apr. 1940.

[60] CR to M. Davis, 10 Jan. 1930.

[61] CR to F. E. Jackson [June 1930].

[62] Max Chapman to author.

Considering both the present demands on his purse and the likelihood of future demands, it was a doubly generous gesture.

He carried on with his writing, and had now completed the memoir of Oscar Wilde which he had started in Tunisia in 1927. This was something which had been in his mind for many years. Before the war he had tried to write about Wilde, but thought the results of value only to himself: 'I found them too personal & not quite the thing', he told Bottomley; but he thought one day he might return to them.[63] In 1921 Ricketts again thought of writing about his old friend, but 'for the moment the war seems to have broken the small remaining sense of contact (with the times as they seem) which I imagined I had before'.[64] However, he had brought Wilde into his other writings; into *Pages from a Diary in Greece* and *Beyond the Threshold*. Now, in these trying and lonely days, Ricketts set out to repay something of his great debt to Wilde. However, after having written a moving tribute to his old friend he found it difficult to get it published. A. J. A. Symons, who had 'queer manners always, and more than the usual number of complexes', hesitated to handle it (though Ricketts was prepared to take the financial risks), and even seemed to have grown unfriendly. Ricketts suspected he was in some personal difficulty. However, they were planning a prospectus as a *ballon d'essai* later in 1931. Ricketts designed a splendid cover for it, appropriately in his earlier *Sphinx* manner, and intended to publish it under his pseudonym, Jean Paul Raymond.

These delays distressed him, as they meant his new book of stories might have to be held up indefinitely. He began writing a new story on Phryne and Praxiteles, and told Moore that these stories 'at the worst ... amuse me & the number of pleasant or amusing things in life do not grow more numerous'.[65] He also produced six silhouette illustrations, which are rather disappointing, and an abstract cover design, similar to that for *Beyond the Threshold*, which was left unfinished at the time of his death and completed by his friend from Vale days, Reginald Savage. It seems likely that Ricketts also intended to publish this under his most recent pseudonym, Jean Paul Raymond, even though one of the stories is signed with his own name in manuscript. He was pleased when an article he had written on Baudelaire was liked,

[63] CR to G. Bottomley, p.m. 20 July 1918.
[64] CR to G. Bottomley, Christmas 1921.
[65] CR to TSM [*c.*Aug. 1931].

and he returned to book design when he did the cover of *Christie's Season* in 1931 at the request of another old friend, Alec Martin.

As the new century progressed, the 1890s legend began to grow and, as a survivor from that period, Ricketts found himself portrayed in memoirs and accounts of that brief golden age. Moreover, many of his friends and associates were writing their autobiographies. When Symons published *An Anthology of Nineties Verse* Ricketts read the poems with interest and found that only some, including those by the Fields, had maintained any interest for him. He read avidly the first volume of William Rothenstein's *Men and Memories*, which came out in February 1931 and of which Sturge Moore commented to R. C. Trevelyan that Rothenstein 'has resolutely determined to be nice to everybody and even nicer to himself'.[66] Ricketts was glad that Rothenstein had mentioned the Fields, as he felt a bit guilty about not having published anything about them himself: 'personalities of their kind are rare'. Now that he had written about Wilde, he hoped that before he died he would also be able to write on Paul Baudry, an artist he greatly admired but who had been forgotten; Ricketts had persuaded the French delegates to add Baudry's work to the large French show at the RA to be held at the end of 1931.[67] Ricketts also wrote a long review of *Men and Memories* in the *Observer*, taking advantage of the opportunity to reminisce about the Nineties himself. As he read of these earlier happier days, especially Rothenstein's lyrical description of him and Shannon at the Vale, he must have fallen prey to sad and bitter thoughts, but he still kept himself as busy as possible. The phrase 'before I die' crept into his letters and conversation, but there were many things he hoped to accomplish before that happened.

'I read a great deal & am absorbed by music. A year ago I could not listen to it', Ricketts told Holst. Even more significant was his admission, 'I find nothing interests me save for the immediate instant. I have almost ceased to work or be interested in Art.' This startling avowal, for the aesthete who had starved in his youth to be able to buy beautiful things, shows how Shannon's condition undermined all his earlier purely aesthetic values. He constantly kept his mind occupied with reading, music, or short-term projects, but was incapable of sustaining any long-term concentration. He still managed to get some painting done (Fig. 52). He described his outlook on life as 'one of

[66] TSM to R. C. Trevelyan, 30 Mar. 1931.

[67] CR to W. Rothenstein [spring 1931].

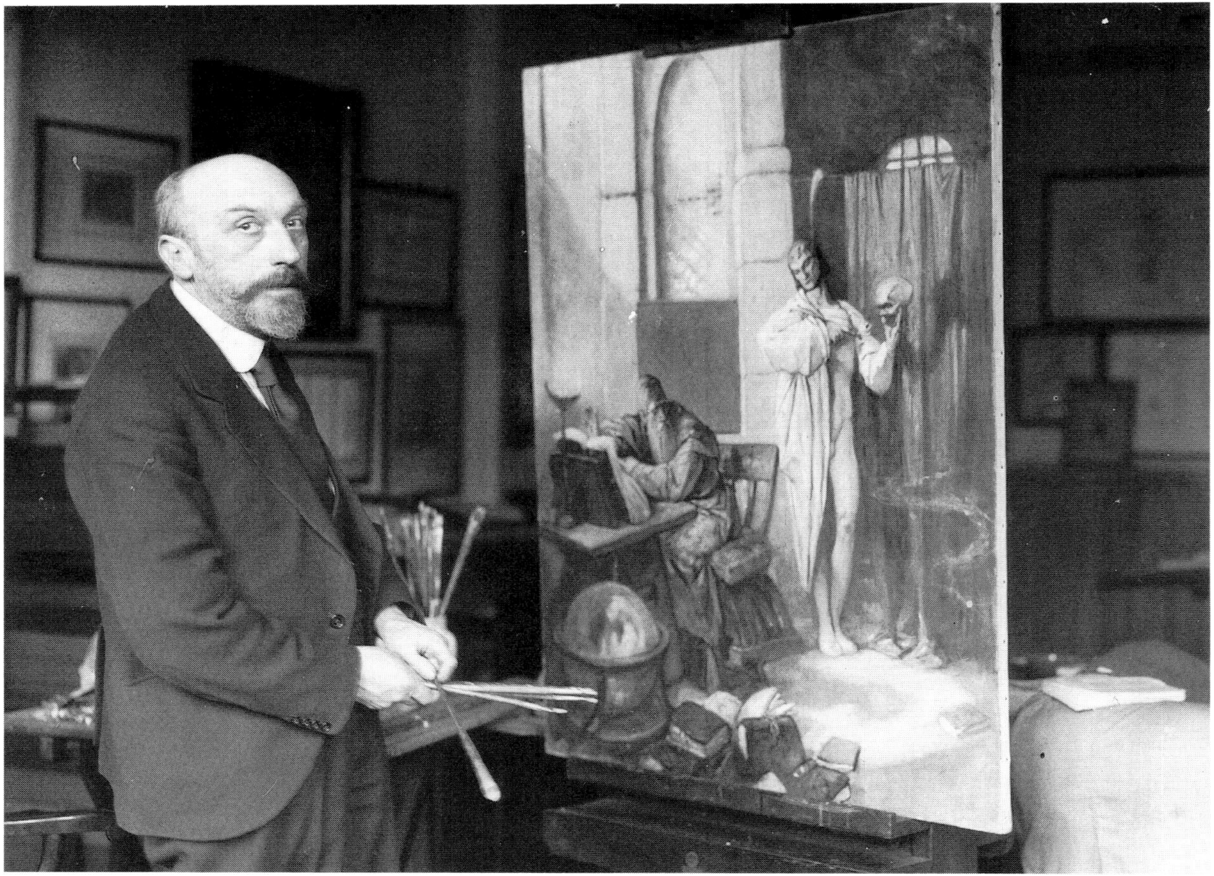

FIG. 52. Charles Ricketts painting *Faust and Mephistopheles* (*c*.1930)

disillusion; my one consolation is that the young seem to have ceased to be interested in post-Impressionism and its following, and I hope to live long enough to find Cézanne forgotten like many other international successes of the past. I have nothing more to say.'[68]

He tried to escape the tragic situation and his own sense of exhaustion and despair by going abroad at every opportunity. Shannon was still giving cause for anxiety: 'recently, things have been a little beyond my staying powers', he confessed to Brown.[69] His own health was failing, but he could not free himself from a frantic round of activities. In the spring of 1931 he decided to make another trip to Italy, his third trip abroad since Shannon's accident. Of course his conception of such a trip, with its arduous round of churches and galleries, was hardly restful. He was to be accompanied again by Nyberg, whom Ricketts still found 'a gentle creature & I wish the house were a little brighter for his sake; though so far he has had a hard & uninteresting life, & has on his side the resilliance of youth to resist dullness or monotony which is inevitable.'[70] No doubt Ricketts felt how much he himself lacked that resilience; but it cheered him to be able to share with Nyberg the artistic splendours of his spiritual home.

On 17 May they set off for Florence, Milan, and Sienna, and for two places he had never seen: Monte Olivetto, which he loved, and San Gimignano, which disappointed him 'as a place, not the art'. Ricketts found himself 'jarred by travel, heat, & noise'.[71] The noise everywhere made him sleepless and anxious. To save him from further bother, they took a taxi through lovely scenery to Assisi, which also saved time and cost only thirty shillings more than their train tickets. They stayed here for a while to recoup and Ricketts spent most of the days 'in a small terraced garden, containing six palms, three arbours, some shrubs, a small pond, and a monkey in a cage to whom I give a pink rose every day'.[72] In a letter from Assisi Ricketts admitted to Lewis, 'I started worn out and grew worse and, I fear, trying to Nyberg, who I fancy has left some feminine interest in London. Once or twice we jarred on each other, probably my fault.'[73] All the beauty and change

[68] CR to RNRH [*c.* Aug. 1930] and [winter 1930] = *SP*, p. 424.
[69] CR to E. Brown [May 1931].
[70] CR to G. Bottomley, p.m. 28 Mar. 1931
[71] CR to M. Davis, 3 June [1931].
[72] CR to C. Lewis, 3 June 1931 = *SP*, p. 423 (in part).
[73] CR to C. Lewis, 3 June 1931 (not in *SP*).

seemed to leave him unmoved, and he even thought of rushing back to London. At some point they also took in Perugia and Arezzo. After a month, he returned, and wrote to Jackson: 'I am not in good fettle. I did not relax in Italy—everything was perfect, but I was bored and depressed and tired. I did not sleep well and am glad to be back.'

No sooner had he got back than Shannon suffered from another attack of gallstones. At first it was thought to be a stroke. The doctors were again considering an operation, as another X-ray had revealed that the fall had opened, or widened, a suture in the skull, which had never closed properly. Shannon spent most of the next months at Chilham, where the large drawing-room had been made into a bedroom and he was in the care of two doctors and two nurses. During the next months the two nurses collapsed in turn, Nicholls was ill and had to go away for three months again, and Shannon fretted over his work and developed a 'phobia' against his studio and against Ricketts.[74]

Undefeated, Ricketts tried to apply himself to designing another play, Bruckner's *Elizabeth of England*, which had had a great success in Germany. 'I don't see it succeeding here', he confessed to Sturge Moore, 'but one never knows.'[75] He dressed the English and Spanish characters in sharp contrast, making the Englishmen look like Isaac Oliver miniatures. He soon found he disliked everyone involved in the play except the German producer, who had done the original production in Berlin. The financial constraints were burdensome: 'as usual economy has to be observed as far as I am concerned, when money is being thrown away on the mechanism of the production, three producers ... an orchestra, etc. etc.'[76] He still thought some of the dresses would be good. During August and part of September he seemed more cheerful than he had been for some time. With the strain of the work, however, he soon fell ill. After a row with the scene-painters over the chief set, he left Nyberg to look after things for him there and at the theatre. Nyberg worked with him daily in the studio, doing the ground-plans and, later, painting the pictures, while Ricketts did the drawings for the scenes and costumes. He soon felt so ill that he offered £50 to Nyberg to complete his work and asked his old friend Bruce Winston to see to the costume part of the production. Though he was not able to oversee this production with his usual

[74] CR to F. E. Jackson [after 30 June 1931].
[75] CR to TSM [Aug. 1931].
[76] CR to L. Binyon [Aug. 1931].

meticulous care, it gratified him to see its success when it opened on 30 September, only a few days before his death.

In his last few days his health had again begun to improve. 'If it is not unlucky to say it', he wrote to Winston, 'I am glad to tell you I am better.'[77] After all the work on the play, Ricketts insisted on Nyberg taking a paid holiday. There may have been another motive for this, since Lewis was in London, and Ricketts may have wanted to prevent any conflict or jealousy between his two protégés. Lewis's presence kept him busy in the evenings and no doubt cheered him up. (He was being treated by his doctor for indigestion, but he had been assured that it was not serious.) Lewis left London for a time and telephoned Townshend House on his return. Nicholls told him that Ricketts was not well and was going to bed early, but suggested Lewis should come round next morning. At 6 a.m. on 7 October 1931 Nicholls gave Ricketts his morning medicine, but when he came to wake him at 10 he was dead. He had died in his sleep of heart failure. Indeed, one might say he had died of a broken heart.

After Ricketts's cremation at Golders Green a memorial service was held at St James's, Piccadilly, on 12 October. Ricketts had wanted the ashes to be scattered to the four winds, so Lewis, Sturge Moore, Lowinsky, and two others went to Richmond Park, where Ricketts had so loved to walk. The five friends began scattering the ashes, which were in a shoe box, but they seemed endless and kept being blown into Sturge Moore's long beard, turning the white into black. With increasing frenzy, they scattered and scattered. In spite of the solemn ceremony, Lewis could imagine Ricketts doubled up with laughter, 'choking and convulsed ... tears streaming down his face, as he laughed when something overwhelmed him, when someone, preferably some "dear beast", had been "planted a sell".'[78] Finally, to everyone's relief, Lowinsky suggested to Lewis that perhaps the remaining ashes should be scattered on the ground in Italy which Ricketts had given him.

This Lewis duly did. A niche was hollowed out of the cliff face, where the bronze head of Ricketts by Wells was placed (Fig. 53). It still looks over the narrow garden towards the beautiful lake.

[77] Bruce Winston to L. McCarthy, 16 Oct. 1931 (Univ. of Texas).

[78] *SP*, p. xiii. The incident of TSM's beard comes from TEL via Sir Brinsley Ford to the author.

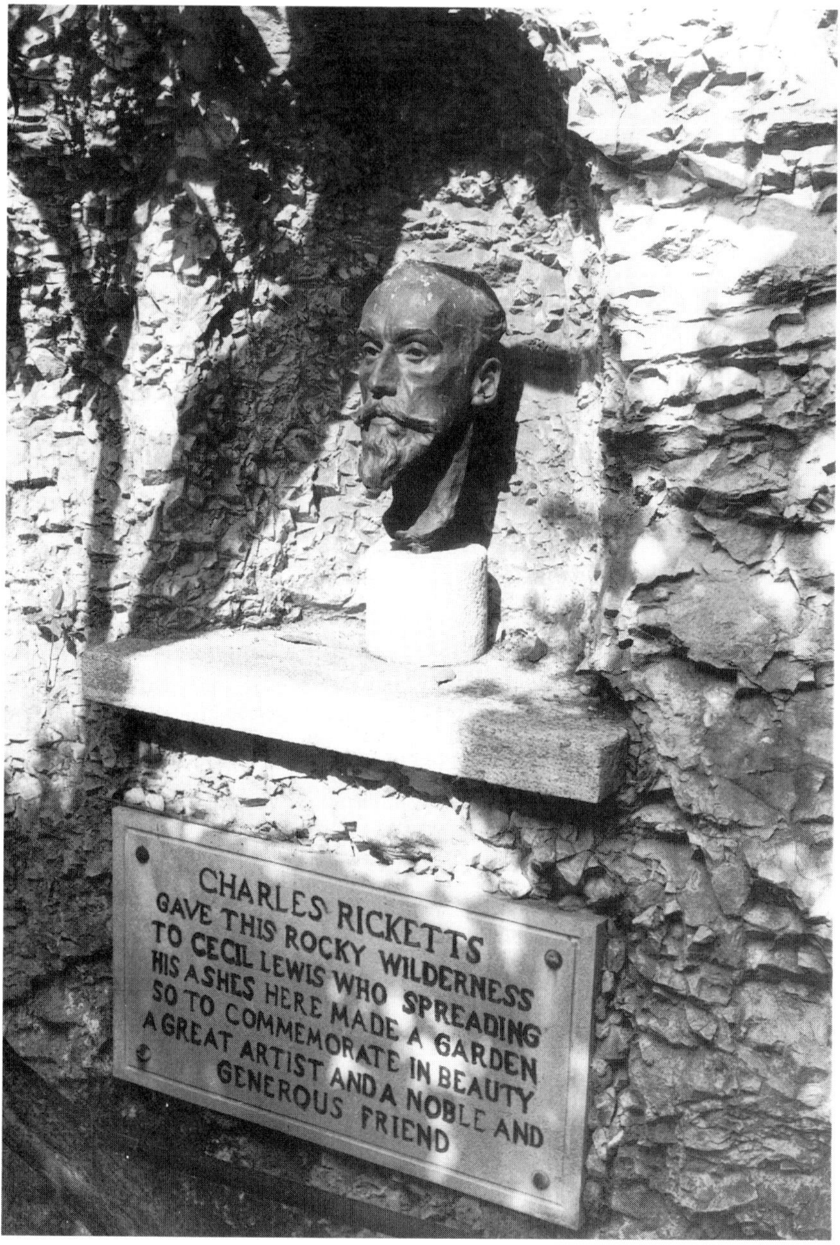

FIG. 53. Charles Ricketts, bust (1902), by F. R. Wells, at Arolo, Lago Maggiore, Italy

Epilogue

In spite of all Ricketts's careful planning for Shannon's future, he left things in rather a mess. In his will, drawn up on 6 March 1929 (some months after Shannon's accident), he had made Ernest Jackson, Bengy Lowinsky, and, surprisingly, Shannon, his executors. Whether this was due to the strain of those months after Shannon's accident, or to bad legal advice, is not clear. It is curious, however, that after his realization that Shannon would never recover he never revised it. As Shannon was the chief beneficiary as well as an executor, the Master in Lunacy appointed the Official Solicitor to act on his behalf. This meant that a public official who had never known Ricketts or Shannon in better days, had no real understanding of their values or priorities, and probably had little knowledge or experience of art was given a dominant position in the arrangements. Moreover, there were many friends, old and new, whose claims and expectations, confused by ambiguities in Ricketts's will, made the situation even more complicated. Through these difficult waters Lowinsky and, in particular, Jackson, steered a sure and even course. In a letter written to Shannon before the accident, Ricketts suggested that Jackson, as his executor, would be a 'tower of strength' in the complex legal and domestic arrangements. So he was.[1]

Before going on his second Italian trip with Nyberg Ricketts gave instructions to his executors in case anything should happen to him. There was £5,000 in his own account and £1,000 in Shannon's. More things from the collection had been put on the market: miniatures by Fragonard and Isabey (valued at £500) given to Ricketts by Princess Alice of Monaco, the Van Dyck portrait of Archbishop Laud (one of their greatest collecting coups, which they had bought in 1919 'at the price of a good frame', and valued by Ricketts at £8,000 in 1931), a painting by Hoppner (valued at £300).[2] All of these appear to have sold except the Van Dyck, which went to Cambridge. There was nearly

[1] CR to CHS [June 1928]. TS from Jackson's papers. The Jackson archive of the Ricketts executorship is the source of this account.

[2] CR to G. Bottomley, p.m. 25 Feb. 1920; *SP*, p. 21. The values are those given by CR to TEL, 15 May 1931.

£2,000 from the value of the house, furniture, and their own work, as well as £1,500 (never paid) owed by Canada. All this Ricketts hoped would bring in enough to secure an income and a cottage for Shannon near his two sisters in Lincolnshire.

However, he recognized that this might not be sufficient, as there was no telling how long Shannon might live or how much care he might require. So, in case further sales might be necessary, he instructed his executors: 'I should like the Greek things and Preraphaelite drawings to be kept to the last. Let Egypt go first, Piero di Cosimo second, and Japan third, and old master drawings fourth'.[3]

Over the problem of caring for Shannon, who meanwhile had been sent to Chilham, Jackson and Lowinsky soon found themselves in conflict with Gilchrist, the Official Solicitor. He thought Shannon should retain Townshend House and, on medical advice, insisted that he should have two nurses in attendance as well as Nicholls and his wife. Shannon could not be left day or night; his medical condition had deteriorated and he suffered from gallstones and a double hernia. Jackson, Lowinsky, Sturge Moore, and others thought this was far more expensive than was necessary, and increased the risk that the collections would have to be sold to pay for his upkeep. Whilst they knew that Ricketts would certainly have wanted them to put Shannon's interests before the collection, they did not want this to be sold unless it was absolutely necessary. In order to reduce the insurance costs, most valuable things had been removed in 1933 and sent to the respective museums to which they had been left in Ricketts's and Shannon's wills: the Piero di Cosimo to the National Gallery, the Japanese prints and drawings divided between the British Museum and Cambridge, the rest to Cambridge. However, not long after the disposal of the collections the Official Solicitor and the Master in Lunacy alarmed Binyon by visiting the British Museum Print Room in order to ascertain the value of the Japanese drawings and prints they had received.[4] Ricketts's lawyers even speculated that the greater part of the collections might have to be sold. Questions were asked about what objects should go first. Jackson insisted that it was the excessive expense of Shannon's upkeep that put the collection at risk. It is probable that the Official Solicitor, seeing such a fine collection, assumed that Ricketts and Shannon had lived luxuriously, and did not

[3] CR to F. E. Jackson and TEL, 16 May 1931 (TS Jackson).

[4] L. Binyon to F. E. Jackson, 24 May 1923.

appreciate how much, for most their lives, they had sacrificed to obtain these objects. A few things were sold, among them three Japanese screens, two Chinese iron lions, all the Japanese wood-block books, and a self-portrait by Sir David Wilkie. Some books, some pieces of furniture including those from Chilham, and works by Ricketts and Shannon were sold privately to friends, and part of the remainder was put up at auction. For £525 the NACF bought a large group of Ricketts's costume designs which it distributed to museums all over the nation. All these brought in money to cover Shannon's expenses.

What finally saved the collections was the decision to sell Townshend House in 1933 for £5,500. A suitable house for Shannon was found at Kew (21 Kew Gardens Road), where a thirty-five-year lease was offered at £1,200, though they probably paid less. The difference between the prices of the two houses finally ensured to Shannon enough money for the foreseeable future. After the new house had been modernized and redecorated, Shannon moved in on 24 June 1933, with his two nurses and Nicholls. A number of things from Townshend House, including some of his own work, had been retained, to give his surroundings some familiarity. At first he had difficulty settling down, but eventually he got used to his new home. For the most part he was good-natured and harmless, though he occasionally suffered from storms, when he got very excited and had to be calmed down. He talked a great deal, mostly nonsense, and spent his time, when alone, looking over pictures in art books and reproductions in portfolios. Early in 1934 he was placed on the list of retired RAs. He seems to have regained some lucidity towards the end; he died on 18 March 1937, just over a month short of his seventy fourth birthday. His memorial service was held at St James's, Piccadilly on 22 March, but no memorial exhibition was organized.

Bibliography

A. Manuscript Sources

The following are the known collections of Ricketts's letters, diaries, and other documents used in this biography. They are arranged in approximate order of importance.

1 The main collection is in the British Library, London: his surviving diaries, 1900–6, 1914–18 (Add. MS 58098–109), his travel diaries (58094–7), drafts of his writings (58092–3, 51724), and his letters to Shannon, 1900–31 (58085), to Thomas and Marie Sturge Moore, 1891–1931 (58086), to Michael Field, 1894–1914 (58086–9), to Sir Sydney C. Cockerell, 1893–1931 (52746), to George Bernard Shaw, 1907–28 (50548), and one to Robert Steele (45850K). In addition, there are transcriptions of letters (61713–20) to Mrs Muriel Lee Mathews, Count Antonio Cippico, Cecil French, W. A. Pye, François Monod, and others, compiled by T. Sturge Moore, who wrote a lengthy introduction to these, part of which, with a selection from the letters, formed the basis for *Self-Portrait* (1939). Shannon's diaries, 1898–1906 (58110–18), supplement Ricketts's at many points, as do the extensive journals of Michael Field (46776–808), which are rich in references to Ricketts and Shannon. Finally Mrs Ursula Bridge's unpublished *Poets and Painters* (61220–3), dealing with the friendship of Ricketts and Shannon with Michael Field, brings together a great deal of material from their letters and diaries.

2 Letters from Ricketts and Shannon to Gordon Bottomley, 1904–31, who carefully noted the postmark on the letters, which Rickets hardly ever dated himself, and to Sir Edmund Davis and Mary, Lady Davis, 1904–31, together with a great deal of material relevant to Ricketts and Shannon collected by Bottomley. (Bequeathed by the late Roger Lancelyn Green to the British Library, London.)

3 Letters from Ricketts and Shannon to Lucien Pissarro, 1890–1914, together with correspondence dealing with Pissarro's books published at the Vale Press, and Lucien's letters to and from his father Camille Pissarro, which contain a lot of material relevant to Ricketts and Shan-

non. (Ashmolean Museum, Oxford.)

4 Letters to Eric Brown, 1923–31, dealing with his purchases for the National Gallery of Canada and including a good deal of personal material. (The National Gallery of Canada, Ottawa.)

5 Letters from Ricketts and Shannon to F. Ernest Jackson, 1907–31, and papers relating to the executorship of the Ricketts estate and the 1933 memorial exhibitions. (Private Collection.)

6 Letters to Robbie Ross, 1910–18, formerly belonging to Mr John Paul Ross. (Mrs Mary Hyde, now Lady Eccles.)

7 Letters to Laurence Binyon, 1905–31. (Mrs Nicolete Gray.)

8 Letters to R. N. Roland Holst, 1910–31, together with a memoir of Ricketts by Holst, seen mostly in transcripts made by David Gould. (Letters formerly in the collection of Mr David Gould.)

9 Letters to T. E. (Bengy) Lowinsky, c.1915–31. (Mr Justin Lowinsky and Mrs Lance Thirkill.)

10 Letters to Florence de Sinner (née Ricketts), 1882–1931. (Dr Vincent von Sinner.)

11 Letters to Martin Birnbaum, his American dealer, 1914–31. (Martin Birnbaum Papers, Archives of American Art, Smithsonian Institution.)

12 Letters to Lillah McCarthy, 1907–31, to Harley Granville-Barker, Sir Hugh Walpole, 1924, to Lady Ottoline Morrell, to George Bernard Shaw, 1907, and to Sir Sidney ?. (Harry Ransom Humanities Research Center, Univ. of Texas, Austin, Texas.)

13 Letters to Sir William Rothenstein, three letters to John Lane, 1892–5, and one each to Fred Holland Day, 1893, and Grenville Winthrop, 1929. (Houghton Library, Harvard Univ.)

14 Letters to W. B. Yeats, 19?–31, kindly supplied by Dr John Kelly. (Senator Michael Yeats.)

15 Letters to A. J. A Symons, 1928–9, Robbie Ross, 1912–13, More Adey, 1897, Holbrook Jackson 1912–13, and Sir Martin Harvey, 1912. (William Andrews Clark Library, Univ. of California.)

16 Letters to Fr. John Gray, 1903–30. (The Beinecke Rare Book and Manuscript Library, Yale Univ.)

17 Letters to Mr Laing, 1906–10, Bruce Winston, and Mr Wilson, formerly belonging to Prof. Carl Woodring. (Columbia Univ.)

18 Letters to D. S. MacColl, 1904–31, with transcripts of some of MacColl's replies. (Univ. of Glasgow.)

19 Letters to Cecil Lewis (TS), 1920–31, probably not complete. (Private Collection.)

20 Letters to Glyn Philpot and Vivian Forbes. (Miss Gabrielle Cross.)

21 Letters to C. J. Holmes, and his wife. (Martin Holmes.)

22 Letters to Laurence Haward. (Manchester City Art Gallery.)

23 Letters to Allan Wade. (Sir Rupert Hart-Davis.)

24 Letters to Huntley Carter, 1910–26, regarding stage design (Victoria and Albert Museum.)

25 Letters to Joseph and Paul Durand-Ruel, 1905–6, together with copies of replies. (Durand-Ruel and Co., Paris.)

26 Letters from Shannon and Ricketts to R. A. Walker, 1921–5. (J. Harlin O'Connell collection, Princeton Univ. Library.)

27 Letters to Sydney Carline. (The late Richard Carline.)

28 Letters to Max Chapman, 1929. (Max Chapman.)

29 Letters to Florence Farr, 1906. (Univ. of London.)

30 Letters to J. M. Synge, 1907–14. (Trinity College Library, Dublin.)

31 Shannon's letters to members of the Dowdall family. (The Huntington Library, San Marino, California.)

32 Shannon's letters to John Quinn. (New York Public Library.)

33 Shannon's letters to Harry Quilter. (Boston Public Library.)

B. *Published Sources*

I have not included the original publication details of articles collected in *Pages on Art* (1913).

Abbott, Claude Colleer, and Bertram, Anthony (eds.), *Poet and Painter: Being the Correspondence between Gordon Bottomley and Paul Nash, 1910–46* (Oxford, 1955).

D'Amat, Roman, *Dictionnaire de biographie française*, 13 (Paris, 1971).

[Anon.] 'A Famous Edwardian Partnership in Art', *The Times* (23 Apr. 1936).

Annigoni, Pietro, *An Artist's Life* (1977).

Archer, William, 'A Rival to Euripides', *Tribune*, (2 Apr. 1906).

Ashwin, Clive, 'Gleeson White: Aesthete and Editor', *Apollo*, 108 (1978), 256–61.

Athill, Philip, 'The International Society', *Burlington Magazine*, 127 (1985), 21–6.

Barber, Giles, 'Rossetti, Ricketts, and Some English Publishers' Bindings of the Nineties', *Library*[5], 25 (1970), 314–30.

Barclay, Michael R., *Catalogue of the Works of Charles Ricketts, RA, from the Collection of Gordon Bottomley* (Stroud, 1985).

—— 'The Scenic Designs of Charles Ricketts', *Apollo* (1985), 184–8.

Beerbohm, Max, [Review of Ricketts's scenery for Norman McKinnel's *Lear*] *Saturday Review* (18 Sept. 1909), repr. in *Last Theatres* (1970), 483–7.

—— *A Peep into the Past*, ed. Rupert Hart-Davis (1973).

BEHMER, MARCUS, 'Charles Ricketts', in *Buchkunst: Beiträge zur Entwicklung der graphischen Künste under der Kunste im Buche* (1935), 35 f.

BINNIE, ERIC A. G., 'The Stage Designs of Charles Ricketts' (Ph.D. Diss., Univ. of Toronto, 1979).

—— *The Theatrical Designs of Charles Ricketts* (Ann Arbor, Mich., 1985).

BINYON, LAURENCE [Review of *Self-Portrait*], *Observer* (10 Dec. 1939).

—— [Review of *Prado*], *Saturday Review* (9 July 1904).

—— 'Theatrical Designs by Charles Ricketts', *British Museum Quarterly*, 8 (1933), 6.

BIRNBAUM, MARTIN, *The Last Romantic* (NY, 1960).

BLANCHE, JACQUES-ÉMILE, *La Pêche aux Souvenirs* ed. Armand Pierhal (Paris 1949).

—— *Portraits of a Lifetime*, trans. and ed. Walter Clement (1937).

BLAND, DAVID, *History of Book Illustration*, 2nd edn. (1968).

BLISS, D. P., *A History of Wood Engraving* (1928).

Books from the Library of John Lane, Publisher (Dulau and Co. [*c.*1936]).

BOTTOMLEY, GORDON, 'Charles Haslewood Shannon', *Dictionary of National Biography, 1931–40* (Oxford, 1949), pp. 805–6.

—— 'Charles Ricketts 1866–1931', *Durham University Journal*, 32, NS (June 1940), 169–84.

—— 'Charles Ricketts, RA', *Theatre Arts Monthly*, 16 (May 1932), 377–95.

BRADSHAW, PERCY V., 'The New *Mikado*: Mr Charles Ricketts's Stage Designs', *Studio*, 92 (1926), 415–19.

BRIDGE, URSULA (ed.), *W. B. Yeats and T. Sturge Moore: Their Correspondence* (1953).

BROOKS, MICHAEL, 'Oscar Wilde, Charles Ricketts, and the Art of the Book', *Criticism*, 12 (1970), 301–15.

BURDETT, OSBERT, *The Beardsley Period* (1925).

Burke's Landed Gentry, i (1952).

CALLOWAY, STEPHEN, '"Tout pour l'art": Charles Ricketts, Charles Shannon, and the Arrangement of a Collection', *Journal of the Decorative Arts Society*, 8 (1973), 19–28.

—— and DELANEY, [J. G.] PAUL, *Charles Ricketts and Charles Shannon: An Aesthetic Partnership* (Twickenham, 1979).

—— *Charles Ricketts: Subtle and fantastic Decorator*, foreword by Kenneth Clark (1979).

CASSON, JOHN, *Lewis and Sybil: A Memoir* (1972).

CIPPICO, COUNT ANTONIO, 'Charles Shannon', *Vita d'Arte*, 5 (1910), 87–102.

CLARK, KENNETH, *Another Part of the Wood* (1974).

COLDWELL, JOAN, '"Images that Yet Fresh Images Beget": A Note on Book Covers', in Robin Skelton and Ann Saddlemyer (eds.), *The World of W. B. Yeats* (Seattle, Wash., 1967), 134–9.

CRAIG, GORDON, letter to the Editor, *Saturday Review* (25 Sept. 1909).

CRANE, WALTER, *Of the Decorative Illustration of Books Old and New* (1896).

DARRACOTT, JOSEPH, *All for Art: The Ricketts and Shannon Collection* (Cambridge, 1979).

—— *The World of Charles Ricketts* (1980).

DELANEY, [J. G.] PAUL, 'Aesthete Valued: Charles Ricketts', *Radio Times* (11–17 Aug. 1979), 12–13.

—— 'Book Design: A Nineteenth-Century Revival', *Connoisseur*, 198 (Aug. 1978), 282–9.

—— 'Charles Ricketts and his unlikely Friendship with George Bernard Shaw', *The Pen: Broadsheet of the International Centre of English Pen*, 15 (1983), 3–5.

—— 'Charles Ricketts: The Decisive Friendship of Oscar Wilde', *Antiquarian Book Monthly Review*, 5 (1978), 290–3.

—— 'Charles Shannon: Master of Lithography', *Connoisseur*, 200 (1979) 200–5; repr. in Garton and Cooke, *Catalogue Thirty* (autumn, 1984), 53–5.

—— 'Father of the well-made Book: Charles Ricketts, Designer', *Country Life*, 97 (1975), 218–19.

—— F. Ernest Jackson: Draughtsman and Lithographer', *Apollo*, 125 (1987), 338–43.

—— 'Heirs of the Great Generation: Yeats's Friendship with Charles Ricketts and Charles Shannon', Warwick Gould (ed.), *Yeats Annual*, 4 (1986), 53–73.

—— (ed.), *Letters from Charles Ricketts to 'Michael Field' 1903–14* (Edinburgh, 1981).

—— *The Lithographs of Charles Shannon* (1978).

—— (ed.), *Some Letters from Charles Ricketts and Charles Shannon to 'Michael Field' 1894–1902* (Edinburgh, 1979).

—— 'Whistler, Shannon, and the Revival of Lithography as Art', *Nineteenth Century*, 4 (winter 1978), 75–80.

DERRY, GEORGES [R. A. Walker], 'The Lithographs of Charles Shannon', *Print Collector's Quarterly*, 5 (1914), 393–406.

DOWSON, ERNEST, *Letters*, ed. Desmond Flower and Henry Maas (1967).

ELLMANN, RICHARD, *Oscar Wilde* (1987).

ENGEN, RODNEY, *Laurence Housman* (Stroud, 1983).

'FIELD, MICHAEL' [i.e. Katherine Bradley and Edith Cooper], *Works and Days* (1933).

FINNERAN, R. J., HARPER, G. M., and MURPHY, W.M. (eds.), *Letters to W. B. Yeats*, 2 vols. (1977).

FLETCHER, IFAN K., 'Charles Ricketts and the Theatre', *Theatre Notebook*, 22 (1976), 6–23.

FLOWER, D., and MAAS, H. (eds.), *The Letters of Ernest Dowson* (1967).

FRANKLIN, COLIN, *The Private Presses* (1969).

FRENCH, CECIL, 'The Wood-Engravings of Charles Ricketts', *Print Collector's Quarterly*, 14 (1927), 191–217.

FRY, ROGER, *Letters*, ed. Denys Sutton, 2 vols. (1972).

—— 'Mr C. H. Shannon's Work at the Dutch Gallery', *Athenaeum* (23 Mar. 1901), 376–7.

FURST, HERBERT, 'Charles Ricketts, ARA, and his Stage Work', *Apollo*, 1 (1925), 329–54.

—— 'The Modern Woodcut', pt. 2, *Print Collector's Quarterly*, 8 (1921), 267–98.

GALLATIN, A. E., 'Introduction', *An Exhibition of Books designed by Charles Ricketts, from the collection of A. E. Gallatin* (The Houghton Library, Harvard Univ. Press, 1946).

[GEORGE, ERIC B.], *Contemporary British Artists: Charles Shannon* (1924).

GERE, CHARLOTTE, *Victorian Jewellery Design* (1972).

GIBSON, FRANK, 'Idylls of Rural Life: A series of Woodcuts by Charles Shannon, ARA', *Studio*, 58 (5 Sept. 1916), 211–15.

GOFFIN, PETER, 'Designing for Gilbert and Sullivan', *Studio*, 149 (1955), 161–7.

GORDON, D. J., and FLETCHER, IAN, *The Poet W. B. Yeats* (Reading, 1957).

GREGORY, LADY, *Hugh Lane* (1921).

GRIERSON, MARY, *Donald Francis Tovey* (1952).

GUTHRIE, JAMES, 'Charles S. Ricketts and Charles Shannon', *Books and Bookplates: A Quarterly for Collectors*, 4 (1903–4), 181–92.

GWYNN, FREDERICK L., *Sturge Moore and the Life of Art* (1952).

HARPER, CHARLES G., *English Pen Artists of Today* (1892).

HARRIES, MEIRION and SUSIE, *The War Artists* (1983).

HART-DAVIS, RUPERT, *The Caricatures of Max Beerbohm* (1970).

—— (ed.) *The Letters of Oscar Wilde* (1962).

HATCH, BENTON L., 'Charles Ricketts's *Hero and Leander*, *Colby Library Quarterly*, 3 (Feb. 1952), 77–9.

—— 'Longus at the Vale', *Colby Library Quarterly*, 3 (Aug. 1952), 103.

HIND, C. LEWIS, 'Charles H. Shannon, Artist and Connoisseur', *Studio*, 46 (1909), 3–16.

—— 'Charles Ricketts: A comment on his Activities', *Studio*, 48 (1910), 259–66.

HODIN, J. D., 'A Swedish Painter in England: Henning Nyberg', *The Anglo-Swedish Review* (June 1954), 208–10.

HOLMES, C. J., 'Original Wood Engraving', *Architectural Review*, 6 (1899) 106–16.

—— *Self and Partners (Mostly Self)* (1936).

HONE, JOSEPH, *The Life of Henry Tonks* (1939).

HONE, JOSEPH, *W. B. Yeats, 1865–1939* (1942).

HOUSMAN, LAURENCE, *The Unexpected Years* (1937).

HUMPHREY, JAMES, 'Books from the Vale Press', *Colby Library Quarterly*[3], 3 (Feb. 1951), 58–68.

HUSSEY, CHRISTOPHER, 'The Keep, Chilham Castle, Kent: A Residence of Mr Charles Shannon, RA and Mr Charles Ricketts, ARA', *Country Life*, 55 (21 June 1924), 1000–6.

JACKSON, F. E., 'Modern Lithography', *Print Collector's Quarterly*, 11 (1925), 205–26.

JACKSON, HOLBROOK, *The Eighteen Nineties* (1922).

JAFFÉ, MICHAEL, 'Van Dyck Studies: i: The Portrait of Archbishop Laud', *Burlington Magazine*, 124 (1982), 600–7.

[KAINS-JACKSON], anon. obituary of J. A. Symonds, *Pall Mall Budget* (27 Apr. 1893).

KENNEDY, DENNIS, *Granville-Barker and the Dream of the Theatre* (Cambridge, 1985).

KENNET, Lady (Kathleen Bruce), *Self-Portrait of An Artist* (1949).

LAGO, MARY, and BECKSON, KARL (eds.), *Max and Will: Max Beerbohm and William Rothenstein: Their Friendship and Letters 1893–1945* (1975).

—— (ed.), *William Rothenstein: Men and Memories* (1978).

[LANE, JOHN], *Books from the Library of John Lane, Publisher* (Dulau and Co., [*c*.1930]).

LAVER, JAMES, 'The Scenic Designs of Charles Ricketts', *Drama*, 10 (1931), 34–6.

LECONTE, GEORGES, 'Lucien Pissarro et ses amis', *Art et Critique* (1892).

LE GALLIENNE, RICHARD, *The Romantic '90s.*, introd. H. Montgomery Hyde (1951).

LEGGE, SYLVIA, *Affectionate Cousins* (Oxford, 1980).

LEWIS, CECIL, *Never Look Back* (1974).

LOCARD, HENRI, 'Le Journal d'Edmond et Jules de Goncourt et *Works and Days* de "Michael Field"', *Confluents: études victoriennes*, 1 (1976), Univ. Lyon 2, 53–77.

LOWINSKY, THOMAS E., 'Charles Ricketts', in *Dictionary of National Biography, 1931–40* (Oxford, 1949), 732–3.

LUCIE-SMITH, EDWARD, *Symbolist Art*, (1972).

MARILLIER, H. C., 'The Vale Press and the Modern Revival of Printing', *Pall Mall Magazine*, 22 (1900), 179–90.

MASON, STUART, *Bibliography of Oscar Wilde: With a Note by Robert Ross* (1914; repr. in 1 vol., 1970).

MCCARTHY, LILLAH, *Myself and my Friends* (1933).

MEADMORE, W. S., *Lucien Pissarro: Un Cœur simple* (1962).

MEYNELL, VIOLA (ed.), *Friends of a Lifetime: Letters to Sydney C. Cockerell* (1940).
MILLER, LIAM, *The Noble Diana of W. B. Yeats* (Dublin, 1977).
MOORE, T. STURGE, 'Charles Ricketts', *Observer* (11 Oct. 1931).
—— *Charles Ricketts, RA: Sixty-five Illustrations* (1933).
—— *Exhibition of the Lithographs of the late Charles Shannon* (Colnaghi, 1938).
—— 'A Note on Pictures by Charles Ricketts', *The RCA Students' Magazine*, 1 (1912), 133–7.
[National Art Collections Fund], *Twenty-Sixth Annual Report, 1929* (1930).
—— *Twenty-Ninth Annual Report, 1932* (1933).
NELSON, JAMES G., *The Early Nineties: A View from the Bodley Head* (Cambridge, Mass., 1971).
NUDELMAN, EDWARD D., *Catalogue*, 9 (Seattle, Wash., 1985).
ORMOND, LEONIE and RICHARD, *Lord Leighton* (1975).
PENNELL, JOSEPH, *Modern Illustration* (1895).
[PERKINS, GEOFFREY], *A Collection of Books designed by Charles Ricketts* (Art Ancien, Zurich, 1972).
PISSARRO, CAMILLE, *Letters to his Son, Lucien*, 4th edn., ed. John Rewald, (1980).
PRAZ, MARIO, *The House of Life*, trans. Angus Davidson (1964).
PRESTON KERRISON (ed.), *Letters from Graham Robertson* (1953).
PURDOM, C. B., *Harley Granville-Barker* (1953).
QUILTER, HARRY, *Preferences in Art, Life, and Literature* (1892).
QUINN, RICHARD, 'Charles Ricketts and the Dial' (Ph.D. Diss., Univ. of Wisconsin, 1971).
R., F. R., 'The Vale Press', *New York Times* (13 Aug. 1904).
REID, B. C., *The Man from New York: John Quinn and his Friends* (New York, 1968).
REID, CHARLES, *Thomas Beecham: An Independent Biography* (1961).
REWALD, JOHN (ed.), *Camille Pissarro: Letters to his son Lucien*, 4th edn. (1980).
—— 'Lucien Pissarro: Letters from London', *Burlington Magazine*, 91 (1949), 188–92.
REYNOLDS, SIMON, 'Sir Edmund Davis: Collector and Patron of the Arts', *Apollo*, 111 (1980), 459–63.
RICKETTS, CHARLES, 'The Abbey and Zeppelin Raids', *Observer* (14 Feb. 1914).
—— 'Antica pittura fiorentina al Burlington Fine Arts Club di Londra', *La Ronda*, vol. 1, no. 5 (Sett, 1919), 81–3.
—— [illustration], *Atalanta*, 3 (Dec. 1889), [1897].
—— [under pseudonym, J. P. Raymond], *Beyond the Threshold* (1929).
—— *A Bibliography of Books issued by Hacon and Ricketts* (1904).
—— 'Book-Printing', *Encyclopaedia Britannica*, 10th edn., 26 (1902), 306.
—— 'The British Medical Association: New Building', *The Times* (24 June 1908).

RICKETTS, CHARLES, 'The Censorship and Foreign Plays', *Observer* (29 Jan. 1928).
—— *Catalogue of Works by Orpen, McEvoy, and Ricketts* (Manchester, 1933).
—— *A Century of Art, 1810–1910* [1911].
—— *A Defence of the Revival of Printing* (1899).
[——] [Designs for Jewellery in *Attila*], *Studio*, 42 (1908), 137–8.
—— [Designs for *Macbeth*], *Observer* (19 Dec. 1926).
—— 'Desultory Musical Recollections', *Music Bulletin*, 5 (1923), No. 1, pp. 10–12; No. 2, pp. 43–6; No. 3, pp. 88–90, 97; No. 5, p. 160.
—— 'Dürer and his Successors', *Burlington Magazine*, 9 (1906), 264–8.
—— 'The Franco-British Exhibition', *Burlington Magazine*, 13 (1908), 192–5.
—— 'Grants for saving the National Treasures', *Observer* (13 Jan. 1918).
—— 'Greek Dress', *Saturday Review* (9 Jan. 1909), 40–1.
—— 'In Praise of the Pianola', *Music Bulletin*, 3 (1921), 60–3.
—— 'Introduction', *Catalogue of the Bliss Collection of Alphonse Legros* (1923).
—— 'Legros', *Saturday Review* (17 Apr. 1897), 406–7.
—— 'The Masterpieces by Velasquez at the Imperial Gallery at Vienna', *Burlington Magazine*, 5 (104), 338–48.
—— *Michael Field: A Memoir*, ed. [J. G.] Paul Delaney (Edinburgh, 1976).
—— 'The National Gallery of Canada', *Observer* (19 Nov. 1930).
—— 'National Treasures and Zeppelins', *Observer* (10 Sept. 1915).
—— 'The New Literary Censorship', *Observer* (4 Oct. 1928).
—— 'A Note on original Wood Engraving', *Pageant*, 2 (1897), 253–66.
—— *Pages from a Diary in Greece*, ed. [J. G.] Paul Delaney (Edinburgh, 1978).
—— *Pages on Art* (1913).
—— *The Prado and its Masterpieces* (1903), reissued in smaller format as *The Art of the Prado* (Boston, 1907).
—— 'Princess Alice of Monaco: A Tribute by Mr Ricketts', *Observer* (10 Jan. 1926).
—— 'The Sculpture and Studies of Mr Ernest Cole', *Country Life* (19 Oct. 1916), 328–30.
—— *Self-Portrait: Taken from the Letters and Journals of Charles Ricketts, RA*, collected and compiled by T. Sturge Moore, ed. Cecil Lewis (1939).
—— [illustrator] *Shakespeare's Heroines* [1926].
—— 'Some Marginal Recollections of the "Nineties"', *Observer* (15 Mar. 1931).
—— Letter to *The Times* regarding the *Butterfly*, *The Times* (22 Apr. 1914).
—— *Titian* (1910).
—— 'The True Significance of Dress', *Saturday Review* (27 Feb. 1909), 268–9.
—— *Unrecorded Histories*, 1933.
—— 'Vladimir Rosing', *Manchester Guardian* (1 Apr. 1919).
—— 'What Removal Means', *Observer* (6 Jan. 1918).
—— 'What Removal Means' *Daily Telegraph* (7 Jan. 1918).

—— Why I Redressed *The Mikado*, *Daily Mail* (18 Sept. 1926).

—— LAVERY, JO and SARGENT, J. S., letter to *The Times* (30 June 1915).

—— and PISSARRO, LUCIEN, *De la Typographie et de l'harmonie de la page imprimée* (1898).

—— 'Of Typography and the Harmony of the Printed Page', trans. from the French of CR by Richard K. Kellenberger, *Colby Library Quarterly*, 3 (Nov. 1953), 194–200.

—— 'William Morris and his influence on the Arts and Crafts', trans. from the French of CR by Richard K. Kellenberger, *Colby Library Quarterly*, 3 (Feb. 1952), 69–75.

—— and RAYMOND, JEAN PAUL (i.e. CR), *Oscar Wilde: Recollections* (1932).

—— and SHANNON, CHARLES, 'The Closing of Museums', *Observer* (16 Feb. 1916).

—— *Daphnis and Chloe*, *Prospectus* (1892).

—— (eds.), *Dial*, 1–5 (1889–97).

—— 'The Future Protection of National Treasures', *Observer* (10 Feb. 1916).

—— 'Governmental Precautions against Zeppelins', *Observer* (16 Oct. 1915).

RINDER, FRANK, 'The Drawings of Mr Charles Shannon', *Art Journal* (1902), 43–6.

ROBERTSON, GRAHAM, *Letters from Graham Robertson*, ed. Kerrison Preston (1953).

ROSS, ROBERT B., *Friend of Friends: Letters to Robert Ross*, ed. Margery Ross (1952).

ROTHENSTEIN, JOHN, *The Artists of the 1890s* (1928).

—— *Summer's Lease: Autobiography, 1901–38* (1965).

ROTHENSTEIN, WILLIAM, *Men and Memories*, 2 vols. (1931–2).

—— *Since Fifty* (1939).

RUTTER, FRANK, *Art in My Time* (1972).

RYSCAMP, CHARLES, *Wilde and the Nineties: An Essay and an Exhibition* (Princeton, NJ, 1966).

SALAMAN, MALCOLM, 'The Sculpture of Ernest Cole', *Studio*, 78 (Jan. 1920), 140–8.

SCARISBRICK, DIANA, 'Charles Ricketts and his Designs for Jewellery', *Apollo*, 116 (Sept. 1982), 163–9.

SCHUCHARD, RONALD W., 'W. B. Yeats and London Theatre Societies, 1901–1904', *RES*, NS, 29 (178), 415–46.

SCOTT, TEMPLE, 'Mr Charles Ricketts and the Vale Press', *Bookselling*, 2 (Dec. 1896), 501–12.

Second Report of the Royal Commission on Technical Instruction, 1882–4, Reports Commissions in the House of Commons Sessional Papers 31 (ii), 184.

SEWELL, BROCARD, *In the Dorian Mode: A Life of John Gray: 1866–1934* (Padstow, 1983).

SHAKESPEARE, WILLIAM, *Macbeth*, illustrations by Charles Ricketts, ARA (1923).

SHAW, GEORGE BERNARD, *Letters*, ii (1898–1910), ed. Dan Laurence (1972).

—— *Letters*, iii (1911–25), ed. Dan Laurence (1985).

—— *St Joan*, illustrations by C. Ricketts (1924).

SHERINGHAM, GEORGE, AND MORRISON, R. BOYD (eds.), *Robes of Thespis: Costume Designs by Modern Artists* (1928).

SKELTON, R., AND SADDLEMYER, A. (EDS.), *The World of W. B. Yeats* (Seattle, Wash., 1967).

Society of Twelve, Catalogue of the First Exhibition (1904).

South London Technical Art School, *Prospectus* for 1882–3 (1882).

SPALDING, FRANCES, *Roger Fry: Art and Life* (1980).

SPEAIGHT, ROBERT, *William Rothenstein: The Portrait of an Artist in his Time* (1962).

SPIELMANN, M. H., 'Original Lithography: The Present Revival in England', *Magazine of Art* (1897), 289–96.

—— 'Our Graphic Humorists: Mr L. Raven-Hill', *Magazine of Art* (1896), 494.

SPRIGGE, ELIZABETH, *Sybil Thorndike Casson* (1971).

STEELE, ROBERT, *The Revival of Printing* (1912).

STURGEON, MARY, *Michael Field* [1921].

SUTTON, DENYS, 'Charles Ricketts and Titian', *Apollo*, 108 (1978), 428–33.

—— (ed.), *Letters of Roger Fry*, 2 vols. (1972).

—— 'A Neglected Virtuoso: Charles Ricketts and his Achievement', *Apollo*, (1966), 138–47.

—— 'An All-Round Aesthete', *The Financial Times* (21 Sept. 1966).

SYMONS, A. J. A., 'An Unacknowledged Movement in Fine Printing', *Fleuron*, 7 (1930), 83–120.

TAYLOR, JOHN RUSSELL, *The Art Nouveau Book in Britain* (1966; reissued 1979).

'THEOCRITUS', 'The Vale Artists: i: Charles Hazlewood Shannon', *Sketch*, 8 (23 Jan. 1895), 617.

—— 'The Vale Artists: ii: Charles Ricketts', *Sketch*, 9 (13 Mar. 1895), 350.

—— 'The Vale Artists: iii: Lucien Pissarro', *Sketch*, 9 (17 Apr. 1895), 615.

—— 'The Vale Artists: iv: Reginald Savage', *Sketch*, 9 (24 Apr. 1895), 683.

THORNDIKE, RUSSELL, *Sybil Thorndike* (1950).

THORNTON, ALFRED, *Diary of an Art Student in the Nineties* (1935).

THORPE, JAMES, *English Illustration: The Nineties* (1935).

'Tis', 'About Charles Ricketts', *Colour* (June 1921), 94–7.

—— *Masters of Modern Art: Charles Shannon, ARA* (n.d.).

TYE, J. R., 'The Letters of Lord de Tabley', *Turnbull Library Record*, NS, 8 (Oct. 1972), 17–24.

UZANNE, OCTAVE, *L'Art dans la décoration extérieure des livres en France et à l'étranger* (Paris, 1898).

Vale Press Prospectuses (1895–8).

WADE, ALLAN (ed.) *The Letters of W. B. Yeats* (1962).

WEBER, CARL J. 'Charles Ricketts and his Books', *Colby Library Quarterly*, 3 (Nov. 1951), 53–7.

—— *The Rise and Fall of James Ripley Osgood* (Waterville, Maine, 1951).

WEDMORE, F., 'The Revival of Lithography', *Art Journal* (1896), 11–14, 41–4.

WEITENKAMPF, RUDOLF, *The Illustrated Book* (1938).

WHITE, GLEESON, 'The Artistic Decoration of Book Covers', *Studio*, 4 (1894), 15–23.

—— 'The Arts and Crafts Exhibition at the New Gallery, 1893', *Studio*, 2 (1893), 57.

—— 'A New Edition of Longus', *Studio*, 1 (1893), 82–3.

—— [The Vale Press] 'Studio Talk', *Studio*, 9 (1896), 62–4.

WILDE, OSCAR, *Letters*, ed. Rupert Hart-Davis (1962).

YEATS, W. B., *Autobiographies: Reveries over Childhood and Youth* (1914, coll. edn., 1926).

—— *The Letters of W. B. Yeats*, ed. Allan Wade (1962).

Index

Abbey, Edwin 30, 32
Académie Julian 33, 49
Academy, the 44, 100, 103
Accademia, Venice 132–5, 160, 232
Adam and Eve (Tintoretto) 361
Adey, More 107–10
Aeneid, The (Morris's and Burne-Jones's attempt to illustrate) 63
Aestheticism and the Aesthetic Movement 26, 45, 142
Agnew, Colin 360–1
Agony in the Garden (Bellini) 271
Aida (Verdi) 341
Ainley, Henry 294
Aitken, Charles 286
Alarum, the 32
L'Album D'Estampes Originales 91
Alexandra, Queen 294
Alice of Monaco, Princess 300, 399
Allen, Maude 321–2
Alma-Tadema, Sir Lawrence 29
Alvary, Max 72
Amiens, France 103, 156, 276
L'Amour Masqué (Guitry and Messager) 369
Anderson, Dr Elizabeth Garrett 187
Anna Ruina (Field) 128
Anthology of Nineties Verse, An (ed. Symons) 393
Aphrodite (Louÿs) 251
Appia, Marie, *see* Mrs Marie Sturge Moore
Archbishop Laud (van Dyck) 399
Archer, William 199, 229
Aretino, Pietro 110, 240–1
Ariosto, Ludovico 110
Arms and the Man (Shaw) 225, 226
L'Art et les Artistes 246
Art Journal 176
Art Nouveau 46, 60, 76, 84, 91, 98, 181
Arts and Crafts Movement 43, 63, 64, 65, 70, 75, 81, 86, 100, 101, 125, 140, 181
Ashbee, C. R. 125
Ashmolean Museum 131, 145, 242
Ashwell, Lena 303
Asnières 12–13
Asquith, Henry, first earl of 270–1
Asquith, Margot 271, 297, 298
Assùnta (Titian) 134
Assisi, CR visits 204, 395
Atalanta 46
Atalanta (Shannon) 103
Athenaeum, The 44, 100, 126, 154
Avory, Rt. Hon. Sir Horace Edmund 368

Bacchus and Ariadne (Titian) 240
Bach, Johann Sebastian 213, 226, 340
Baedeker, Karl 162, 255
Bakst, Leon 259, 260–1, 304
Baldovinetti, Alesso 163
Balfour, Arthur, earl of 287
Ballantyne Press 69, 91, 95, 141, 150, 173, 174
Ballets Russes 258–61, 265, 267, 298, 304–5
Balzac, Honoré de 234, 365
Barbari, Jacopo de 310
Barrie, J. M. 231, 270–1, 297
Bartók, Béla 379
Barye, Antoine-Louis 190
Basle 136, 161
Bassano 10
Bastien Lepage, Jules 13
Baveno, CR visits, 335, 346
Bath of Venus, The (Shannon) 291, 384
Baudelaire, Charles 13–14, 51, 54, 57, 207, 216, 291, 392
Baudry, Paul 30, 263, 264, 393
Beardsley, Aubrey 2, 66, 82–5, 88, 97, 126–7, 272, 318
Beardsley, Mabel 85, 231, 272, 318
Beardsley, Mrs 85, 318
Beatrice denying her Greeting to Dante on the Public Stairs (Rossetti) 119
Beaufort Street 31, 92–3, 94, 98, 113, 122, 153
Beaumont, Cyril 304
Beecham, Sir Thomas 1, 299, 300–1
Beerbohm, Sir Max 49, 100, 162, 201, 237–8

Beethoven, Ludwig van 142, 301
Beit, Sir Otto 298
Bell and Sons, George (publishers) 63
Bellini, Gentile 117, 132, 133, 335
Bennett, Arnold 295
Benson, Robert Henry 361, 364, 365, 376
Bensusan, Esther (Mrs Lucien Pissarro) 72, 107
Berenson, Bernard 1, 87, 88, 127, 146, 157, 158, 162–3, 171, 195, 205, 222, 242, 340, 360–1
Berenson, Mary 195, 196
Bergamo 161
Berlin 119, 279, CR visits 166–7, 266, 386
Berlioz, Hector 241
Bernard, Émile 248
Bernhardt, Sarah 31, 96
Bethlehem (Housman) 197
Bigio, Francesco (Franciabigio) 118–19
Billing, Pemberton 321–2
Bing, Samuel 91
Binyon, Laurence 19, 54–5, 121, 170, 178–9, 186, 197, 199, 222, 226–8, 229, 242, 286, 290, 312, 318, 340, 376, 400
Biondina (Shannon) 91
Birnbaum, Martin 269, 376
Bishop, Gwendolyn 198, 199
Black and White 46
Blackwood's Magazine 45
Blake, William 59, 104, 207, 263, 286, 364, 376
Blanche, Jacques-Émile 58, 185, 188
Boccaccio, Giovanni 141
Bodley Head 76–80, 86, 95
Boer War 145, 175
Bone, Sir Muirhead 193, 291
Books and articles written or proposed by CR: 'Alphonse Legros' 52; *Beyond the Threshold* 357, 367, 392; 'Charles Shannon' 246; *Dial* (founds with CHS) 41–4; 'Dictionary of the Italian Masters' (unrealized) 164, 167; 'Handbook on Italian art' (unpublished) 381; 'History of the Art of the World' (unrealized) 381; 'Maximes en Marge' (unpublished) 366–7; *Pages from a Diary in Greece* 392; *Pages on Art* 81, 246, 249, 262; *Prado, The* 2, 164–71, 174, 219, 234, 239, 242; 'Raphael' (unrealized) 170; *Recollections of Oscar Wilde* 144, 357, 389, 392; Revival of Printing in *Encyclopaedia Britannica* 151; *Titian* 201, 219, 238–42, 243, 262; *Unrecorded Histories* 389; 'Watteau' (unrealized) 170; 'Why I Redressed the Mikado' 352
Books with CR designs: *Ballad of Reading Gaol, The* (Wilde) 109; *Beyond the Threshold* (Ricketts) 367; *Book of Thel: Songs of Innocence and Songs of Experience, The* (Blake) 104; *Borgia* ('Field') 272; *Christie's Season, 1931* 392; *Collected Works by W. B. Yeats* 318; *Cupid and Psyche*, see *The Marriage of Cupid and Psyche*; *Danaë* (Sturge Moore) 174; *Daphnis and Chloe* (Longus) 73–6, 80, 85, 124, 315; *De Cupidinis et Psyches Amoribus* 113; *Dedicated* ('Field') 275; *Defence of the Revival of Printing, A* (Ricketts) 126–7; *De la Typographie* (Pissarro and Ricketts) 107; *Dial, The* 41, 43, 48, 55, 65, 69–70, 81–2, 85, 98, 100, 101, 113, 151, 262; 'Dialogues of the Dead', see *Beyond the Threshold*; *Early Poems of John Milton, The* 102; *Ecclesiastes* 173; *Empedocles on Etna* (Arnold) 104; *Epicurus, Leontion, and Ternissa* (Landor) 104; *Fair Rosamund* ('Field') 104, 128; *Faustus, Dr* (Marlowe) 198; *Fifty Songs by Thomas Campion* 102; *Golden Treasury of Modern Lyrics, The* (ed. Binyon) 318; *Group of Noble Dames, A* (Hardy) 62; *Gruach and Britain's Daughter* (Bottomley) 316; *Hamlet* (Shakespeare), Vale Press 141; *Hero and Leander* 85–6, 124, 316; *House of Pomegranates, A* (Wilde) 59–62; *Intentions* (Wilde) 59; *In the Key of Blue* (Symonds) 76–7; *Julia Domna* ('Field') 128–30; *Kingis Quair, The* 98–9, 101; *Lord Arthur Savile's Crime* (Wilde) 59; *Marriage of Cupid and Psyche, The* 124, 126, 179; *Mystic Trees* ('Field') 275; *Nimphidia* (Drayton) 104–5; *Pageant, The* 100, 103; *Parables, The* 171–2, 193; *Picture of Dorian Gray, The* (Wilde) 45, 59; *Poems* (Wilde) 45, 76, 81; *Poems and Songs of Sir John Suckling, The* 102, 104; *Poems Dramatic and Lyrical* (de Tabley) 78–81, 363; *Poems in Prose* (Wilde) 318, 376; *Poems of Adoration* ('Field') 275; *Poems of Sir John Suckling, The* 104; *Poems of Thirty Years* (Bottomley) 316; 'Portrait of Mr W. H., The' (Wilde) 45, 102, 109; *Recollections of Oscar Wilde* 392; *St Joan* (Shaw) 344–5; Shakespeare, Vale Press edition 171, 173; *Shakespeare's Heroines* 353; *Silverpoints* (Gray) 1, 76–7, 81; *Sonnets of Sir Philip*

Sidney, The 102, 106; *Sphinx, The* (Wilde) 2, 82–3, 91, 104, 115, 143–4, 145, 230, 252, 290–1, 392; *Sphinx* (2nd set) 318, 364, 376; *Spiritual Poems* (Gray) 104; *Tess of the D'Urbervilles* (Hardy) 62; *Troy* (Wolfe) 367; *Unrecorded Histories* 392; *Vision of Giorgione, A* (Bottomley) 316; *Wild Honey* ('Field') 274; *World at Auction, The* ('Field') 106, 128; founts designed by CR 95–6, 172–3, 175; unrealized projects for books and magazines 85, 102, 176

Bordone, Paris 370

Boris Godunof (Mussorgsky) 267, 275

Botticelli, Sandro 157, 234, 338–9, 357, 380, 385

Bottomley, Gordon 22, 52, 57, 151, 193, 210, 237, 260, 316–18, 333, 346–7, 390–1

Boudin, Eugène 361

Boulogne 12

Bourdelle, Émile Antoine 260

Boutet de Monvel, Bernard 341

Bradley, Katherine, *see* 'Field, Michael'

Brahms, Johannes 302

Brayton, Lily (Mrs Asche) 227, 231

Bréal, Auguste 155

Bremen Kunsthalle 151, 246

Brescia 136, 164

Breugel, Pieter 166

Bridges, Robert 100

Brighton 250

Brinkley, Capt. Frank 115

British Museum 121, 156, 175, 277, 281, 302–3

Britomart (Watts) 262

Broadbent, Edith, *see* Hacon, 'Ryllis

Broadstairs 92, 159, 166

Brooke, Rupert 270, 282, 297

Brown, Arnesby 337

Brown, Eric 337–40, 361–3, 369–70, 380–1, 387

Brown, Ford Madox 37

Brown, Fred 26

Bruce, Kathleen (Lady Scott and Lady Kennet) 220–1, 230, 232, 253, 298

Brussels 21, 264

Buckman, Rosina 300

Burlington Fine Arts Club 116

Burlington Magazine, The 155, 170, 176–7, 246, 262

Burne-Jones, Sir Edward Coley 56, 63, 64, 96, 100, 110, 115, 141, 214, 263, 269, 310, 326, 364, 373, 374

Butler, Samuel 135

Buzzard, Sir Farquhar 372, 374

Byam Shaw School 359, 381, 383

Cambridge 156

Camel, The (Persian) 117

Cameron, Sir D. Y. 193, 291, 327

Campbell, Mrs Patrick 199, 228, 233, 246, 268

Canaletto, Antonio 285

Carfax Gallery 119, 121, 151, 154, 201, 229

Cariani (Giovanni Busi) 370, 380

Carmichael sale 116

Carnaval (Schumann) 258–9, 302

Carpaccio, Vittore 118, 132

Carrots (squirrel) 265

Cassatt, Mary 66

Casson, Lewis 341, 344, 347

Castiglione, Giovanni Benedetto 338

Catena, Vincenzo 132

Catullus 13, 102

Cellini, Benvenuto 241

Centaur theme in CR painting 146–7, 175

Century Guild Hobby Horse 43, 44, 64, 70, 151

Cézanne, Paul 148, 177, 203, 246, 248–9, 386, 395

Chaliapin, Feodor 267, 298

Chamberlain, Sir Austen 178

Chamberlain, Lady 384

Chapman, Max 359, 391

Charley's Aunt (Thomas) 258

Chassériau, Théodore 291

Chesterton, G. K. 229, 270, 318

Chifflart, Nicolas François 264

Chopin, Frederic 15, 301–2, 314, 379

Christ and St John (Botticelli) 338–9, 380

Christ Carrying the Cross (Rubens) 370

Christ in the Garden of Olives (Gauguin) 248

Christchurch, Hampshire 107

Christmas Carol, The (Rossetti) 114

Cima da Conegliano, Giovanni Battista 132

Cippico, Count Antonio 240, 246, 383–4

City and Guilds Technical Art School 27, 51

Clark, 'the dark' 38

Clark, Kenneth, Lord 1, 2, 181, 283, 385

Clausen, Sir George 147, 193, 242, 326

Cleopatra (Massenet) 258–9, 304–5

Clouet, François 45

Cockerell, Sir Sydney 122, 125, 131, 140, 242, 250, 251, 290, 310, 327, 342

Cole, Ernest 288–9, 302, 306, 308

Coles, W. C. 26

Collier, Constance 199
Colvin, Sir Sidney 52, 53, 118
Conder, Charles 92, 163, 193
Connard, Philip 326–7
Constable & Co. Ltd, Archibald (publishers) 165
Constable, John 102, 263, 290
Constable, W. G. 384
Contemporary Review 126–7
Cook, Sir Herbert 285
Coombs, Mrs 388
Cooper, Amy (Mrs Ryan) 137, 222
Cooper, Edith, *see* 'Field, Michael'
Cornforth, Fanny 64
Corot, Jean Baptiste Camille 263
Correggio, Antonio 166
Cotton, Albert Louis 126
Countess Cathleen (Yeats) 197
Courbet, Gustave 249, 361
Craig, Edward Gordon 193, 197–8, 238, 262, 328, 341–2
Cranach, Lucas 387
Crane, Walter 80, 103, 198
Craven St, Strand, no. 17 137
Credi, Lorenzo di 27, 118–19
Crivelli, Carlo 8, 335
Croydon Art School 31
Crystal Palace 6, 119
Cunard, 'Emerald', Lady 284, 299–300
Cunninghame Graham, R. B. 100
Curzon, George Nathaniel, Marquess 289, 303
Cust, Sir Lionel 242–3

Daily Mail 352
Daily Telegraph 196, 302
Daniel Barbaro (Titian) 369
D'Annunzio, Gabriele 270
Dante Aligheri 216, 252, 286
Daphne (Watts) 339
Darling, Charles, 1st Baron 368
Darragh, Letitia Marion Dallas 199
Daubigny, Charles François 263
Daumier, Honoré 147, 185, 210, 332, 364
David, Jacques Louis 168
Davis, Sir Edmund 52, 121, 154–5, 163, 176, 184, 185, 186–7, 191–2, 232, 262, 268, 289, 290, 309, 311, 312, 335, 339, 346, 375
Davis, Mary, Lady 52, 163, 184, 186, 191–2, 199, 290, 308, 310, 311, 312, 330, 334, 335, 346
Deacon, Esther (Hetty) 159–61, 179, 187, 220
Dead Christ (Mantegna) 335
Dead Poet and Centaur, A (Moreau) 147
Death of a Doe (Stubbs) 338
Degas, Edgar 65, 96, 152, 247, 249
Delacroix, Eugène 29, 46, 51, 115, 156, 207, 210, 211, 240, 263, 291, 378
Delaunay, Robert 304
Denis, Maurice 248–9, 260
Dent & Co., J. M. (publishers) 63
Descent of the Holy Ghost, The (Titian) 133
Desiderio da Settignano 337
Deslys, Gaby 270–1
Diaghilev, Sergey 258, 260
Dicksee, Sir Frank 29
Die Walkürie (Wagner) 15, 345
Dispute of St Catherine, The (Pinturicchio) 222
Dodgson, Campbell 242
Dolmetsch, Arnold 140
Don Giovanni (Mozart) 7–8, 207
Don Juan theme 7–8, 53, 205–10, 216, 224, 230, 316, 321, 328–9, 353
Donatello 203
Dossi, Dosso 360
Douglas, Lord Alfred 229
'Dove and Fig', textile design by CHS 311
Dowdall, Harold Chaloner 131
Dowdall, Hon. Mrs Chaloner 150
Dowson, Ernest 54, 62, 103
Dresden 246, 279; CR visits 166, 267
Duccio da Siena (di Buoninsegna) 203
Du Cros, Sir Arthur 284
Dulac, Edmund 277, 299, 301
Dulac, Mrs 294
Duncan, Isadora 231–2
Durand-Ruel, Paul 91, 146, 157, 201, 332
Dürer, Albrecht 118, 164
Durlacher, Mr (a dealer) 387
Duse, Eleanora 201, 298
Dutch Gallery, The 123
Duveen, Joseph Baron 360, 362, 363, 365, 387
Dyck, Sir Anthony van 118–19, 164, 165, 202, 229, 234, 247, 308, 386, 399

Eastlake, Sir Charles 36
Egyptian art 251–4, 257, 264–5, 400
Ellis, Havelock 140
Entombment (Titian) 168
Epstein, Sir Jacob 229

Eragny Press 151, 204
Espérance (Puvis de Chavannes) 276
L'Estampe Originale 91
Ethel (servant) 334
Etty, William 24, 25, 115
Eworth, Hans 338, 362, 363
Exhibitions of CR's work: Agnew's (1906) 221; American Exhibition (1914) 269; L'Art Nouveau Bing 91; Bradford 221; CR paintings, Dutch Gallery (1906) 201, 205, 218–19; CR sculpture, Carfax Gallery (1906) 201, 218–19; International Society 187, 230, 268; Le Livre Aesthetique (Brussels) 91; Manchester 234; Nederland Etsclub 91; Society of Twelve 193; Van Wisselingh Gallery (Amsterdam) 91; Wolverhampton (1902) 147–8; Exhibitions organized in whole or in part by CR: Century of Art Exhibition 262–4; Decorative Arts Exhibition (RA) 328; Exhibition of Fair Women 234–5; *Exhibition of Italian Art,* 1930 380–1, 384–6; International Theatre Exhibition (V & A) 328; Old Master Exhibition 234–5; RA Summer Exhibition, 1929 379
Eyck, van 342

Fantin-Latour, Henri 155
Farquharson, Robert 201, 222
Farr, Florence 198, 224
'*Femme aux Chats, La*' (= *Biondina*) 91
Fêtes Galantes (Verlaine) 57
'Field, Michael' (Katherine Bradley and Edith Cooper) 52, 86–91, 98, 102–3, 113, 122–3, 127–31, 137–40, 141, 142–3, 146, 151, 157, 160, 171, 176, 180, 186, 187, 191–2, 196–7, 204–5, 220, 222–3, 228, 244, 246, 310, 312, 316, 319, 333, 390, 391, 393; illness and deaths of the Fields 272–6
Fight between the Lapiths and the Centaurs, The (di Cosimo) 119–21, 147, 381
Fine Art Society 152
Fingal's Cave, Staffa 252
Finney, Dr 390
Finny, Richard 359, 383, 391
Firebird, The (Stravinsky) 259
Fisherman and his Soul, The 259
Fitzwilliam Museum 2, 122, 276, 286, 399–400
Flaubert, Gustave 13–14, 31, 57, 141, 155, 291, 356
Florence 232; CR visits: 161–2, 195, 201, 202, 204, 257, 346, 369, 395
Fokine, Michel 304
Forbes, James Staats 150–1, 153
Forbes, Vivian 312–14, 334–5
Form 290
Fra Angelico 157, 234, 384, 385
Fragonard, Jean Honoré 399
Frampton, George 125
France, Anatole 13
Franklin, Maud 38
French, Cecil 288
Fromentin, Eugène 14, 240, 292
Fry, Roger 1, 52, 121, 144, 153–5, 170, 176–7, 186, 195, 201, 242–3, 246–8, 263, 381
Furse, Mrs Charles 140

Gainsborough, Thomas 116, 185, 309, 338, 370
Garden of Loves, The (Titian) 168
Gardner, Mrs Isabella 360, 365
Gate of Smaragdus, The (Bottomley) 193
Gauguin, Paul 203, 248
Geddes, Patrick, Professor 281
Gelder, Aert de 370
Goncourt, Edmond and Jules de 26, 43
Genoa 14, 194
Germ, The 41, 151
Gilbert, Sir Alfred 124
Gilchrist, Mr, Official Solicitor 400
Giorgione 234, 327, 365, 385
Giotto 136, 203, 278
Girton College, Cambridge 116
Giuliano (jeweller) 179
Good Samaritan theme in CR's painting 147, 230, 274, 287, 290
Gorki, Maxim 192
Gosse, Sir Edmund 96, 103, 289, 291
Goya y Lucientes, Francisco de 65
Gozzoli, Benozzo 387
Grafton Gallery 36, 234
Granville-Barker, Harley 1, 222, 228, 244, 261, 292, 297, 299, 340, 344
Graphic, the 366, 378
Gray family 15, 16
Gray, Canon John 1, 43–4, 49, 52, 54, 55, 64, 68, 69, 76–7, 104, 113, 319
Gray, Ronald 25
Greco, Domenikos Theotocopolous, El 165, 292
Greek art 116–17, 123, 202, 204, 357, 400
'Green Gaffer, The' (Housman) 51

Green, Mrs J. R. 140, 316
Gregory, Augusta Lady 270, 294
Grosvenor Gallery 36
Gruach (Bottomley) 346–7
Guérin, Maurice de 69, 291
Guilbert, Yvette 196–7
Guthrie, James 124

Hacon and Ricketts Ltd 95, 104, 174–5
Hacon, 'Ryllis (Edith Broadbent) 95, 140, 177, 316
Hacon, William Llewellyn 94–5, 98, 106, 174–5, 176, 177, 272
Hakart (i.e. Hackert), Jacob Philip 386
Hall, North 58
Happy Prince, The (Wilde) 45
Hardy, Thomas 62, 96, 201, 292
Harker, Joseph 237
Harris, Frank 109
Hart, Dr Bernard 380
Harunobu 115, 116
Hedmondt, C. E. 377
Hellwag, Rudolph 279–80
Henley, W. E. 40, 52
Henry VIII (Holbein) 119
Henry VIII (Shakespeare) 347, 352, 353, 386
Herrera, Francisco 167
Hewart, Gordon, 1st Viscount 368
Hignett, H. R. 237
Hiroshige 40, 114
History of Greece (Grote) 251
Hodson, Lawrence 98, 147, 203
Hofer, Philip 376
Hofmannsthal, Hugo von 228
Hokkei 234
Hokusai 115, 116, 131, 185, 234, 334
Holbein, Hans the Younger 119, 136, 166, 169
Holford, Lt.-Col. Sir George 376
Holland, Cyril 282
Holland, Vyvyan 229
Hollway, Robin 288
Holmes, Sir Charles J. 37, 55, 56, 57, 68–9, 75, 82, 106–7, 114, 115, 121, 122, 134, 138, 143, 154, 164–5, 170, 173–6, 177–8, 184, 211, 229, 234, 242, 282–3, 285, 314
Holmes, Dr Gordon 374, 390
Holroyd, Sir Charles 280–1, 282
Holst, Gustav 366
Holst, Richard Nicolaas Roland 49
Hopkins the jerboa 249
Hoppner, John 310, 399
Horne, Herbert 43, 56, 64, 65, 95, 121, 162–3, 195, 202
Horner, Frances Lady 284
Houdon, Jean-Antoine 184
Houghton, A. Boyd 28, 123
House of the Wolfings, The (Morris) 63
Housman, Clemence 28
Housman, Lawrence 51, 68, 113, 125, 197
Howard, Francis 187, 268
Howard de Walden, Thomas Evelyn Scott-Ellis, 8th Baron 233
'Hughes, Willie' 45, 102
Hunt, William Holman 36
Huysmans, Joris Karl 43
Hypnerotomachia (Poliphilus) 73

Ideal Husband, An (Wilde) 94, 142
Illustrated London News 366
Image, Selwyn 43, 64, 65, 86, 186, 242–3, 284, 288
Importance of Being Earnest, The (Wilde) 318
In the House of Delia (Shannon) 95
Ingres, Jean Auguste Dominique 48, 51, 291, 364
International Society 124, 186–7, 192, 221–2, 230–1, 234, 262–4, 268, 326
Irish National Theatre Company 198
Irving, Sir Henry 201
Irving, Lawrence 214, 312
Isabey, Eugène-Gabriel 399
Israels, Josef 264
Italian Poets, The (unused frontispiece by Rossetti) 310

Jackson, Ernest 289, 304, 312, 359, 383, 391, 399–400
Jacob's Room (Woolf) 368
James, Henry 25
Japanese art 41, 84, 114, 115, 116–17, 349
Jazz and Jasper (Gerhardi) 368
Jesus Christ theme in CR's work 55–6, 87, 206–8, 216
Joachim, Joseph 195
John, Augustus 147, 152, 193, 368
Johnson, Lionel 54, 84, 103
Johnston, Alec 378, 381
Julius II, Pope 241
Julius Caesar (Shakespeare) 383

Kains-Jackson, Charles 25
Karsavina, Tamara 259
Keats, John 13, 88, 102, 142, 260

Keep, The, Chilham Castle 309–12, 379–80, 389, 396, 400
Keller, Ferdinand 386
Kelmscott Press 63, 91, 122, 126, 151, 176, 291
Kennington, Eric 330
Kessler, Count Harry 157, 192
Kew Gardens 137
Keyserling, Count 157
King, Sir William Mackenzie 361, 362–4, 387
King Lear's Wife (Bottomley) 316
Kingsley, Miss Mary 140
Kipling, Mrs Alice 141
Kipling, Rudyard 141
Kitchener of Khartoum, Earl 265, 289
Kiyonaga 115, 238
Klafsky, Katharina 72
Klinger, Max 67
Kohitsu 157–8
Korin 158, 351

Lady Chatterley's Lover (Lawrence) 368
Lady Windermere's Fan (Wilde) 85, 108
Lady with a Cyclamen, The (Shannon) 150
Lady with a Feather, The (Shannon) 160
Lamb, Henry 102, 330
Lambert, Constant 312
Land, The (Sackville-West) 368
Lane, Sir Hugh 188–90, 282
Lane, John 80–1, 85, 95, 96, 97–8, 109, 167, 290
Lansdowne, Henry, 5th Marquess of 376
Lansdowne House, Lansdowne Rd, Holland Park, London 179, 181–6, 195, 309, 332, 334–5
Lanteri, Edward 152
Lapin, Le (Manet) 201
Lara, Isidore de (Cohen) 300
Lavery, Sir John 124, 291
Lawrence, A. K. 391
Lawrence, D. H. 368
Lawrence, Sir Thomas 25, 115
Lecomte, Georges 65, 66
Lee, T. Stirling 38
Legros, Alphonse 51–2, 101, 124, 152, 193
Leighton, Frederic Lord 16, 25, 46, 70, 115
Lenbach, Franz 386
Leonardo da Vinci 27, 136, 185, 288
Lethaby, W. R. 242–3
Lewis, Cecil 314–17, 328, 344, 353–8, 367, 375–6, 379, 391, 397
Lewis, Sinclair 368
Licinio, Bernardino 338
L'île Lacroix (etching by Camille Pissarro) 71
Limburg, Pol de 342
Lippi, Filippino 118
Lippi, Fra Filippo 234, 385
Lippman, Walter 118
Liszt, Franz 302
Literary Theatre Society 197–8, 222–3, 224–5
Livre Aesthetique, Le 91
Llewellyn, Sir William, PRA 327, 378, 384
Lopokova, Lydia 304
Loraine, Robert 225
Lorraine, Claude 115
Lotto, Lorenzo 118
Louvre, Musée du 33–4, 119, 335, 338, 358, 359
Lowinsky, Thomas E. 'Bengy' 24, 288, 383, 397, 399–400
Lucerne 161
Lucia di Lammermoor (Donizetti) 384
Lucrezia Borgia (Rossetti) 276, 284
Ludovici, Albert 124, 186
Luxembourg Gallery 136, 268, 276, 335–6
Lykiardopoulos, Michael 277

Mabuse (Jan Gossaert) 281, 338, 361
McCarthy, Lillah (Mrs Granville-Barker) 1, 225–6, 261, 270, 294, 295–7
McIlvaine, Clarence 59, 62
McKinnel, Norman 237
MacColl, D. S. 52, 53 4, 144, 170, 285, 303
MacGregor (office boy) 100
Madonna (Crivelli) 335
Madonna and Child with Angels (Masaccio) 284
Madrid 164–5, 239
Maeterlinck, Maurice 196, 261–2
Magazine of Art 44, 46, 62
Magdalen (Veronese) 365
Maillol, Aristide 284
Mallarmé, Stéphane 44
Man and Superman (Shaw) 204, 224–6, 258, 344
Man in a Black Shirt, The (Shannon) 110, 111, 155
Man in the Inverness Cape, The (Shannon) 110, 112

Man Wearing a Collar of the Golden Fleece (Flemish School, attrib. by CR to Mabuse) 361
Man with the Greek Vase, The (Shannon) 306–8
Man with the Yellow Glove, The (Shannon) 110
Manchester 234
Manchester Guardian 299
Manet, Édouard 13, 40, 51, 155, 201, 247, 263, 364
Mantegna, Andrea 234, 335, 384, 385
Manutius, Aldus 73, 76, 85
Marchesi, Blanche 157
Marie (servant) 180, 186
Marks, H. Stacy 29
Marlowe, Christopher 85, 88, 198
Marseilles 251, 265
Martin, Sir Alec 392
Martin, Mr E. LeBreton 100
Marty, André 91
Mary Magdalene at the Door of Simon the Pharisee (Rossetti) 100, 114, 123
Masaccio, Tommaso Giovanni di Mone 10, 121, 385
Masefield, John 221, 290, 292, 366
Massine, Leonid 304–5
Mathews, Mrs Muriel Lee 301
Matisse, Henri 148
Matthews, Elkin 80
Maupassant, Guy de 141
Mauve, Antoine 361
Meier-Graefe, A. Julius 328
Melchior, Lauritz 345–6
Memmi, Lippo 115
Men and Memories (Rothenstein) 393
Menzel, Adolf 30
Meredith, George 96
Meštrovič, Ivan 260, 283–4, 304, 370
Metropolitan Museum, New York 154, 364
Michael Angelo Buonarrotti 118, 154, 161–2, 190, 202, 241, 280, 288, 289
Milan, CR visits 131, 161, 395
Mill Pond, The (Shannon) 234
Millais, Sir John Everett 28, 36, 64, 100, 123, 152, 171
Millet, Jean François 51, 123–4, 185
Milward, Dawson 237
Monet, Claude 361
Monna Vanna (Maeterlinck) 196
Monna Vanna (Rossetti) 100
Montaigne, Michel de 368
Monte Olivetto, CR visits 395
Montezuma theme 207, 216, 268, 314–15
Moore, G. E. 31
Moore, George 25, 201, 233, 318
Moore, Hilda 221
Moore, Mrs Marie Sturge (née Appia) 178, 333
Moore, Thomas Sturge 2, 21, 30, 31–2, 46, 51, 52, 57–8, 67, 69, 75, 92, 101, 110, 113, 117, 123, 131, 142, 151, 154, 155, 156, 160, 166, 169, 171, 173, 174, 178–9, 193, 197, 198–9, 222, 248, 290, 294–5, 333, 340, 377, 393, 397, 400
Mordkin, Michael 258
Moreau, Gustave 13, 30, 33, 36, 41, 48, 81, 103, 147, 148, 155–6, 198, 211, 214, 249, 260, 263, 295, 335–6
Morelli, Giovanni 118, 171
Morgan, Pierpont 234, 334
Morgan, William and Evelyn de 38, 49
Morice, Charles 248
Morisot, Berthe 66
Morland, George 370
Morning Post 262
Moro, Antonio 337, 338, 370
Moroni, Giovanni Battista 337–8
Morris, Miss May 140, 162
Morris, William 1, 63, 70, 88, 95, 96, 115, 121, 126, 151, 175, 181, 284
Mozart, Wolfgang Amadeus, 7–8, 207
Mr Ricketts and Mr Shannon in the Enjoyment of Popular Success (Beerbohm) 49–50
Mrs Warren's Profession (Shaw) 369
Munich 110; CR visits 164, 386
Munnings, Sir Alfred 328
Murder in the Cathedral (Eliot) 366
Murillo, Bartolomé Estebán 168
Murray, Charles Fairfax 64, 114, 163
Mussolini, Benito 346, 365, 380, 384
Mutach, Alfred de 17
Myers, Capt. W. J. 117
Myriam (Shannon) 372

Naples 1; CR visits 201, 203–4, 239, 257
National Arts Collection Fund 119, 121, 192, 284, 286–7
National Gallery 12, 121, 280–1, 282–3, 400
Nash, Paul 328–9
Nederland Etsclub 91
New English Art Club 110, 144, 147, 192
New York 13, 154, 224, 267; CR visits 364

Newton, Thomas Woodhouse Legh, Lord 289
Nicholls, Percy 334, 372, 373, 374, 380, 388–9, 396, 400, 401
Nicholson, Sir William 123, 147, 193
Nietzsche, Friedrich Wilhelm 249, 265
Nijinski, Vaslav 259–60, 304–5
Nineteenth Century, The 249
Northcliffe, Sir Alfred, 1st Viscount 291
Nyberg, Henning 381–4, 386, 388–9, 390, 395–7, 399

Observer 285–6, 300, 302, 352, 381
Oedipus Rex 369
Offenbach, Jacques 192
Ophelia (Millais) 360
Orchardson, Sir William 155
Orpen, Sir William 147, 214, 291, 368
Orvieto, CR visits 346
Osgood, James Ripley 59, 62
Osgood and McIlvaine 59, 62, 63
Oxford, CR visits 16, 131, 228, 249–50

Padua 135–6
Paguin (Mme) 270
Paintings by CR: *Argonauts, The* 147; *Ariadne* 233; *Betrayal, The* 216, 221; *Burial, A* 187; *Calvary* 221; *Centaurs Quarrying* 146; *Centaur Swimming* 146; *Chimaeras* 216, 217; *Danaïds* 244, 321, 361–2; *Death in a Picture Gallery* 219, 221, 244; *Death of Cleopatra, The* 216, 221; *Deposition from the Cross, The* 291, 321; *Descent from the Cross* 206, 207–8; *Don Juan* 53, 216, 224, 353; *Don Juan* (Tate) 321; *Don Juan and the Commendatore* 205–6, *Don Juan Challenging the Commander* (RA) 328–9; *Descent from the Cross* 305; *Don Juan in Hell* 53–4, 208–10, 230; *Ecce Homo* 216; *Faust and Mephistopheles* 216, 394; *Faust Riding the Centaur* 146; *Flight of Cleopatra, The* 216; *Good Samaritan, The* 147, 230, 290; *Head of John The Baptist, The* 215; *Heliodorus driven from the Temple* 211–12; *Holy Women and the Angel of the Resurrection, The* 216, 221, 321; *Icarus* 137–8; *Jephthah's Daughter* 216; *Judith Returning to Bethulia with the Head of Holofernes* 216; *Leda and the Swan* 230; *Messalina leaving the Pincian* 230; *Montezuma* 207, 210, 216; *Montezuma Sacrificing to the Sun* 268, 319–21; *Orpheus and Eurydice* 216, 322–5; *Parable of the Vineyard, The* 190, 269; *Plague, The* 244–5, 268; *Resurrection, The* 207, 230; *Salome* 321–2; *Siegfried and the Water Maidens* 321; *Tobias being Comforted by the Angel* 145, 146; *Triumph of Bacchus, The* 221; *Wise and Foolish Virgins, The* (Cowdray) 322–3; *Wise and Foolish Virgins, The* (Matsukata) 321
Pall Mall Gazette 44, 144
Palmer, Samuel, 213
Panel Society, 193
Paragon, 1 The 130–1, 137, 163, 275
Paris 13, 49, 53, 60, 65, 91, 107, 108, 109, 142, 246, 257, 259; CR visits 33–5, 103, 136, 155, 161, 164, 201–2, 204, 276, 335–6, 346, 358, 359
Parma 161
Parsifal (Wagner) 249, 341
Passage to India, A (Forster) 368
Pastel Society, The 124
Pater, Walter 56
Patience (Gilbert and Sullivan) 142
Patricia of Connaught, Princess 302, 362
Paul III, Pope 241
Pavlova, Anna 258
Pêcheur, Le (Puvis de Chavannes) 332
Peer Gynt (Ibsen and Greig) 299
Pennell, Joseph 40, 124, 186
Perseus Slaying Medusa (Burne-Jones) 310
Persian miniatures 117, 376
Perugino (Pietro Vannucci) 118
Pesaro Madonna (Titian) 133
Pesellino 10
Peter Pan (Barrie) 231
Philips, Sir Claude 259, 281
Philpot, Glyn 312 14, 326, 334–5, 366, 368, 391
Piero Della Francesca 87, 157, 359, 385
Piero di Cosimo 119–21, 360, 365, 381, 400
Pietà (Bellini) 335
Pietà (Veronese) 361
Pike, Bernard 359, 383
Pile, Howard 30
Pinturicchio, Bernardino 119, 222
Pisa 194
Pissarro, Camille 52, 64, 65, 67, 70–2, 92, 107, 136, 361
Pissarro, Lucien 24, 35, 54, 65–7, 73, 75, 84–5, 86, 91–2, 97–8, 101, 103, 107, 123, 137, 204
Place de la Republique, La (etching by Camille Pissarro) 71

Plage de Boulogne, La (Manet) 201
Plays, operas, etc., designed in part or in whole by CR: *Admirable Crichton, The* (Barrie) 297; *Annajanska* (Shaw) 297; *Aphrodite against Artemis* (Moore) 198; *Attila* (Binyon) 226–8, 229, 231, 260; *Bacchae, The* (Euripides), 294–5; *Betrothal, The* (Maeterlinck) 340–1, 344; *Bride of Dionysus, The* (Tovey) 377–8; *Coming of Christ, The* (Masefield) 2, 366; *Dame de Pique, La* (Tchaikowsky) 298–9, 340; *Dark Lady of the Sonnets, The* (Shaw) 243–4; *Death of Tintagiles, The* (Maeterlinck) 261–2; *Deirdre* (Yeats) 270; 'Don Juan in Hell' from *Man and Superman* (Shaw) 206, 224–6, 237; *Door must be kept open or shut, A* (de Musset) 297; *Dynasts, The* (Hardy) 292–3; *Elektra* (Hofmannsthal) 228, 233; *Elizabeth of England* (Bruckner) 396; *Fanny's First Play* (Shaw) 257–8; *Florentine Tragedy, A* (Wilde) 199–201, 233; *Gondoliers, The* (Gilbert and Sullivan) 378; *Joan of Arc* (a pageant?) 305; *Judith* (Bennett) 295–6, 351; *Judith* (Moore) 294–5, 351; *King Lear* (Shakespeare) 235–8, 260; *King's Threshold, The* (Yeats) 269–70, 294; *Lamia* (Keats) 260; *Lanval* ('Ellis', i.e. Lord Howard de Walden) 233; *Lithuania* (Brooke) 297; *Macbeth* (Shakespeare) 261, 352, 353, 366, 376, 385; *Man of Destiny, A* (Shaw) 225; *Man who married a Dumb Wife, The* (France) 295; *Manfred* (Byron) 299–300; *Medea, The* 228; *Merchant of Venice, The* (Shakespeare) 303–4; *Mikado, The* (Gilbert and Sullivan) 2, 349–52, 353; *Miracle, A* (Vollmöiller) 222–3, 260; *Montezuma* (Lewis) 314–15; *Nail* (de Lara) 300, 340; *Oedipus Rex* (Reinhardt production) (Sophocles) 261; *On Baile's Strand* (Yeats) 294; *Paris and Oenone* (Binyon) 198; *Parsifal* (Wagner, unused) 312–13; *Persians, The* (Aeschylus) 222–3, 231, 341; *Philip the King* (Masefield) 292, 293; *Prince Marcocana* (de Lara) 357; *Retribution, The* (pageant) 228; *St Joan* (G. B. Shaw) 2, 312, 341–5, 347, 351; *Salome* (Wilde) 197, 199–201, 222, 260, 261–2, 351; *Tanta: or The Impossible Woman* (Haddon Chambers) 270; *Twelfth Night* (Shakespeare) 261, 303–4; *Two Gentlemen of Verona, The* (Shakespeare) 303–4; Wagner's operas 351; *Well of the Saints* (Synge) 294; *Winter's Tale, The* (Shakespeare) 261
Plotinus 253
Poems (Dowson) 62
Poems (Rossetti) 274
Poiré (i.e. Poiret), Paul 270
Pollajuolo, Antonio 118
Pompeii, CR visits 203–4
Post Impressionist Exhibition, The 246–9, 262, 263, 330
Poussin, Nicolas 34, 51
Poynter, Sir Edward, PRA 51, 289
Prado, Museo del 155, 165
Pre-Raphaelites 36, 43, 48, 54, 56, 60, 68, 81, 84, 100, 148, 262, 400
Prince Igor (Borodin) 330
Prints and Drawings by CR: *Athaliah* (pen drawing?) 29; *Autumn Muse, The* (pen drawing with white bodycolour) 103, 128, 145; *Defeat of Glory, The* (pen illustration) 80; *Deluge, Le* (woodcut) 91; *Dynasts, The* (lithographic poster) 292–3; *Finis* (lithograph) 290; 'Had Zimri Peace Who Slew His Master?' (after a pen drawing) 42; International Society (poster) 230; *Italia Redenta* (colour lithographic poster) 304; lithography 191; *Oedipus and the Sphinx* (pen drawing) 48, 84; *Parables, The* (woodcuts) 171–2, 193; *Persians, The* (lithographic poster) 222; *Prodigal, The* (pen illustration) 78; self-portrait (chiaroscuro woodcut) 124; *Two Old Kings, The* (pen illustration) 78–9, 81; Wood-engraving 28–9, 69–70, 73–5, 86, 95–6, 104–5, 124, 126, 141, 171–2, 174, 175, 193, 269; *Worm, The* (lithograph: design by CR) 43
Proust, Marcel 357, 368
Pryde, James 326
Punch 308
Puvis de Chavannes, Pierre 13, 30, 33, 34, 36, 43, 103, 143, 148, 152, 155–6, 185, 249, 263, 265, 288, 291, 332
Prunella (Housman) 196
Pye, William 186, 198, 219

Queensberry, eighth Marquess of 96
Quilter, Harry 41, 55
Quinn, John 267–8

Rackham, Arthur 267
Raffalovich, Marc André 44

Raphael (Raffaello Sanzio) 118, 162, 169, 170–1, 211, 234, 385
Raven-Hill, Leonard 22, 30
Rebirth of the Arts, The (Shannon) 304
Redon, Odilon 248
Regnault, Henri 13
Regnier, Henri 14
Reine des Poissons, La (ill. Lucien Pissarro) 73, 91, 97
Reinhardt, Max 261
Repose (C H S sculpture) 230
Rembrandt van Ryn 71, 118, 169, 234
Renoir, Auguste 247
Requiem (Verdi) 7
Reynolds, Sir Joshua 25, 36, 134, 309, 338
Reynolds, Victor 287
Richmond 130–1, 140, 179–80, 333
Richmond Park 122, 137, 397
Ricketts, Arthur (uncle) 6, 16, 328
Ricketts, Blanche Isabella 11, 17–18, 131
Ricketts, Charles de Sousy: birth 5; death 397; ability to laugh at himself 22, 56, 76, 151, 232, 270; appearance 52, 55–6, 64, 87, 192, 358; artistic ideals 31, 34, 64–5, 84–5, 148–9, 169–70, 205–18, 263–4; attitude to Impressionism 13, 40, 52, 66, 71–2, 136, 169, 276; attitude to Post-Impressionism 2, 52, 148–9, 246–9, 265, 304, 306, 386; attitude to religion 15, 162, 221, 272, 275, 285; Canada and U S A, C R's visit to 359–65; Canadian advisership 337–40, 359–65, 380, 386–8, 399; Collections: 2, 8–9, 40, 100, 114–21, 156, 181, 185–6, 202, 251, 264–5, 276, 310, 332–3, 376, 400–1; the Cole episode 288–9; contacts with art movements abroad 66–7; contributions to the war effort 290–4, 303–4; critical ability 67, 127, 149–50, 170, 191; difficulties with different media 191, 210, 218, 230, 353; difficulties with publishers 76, 78–81, 97–8; disingenuousness 42–3, 82, 151, 339; early book designs 59–63; early 'hack' work 32, 41, 46; early visits to Paris (1885 and 1887) 33, 34; education and training 10, 12, 14, 18, 22, 27–30, 36; 'enemies' 52–4, 124, 144, 153–5, 171, 192, 222, 328; Friday evenings 49, 69, 141, 157, 228; friendship with Bottomley 193; friendship with 'Michael Field' 86–91, 127–31; friendship with Charles Holmes 69, 106–7; friendship with Cecil Lewis 314–17; friendship with Henning Nyberg 381–4; friendship with Lucien Pissarro 64–5, 70–2, 91–3; friendship with G. Bernard Shaw 223–6; friendship with T. Sturge Moore 21, 31; friendship with Oscar Wilde 44–5, 56–7, 59–62, 94–8, 107–10; friendship with Yeats 141–2; generosity 1, 45, 67, 82, 109–10, 117–18, 121, 151, 173, 179, 184, 197–8, 233, 258, 272, 274–5, 280, 290–1, 297–8, 315–16, 318–19, 339–40, 375–6, 390–1; homosexuality 24–6, 97, 109, 203–4, 315–16, 356–7; how his mind worked 21–2, 31, 57–8, 132–3; jewellery designs 137–40, 178–9, 316, 377; love of flowers and gardening 7, 40, 311, 333–4; love of influencing people 44, 52, 67–9, 127–8, 288–9, 298–9, 312–15, 330; on marriage 177–9; and money 38, 69, 81, 97–8, 106, 110, 114, 116–17, 175, 202, 219, 268, 282, 337–40, 344–5, 355, 369–70, 376, 379, 386–8; move to 31 Beaufort St 91–2; move to England (1868) 6; move to France 12–14; move to The Keep, Chilham 309–12; move to Lansdowne House 181–6; move to Townshend House 331–3; move to the Vale 38–41; and music 7–8, 10, 11, 14, 72–3, 196, 258–9, 267, 279, 281, 298–9, 301–2, 304–5, 363, 379, 383, 393; obtaining works of art for the nation 284–7; pessimism 10, 13, 14, 145, 181, 216–17, 282, 295; plans for setting up exhibiting society 17, 65–6; pseudonyms 81, 102, 113, 222–3, 270, 345, 357, 367, 392; refuses Directorship of National Gallery 282–3, 337; relationship with Charles Shannon 2–3, 19, 22–7, 34–5, 37, 46, 125, 149–50, 151, 153, 155, 158–61, 203–4, 219–21, 246, 330–1, 336, 373–5, 380, 384, 389–90, 400; and the Royal Academy 322, 325–8, 328, 365–6; row with Holmes 173–4; secretiveness 6, 25, 76, 82, 127–8, 179, 204, 356; takes up painting 145–7, 205–20; takes up sculpture 190–1, 233; timidity 16, 18, 33, 165; Toronto Gallery advisership 370; touchiness 41, 44, 125–6, 128–30, 137, 173–4; trip to Belgium and Holland (1911) 264; trip to Berlin and Munich with Nyberg (1930) 386, 388; Egyptian trip (1911) 250–4, 264; second Egyptian trip (1912) 265–7, 275; trip to Germany (1903) 165–7; trip to Greece

Ricketts, Charles de Sousy (*cont.*): (1911) 254–7; trips to Italy; Venice (1899) 131–6, and Spain (1903) 161–7, 220, Rome and Florence (1905) 194–5, for *Titian* (Florence, Rome, Naples, 1906) 201–4, 220, Venice (1908) 232, Baveno (1923) 335–6, Baveno, Rome, Florence (1924) 346, to Florence with Brown (1928) 369; to Rome with Nyberg (1929) 383–4, second trip to Italy with Nyberg (1931) 395–6; trip to Paris (1913) 276; Tunisia, CR's visit to 355–8, 362, 366, 392; turns to theatre work 197–201, 225, 227–8, 260–1, 348–9; visits to Ryde 8, 10, 16, 17
Ricketts, Charles Robert (father) 5, 11–12, 15, 17, 177, 327–8
Ricketts, Edward Woodville (grandfather) 8, 16, 17, 32, 95
Ricketts, Eleanor (aunt) 8, 17
Ricketts, Flora (aunt) 32
Ricketts, Florence (cousin), *see* Baroness de Sinner
Ricketts, Hélène Cornèlie (née de Soucy) 5–6, 7, 12, 14, 177
Ricketts, Isabella (née Gibson) (grandmother) 17
Riley, Henry James 69, 75
Rimbaud, Arthur 87
Ring, The (Wagner) 14–15, 72–3
Roberts, Charles, wood engraver 28
Robertson, W. Graham 55, 57
Robinson, Sir Charles 118
Rodin, Auguste 51, 186–8, 190–1, 192, 220, 249, 289
Roerich, Nicholas 330
Rokeby Venus (Velasquez) 271
Romano, Giulio 171
Rome 14; CR visits 161–2, 194–5, 201, 204, 230, 346, 383–4
Romney, George 134
Ronda, La 257
Rooke, Noel 125
Roots of the Mountains, The (Morris) 63
Rose, J. Andrew, sale 114
Rose, Madame 305
Rosenkavalier (Strauss) 267
Rosing, Vladimir 298–9, 314, 379
Ross, Robert Baldwin ('Robbie') 24, 25, 49, 53–4, 55, 107–10, 121, 140, 151, 178, 186, 196, 200–1, 203–4, 216, 222, 229–30, 243, 259, 282, 286, 303, 312
Rossetti and His Circle (Beerbohm) 49
Rossetti, Dante Gabriel 28, 30, 33–6, 41, 51, 56, 59, 64, 68, 87, 95, 100, 102, 114, 119, 122, 169, 193, 211, 213, 263, 269, 274, 284, 287, 310, 326, 327, 360, 364, 373, 374
Rossini, Gioacchino 7
Rosso Fiorentino (Giovan Battista di Jacopo) 202
Rothenstein, Sir John 82
Rothenstein, Sir William 24, 40, 49–50, 87, 94–5, 101, 103, 115, 116, 124, 142, 151, 153–4, 192, 193, 368, 393
Rowlandson, Thomas 310
Royal Academy 36, 148, 155, 166, 325–8, 331, 351, 365–6, 378–9, 380–1, 393
Rubens, Sir Peter Paul 118–19, 164, 167, 169, 234, 247, 338, 370, 386
Rubenstein, Anton Grigorevitch 15
Rue de l' Épicerie (etching by Camille Pissarro) 71
Ruskin, John 32, 41, 132, 134, 203
Russell, A. G. B. 342
Rutherston, Albert 259, 261
Rutter, Frank 144, 147–8
Ruysdael, Jacob van 370
Ryan, Dr John 137, 222

Sachs, Dr Wilhelm (brother-in-law) 17–18
Sacre du Printemps, Le (Stravinsky) 267, 330
Sainte Courtisane, La (Wilde) 109
St John (Rodin) 187
St Mark enthroned with four Saints (Titian) 133
St Mark Rescuing a Slave (Tintoretto) 134
Sala, Count 360–1
Salammbô (Flaubert) 356
Salome theme in CR paintings 214, 321–2
Salome (Moreau) 214
Salome (Wilde) 1, 108, 166, 196
Salome (Wilde, ill. by Beardsley) 82, 85
San Gimignano, CR visits 395
Sargent, John Singer 25, 35, 155, 328
Saturday Review, The 52, 104, 144, 170, 238
Satyricon (Petronius) 367
Saumarez, Lady 116
Savage, Reginald 30, 38, 43, 54, 65, 69, 72, 75, 101, 122, 123, 392
Savoy, The 95
Scheherazade (Rimsky-Korsakov) 259

Schubert, Franz 302
Schumann, Robert 15, 258, 299, 301–2, 362
Schuster, Adela 229
Scott, Capt. Robert Falcon 220
Scott, William Bell 80
Sculpture by CR: *Deposition, The* 233; *Good Samaritan, The* 233; *Mother and Child* 205; *Orpheus and Eurydice* 268, 315, 322, 324; *Paolo and Francesca* 233; *Silence* 272–3; *Tragic Actor, The* 233
Sebastiano del Piombo 154, 338
Setting Sun (etching by Camille Pissarro) 71
Seven Characters (Pirandello) 369
Shakespeare, William 55, 102, 109, 141, 171
Shannon family 113, 131, 305, 380
Shannon, Charles Haslewood: accident 372–5, 379–80, 384, 388, 389–90, 396; background and appearance 19, 87; character 19–20, 22–4, 26–7, 31, 34–5, 58, 142, 179, 253, 331, 336, 370, 373, 389; confusion with J. J. Shannon 113, 232, 362; death 401; early work 29, 41, 60; edits *Pageant* 100, 103; given overdose of strychnine by CR 194–5; illnesses abroad 134–6, 194–5, 232; joint show with Rothenstein (1894) 95; lithographs 40, 43, 48–9, 65, 70, 71, 73, 91, 101, 103, 110, 147, 154, 193, 234, 279, 304; pact with CR 37; painting 36–7, 103, 110–13, 145, 147, 150–1, 155, 186, 187–90, 207, 224, 225, 230, 231, 233, 234, 246, 267–9, 287, 291, 295, 302, 306–7, 312, 331, 362, 370–2, 374; portraits of CR 21, 110, 112, 306–8; relations with women 22–4, 158–61, 178, 219–20; reputation 43, 55, 67, 246; and Royal Academy 264, 325–6, 366; teaching 31–2, 358–9; trip to Antwerp 359; twelve roundel woodcuts 124; visits Madrid 164
Shannon, James Jebusa, confusion with CHS 113, 232, 362
Shaw, G. B. 1, 52, 151, 201, 223, 224–6, 257–8, 270, 297–8, 315, 341–5
Shell-Gatherers, The (Shannon) 151
Shepheard, Dr 361
Shields, Frederick 56
Sickert, Walter 40, 49–50, 66
Sienna, CR visits 202–3, 395
Signac, Paul 66, 67
Sinner, Edward, Baron de 17
Sinner, Baroness de (Florence Ricketts) 16, 17, 327, 375
Sketch, The 54
Slade Professorship 239, 242–3, 337
Société des Vingt 66
Society of British Artists 36
Society of Painters in Watercolour 36, 38
Society of Twelve 193, 288
Soucy, Armand Louis de Fitte, Comte de 5, 6
South Wind (Douglas) 368
Spanish Singer, The (Manet) 201
Spare, Austin 290
Sparkes, John 28, 29
Speaker, The 60
Spencer, Sir Stanley 330
Sphinx Parisien, Le (Stevens) 264
Spring Terrace, No. 8 122–3, 130, 140
Squire (barrister) 176
Stage Society, The 294–5
Stannus, Hugh 29
Stations of the Cross (Tiepolo) 133
Steed, Wickham 279
Steer, Philip Wilson 25, 49, 65, 66, 103, 147, 214
Stevens, Alfred (English artist) 25, 29, 119, 285, 289, 326, 359, 376
Stevens, Alfred (Belgian artist) 264
Stevenson, R. A. M. 144
Stone, Miss 227
Stonehenge 235, 252
Strang, William 65, 103, 124, 193, 234, 248, 326
Strauss, Richard 267, 279
Strong, Arthur 121
Studio, The 59, 62, 81, 104, 227, 328
Sucher, Rosa 72–3
Suetonius 368
Summer (Shannon) 269
Swan, J. M. 110
Swinburne, Algernon 25, 26, 55, 56, 174, 310
Switzerland 5, 17, 131
Symbolism 2, 33, 43, 44, 48, 84, 214–18, 321
Symonds, John Addington 25, 76, 276
Symons, A. J. A. 367, 392, 393
Symons, Arthur 43, 54, 95, 140, 201
Synge, J. M. 294

Tabley, John Byrne Leicester Warren, Lord de 78–81
Tanagras 115–16, 123, 125
Tempesta, La (Giorgione) 327

Terry, Ellen 108, 232
Thompson, Francis 276
Thorndike, Dame Sybil 1, 2, 344–5, 347–9, 353
Thornley, George 73
Tiepolo, Giovanni Battista 121, 133, 278, 378, 386
Tiepolo, Giovanni Domenico 133
Times, The 201, 229, 280, 284, 366
Tintoretto, Jacobo 132–5, 361, 386
Tintoretto's Servant (Tintoretto) 338, 361, 380
Titian (Tiziano Vecelli) 2, 36, 95, 110, 116, 132–4, 148, 155, 164, 165, 166, 167–8, 169, 170, 203, 204, 213, 234, 238–42, 285–6, 363, 365, 385
Tonks, Henry 25, 147, 214
Toorop, Jan 67
Toulouse-Lautrec, Henri de 67, 157
Tovey, Sir Donald 377–8
Townshend House, Regents Park 331–3, 372–3, 383, 400–1
Tragic Mary, The ('Michael Field') 86
Travaux des Champs, Les (Lucien Pissarro after Camille Pissarro) 73
Trench, Herbert 259
Trevelyan, R. C. 186, 198, 377–8
Tribune, The 199
Tura, Cosmé 339
Turin 202
Turner, J. M. W. 25, 134, 153, 185, 263, 285–6
Turner, Reginald (Reggie) 162, 195
Two Sisters, The (Philpot) 326

Universal Review 41, 51
Utamaro 113, 114, 116, 238
Utopia Ltd (Gilbert and Sullivan) 349

Vale Artists, The (The *Sketch*) 54
Vale, Chelsea, The 38–41, 44, 58–9, 69, 82, 92–3, 94–5, 114, 123, 181, 182, 311, 312
Vale Press 1, 73, 86, 94–6, 100–3, 104–7, 113, 126–30, 137, 141, 151, 171–7, 184, 198, 242, 246, 279, 328, 337, 377
Valloton, Felix 67
Van de Velde, Henri 66, 91
Van Gogh, Vincent 248
Van Horne, Sir William 361
Vanity and Sanity (Shannon) 312
Van Nu en Straks 84, 91
Van Wisselingh, E. J. 91–2, 101, 115, 116, 150, 151, 153, 185, 201, 205, 272
Vasari Society 154, 192
Vatican 161–2, 203, 221
Velázquez, Diego Rodríquez de Silva y 144, 165, 166, 169, 170, 234
Velaya Costa, Ann 315–16
Veneziano, Domenico 163
Venice 160, 195, 246; CR visits 118, 131–6, 163–4, 232
Verdi, Giuseppe 7
Verhaeren, Émile 113
Verlaine, Paul 13–14, 56, 100, 108, 231, 291
Verona 136, 164
Veronese, Paolo 115, 132, 135, 361, 365
Vermeer, Jan 280
Verrocchio, Andrea del 27, 123
Vicenza 136
Victoria, Queen 142
Vienna, CR visits 166–7, 266–7
Villiers de l'Isle-Adam, Philippe Auguste Mathias, Comte de 56
Vintage, The (Shannon) 267
Vita d'Arte 246
Vivarini family 8, 133–4
Vollard, Ambroise 91, 103

Wagner, Richard 14–15, 77–3, 142, 214, 225, 249, 279, 298, 301, 314
Waller, Pickford 186
Waller, Sybil 26–7
Walpole, Sir Hugh 345–6
Warham, Archbishop William 169
Warren, Edward 26
Warwick St., no. 52, 100, 123, 137
Watteau, Antoine 119, 145, 170, 185, 228–9, 234, 268, 291–2
Watts, G. F. 30, 37, 141, 214, 262, 263–4, 269, 287, 339, 360
Watts-Dunton, Theodore 26
Well of Loneliness, The 369
Wells, H. G. 229, 381
Wells, Reginald F. 152–3, 315, 316, 397–8
Wheeler, Penelope 199, 294, 304
Whistler, J. A. M. 26, 37, 38–41, 59, 64, 66, 69, 84, 92, 100, 101, 114, 124, 186, 262
White, Gleeson 59, 81, 100, 103
Whym Chow, Flame of Love (Field) 204
Wilde, Oscar 1, 25, 40, 43, 44–5, 49, 52–3, 55, 56–7, 58, 59–62, 76, 82–5, 94–8, 102, 107–10, 151, 192, 197–201, 207, 219, 225,

229–30, 234, 272, 278, 282, 290, 305, 312, 318–19, 321, 367, 392; death of Oscar Wilde 142–4
Wilhelm II, Kaiser 166
Wilkie, Sir David 152, 264, 310, 400
Wilkinson, Norman 261
Willette, Adolphe Léon 30
Willingdon, The Viscount 362–4
Wilson, Henry 234
Wilson, Mona 287
Wilson, Richard 121
Wilton Diptych 285
Winslow, Henry 375
Winston, Bruce 347, 366, 396
Winthrop, Grenville 364, 376
Witt, Sir Robert 282, 384–5
Wolverhampton, 1902 exhibition 147
Woman's World 44
Wood, Derwent 187
Wood, Mr 363
Work (Puvis de Chavannes) 156
Wounded Amazon, A (Shannon) 103, 110
Wreckers, The (Morland) 370

Yeames, W. F. 29
Yeates [i.e. Yates] Thompson, Henry 376
Yeats, W. B. 1, 30–1, 35, 52, 71, 140–1, 151, 197–8, 201, 207, 210, 269–70, 290, 294, 318, 340
Yellow Book, The 84–5, 97, 100, 196

Zola, Émile 13, 31, 155